Y0-BPZ-035

Data Communications

A Comprehensive Approach

Data Communications

A Comprehensive Approach

Gil Held

Ray Sarch

Third Edition

McGraw-Hill
New York San Francisco Washington, D.C. Auckland Bogotá
Caracas Lisbon London Madrid Mexico City Milan
Montreal New Delhi San Juan Singapore
Sydney Tokyo Toronto

Library of Congress Cataloging-in-Publication Data

Held, Gilbert, 1943–
Data communications : a comprehensive approach / by Gil Held, Ray Sarch. — 3rd ed.
p. cm.
Includes index.
ISBN 0-07-028049-5
1. Data transmission systems. I. Sarch, Ray. II. Title.
TK5105.H425 1995
004.6—dc20 95-9199
CIP

McGraw-Hill 
A Division of The McGraw·Hill Companies

hc 1 2 3 4 5 6 7 8 9 DOC/DOC 9 0 0 9 8 7 6 5

ISBN 0-07-028049-5

The editors of this book were Acquistions Editor, Steve Chapman; Executive Editor, Joanne Slike; Managing Editor, Andrew Yoder; and Book Editor, Andrew Yoder; and the production supervisor was Katherine G. Brown. This book was set in ITC Century Light. It was composed in Blue Ridge Summit, PA.

Printed and bound by R. R. Donnelley & Sons Company, Crawfordsville, Indiana.

MH95
0280495

Contents

Foreword to the Third Edition

Again, fast-evolving technologies have prompted the authors to create this third edition. The basic style and format remain the same.

The chief differences with the previous edition are again represented by the additions—especially those covered in some depth. Other than the effects of mergers and sell-offs, the topics that represent the major additions include: expansion of the discussion on the PC (Chapter 3); an extended commentary on the universal asynchronous receiver transmitter (UART) in Chapter 4; an extended update to modem standards and associated developments (including details on commands, data compression, and the MNP protocol), and the latest on the channel service unit (Chapter 5); the new lineup of the ITU-T (formerly CCITT) study groups (Chapter 6); the most recently available common carrier offerings and costs (Chapter 8); an extensive revision of the LAN material, now Chapter 10; the relationships of servers, bridges, routers, and gateways to network operations (Chapter 11); extensive additions to Chapter 12's PC software material; an introduction to the subject of sizing (Chapter 14); an introduction to ATM, which appears to have the potential of becoming the dominant networking technology (Chapter 15); a new chapter on the ongoing developments of videoconferencing (Chapter 16); a detailed examination of LAN monitoring (Chapter 17); coverage of trends in areas such as the Internet, ISDN, and networked multimedia (Chapter 20).

Some earlier topics are losing their value, such as OSI, because of TCP/IP's hardiness. Others needed minimal updating. The authors have attempted, to the best of their abilities, to make all the contained information as current as possible. Keep up-to-date by reading the trade press, following vendor and standards activities, and attending major conferences and exhibitions.

January 1995

Introduction

The following overview defines data communications, its functions, and its scope. Its purpose is to provide an understanding of the significance of data communications technology in today's business-communications environment. It is adapted and revised from an article written by Raymond Sarch, executive technology editor of *Data Communications* magazine, that then appeared in the 1982 edition of *The Encyclopedia of Management*, edited by Carl Heyel.[1]

Data Communications is, as the name indicates, the means of communicating data—as opposed to communicating analog (telephone) voice. It comprises communications between digital computers, between terminals and these computers, and between terminals themselves. Thus, it is the means of tying together coherently all local and far-flung computerized devices.

Strictly speaking, data communications has existed as long as data processing, but its growth was severely restrained by the tariff restrictions imposed on and by the telephone industry. Telephone networks were the major means of connecting a remotely located terminal user to a centrally located mainframe computer. However, on June 26, 1968, the ground rules were drastically changed, and a "new" industry was spawned. On that date, the Federal Communications Commission rendered its landmark Carterfone Decision permitting direct connection to the telephone network of non-telephone company devices. Although the telephone companies remained free to maintain standards for the protection of their networks, this proved no deterrent to the growth of the newly liberated data communications applications. This growth is a major reason for the increasing computer orientation of our society.

Besides the telephone companies—especially AT&T—the other major influence affecting this growth has been and is IBM. This company is primarily known for its computers, typewriters, and other business machines, but its impact on data communications is twofold: in the interconnection of its own products, and in the connection of other manufacturers' "plug-compatible" products to IBM computers and other devices. The company's extremely strong position in the large-computer industry,

---

[1] *The Encyclopedia of Management*, edited by Carl Heyel. Copyright © 1982 by Van Nostrand Reinhold Company. Reproduced by permission of the publisher.

and its strong positions in other business-equipment areas, have enabled it to establish de facto standards for both the EDP and data communications fields.

Networks. The user realizes the full potential of data communications in the application of networks. The most elementary connection method of a remote terminal to a computer is via a leased line, point-to-point. When there are several terminals at a remote location requiring connection to the computer, instead of leasing an individual line for each terminal, it is possible and less costly to use only the one line in combination with a "multiplexing" (line-sharing) device. The multiplexer, representing an elementary form of networking, is commonly available in several forms and variations of "intelligence."

As terminals from different remote sites require access to the centrally located computer, the network grows in complexity. Waystations, called nodal processors or nodes, become necessary within the network to interact with the terminals and the computer, storing, forwarding, and controlling the network's data flow. As more computers and terminals are added, the network design becomes ever more complex to provide the capability for any terminal to reach any computer or any other terminal.

The user firm is faced with a major design and implementation effort to provide a private network—which yields many control and security benefits—or it can avail itself of a public network to move its data. Among the different types of network designs is circuit-switching, which involves a straightforward transfer of data from one path onto another. Message-switching, on the other hand, adds a store-and-forward function: the message is stored at a network node, possibly checked for errors, then transferred onto another circuit (path). Depending on the distances involved and the network complexity, these nodal "stops" may occur from one to several times.

A popular variation of the message-switched network is the packet-switched network. Here, as the message enters the network, it is divided into discrete segments, called packets. Each packet travels the network independent of the others, with suitable identification. A packet is routed over the "free-est" path, node by node, until it reaches its exit node. There, all the packets are reassembled in their proper order, and presented to the destination device as the entered message.

An important element of the data communications network is the satellite—particularly, the geostationary satellite. Its orbit matches the earth's angular rotation, and it is positioned over the equator. To an observer on the earth, it appears stationary in space. In this position, it acts as a relay and disseminating station for all forms of electrical signals. When a firm's network requires satellite use, "dish" antennas may be positioned on building roofs, to minimize land-line use.

One drawback in the application of satellites to data transmission is the inherent propagation delay of this technique. There are several schemes to negate the delay's effect—especially noticeable in interactive (inquiry-response) applications—but the user must balance their costs against those of transmission via cable and microwave links. For non-interactive transmission, such as batch data, video, and facsimile, the delay is more readily absorbed.

Among the early established public packet-switched networks in the United States were those of Telenet and Tymnet. Each interconnected internationally to Canada's Datapac, and, via international record carriers to Europe and Japan. A private network that met the interface requirements of one of the public ones could establish an economical combination of the two network types. International and overseas remain available to the private network, with the proper interface.

The Terminal. The most ubiquitous data communications device is the operator's terminal. Commonly, it has a typewriter-like keyboard, and a cathode-ray-tube (CRT) screen or a hard-copy capability (the two are not mutually exclusive). It permits the user to gain access to a wide range of databases (information sources), including business, training, entertainment, and interactive applications. These services are usually available on a timeshared basis—meaning accessible to many users "simultaneously" (i.e., it appears so to the user).

The business office is where the terminal and all its attendant peripherals find their potential increasingly realized. Office automation is driven by the need for higher office productivity. This drive had been fueled by the replacement of discrete pieces of mechanical office equipment with their electronic counterparts. But significantly higher productivity is gained by the application of data communications concepts.

One of these concepts is the local area network (LAN). In its most common form, it consists of a loop of coaxial cable interconnecting offices in adjacent buildings, operating at megabit/second data rates. Ideally, the LAN enables the user to attach any communicating office device to the loop, automatically gaining access to all the other devices so connected. This idea is exemplified by the Ethernet offering. The user thus gains one of the most significant elements toward productivity improvement: resource sharing. Each office terminal device may readily access common databases, rather than having to maintain its own. Also, with the advent of digitized voice transmission, the office network becomes the common shared medium for all office transmissions, thus avoiding the need for separate, costlier facilities.

LAN standards are established primarily by the Institute of Electrical and Electronics Engineers (IEEE). The increasing application of optical fiber cable and its 100 Mbit/s (and up) data rates has led to newer LANs and developing standards. One offshoot, the metropolitan area network (MAN), has led to the fiber distributed data interface (FDDI).

To enable resource sharing on a global scale, by allowing LANs to connect to each other, standards bodies specify the functions of the internetwork interface, called a gateway. These functions include responsibility for end-to-end accountability, data routing, and traffic flow control. With gateway availability, LANs interconnect and also gain access to existing long-haul networks.

Increasingly, the business-office terminal is a multifunctional workstation device: a microcomputer or personal computer (PC). Depending on the amount of intelligence its owner is willing to purchase, it can not only handle interoffice message communications—such as electronic mail—but also do a considerable amount of data processing. With today's integrated-circuit chip technology, the PCs include microprocessors, each with its own designated function. The information processing power is limited only by cost and the abilities of its programmers.

Facsimile terminals are also evolving. New digital techniques enable scanning in seconds the material to be transmitted. Among these techniques is the terminal scanner's ability to skip over or "compress" redundant characters, including "white space." Of course, the receiving machine interprets the received compression codes and restores the original material in its entirety. Adding a fax board to a PC adds facsimile-sending capability.

The Network Minicomputer. One "older" device still in evidence in network applications is the minicomputer. As a communications processor, it "front-ends" a mainframe computer to handle the network traffic, thus freeing its host for the "number-crunching" tasks it handles more efficiently. The communications processor also functions as a switch, connecting terminals to each other as required. When a user desires access to the host's database, the front end opens a path to the mainframe.

Other network functions of the minicomputer are as a stand-alone message switch, as a cluster controller (directly controlling a group of terminals), as a gateway processor and as a distributed processor. In the latter roll, the mini may operate totally independent of a host computer, in conjunction with other minis. The resultant distributed data processing (DDP) network thus has no one element—such as a mainframe—that can make the network fail. Instead, if any of the DDP units becomes inoperative, only those functions it controls become unavailable. Of course, the coordination and programming tasks required by a DDP network are considerable, and the distributed database techniques continue to evolve.

Network Control. A network does not run itself. For greatest efficiency, a control function is part of the basic network design. And the primary purpose of network control is to minimize—or better, avoid—downtime. Network control requires some specialized equipment, proven cost-effective, to acquire and interpret operational statistics. In this way, as a variable such as response time approaches its tolerance limit, the preprogrammed control mechanism apprises the network supervisor terminal of the condition. Depending on the control's complexity and the level of artificial intelligence, corrective action may be automatic as the tolerance limit is reached, or the supervisor may key in suitable instructions to alleviate the condition.

When a network component experiences an outage, before the failure can be corrected it must be located and diagnosed. A properly designed network, with sufficient detection devices and diagnostic routines, and with built-in redundancy of critical components, will have minimum downtime. The problem will be located quickly, its extent readily evaluated, and the proper corrective measures promptly taken. The lesson to be learned is not to wait for trouble to decide how to handle it, but to anticipate and design for it at a very early stage of the overall project. One example of a popular set of network management packages is IBM's Netview.

Protocols. The rules under which data communications devices communicate with each other are called protocols. They range from rules for simplex (one-way-only) transmission to those packet-network high-level data link control (HDLC). They cover transmission modes such as half duplex (one direction at a time) and full duplex (both directions simultaneously), as well as operations of multilayered interfaces. The more complex the protocol, the more intelligence required of the terminal/computer device and of the network. To the operator, protocols should be "transparent," i.e., they should require minimal or no user action. As protocols become standardized, the programming efforts needed to formulate them become susceptible to production methods such as those of solid-state chip technology, which makes the protocol a less-expensive hardware function. An example of this result of standardization is the protocol for access to packet-switched networks. After much negotiation, the International Consultative Committee for Telegraphy and Telephony (CCITT) formulated its Recommendation X.25. Shortly thereafter, chips incorporating that protocol became available for packet-network access. As the standard is further refined, access requires even less software as more functions are handled by hardware.

Software. Some firms specialize in software packages to optimize network design. The number of network alternatives is growing so fast that the demand for optimization—both in private-network design and in the interface to, and use of, public-network facilities—can be satisfied only through computer aids. These aids range from simple memory devices that compile and store network statistics to complex data traffic simulators, network design configurators, and software for data-distribution modeling. Much time and effort can be spared by computer modeling of a proposed network prior to implementation. The effects of traffic variables and equipment and tariff changes may be examined in the model, leading to network design optimization.

Other data communications software is concerned with accessing of computers by remotely located terminals. For example, among the software packages implementing this function are IBM's Virtual Telecommunications Access Method (VTAM) and Telecommunications Access Method (TCAM), two of the more common early ones. A software product that is designed to manage access methods and protocols is the teleprocessing (or telecommunications) monitor. The TP monitor optimizes a computer's processing time by off-loading the data communications software functions from the operating system. The monitors are available from many software firms, as well as from computer vendors.

The increasing networking application of PCs has given rise to PC-specific programs for controlling and communicating over a variety of networks. In many instances, PCs have replaced terminals and other networking computers.

A software-related function of increasing importance to the network planner and user is database management, especially in DDP networks. A database management system (DBMS) provides users with a method of readily accessing data no matter where the data resides in a network. The aim of DBMS designers is to approach the ideal concept of totally distributed network processing. In this DDP environment, the user accesses a database by subject, and the network connects the user terminal to the proper computer. The user is completely unaware of this network operation. Common communications-related DBMS problems are contention, deadlock (simultaneous access), and recovery in a distributed environment.

To facilitate DBMS development, computers are needed that are capable of storing and retrieving information in an efficient manner. Ideally, they would handle these functions by information content, rather than by the customary techniques of physical addressing. To achieve the greater efficiencies of a computer architecture dedicated to information storage and retrieval, a database computer is needed. In its most efficient application, the database computer acts as a network node, totally decoupled from the data processing functions of the other network computers.

Current Trends. Spurred on by technological advancements, and enjoying an increasingly competitive climate, the data communications industry's influence on business is considerable. With the FCC and Justice Department rulings, AT&T is in the computer field, and that other industry giant, IBM, is well into the data communications field. The considerable research and development capabilities of the two companies are expected to respond to the competitive pressures of each other and of the growing number of independents with more and more technical breakthroughs.

Enhanced electronic document interchange (EDI) services, for example, will be far ranging. Such services provide computer-to-computer links and high-speed document communications.

The pace of local area network growth is accelerating. As expected, the key to this growth—as to so many recent computer-related developments—is the microprocessor. The network interface unit is microprocessor-equipped to manage data-packet processing, circuit connection, and error detection.

Another notable business-office development is the digital private branch exchange. As an office's communications interface to the outside world, the PBX is undergoing considerable transformation. Besides handling both data and voice, more recent versions of the device integrate the two by digitizing the voice signals. Digitized information from PCs, facsimile equipment, and other devices are all funneled through the digital PBX. Its own memory and switching functions, and interfaces to analog and digital networks (including packet-switched types), are all made possible by the application of microprocessor and chip technology.

The Future. One data communications technology that is fast maturing is lightwave transmission. The medium in greatest use is glass, in the form of optical fibers. Using a light-emitting diode (LED) or a laser as the light source, data rates in the multimegabit-per-second range have been achieved. Other advantages are the material's ready availability, light weight (lighter than copper), and narrow gauge (less than coaxial cable)—all made available in a sufficiently strong packaged product. Connectors and field-splicing techniques are available.

Speech recognition and voice response technologies are being applied to data communications devices in several areas. One is voice mail, where a computer synthesizes messages for the user who dials up a "mailbox." Another is the voice-activated workstation (both with and without a data communications interface), which at this writing is still in an experimental stage. In its ultimate form, the operator will speak to it, and the machine will print and transmit the words as spoken and correctly spelled, and the text will appear as typical typed material. The first commercial versions are expected to recognize correctly about 95% of "typical" business English as spoken by the "average" executive.

A growing phenomenon outside the United States, but appearing here in several limited versions, is interactive home TV, called videotex. Videotex services include electronic mail, timeshared computing, remote shopping and banking, and travel and event reservations, besides the already vast array of home entertainment modes.

Of less visibility outside the data communications industry, but of tremendous importance, are the efforts of standards bodies to formulate specifications permitting interconnection of devices no matter which company is the manufacturer. Called open systems, considerable progress is reported.

The increasingly competitive climate that evokes more and more advanced versions of data communications devices is starting to sharpen the productivity needs of budget-plagued corporations. Just as standalone computers were seen as an aid in earlier troubled times, so today are distributed data processing and allied data communications techniques and networks seen as the vehicles to help overcome the problems of more recent and near-future financially anxious eras.

Chapter

1

Historical Development of Communications

1.1 Introduction

Man's inability to make his voice heard beyond a very limited distance was perhaps the governing factor in the early development of communications techniques. In the 11th century B.C., Homer, describing the fall of Troy, wrote about a chain of fires that were used to transmit news of the capitulation of that city. In the 18th century, as the pioneers moved westward across the Missouri, two shots from the scout's revolver were used as the "all clear" signal to the wagon train several miles to the rear.

As man's level of technological development progressed, a corresponding advancement in the level of communications techniques occurred. In the 18th and 19th centuries, complex messages of war and peace, trade and art, science and history were transmitted in written form over distances of hundreds or thousands of miles, with the Pony Express linking the North American continent from shore to shore in a matter of days.

Wanting to communicate between greater distances on a faster scale, man experimented to develop more rapid and reliable communications systems. The development of these systems was based upon the utilization of a variety of machines that transmit information electronically to other machines located in another room or in a remote part of the world.

No matter what technology or type of communications is used, three common items are required in order to communicate. First, a source or transmitter must exist. The transmitter can range from the tom-tom, to the more modern telegraph key, to a present-day cellular telephone or a personal computer using a built-in or external modem. Second, for the transmission to reach its destination, it must travel over an appropriate medium. This medium can be the atmosphere for the signal of the tom-tom, a wire for the transmission of telegraph signals, or a combination of media.

An example of the latter could be outer space and wires within a building to connect an antenna receiving telemetry information to a computer. Third, a sink or receiver must be present to capture the signals that are transmitted.

1.2 Telegraph Systems

The era of electronic communications began on May 24, 1844, when Samuel Morse transmitted the well-known "What hath God wrought?" from the United States Capitol in Washington to his partner, Alfred Vale, in Baltimore. Shortly thereafter, the telegraph line was extended to New York City and additional lines were rapidly installed throughout the United States. The explosive growth of the telegraph system can be attributed to the simplicity of the code Morse developed as well as to the expansion of the railroad system across the North American continent. A major portion of telegraph message volume consisted of information concerning train dispatching. That telegraph lines could share a right of way with railroad lines resulted in the growth of each system complementing the other.

Because of the lack of a satisfactory system for sending and receiving information automatically, the use of the telegraphy for many years was limited to the hand keying of the Morse code. The transmitter in the Morse telegraph system was a telegraph key, which is simply a switch with a handle so that it can be operated by trained personnel. This switch was used to open or close an electrical circuit whose power was provided by a battery or another source of direct current. The receiver consisted of an electromagnet with a moving armature.

When the transmitting operator depressed the telegraph key, he would close the circuit, which in turn would cause current to flow, making the movable armature click at the receiving end as it was pulled by the electromagnet to its stop position. These clicks of the armature could be "read" by a trained ear, and the sequence of clicks—a series of short and long contact closures—was used to represent different characters in Morse code.

A simple one-way telegraph circuit is illustrated in Figure 1.2.1. This type of circuit is also referred to as a *simplex circuit* because information flows in one direction. When information can travel in either direction, but not at the same time, the circuit is known as *half duplex*; and if it flows in both directions simultaneously, it is *full duplex* (see Sec. 4.3).

Although the Morse code was extensively utilized on telegraph systems, two factors hampered the development of automatic transmission with the use of that code. First, the structure of Morse code is not suitable for automatic reception of information because different characters have different numbers of elements, each a series of short (dot) and long (dash) contact closures that identify a character. Secondly, because there were no prescribed time intervals between characters, a method to synchronize the transmitting and receiving units was lacking.

These problems persisted until 1874 when a Frenchman, Emil Baudot, devised a constant-length code, which was an important step toward the development of automatic transmission systems. In the constant-length code, the number of elements or signals used to represent a character is the same (constant) for every character, and the duration of each element is also a constant.

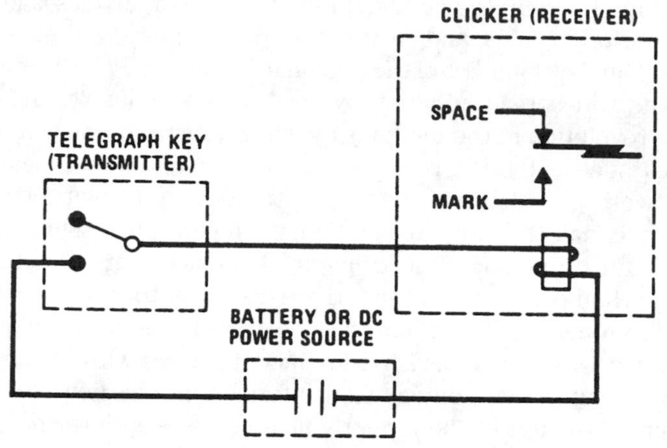

Figure 1.2.1 Simplex Telegraph Circuit.

The Baudot code is commonly referred to as a *five-level code* because five elements or signals are used to represent a character. It is interesting to note that the Baudot code is considered by many to be the forerunner of most data processing codes in use today.

In 1910, an American, Howard Krum, devised a technique for synchronizing the transmission and reception of each character by appending a standard element to be used as an identifier to denote the beginning and end of each character. Krum's technique, known as *start/stop synchronism*, combined with the constant-length Baudot code, resulted in the development and widespread utilization of automatic telegraph equipment.

Commercial operation of automatic telegraph equipment, normally referred to as *teleprinter service*, began in 1910 with the Postal Telegraph System and later with the Western Union System. From a handful of teleprinters in 1910, rapid expansion resulted in tens of thousands of teleprinters serving private, commercial, and governmental users. Two of the largest and best-known teleprinter networks were TWX and Telex.

Telex service originated in Germany in 1934 and was subsequently expanded into an international message-switching communications service. In 1956, Western Union introduced Telex service in the United States. A similar service, TWX, was developed by AT&T and sold to Western Union in 1971. In 1989, Western Union sold its Easy-Link service—which then included both Telex and TWX—back to AT&T. The service, now spelled *Easylink*, encompasses Telex, facsimile, and database research applications. The TWX service is no longer offered. The user equipment recommended by AT&T for Easylink is a personal computer (discussed next) and a modem, effectively obsoleting the teleprinter. Telex subscribers use the 8-level American Standard Code for Information Interchange (ASCII), replacing the earlier 5-level Baudot code.

Within the Department of Defense, the Army, Navy, and Air Force operated very large worldwide teleprinter networks through the 1980s. Many commercial organizations also developed corporate-wide teleprinter networks for message switching by integrating teleprinters, lines, and computers to form specialized data communications networks. The development of the personal computer (PC) during the 1970s was followed by its widespread adoption by government, industry, and academia. This resulted in a revolution in the methods by which organizations communicated. Intelligent devices, such as the PC, replaced dumb terminals at user locations. Software developers were provided with new opportunities to develop programs that supported electronic mail (e-mail), calendaring, document processing and storage, and other related office functions. Electronic mail enables users to compose messages on screens at their desks and transmit these messages to another person in the same building or to a mailing list of hundreds scattered around the globe. E-mail, stand-alone facsimile (fax) machines, and fax modems used with PCs have drastically changed how organizations communicate information. This new technology has practically eliminated the use of the formerly ubiquitous teleprinter networks. The transition is an example of how advances in technology—in this case, communications technology—enable new industries to replace enterprises that are based on older technologies.

1.3 Telephone Systems

In 1874, 30 years after the first telegraph transmission, Alexander Graham Bell built a device that permitted an electrical current to vary in intensity within the device as the density of the surrounding air varied in passing through the device. Using a rudimentary diaphragm, Bell was able to pick up the sound waves generated by the twang of a clock spring, convert the waves into an electric current, and then convert that current back into sound at the distant end of the electrical circuit.

Bell continued experimenting with his device, which was known as a *harmonic telegraph*, until, on March 10, 1876, he transmitted the now famous sentence over a wire from his laboratory in Boston to his associate in the adjoining room: "Mr. Watson, come here, I want you!"

A simplified one-way telephone circuit is illustrated in Figure 1.3.1. Here, the sound waves from the speaker cause the diaphragm to vibrate, thus causing the carbon grains behind it to be compressed or released, according to the speaker's inflection. The change in the carbon grains causes varying resistance in the circuit, which in turn causes the current flow to vary. The resulting alternating current is similar in wave shape to that of the sound waves producing it. The alternating current, in turn, is passed into the primary winding to induce current into the secondary winding, from which the current is passed onto the telephone line.

At the receiver, the coils in the electromagnet are connected across the pair of wires that forms the telephone line and alternately attract or repel the iron diaphragm. The diaphragm in turn acts on the air next to it, to create sound waves similar in form to those that the speaker originated. Electrical power for the circuit is supplied by a battery or other direct current source. The permanent magnet in the receiver holds the diaphragm in a central or neutral position when no alternating current is flowing, permitting the diaphragm to move either backward or forward.

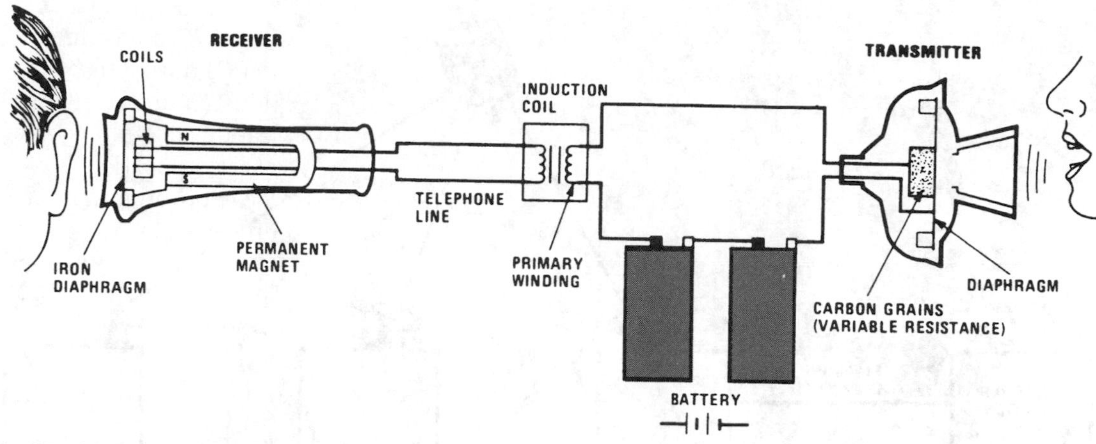

Figure 1.3.1 Elementary Telephone Circuit.

In 1877, a telephone line was constructed between Boston and Somerville, Mass. Over the next three years, additional lines and more than 50,000 telephones were installed in the United States. Approximately 100 years after the invention of the telephone, the United States was covered by a network of about one-half billion miles of telephone circuits, with 100 million telephones in use. Also in 1877, the first manual telephone switchboard was introduced; in 1882, the first dial-system switchboard (no manual intervention) was placed into operation. Today, almost 100 percent of the telephones in the United States have direct distance dialing (DDD) capability.

1.4 New Technological Developments

Two recent technological developments have resulted in significant advances in the state of communications: satellites and lasers.

A communications satellite is basically a microwave radio relay station apparently suspended (geostationary) at a point above the earth. First used by the military in the 1960s, satellites offered greater reliability and flexibility than conventional methods of communications; this prompted a number of commercial satellite offerings. By the early 1990s, thousands of companies had earth stations installed on their premises to enable them to communicate directly via satellite to other company locations. There, connections were provided by other ground stations or by a telephone line connecting a satellite communications carrier's ground station to the customer's premises.

New developments in satellite technology permit thousands of simultaneous connections from different locations to be retransmitted via the satellite to ground stations distributed over a wide geographical area. The communications satellites' key limitation is that they are required to be in a line-of-sight position to both receive transmitted data and retransmit that data to the receiving ground station. One typical customer's satellite network is illustrated in Figure 1.4.1.

During the 1970s, many industry analysts predicted that the use of communications satellites would produce a significant reduction in the growth of terrestrial

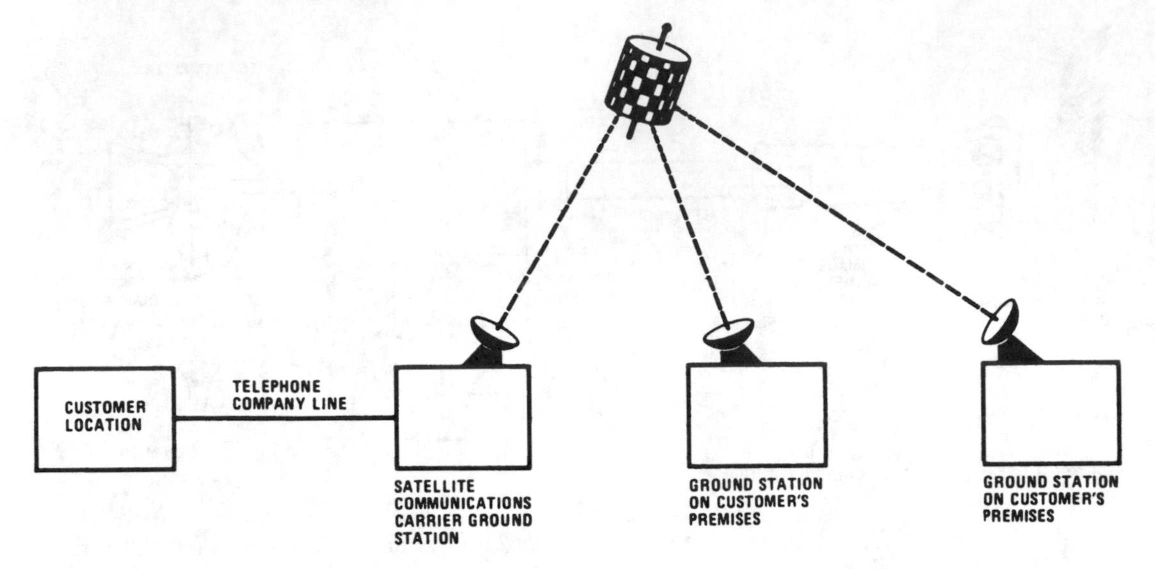

Figure 1.4.1 Satellite Network.

communications. However, because of a communications satellite's associated propagation delays, its usefulness was found to be greatest in transaction applications. (The propagation delays result from a round-trip communications path of approximately 50,000 miles between the satellite and its earth stations.) Today, you will probably notice a satellite dish antenna on top of your bank, drug store, and department store. Instead of using relatively expensive terrestrial circuits, transactions are transmitted via satellite to corporate data centers for processing. Unlike voice interchanges, which are ill-suited to ½- and 1-second delays, transactions of retail stores and banks are ideal for satellite transmission. This communications application enables modern organizations to rapidly track the effect of advertising, be alerted to potential inventory problems, and perform other financially related tasks in time to stay in step with the competition.

The second technological development that has resulted in significant advances in communications is the application of lasers. Coherent light of the laser can transmit tremendous quantities of data when compared to conventional communications systems. Fiber optics, which is a grouping of glass fibers bundled together, permits light to be transmitted in other than a straight path. When combined with lasers, fiber optic transmission networks have been developed that permit data transmission rates tens of thousands of times greater than that obtainable on conventional twisted-pair telephone wire. Because of this high transmission capability, communications carriers were among the first users of fiber optic transmission. The equipment incorporated lasers to generate the required light energy, thus obtaining the capacity to combine thousands of telephone and data calls routed between cities on a common fiber optic cable. As the technology has evolved, it has gained acceptance as a transmission mechanism for many local area networks, linking computers and terminals within close proximity of one another as well as for underseas cables transmit combined digitized voice and data between continents. In fact, because of the low cost associated with transmitting high data rates via fiber optics, this technology competes directly with satellite transmission. In many cases, fiber is more economical for a communications carrier, which must obtain a capacity measured in thousands of circuits between two locations.

1.5 Summary

As man developed, he sought to transfer information between persons and locations by meaningful sounds and symbols. When the requirement arose to communicate at a more rapid rate between locations separated by great distances, new communications systems were developed, resulting in a series of progressively more sophisticated electronic communications systems. Man is both a communications system and a manipulator of machines that transfer information by wire, radio, and optical signals.

Questions

1.1 What three items are required for all communications? Discuss two different types of communications systems and how the three items required for communications relate to one another.

1.2 Why was the Morse code not suitable for utilization by an automatic transmission system?

1.3 What two developments resulted in the development and widespread utilization of automatic telegraph equipment?

1.4 What was the effect of electronic mail and facsimile machines upon the use of teleprinter networks?

1.5 Discuss differences in usage between satellite communications and fiber optic transmission.

Chapter

2

Development of Computational Machinery

2.1 Introduction

During the last three decades, we have witnessed two of the most incredible technological revolutions in the history of the world. The first (described in Chapter 1) was the advent of a truly global communications system, linking remote locations via satellites. The second revolution was the advent and evolution of the digital computer. In a short period of time, the advance of computer technology has become so profound that the term "generations" has been applied to the sequences of machines that have been developed.

The first device used to help man perform calculations was the abacus, which was developed in China around 450 B.C. and is still used today in many areas of the world. The next significant development occurred more than 2000 years later when man started to design and build mechanical calculators. In 1812, a British mathematician, Charles Babbage, designed a machine to aid in the computation of mathematical tables. This machine, which he called a "difference engine," was completed in 1822. Other mathematicians of that period also attempted to build calculating machines; however, a calculator is not a computer. In 1833, Babbage designed an "analytical engine," which was to be completely automatic. This heralded the dawn of the computer era. Although the machine was not constructed because the parts it required could not be built with sufficient precision in those days, the device had all the elements of a modern digital computer. It had a storage area for holding information (memory), components to perform calculations (arithmetic unit), and a means of controlling the operation (control unit).

The modern history of digital computation began in 1939 with Howard Aiken and his associates at Harvard University. Their work on an automatic-sequence-controlled electromechanical calculator, called the *Mark I*, was completed in 1944. This machine was primarily designed to solve differential equations and followed a sequence of programmed instructions stored on punched paper tape.

Like Babbage's analytical engine, the Mark I used binary arithmetic and had the same three basic elements of memory, arithmetic, and control units. An input/output unit, however, was added to the Mark I to enter data and return answers, providing a firm foundation for further computer development. The Mark I could perform any specified sequence consisting of the four basic arithmetic operations (addition, subtraction, multiplication, and division) and references to tables of previously computed results. Information was supplied on punched cards and by the setting of switches on the machine, while answers were punched on cards or printed by a typewriter. The Mark I was in use for over 15 years, and a typical multiplication on that machine required about 3 seconds.

In 1946, the second key development of modern computer technology occurred. J.P. Eckert and J.W. Mauchley at the University of Pennsylvania designed the *Eniac (electronic numerical integrator and computer)*. This machine was the forerunner of the first generation of electronic digital computers and contained 18,000 vacuum tubes. Because the device had no moving parts, its development represented a significant advance in computer technology. Addition required 0.2 milliseconds and multiplication took 2.8 milliseconds. The Eniac was programmed by the wiring of circuit boards on the machine, which was a slow process that limited its versatility.

In 1946, Dr. John von Neumann at the Institute of Advanced Study in Princeton introduced the concept of the stored program, which is used in all modern computers. In a stored program, a sequence of instructions maintained in memory are examined by the device to determine what sequence of computations is to be followed. The computer proposed by von Neumann was completed in 1950. Called an *Edvac (electronic discrete variable automatic computer)*, it had 12 different instructions and performed additions in 0.9 milliseconds and multiplications in about 2.9 milliseconds.

The first commercially produced electronic digital computer was the Univac I. This computer was used by the United States government to perform all the calculations of the 1950 national census and was sold to commercial organizations in 1951. The Univac I and most of the following computers developed during the 1950s used vacuum tubes.

Vacuum tube technology, while permitting computation to occur at a rate thousands of times that expected from Babbage's analytical engine, had a number of drawbacks. First, tubes were bulky devices that consumed large amounts of power and correspondingly produced a large amount of heat that had to be removed from the environment. Second, the vacuum tubes were short-life devices; at any given time, the probability was quite high that at least one tube had failed in the computer.

These early machines were predominantly used for scientific work and there was very little desire to add to the hazards of computing by transmitting information to remote locations via telephone or telegraph lines. As the volume of information from computing grew, and as decisions became more time critical, new developmental efforts turned toward the existing telegraph lines and the Telex message-switching network, fostering the beginning of data communications technology as we know it today.

By 1960, the semiconductor industry had been formed, and during the early 1960s, a second generation of computers was developed that used discrete solid state electronic components in place of vacuum tubes. These computers performed operations hundreds of times faster, with instructions executed in microseconds, and were less expensive than the machines they replaced.

By the mid-1960s, advances in technology resulted in the replacement of discrete electronic components (transistors) by integrated components that combined hundreds of functional units on one chip. A dramatic price/performance breakthrough in computational capability occurred, fostering the third generation of computers, with computer instructions now executed in nanoseconds (billionths of a second).

Advances in semiconductor technology have led to chips replacing tens of thousands of discrete units, resulting in computer developments that some consider to be the fourth generation of computer technology. The LSI (large-scale integrated) and VLSI (very-large-scale integrated) chips formed the foundation for the development of the microcomputer, also known as the *personal computer (PC)*.

The first PCs were marketed in kit form during the mid-1970s. They were followed by the "ready-made" Apple I and Apple II in the late 1970s, and by the IBM PC in 1981. Although those devices are primitive when compared to modern PCs, their introduction began a computer-communications revolution. For the first time, true computer intelligence and processing capability was available at the user's desktop.

As the use of PCs within business organizations began to grow, so did their users' need to communicate with each other and with the corporate mainframe computer. Eventually, this need resulted in the development of different types of local area networks (LANs) and gateways to connect LAN-based computers to mainframes.

The remainder of this chapter focuses on how computers operate and the evolution of personal computing and its communications requirements. The later chapters feature in-depth coverage of PC communications software, LANs, and other PC-communications-related topics.

2.2 Computational Numbering Concepts

The basic building block mathematics of a digital computer is Boolean algebra. Named after the English mathematician, George Boole, who pioneered in the field of symbolic logic, it is an important tool in the design and analysis of computers.

The simple circuits illustrated in Figure 2.2.1 show why mathematical logic is important in the design of digital computers. Assign the value of 0 to the switch when it is open and the value of 1 when it is closed. These two switch states then correspond to the value of false (0) and true (1) of the symbolic logic of Boolean algebra.

For the simple series circuit illustrated in Figure 2.2.1, there will be continuity to the output point C only when both switch A and switch B are closed. This situation can be described by the Boolean statement A AND B, which is written symbolically as $A \bullet B$.

In the lower portion of Figure 2.2.1, a parallel circuit is represented. This is a circuit where there will be continuity to point C if either A or B is closed and is described by the Boolean statement A OR B, which is written symbolically as $A + B$.

Although George Boole's book, *An Investigation of the Laws of Thought*, concerning mathematical logic, was published in 1854, it was not until the 1930s that bistable electronic components were available from which computers could be con-

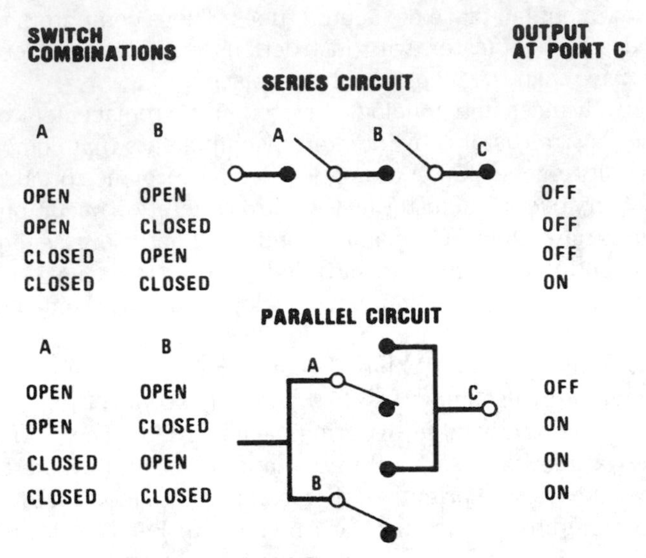

Figure 2.2.1 Switching Circuit Analysis.

structed. A bistable electronic component is so named because it exists in one of two states. Consider the action of a bistable electronic component illustrated in Figure 2.2.2. This component could be a vacuum tube, as found in a first generation computer, a semiconductor (such as a diode) common in a second generation computer, or a one functional unit on a chip in a third generation computer.

As illustrated in Figure 2.2.2, the bistable component is initially at state A; then an energy pulse drives it to a new state B. If the component is in state B, the removal of the energy pulse will return it to state A. This can be thought of as being similar to turning on and off a transistor. If you define state A to represent the number 0 and state B to represent the number 1, you can—through the binary numbering system—use bistable components to represent any number.

Prior to examining the binary numbering system, review a commonly used, but infrequently discussed numbering system, the decimal system.

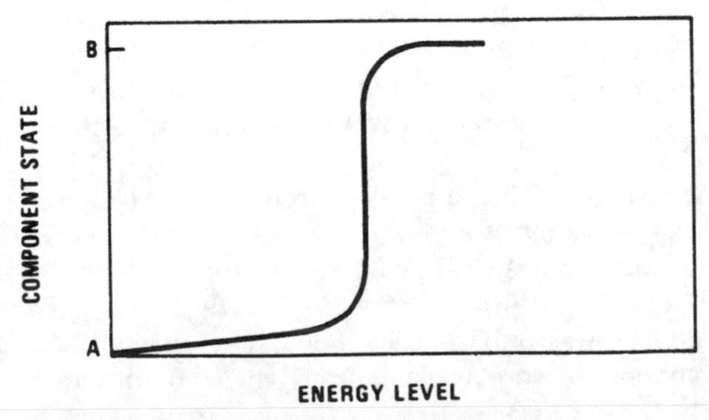

Figure 2.2.2 Action of bistable electronic component.

TABLE 2.2.1 Decimal System Powers of the Base

$1 =$	$= 10^0$
$10 = 10$	$= 10^1$
$100 = 10 \times 10$	$= 10^2$
$1{,}000 = 10 \times 10 \times 10$	$= 10^3$
$10{,}000 = 10 \times 10 \times 10 \times 10$	$= 10^4$
$100{,}000 = 10 \times 10 \times 10 \times 10 \times 10$	$= 10^5$

The decimal system

The decimal system is also known as the *base-10 number system*. It has 10 different symbols that make up all its numbers—0 through 9. Consider the number 575. This number really means 5 hundreds plus 7 tens plus 5 ones. Each digit corresponds to a place value that is obtained by multiplying 10 by itself over and over, resulting in the 10s place, the 100s places, and so on. The place values can thus be represented as powers of the base, meaning the number of times the base is multiplied by itself, as shown in Table 2.2.1. From Table 2.2.1, notice that 10 to the 0 power (10^0) is 1. In fact, any number to the zero power is 1. Note also that 10 to the first power (10^1) is 10. Returning to the decimal number 575, it can also be read as follows:

$$
\begin{aligned}
575 &= 5 \times 10^2 + 7 \times 10^1 + 5 \times 10^0 \\
&= 5 \text{ hundreds} + 7 \text{ tens} + 5 \text{ ones}
\end{aligned}
$$

The binary system

In the binary or base-2 system, only the digits 0 and 1 are used. When using binary numbers, the place values are found by multiplying 2 by itself over and over to obtain the 2s place, the 4s place, the 8s place and so on. The place values can be represented as powers of the base, as shown in Table 2.2.2. Using the information in Table 2.2.2, the binary number 11010 can be expressed as its decimal equivalent:

$$
\begin{aligned}
11010 &= 1 \times 2^4 + 1 \times 2^3 + 0 \times 2^2 + 1 \times 2^1 + 0 \times 2^0 \\
&= 16 + 8 + 0 + 2 + 0 \\
&= 26
\end{aligned}
$$

TABLE 2.2.2 Binary System Powers of the Base

$1 =$	$= 2^0$
$2 = 2$	$= 2^1$
$4 = 2 \times 2$	$= 2^2$
$8 = 2 \times 2 \times 2$	$= 2^3$
$16 = 2 \times 2 \times 2 \times 2$	$= 2^4$
$32 = 2 \times 2 \times 2 \times 2 \times 2$	$= 2^5$
$64 = 2 \times 2 \times 2 \times 2 \times 2 \times 2$	$= 2^6$

To alleviate the confusion that can occur when working with different number systems, we can employ the use of a subscript to indicate the numbers base. Thus, $11010_2 = 0.26_{10}$, which is read as "11010" in base 2 is equal to "26" in base 10.

To go in the opposite direction and convert a decimal number to binary, two methods can be used. The easier, but more laborious, method is to write out all the binary place values up to the highest one, you will require for the decimal number. Thus, for converting the number 26_{10}, you would write:

16 8 4 2 1

The next higher place number, which would be 32, is not required because the decimal number to be converted (26_{10}) is smaller than that one.

Now all that remains is to put a 1 under each of the terms via trial and error until they add up to 26. After some fast addition, the result would be:

16 8 4 2 1

1 1 0 1 0

A second method to convert decimal numbers to binary is by continuous division by 2, where the remainder is used to indicate the binary place value while the quotient is divided again and again by 2 until it becomes zero, as illustrated in Table 2.2.3.

Although binary numbers are the basis of computer technology, they are difficult for people to use—especially when such numbers are long. This makes the probability of an error occurring relatively high when they are written or read. Because binary numbers tend to be long, it is often convenient to group the binary digits in sets

TABLE 2.2.3 Decimal to Binary Conversion by Repeated Division

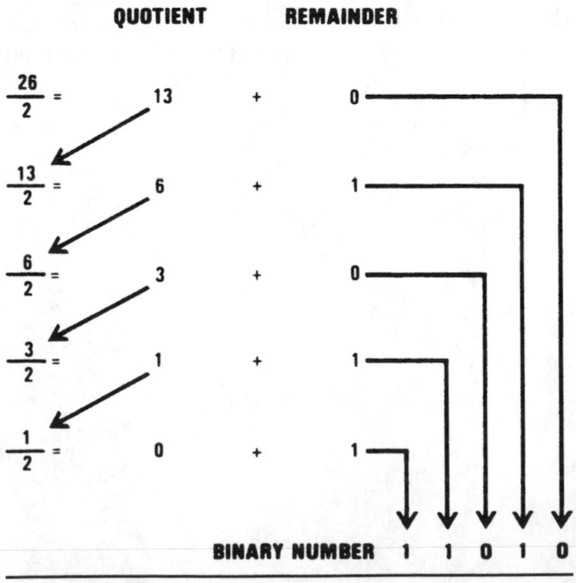

	QUOTIENT		REMAINDER
$\frac{26}{2}$ =	13	+	0
$\frac{13}{2}$ =	6	+	1
$\frac{6}{2}$ =	3	+	0
$\frac{3}{2}$ =	1	+	1
$\frac{1}{2}$ =	0	+	1
		BINARY NUMBER	1 1 0 1 0

TABLE 2.2.4 Octal System Powers of the Base

1 =	$= 8^0$
8 = 8	$= 8^1$
64 = 8 × 8	$= 8^2$
512 = 8 × 8 × 8	$= 8^3$

of digits. These sets of digits now produce different bases with different numbering systems and provide a shorthand method of binary representation.

Most digital computers are designed for one or another of these shorthand systems. As an example, a computer can be designed so that humans can easily use the octal number (base-8) system with it, as did the Honeywell Series 700. The IBM System/360 and 370 computers were designed so that the hexadecimal (base-16) system can be easily used with it. In any case, the computer itself always uses the binary system for internal data manipulation, and the shorthand systems based upon the octal and hexadecimal number systems are only for human convenience.

Octal and hexadecimal number systems

By grouping the binary digits into sets of 3, the numbers are covered into octal format. The octal number system is based on 8, as illustrated in Table 2.2.4.

To convert an octal number to decimal is just like converting a binary number to decimal, with the exception that one uses powers of 8 rather than powers of 2. Thus, the octal number 512 can be represented as:

$$5 \times 8^2 + 1 \times 8^1 + 2 \times 8^0 = 330_{10}$$

To convert a decimal number to an octal number, one may use either of the two methods for converting a decimal number to a binary number, which were previously explained. As an example, consider the decimal number 330. To convert to octal, we divide the decimal number by 8 and save the remainder, continuing the process by dividing the quotient by 8 until no further division is possible, with the remainders forming the octal as a number as illustrated in Table 2.2.5.

Now that we have examined octal as a number system, let us explore how it can be used as a shorthand for the binary number system. The first eight octal numbers and their binary equivalents are illustrated in Table 2.2.6.

From Table 2.2.6 we can see that the highest number that can be written with a single digit in octal is 7, and its binary equivalent requires three binary digits. By grouping binary digits into sets of 3 we can therefore convert the number into octal format. Thus, the binary number

$$1\,1\,1\,0\,1\,1\,0\,1\,0\,1_2$$

can be grouped into 3 sets of 3 digits and a most significant digit.

$$1\ \ 1\,1\,0\ \ 1\,1\,0\ \ 1\,0\,1$$

Converting to octal, the binary number becomes 1665_8.

The hexadecimal number system, whose base is 16, uses the symbols 0 through 9 for the first ten digits and the letters A through F to stand for 10, 11, 12, 13, 14 and

TABLE 2.2.5 Decimal to Octal Conversion by Repeated Division

	QUOTIENT		REMAINDER
$\frac{330}{8}$ =	41	+	2
$\frac{41}{8}$ =	5	+	1
$\frac{5}{8}$ =	0	+	5

OCTAL NUMBER 5 1 2

TABLE 2.2.6 Octal and Binary Equivalents

Octal	Binary
0	0
1	1
2	10
3	11
4	100
5	101
6	110
7	111

15. By grouping the binary digits into sets of 4, the number is converted to hexadecimal format. Thus, the binary number 1 1 1 0 1 1 0 1 0 1 can be grouped into 2 sets of 4 digits and 1 set of 2 digits to produce:

$$\underbrace{1\,1}_{3}\quad \underbrace{1\,0\,1\,1}_{B}\quad \underbrace{0\,1\,0\,1}_{5}$$

the hexadecimal number $3B5_{16}$. In Table 2.2.7, the relationships between the first 17 numbers of the hexadecimal, decimal, octal, and binary number systems are listed.

2.3 Binary arithmetic

Now that several numbering systems have been examined, investigate how one such system can be used. Binary numbers can be operated on in the same manner as decimal numbers, with a carry occurring when 1 is added to 1, which is similar to the carry when 1 is added to 9 in the decimal system. The four possible combinations in binary addition are:

AUGEND	+	ADDEND	=	RESULT	+	CARRY
0	+	0	=	0	+	0
0	+	1	=	1	+	0
1	+	0	=	1	+	0
1	+	1	=	1	+	1

Like decimal arithmetic, the carry is added to the next higher position. As an example, consider the following:

1	carries	1	
6		0110	
+ 5	in decimal becomes	0101	in binary.
11		1011	

Binary subtraction can be performed by following decimal system rules where the next position of the minuend is borrowed when the present value of the subtrahend

TABLE 2.2.7 Numbering System Relationships

Hexadecimal	Decimal	Octal	Binary
0	0	0	00000
1	1	1	00001
2	2	2	00010
3	3	3	00011
4	4	4	00100
5	5	5	00101
6	6	6	00110
7	7	7	00111
8	8	10	01000
9	9	11	01001
A	10	12	01010
B	11	13	01011
C	12	14	01100
D	13	15	01101
E	14	16	01110
F	15	17	01111
10	16	20	10000

is greater than the present value of the minuend. Because many computers perform subtraction by doing a special type of addition operation to minimize circuitry, examine how this process can occur.

Subtraction can be performed by the implementation of complement binary arithmetic. The ones complement of a binary number is formed by changing every one to zero and every zero in the number to one. For example, the ones complement of 0110 is 1001. The twos complement of a binary number is obtained by adding 1 to the ones complement of that number, as shown:

binary number	0110
Ones complement	1001
Twos complement	+ 1
	1010

In the binary number system, subtraction can be performed by adding the twos complement of the subtrahend to the minuend and discarding the final carry. For example, to perform the binary subtraction

Minuend	0110
Subtrahend	0101

the twos complement of the subtrahend is formed, and it is obtained by adding 1 to the ones complement (1010 + 1 = 1011). Next, the twos complement of the subtrahend is added to the minuend and the final carry is discarded as shown:

Minuend	0110	
Twos complement of subtrahend	+ 1011	
	1) 0001	Difference
discard final carry ↙		

The discarded final carry provides the sign of the answer. If the final carry is 1, then the answer is positive. Conversely, if the final carry is 0, then the answer is negative.

In many computers, binary multiplication is performed by repeated additions because each partial is either zero or exactly the multiplicand. For example:

$$\begin{array}{r} 6 \\ \times 5 \\ \hline 30_{10} \end{array} \quad \text{in decimal becomes} \quad \begin{array}{r} 110 \\ \times 101 \\ \hline 110 \\ 000 \\ 110 \\ \hline 11110_2 \end{array} \quad \text{in binary}$$

The last operation to be considered in this section, binary division, is performed in most computers by a series of repeated subtractions. For example:

```
                     110     Quotient
(Divisor)    101 ) 11110     Dividend
                   101
                   101
                   101
                   0000
                   0000
                      0
```

2.4 Computer Hardware Organization

Architecturally, computers generally resemble one another to a very high degree. In Figure 2.4.1, a simplified block diagram is used to show the hardware organization of a typical digital computer.

Today, most computer memories are implemented using high-speed semiconductors or magnetic core elements. They are random-access devices, where data can be written to or read from any addressed section of memory.

As the name implies, the arithmetic and logic unit performs arithmetic operations on data transferred within the computer, the memory, and from and to input and output devices.

Referred to as the computer's "brain" because it coordinates all the computer units in timed logical sequence, the control unit receives sequences of instructions from memory. These sequences, called *programs*, reside in memory and are best known as *software*.

Data transfers for communicating with a wide variety of devices occur via the input or output sections of a computer. The input/output unit of a computer is bidirectional and referred to as the *I/O unit*. Devices connected to a computer's I/O unit

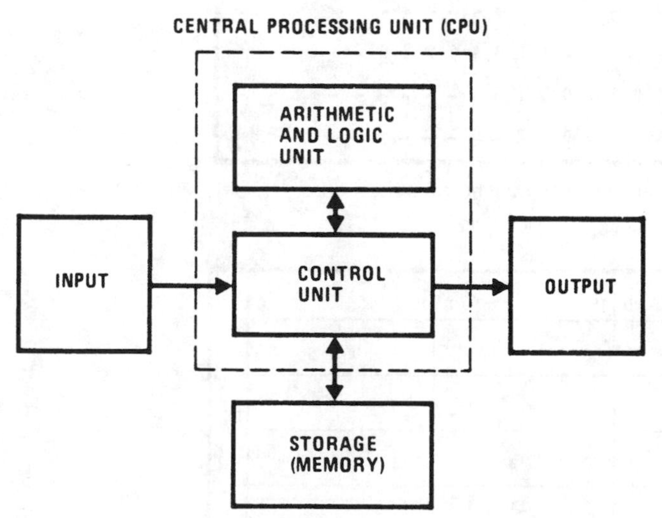

Figure 2.4.1 Hardware organization of a digital computer.

are known as *computer peripherals* and include printers, cathode ray tube displays, card readers and punches, magnetic storage disks and drums, paper or magnetic tape units, and assorted communications devices and lines.

In addition to the five functional units illustrated in Figure 2.4.1, every computer contains a set of registers, and circuits linking the five functional units to the registers. A register can be thought of as a box that provides intermediate storage of data and instructions. As an example, when two numbers are to be added, they might first be loaded into two registers. Then the circuitry to add the contents of the two registers would be activated to store the result in a third register.

Memory organization

At any instant during the execution of a program, the memory is equivalent to a grid of binary digits (bits) consisting of a pattern of ones and zeros as illustrated in Figure 2.4.2. Because a single bit can represent only two states, 0 or 1, you must code larger numbers by grouping bits into "words" to convey greater amounts of information.

Common computer word sizes include 8, 12, 16, 18, 24, 32, and 64 bits. A computer that groups bits into 16-bit words, for example, is referred to as a *16-bit machine*, and a computer whose logic is based on 32-bit words is referred to as a *32-bit machine*. Memory in a 16-bit computer can be visualized as illustrated in Figure 2.4.3. By convention, the bits of a computer word are numbered from right to left. In Figure 2.4.3, the low order or least significant bit is 0 and the most significant or high-order bit is 15.

Within the computer's memory, every word has a unique address. In addition to consisting of a grouping of bits, a computer word is also a group of bits that represents the largest addressable unit of information in memory.

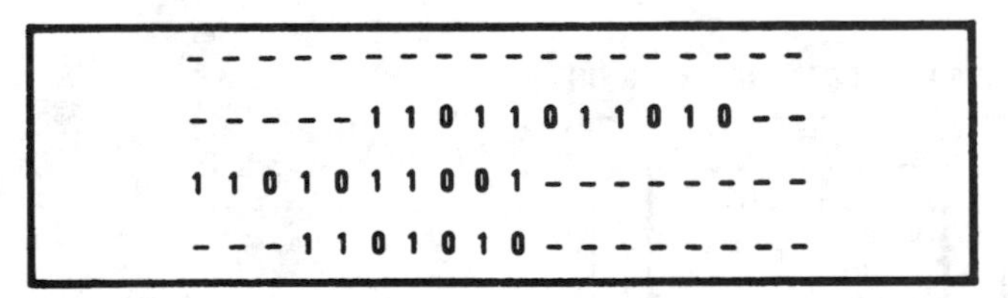

Figure 2.4.2 Computer memory during program execution.

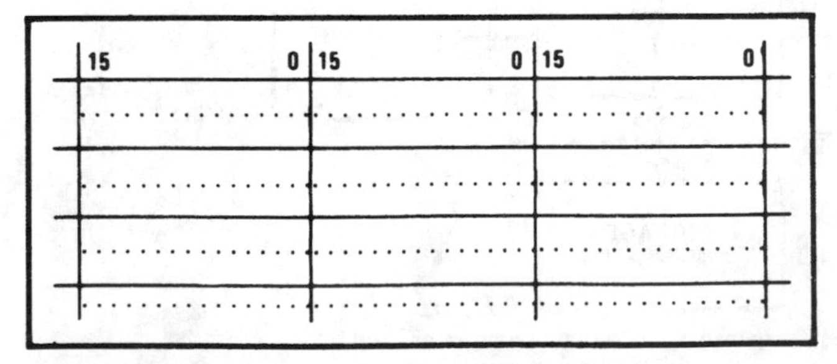

Figure 2.4.3 Organization of 16-bit-word computer memory.

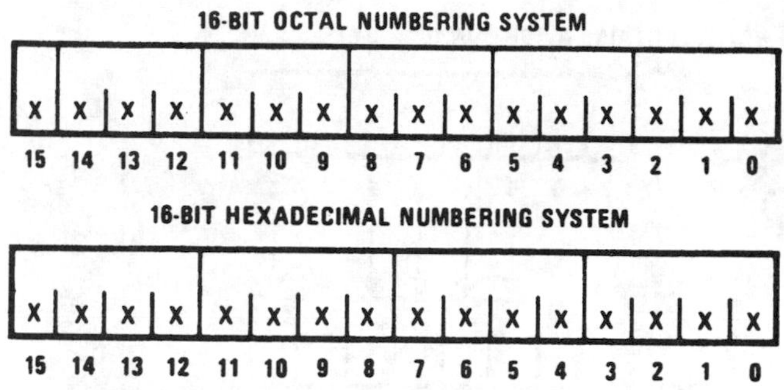

Figure 2.4.4 Numeric representation based on word structure.

Numeric representations

Depending on the architecture of the machine, numeric data can be represented in one of several ways. In a 16-bit computer, each word might be subdivided as shown in the top portion of Figure 2.4.4. Here, the 16-bit word is subdivided to contain 5 octal digits, and the 15th bit represents the sign of the data. Thus, numeric data would fall within the range:

$$177777_8 = -32767_{10}$$

to

$$077777_8 = +32767_{10}$$

In the bottom portion of Figure 2.4.4, the 16-bit word is subdivided into four hexadecimal digits. If the high-order bit is used to represent the sign (one convention), then the numeric data would fall within the range:

$$FFFF_{16} = -32767_{10}$$

to

$$7FFF_{16} = +32767_{10}$$

Character representation

In addition to entering, processing, and outputting numeric data, the computer must interpret letters of the alphabet and special characters, such as $, + , –. Then, by having appropriate software and peripheral devices, a payroll program (for example) could be executed and employee checks with names, addresses, and remuneration could be printed. To represent characters, a set of bits are grouped together to form what is referred to as a *byte*.

Two of the most common codes used to represent data with bytes are the American standard code for information interchange (ASCII) and the extended binary coded decimal interchange code (EBCDIC). If a character is represented by an 8-bit byte, then a computer will relate its word to character bytes as illustrated in Figure 2.4.5.

6-BIT COMPUTER WORD CONTAINS A PORTION OF A BYTE

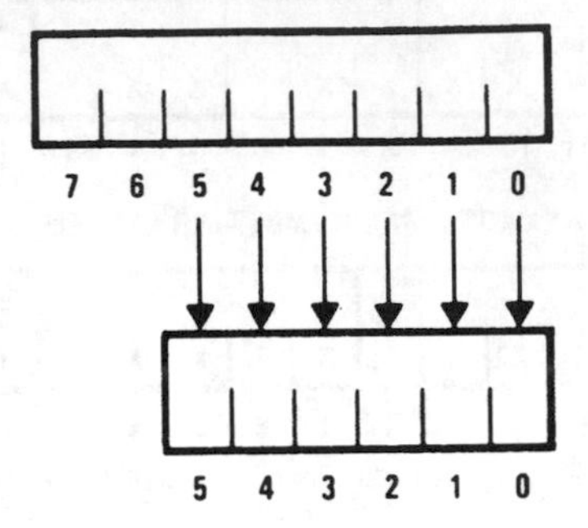

8-BIT COMPUTER WORD CONTAINS 1 FULL BYTE

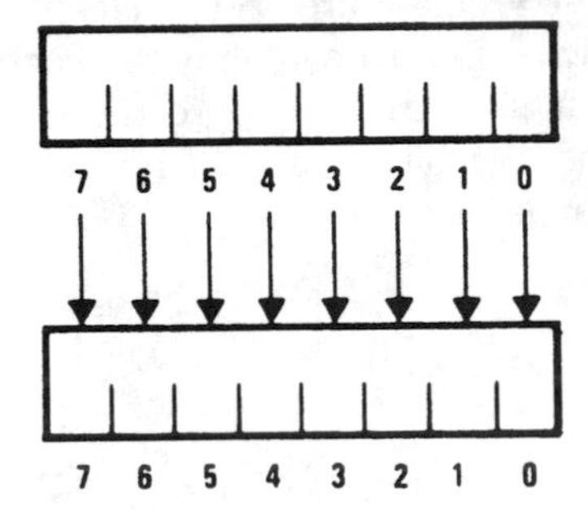

16-BIT COMPUTER WORD CONTAINS 2 FULL BYTES

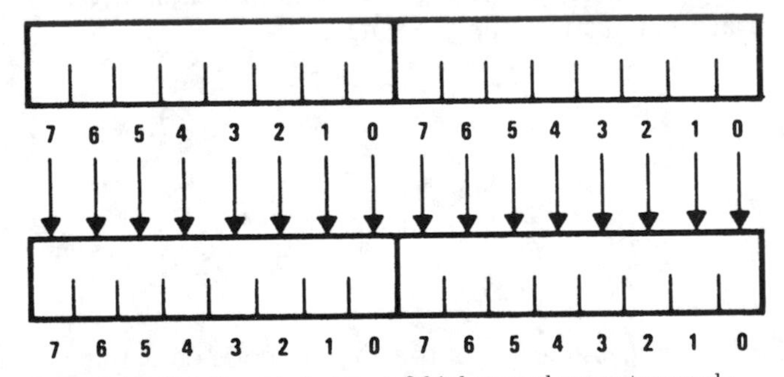

Figure 2.4.5 Relationship between 8-bit byte and computer words.

If the computer's word size is 6 bits, the two high-order bits from the 8-bit byte are eliminated, which reduces the number of different characters that can be represented on such a machine. In the middle portion of Figure 2.4.5, an 8-bit-word computer directly represents one byte with one 8-bit word. When the computer's word size is 16 bits or greater, more than one byte may be packed into the computer word. This is illustrated in the bottom portion of Figure 2.4.5, where two 8-bit bytes have been packed into one 16-bit computer word. Although you have seen that at its lowest level a computer's memory consists of a sequence of binary digits, you can now recognize that both numbers and characters can be read, manipulated, and transmitted to peripheral devices by the computer.

2.5 Computer Software Concepts

A digital computer is constructed to perform calculations by executing a sequence of operations called *instructions*. The sequence of instructions is called a *program*.

In early digital computer applications, programming was accomplished by logically and electrically inserting binary digits into the computer's memory. Here, a particular string of bits (for example, 8 or 16) constitutes a language instruction, such as "store the results." The computer hardware then interprets each coded instruction for program execution. Such numeric coding, called *machine language programming*, makes excellent use of computer resources. But the effort is so tedious and time consuming that few computers are now programmed this way.

Assembly-language programming uses easy-to-remember symbolic notations, or mnemonics, instead of strings of 1s and 0s, to define an operation or instruction. The mnemonic ADD (for addition), for example, when inserted by the programmer, is interpreted by the computer to generate the corresponding machine-language instructions. That is, there is a one-to-one correspondence in that each assembly-language mnemonic symbol produces an equivalent machine-language instruction. The result of the assembly translation is called an *object program*. The object program is then executed by the computer as illustrated in the top portion of Figure 2.5.1.

The chief advantage of the assembly language is that the programmer only needs to remember and use simple symbols instead of a long string of 1s and 0s. The net result is that assembly-language programming is easier, faster, and more accurate than machine-language programming.

Applications programmed in assembly language appropriate to a particular computer are, however, usually valid only for that computer or generic family of computers. So, if circumstances warrant a changeover to another vendor's computer, the application must be reprogrammed for the new computer. In other words, machine-language and assembly-language programs are both "machine-dependent."

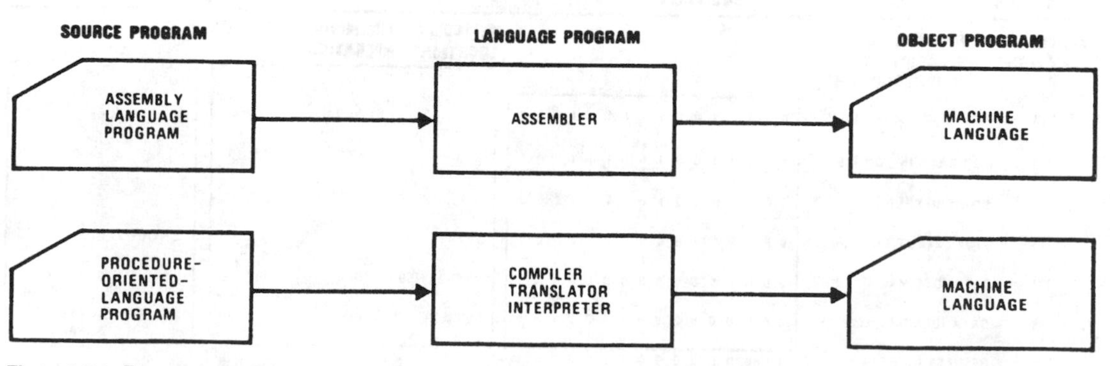

Figure 2.5.1 Program processors.

With a procedure- or higher-level-oriented language, the programming of a particular application becomes independent of the computer—provided that the computer vendor can supply a compiler, translator, or interpreter that converts the procedure-oriented language program into a machine-language object program suited to that computer. This is illustrated in the bottom portion of Figure 2.5.1.

With a procedure-oriented language, programmers can produce, with one statement, a segment of a program that in assembly language would require many more individual statements. In Figure 2.5.2, the relationship between machine-language, assembly-language, and procedure-oriented-language programming for a simple application involving the summing of two numbers is illustrated.

What the programmer has to code for each of these methods is enclosed in boxes. Note that in both the machine-language and assembly-language methods, the programmer writes five "lines" of instructions and two lines of data. The procedure-oriented language requires the same two lines of data, but just one statement, which the computer's translator automatically converts to the equivalent machine-language instructions.

In more complex programs, five procedure-oriented language statements might perform the equivalent of 50 or more assembly-language instructions. Thus, speed of programming in a procedure-oriented language is one of the key advantages over programming in assembly language.

Early procedure-oriented languages were certainly less efficient and slower in program execution than equivalent assembly-language programs because the procedure-oriented language invokes a greater number of machine instructions than would be required by the equivalent assembly-language program. This is mainly because the required compiler, translator, or interpreter is designed to perform generalized translations and the assembly language program can be code optimized, reducing the size of the object program and thus permitting the assembly-language program to execute in less time.

In general, procedure-oriented languages are easy to learn and use, hence they reduce training and program writing time. They also simplify program checkout, main-

LANGUAGE COMPARISON	MACHINE LANGUAGE	MACHINE LANGUAGE EQUIVALENT (BINARY)	ASSEMBLY LANGUAGE			PROCEDURE - ORIENTED LANGUAGE	
MEMORY LOCATION	FUNCTIONAL DESCRIPTION OF INSTRUCTION		SYMBOLIC LOCATION	MNEMONIC OPERATION	OPERAND		
				LDA	DATA1		
770	PICK UP DATA IN LOC 774	0 0 0 0 1 0 0 1 1 1 1 1 1 1 1 0		ADD	DATA2		
771	ADD DATA IN LOC 775	0 0 0 1 1 0 0 1 1 1 1 1 1 1 0 1		STA	RESULT		
772	STORE RESULT	0 0 0 1 0 0 0 1 1 1 1 1 1 1 1 0		STOP			
773	STOP EXECUTION	0 0 0 0 0 0 0 0 0 0 0 0 0 0 0 0	DATA1 (=1)	DEC	1	DATA (=1)	
774	DATA (DECIMAL 1)	0 0 0 0 0 0 0 0 0 0 0 0 0 0 0 1	DATA2 (=100)	DEC	100	DATA (=100)	
775	DATA (DECIMAL 100)	0 0 0 0 0 0 0 0 0 1 1 0 0 1 0 0	RESULT	BSS	101	RESULT = DATA1 + DATA2	101
776	RESULTS GO HERE	0 0 0 0 0 0 0 0 0 1 1 0 0 1 0 1					

Figure 2.5.2 Language relationships.

tenance, and documentation, and speed program conversion from one computer to another. Overall, procedure-oriented languages save expensive programming manpower, the cost of which keeps going up.

However, procedure-oriented languages are inefficient on object (machine-language) code. They also require more memory for program execution and extend program run time. In addition, sometimes they cannot express all the operations required in specialized applications, such as bit and byte manipulation, which are extensively used in data communications programming.

Since the introduction of procedure-oriented languages, substantial improvements have occurred in their efficiency, speed, and scope. Such popular procedure-oriented languages as Cobol (common business-oriented language), Fortran (formula translation), Basic (beginners all symbolic instruction code), RPG (report program generator), and others have been developed to meet specific business and scientific requirements. However, none is specifically designed for programming data communications applications.

Currently, computer programming for data communications must rely on an assembly language or an extended version of a language specifically tailored for other applications. Thus, a substantial portion of the effort expended in programming a data communications application on one computer might be lost if for some reason a new computer is acquired. Many persons in the industry view communications programming as a frontier for further progress because even today it is comparatively primitive and expensive, a situation that belies the sophisticated design and advanced technology of modern communications networks.

Part of the problem might be that the growth of data communications followed the introduction and initial expansion of scientific and business data processing. Thus, while specialized procedure-oriented languages were being developed for scientific and business use, communications programmers were just starting to use assembly language, which permits a large degree of bit and byte manipulation that is well suited to such programming.

2.6 Data Processing Developments

Until the early 1960s, most computers were used by persons at the computer's location. Programmers had their programs keypunched and the resulting cards were sent to the computer center, where they were batched together with other programs and executed one at a time. This process is known as *batch data processing*.

The development of magnetic disks permitted millions of additional bits of storage to be rapidly loaded into and transmitted from the computer's memory. Coupled with the development of multiprogramming, which allowed a computer to process a number of programs simultaneously by interleaving their execution, more efficient utilization of computer resources was possible. One job could now be executing while a second job was perhaps having its output listed on a peripheral device.

This resulted in the development of remote-access data processing systems. Here, data is transmitted from many locations via communications links to and from a computer performing the required data processing functions. This permitted many users to simultaneously access the computer. Because the computer's processing speed was many millions of operations per second, to each user it appeared that he was the only one using the computer's facilities.

The variety of remote-access data processing systems is already large and rapidly growing in both number and scope of operation. Through the mixture of computers and communications, many users can now obtain the immediate use of computational power regardless of location, while preserving on one site the expertise required to operate the system. A few of the more commonly used remote-access data processing systems include:

- *Conversational timesharing systems.* These systems permit the simultaneous sharing of a central computer among a group of users located at remote terminals and connected to the computer via a variety of communications lines. These systems permit a user to "interact" with a computer. The user develops, debugs, and executes programs "on-line," and receives the program results at his terminal. When conversational timesharing systems were first developed during the 1970s, the PC did not yet exist. Terminals were hard-wired to fixed logic devices that lacked intelligence. Today, the PC has mostly replaced dumb terminals, and the term *terminal* is now considered to be synonymous with PC.
- *Remote batch processing systems.* These systems permit the computer processing of tasks transmitted from distant locations. In contrast to conversational systems where the quantities of data input or output are limited, remote batch processing provides a mechanism for transmitting large quantities of data to and from the central computer. Because of the large amount of data that is transmitted in comparison to conversational systems, remote batch terminals usually operate at higher data rates than timesharing terminals.
- *Inquiry systems.* These systems are usually limited in scope because they are designed to provide information on a particular subject. One widely used inquiry system is a stock quotation service, where many terminals are connected to a computer via communications lines. This system permits remote users to enter a stock symbol and obtain such data as the stock's current price, its latest earnings per share, and other financial data.
- *Data collection systems.* These systems involve the transmission of information—gathered at many remote points—to a central computer for processing and storage. One common data collection system is a series of badge readers located at many entry points in a building. As workers enter the building after normal working hours, coding on their badges is read by the reader and transmitted to the computer.
- The computer can then compare the badge numbers against a list of authorized badge numbers, and, if a match occurs, transmit a signal to unlock the door. If required, the computer can also be programmed to prepare a report of all after-hour entries and other information required by management.

For each of these four systems, PC hardware and software enable the PC to function as a ubiquitous type of terminal. The PC provides a computing platform that, when tailored via appropriate additional hardware and software, enables it to perform a variety of processing and communications-related tasks.

In order to connect the computer to communications lines, it is necessary to add compatible with the particular speed and code of the terminal device connected to the remote end of the line. In addition, it is necessary to indicate to the computer that a complete unit of information has been received or is ready to be transferred to the terminal. Such a unit of information could be a bit. However, in most cases it is a byte, which is a grouping of bits that represent a character. By acting on bytes of information instead of bits, the computer can spend more time processing information than keeping track of communications.

Early large-scale data processing computers usually interfaced one, or at most a few, terminals by the use of single-line controllers. In addition to providing the clocking and circuitry necessary to interface the communications line, these controllers buffered incoming bits into bytes of information that were then passed to the computer. This is illustrated in Figure 2.6.1.

As remote-access networks evolved, more and more terminals were linked to an organization's large-scale data processing computer. To minimize the effect of interfacing a large number of communications lines, multiline controllers were designed. They provided the buffering capability and clocking circuitry to interface a large number of communications lines to the computer, as illustrated in Figure 2.6.2.

These controllers relieved some of the computer's communications processing functions. At the same time, they not only required the execution of computer instructions for their operation but also placed a large memory buffering allocation burden on the computer. This was because incoming and outgoing information had to be buffered to compensate for the difference between the computer's internal operating speed and the slower data rate permissible over the communications line.

With remote-access systems growing in size, number of terminals, and complexity, it became obvious that a computer with one or more multiline controllers would be able to handle only a limited number of lines and still maintain efficient processing of

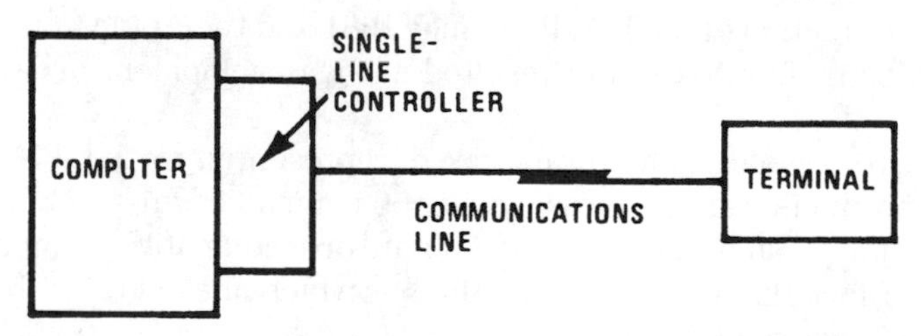

Figure 2.6.1 Single-line controller.

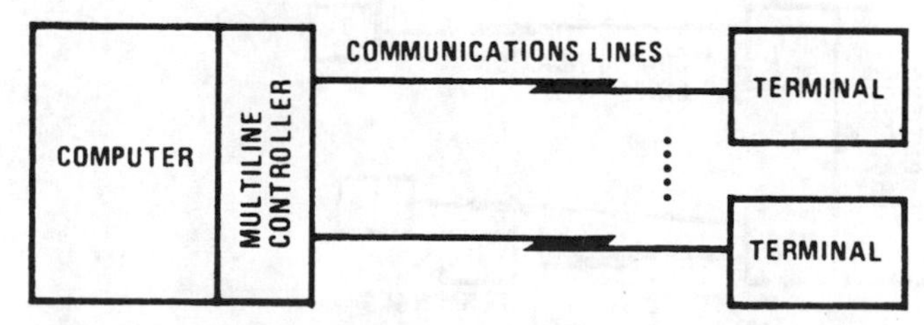

Figure 2.6.2 Multiline controller.

data. Beyond that point, the overhead in processing instructions for data transmission would tend to decrease significantly the computer's capability and throughput that remained for processing data. In recognition of these facts, computer manufacturers realized that if the communications function was to be performed efficiently and the overhead on the computer was to be minimized, it would have to be removed from the computer and performed elsewhere. This resulted in the development of data communications processors—small-scale computers with a mixture of single and multiline controllers and the necessary hardware to interface with large numbers of communications lines.

The communications processor makes it possible to remove most of the data communications overhead from the central or host computer by processing the communications in front of the host—hence the name *front-end communications processor*. A modern remote-access computer network is illustrated in Figure 2.6.3. Through the use of a front-end processor, the memory space and communications-instructions executions previously required for communications in the host computer were removed. This permits the host to concentrate its effectiveness on information processing.

Evolution of personal computing

In the late 1970s, a new type of computer, based on the use of microprocessor technology, arose from the basements and garages of hobbyists and made its way into corporations, government agencies, and homes. Commonly called the *PC*, the personal computer has had as profound an effect on the workplace as the office copier and telephone. Software designed to facilitate word processing functions, local database construction and retrieval, and electronic spreadsheet calculations provided the PC user with the ability to quickly perform operations that previously required terminal access to mainframe computers.

Although the PC field was initially fragmented because of the lack of a dominant computer, the introduction of the IBM PC in mid-1981 and the Apple Computer Corporation's Macintosh a few years later resulted in the development of two de facto standards.

The first de facto standard is hardware-based, represented by the IBM PC Series and compatible products manufactured by over 50 companies. In general, programs written for operation on an IBM PC will also operate on a compatible computer. However, the user interface that governs the method by which one operates the program can vary drastically between programs.

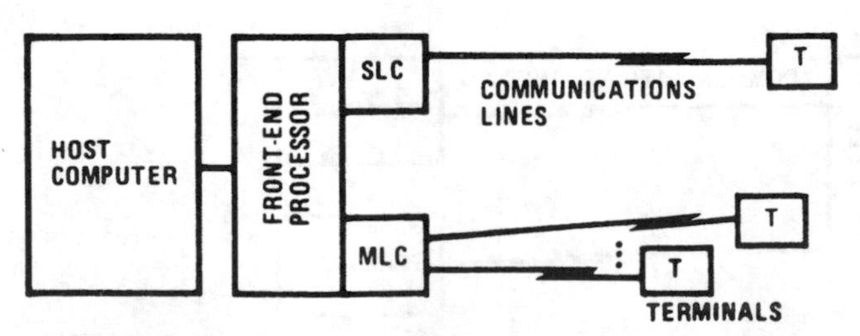

Figure 2.6.3 Using the front-end processor.

The second de facto standard is software based, represented by the Apple Macintosh. Unlike IBM, which used numerous commercially available parts in developing the PC series, the Macintosh was essentially a proprietary design and could not be legally duplicated. However, Apple developed the Macintosh user interface based on "pull-down windows" and icons and provided software developers with a mechanism to design similar-appearing user interfaces. The ability of programs to operate in a similar manner is probably the reason that user surveys indicate that the time required to learn the operation of a program on a Macintosh is less than that required on an IBM PC or compatible.

In 1987, IBM introduced a new family of computers under the PS/2 (personal system) label, which made its computers similar in many ways to the Macintosh. First, IBM based its new series primarily on proprietary technology, which has made it more difficult for other vendors to duplicate. Secondly, IBM introduced several new operating systems to include a "presentation manager," which is based on the use of icons and pull-down menus and permits a common user interface to be developed between application programs and the computer user.

IBM's presentation manager software products have not achieved a notable market success. However, a competitive product, Windows, developed by the Microsoft Corporation, has greatly facilitated the use of IBM PC and compatible computers. Windows can be considered a graphics environment as it replaces the command-line prompts—associated with the Disk Operating System (DOS)—previously required for use with icons and menus. The first versions of Windows were noted for their sluggishness and tendency to cause unrecoverable errors. By version 3.1, prior problems had been resolved. At the time of this writing, Microsoft was shipping over a million copies of Windows each month.

Similar to the requirement to connect terminals to computers, a multitude of applications have been developed that resulted in the requirement for PCs to access mainframe computers, minicomputers, and other PCs. In some situations, the PC has replaced the nonintelligent terminal previously connected to the mainframe, enabling users to access the larger computer to operate extensive databases of information and to extract the data required for local processing. Then, PC users can download data from the other computer onto their PC's floppy or fixed disks and use software on their PCs to analyze and manipulate the data. Other PC applications that can require a communications capability include such diverse functions as electronic mail, budgeting, accounting, payroll, and similar common organizational activities. Here, the communications capability required for budgeting, accounting, and payroll operations can range from the transmission of regional office data to a central computer site to the use of PCs to communicate with other corporate PCs to pass appropriate information between locations.

As the use of PCs proliferated, the methods developed to permit communications between PCs and other computers correspondingly increased. Initially, PCs communicated with other computers by transmitting and receiving data one line at a time. This line-by-line transmission method is also known as *TTY* for teletype transmission because it duplicated the method by which the common teletype terminal communicated. Later, the addition of different software and hardware enabled personal computers to emulate the video attributes and operational characteristics of a variety of terminals, manufactured by different vendors, which transmit and receive full screens of data. This capability is more commonly known as *full-screen terminal emulation* and it permits a PC to access different types of computers. Figure 2.6.4 illustrates how one personal computer with appropriate hardware and software could access three different computers, emulating a different type of full-screen terminal when communicating with each computer. In comparison, three separate terminals might be required to provide a similar communications capability.

Because of the extensive penetration of personal computers into most business organizations, universities, and government agencies, it was not long until additional requirements arose to transmit data between PCs, and to share common files and application programs, as well as high-cost peripheral devices. Such requirements were satisfied by the development of local area networks, which enable PCs to be linked to one another, as well as to mainframe computers.

The structure by which computers are connected to one another on a local area network (LAN) is known as the network's *topology*. Figure 2.6.5 illustrates three common LAN topologies. Detailed information concerning LANs is covered later in this book.

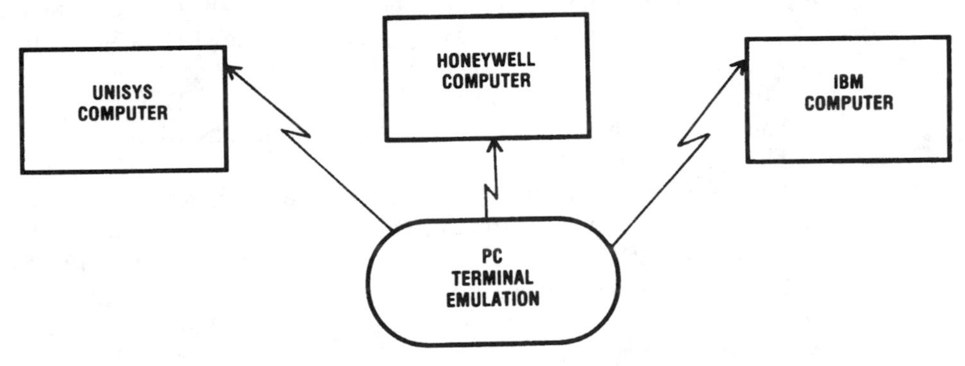

Figure 2.6.4 Terminal emulation.

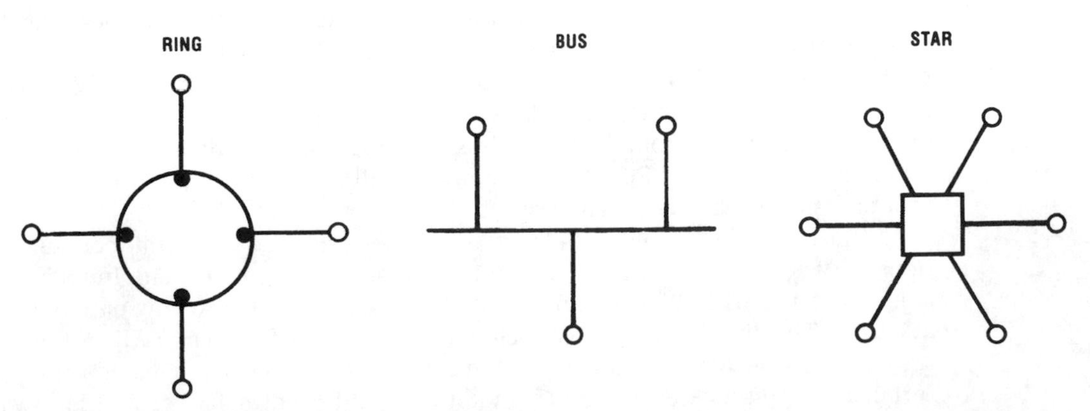

Figure 2.6.5 Common LAN topologies.

2.7 Summary

Although the first device used by man to perform calculations, the abacus, was developed around 450 B.C., it was not until 1833, when Charles Babbage designed an "analytical engine," that the computer era dawned.

The modern history of digital computation commenced in 1939 at Harvard University and resulted in the development of the Mark I automatic sequence-controlled electromechanical calculator in 1944. In 1946, Dr. John von Neumann introduced the concept of the stored program, which is the basis of all modern computers. Four years later, the world's first commercially produced electronic digital computer was used by the United States government to perform all the calculations of the 1950 national census. Computers' computation power and labor-saving attributes were recognized. The machines rapidly gained acceptance as management tools as companies began to apply them to both scientific and business problems.

In spite of the rapid growth in the number of operating computers, their cost prevented every needful location within a company from having one. This resulted in the centralized location of computational equipment and the development of data communications networks with terminals at remote locations connected to the computer. In order to extend computational power to remote users, techniques and specialized devices were developed that permitted rapid data transfer to the computer for processing. The appropriate report and data were then sent to their designated receiving terminals.

Among the first places that the computer pioneers turned to for developing networks were the common carriers: Western Union, the Bell System and associated telephone companies, and others, mainly because they already had large operating networks in place. Although few of these networks were designed to transmit data, they were available to almost all locations. Through the use of specialized components, they could be used for data transmission. Another advantage in turning to the common carriers was that they already had terminal devices available that sent messages between devices via their message-switching services.

These terminals were essentially typewriters coupled or connected to communications lines and were a natural evolution from the original telegraph keys. The terminals, appropriately known as *teletypewriters*, were easily adapted for communicating with computers and formed the foundation of early data communications networks.

Beginning in the late 1970s, the rapid acceptance of PCs resulted in a corresponding increase in communications between computational devices. Initially, personal computers communicated on a line-by-line basis in a manner similar to teletype terminals. Later, the development of specialized hardware and software enabled PCs to emulate the operation and functionality of different types of terminal devices. This permitted computer centers containing equipment manufactured by two or more vendors to replace—in many instances—single-computer-access terminals with PCs that could emulate two or more terminals and provide access to each computer.

Because of the growth in personal computing, local area networks were developed to permit each computer on a network to send and receive electronic messages, share access to common programs and data files and to expensive peripheral devices, and, in some instances, to provide a common method of access to a mainframe computer. Today, the prevalence of PCs in business organizations, government agencies, and educational institutions has resulted in a large increase in the requirement to share and exchange information. This, in turn, has resulted in the development of specialized hardware and software products to enable PCs to communicate with one another as well as with other types of computational devices.

Questions

2.1 Discuss some of the similarities between Babbage's "analytical engine" and modern digital computers.
2.2 What were some of the disadvantages of vacuum tube technology employed in the first generation of computers?
2.3 What are the decimal equivalents of the following binary numbers?

A. 10110110 C. 10110111
B. 01101011 D. 11111111

2.4 Convert the following decimal numbers to their binary equivalents:

A. 27 C. 410
B. 186 D. 206

2.5 Convert the following binary numbers to their octal equivalents:

A. 11111111 C. 11000101
B. 10101011 D. 10001100

2.6 Convert the following binary numbers to their hexadecimal equivalents:

A. 11111111 C. 11000101
B. 10101011 D. 10001100

2.7 Construct a table from decimal 17 to 26 and their binary, octal, and hexadecimal equivalents.
2.8 Perform the following subtractions by using twos-complement addition:

A. 101110 − 100111 C. 100011 − 011010

B. 1101110 − 1111001 D. 011111 − 100000

2.9 How might data be represented octally in an 18-bit computer, 24-bit computer, and 32-bit computer?
2.10 What is the relationship between a procedure-oriented language, an assembly language, and machine language? Discuss ease of usage, machine transferability, documentation, and maintenance aspects.
2.11 Why is assembly language today the primary means of programming communications applications?
2.12 Why are bytes the preferred unit of information to be acted upon instead of bits?
2.13 What role do single-line and multiline controllers play in computer communications?
2.14 What are some of the advantages of employing a front-end processor to control computer communications?
2.15 Discuss a possible relationship between processing data on a large computer and a personal computer with respect to communications.
2.16 How could one personal computer be used to provide access to two or more different types of mainframe computers?

Chapter

3

Role and Utilization of Terminals and Personal Computers in Data Communications

3.1 Introduction

During the last decade, a large number of terminals have been developed that can be attached to communications lines for transmitting data to computers. These devices can be categorized into two specific types: terminals into which data is entered by human operators and terminals that perform the automatic collection of data from instruments. Prior to discussing the types, operation, and functions of terminals, let us first define what we mean by a terminal in the context of the data communications field.

A *terminal* is a device that permits information to enter or exit a data communications network, where the network consists of one or more terminals linked via a communications line to computers. The terminal can be physically located anywhere communications facilities are available to link it to the computer: in the room next to the computer, on the opposite side of the continent from where the computer is located, or even, perhaps, on a different continent. Because any device that is used to input or output data can be classified as a terminal, the computer itself can be considered as a terminal device.

When terminals were first designed for data communications applications, they were essentially "dumb" devices. That is, they consisted of fixed logic developed through wiring that connected a mechanical part, such as a print head, to a serial port that transmitted and received a sequence of bits. As a user typed at the terminal's keyboard, the print head would move laterally, printing the keystrokes' results. At the same time, control circuitry would convert the keyed character to a sequence of bits, which would then be transmitted.

Advances in logic design added some intelligence to terminals, especially to cathode-ray-tube (CRT) devices. Through the use of logic modules, bit streams would be examined for special codes. These codes could, for example, position the next character two lines down and centered on the screen, clear (blank) the screen, or move the cursor to the first (usually upper left) screen position. The addition of such intelligence enabled terminal devices to be programmed to better support a variety of data processing applications. One such application was the display of a form to facilitate the entry of an individual's tax-return information by a government operator.

Further advances in logic design considerably improved the utility of terminal devices. The terminal had been designed with logic that would recognize the display codes generated by a particular computer. Therefore, the use of a specific terminal was essentially tied to a specific type of computer.

The development of the microprocessor and its incorporation into PCs provided a large degree of flexibility in using the PC as a terminal device. Because the PC had an open architecture, vendors were able to develop communications hardware and software products that enabled a PC to support the terminal-control codes of computers manufactured by different vendors. In effect, this development resulted in the PC becoming a ubiquitous terminal device.

The use of PCs increased; they essentially replaced typewriters in office popularity. There then rose the need for organization employees to exchange and transfer information. This need resulted in a new method of communications infrastructure, referred to as a *local area network (LAN)*. Using the PC as a platform, vendors developed hardware and software that enabled a PC user to communicate with other users connected to the same LAN. This activity brought about a change in nomenclature: the LAN-connected PC was no longer a terminal device, but—according to many vendors—was a node or workstation.

Unless a specific reference is made to the use of a personal computer, the term *terminal* is used to refer to any device that permits information to be communicated. In certain instances, where the method of access into a communications network or the hardware and/or software required to obtain a specific type of communications capability is unique to a personal computer, a distinction is made by specifically referring to that device.

In normal practice, a user's input/output (I/O) device is called a *terminal* and it is considered to be communicating with a computer at the other end of the line. This arrangement forms a terminal-to-computer link or path. When a computer communicates with another computer, the link or path is referred to as a *computer-to-computer link* even though one or both computers might function as terminals. With the terminal-to-terminal link, these are the basic communications links (illustrated in Figure 3.1.1). Many computer peripherals can be removed from the computer room and attached to a communications line, thereby becoming terminal devices. Thus, such computer peripherals as card readers, card punches, and even the computer's control console can be considered terminal devices.

Types of terminals

Some terminals can be used only to input data, some only to output (or receive) data, while some terminals and all personal computers permit data to be both inputted and outputted.

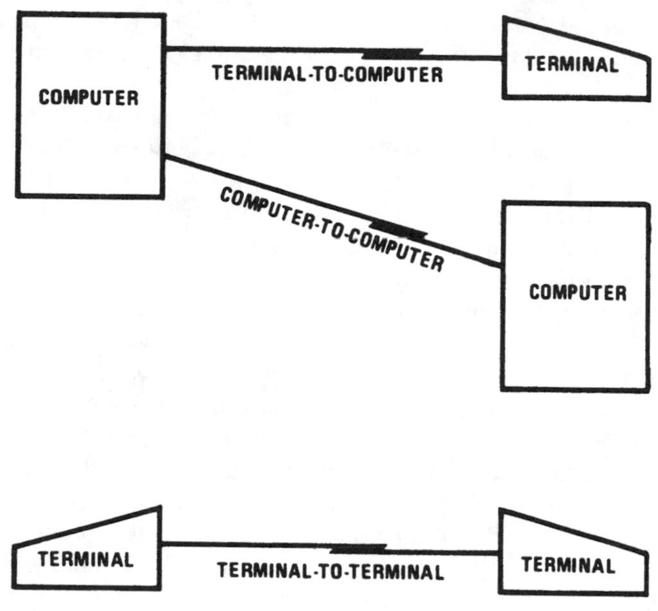

Figure 3.1.1 Basic communications links.

A keyboard-only device is generally used to transmit data to a data-collection device such as a computer. One example of a keyboard-only device is the push-button telephone, which, after a call has been established, can be used to send data by pressing the buttons, each of which transmits a different tone or frequency.

Quite similar to a normal typewriter, a keyboard-printer or teleprinter is a terminal that has a keyboard consisting of alphabetic, numeric, and function keys, as well as a printing mechanism. As a key is pressed, an internal signal generator transmits an electrical signal that defines that character. The signal is then transmitted over a communications channel to the computer, where the signal is interpreted and the character reproduced.

The transmit and receive capabilities of such terminals can be described as either keyboard send and receive (KSR), receive only (RO), or automatic send and receive (ASR). The KSR terminal has an alphanumeric keyboard for data entry at the operator's typing rate, and a character printer. As data is entered on the keyboard, a "hard" copy of each character is printed. Thus, data can be transmitted manually by keyboard and a local printed copy produced.

A receive-only terminal consists of a printer (and no keyboard). Here, data is received as printed copy. This device is normally used when there is no requirement at the terminal to transmit information to the computer. The RO can also be slaved to a video display to provide a local hard copy on an as-required basis (detailed later).

ASR machines use local storage and have automatic send and receive capability. Data can be transmitted manually by keyboard or automatically via such storage devices as punched paper tape, magnetic tape, cassettes, or diskettes. As data is transmitted or received, a local copy is printed for visual reference with or without using the storage device. With the storage device used, data can be batched off-line at the terminal operator's typing rate, then transmitted from storage at the communications line data rate in one continuous stream, to reduce communications costs. One of the earliest and most-utilized teleprinters is the Model 33, once produced by an AT&T subsidiary, the Teletype Corp.

Operating rates of keyboard terminals typically range from 10 to 120 characters per second. These data rates include both KSR and ASR terminals. Receive-only terminals have a much wider range of data rates, varying from 10 to 1,200 characters per second. This wider range is caused by the greater variety of RO printers on the market. They range from simple teleprinter devices (without keyboards) to line printers specially adapted for data communications.

Although first used in telegraph networks, teleprinters eventually evolved dramatically to the point where, during the 1970s and early 1980s, several million such devices were in use connecting users to timesharing networks via communications lines or directly to computers. The vast majority of teleprinters have been replaced by PCs operating with appropriate communications programs. However, by the mid-1990s, there were still tens of thousands of those antiquated devices still in use. The ASR-33 teleprinter has 63 printing characters plus the nonprinting graphic, "space," and control characters formed by depressing the control key and the associated character key at the same time. This type of terminal can print up to 72 characters per line, as well as respond to such control keys as carriage return, line feed, and signal bell.

The ASR-33 is one of many terminals that transmit and receive data encoded according to the American National Standard Code for Information Interchange (ASCII). A family of terminals evolved based on the functional capability of the ASR-33. These devices, commonly referred to as *TTY (Teletype)-type terminals*, transmit and receive data encoded in ASCII, like the ASR-33. However, they usually operate at higher data rates, have more printing characters available, and can print more characters on a line than the ASR-33. Some of these devices include:

- GE TermiNet 300
- Computer Transceiver Execuport 300
- Texas Instruments 700 Series

With respect to communications, the ASR-33 and compatible TTY terminals transmit and receive information on a line-by-line basis. Personal computers that transmit and receive data on a line-by-line basis as opposed to transmitting and receiving a full screen of information at one time are said to be communicating as a *TTY-compatible device*.

One terminal still in use, but rapidly being phased out of existence because of advances in technology, is the punched-paper-tape type. This device consists of a transmitter and a receiver, and permits data recorded as holes punched in paper tape (one of the storage methods mentioned previously) to be transmitted and received over communications facilities. The transmission unit of the device consists of a paper-tape reader and a signal generator. As the paper tape passes through the reader, the holes previously punched in the tape are read by a sensing mechanism connected to the signal generator. The signal generator produces a signal according to the presence or absence of a hole on the paper tape, while at the receiver the operation is reversed with a paper-tape punch interfaced to a signal receiver. The presence of an electrical signal causes a hole to be punched, and in the absence of a signal, no hole is punched. These terminals operate at speeds ranging from 10 to 300 characters per second.

One type of terminal that has gained widespread acceptance is the visual display or cathode ray tube (CRT) terminal. This terminal consists of a keyboard, a cathode-ray tube for character display, a signal generator and interpreter, and a buffer area. These terminals are widely used in data communications applications where a hard-copy (printed) output is normally not required. (When a hard-copy output is required, an RO printer is normally slaved to the CRT so that the operator can depress a print button on the CRT's keyboard and have the current image on the tube transferred to the printer.)

The personal computer in many ways is very similar to a CRT terminal. Most PCs are modular devices that consist of a keyboard unit, a system (control) unit, monitor, and such peripheral devices as printers and plotters. The control unit is the key to the functionality of the personal computer, containing a microprocessor, memory, and adapter cards that are installed into expansion slots to perform specific functions. In many PCs, an added feature of the control unit is housing "floppy" diskettes and one or more fixed disks, providing the computer with an online storage capability. The keyboard unit plugs into the control unit and a video adapter in an expansion slot is normally cabled to a CRT-based monitor. When a printer is cabled to another adapter board in the control unit, a print button on the PC's keyboard can be used to transfer the image on the monitor to the printer.

When the CRT operator queries the computer by pressing the appropriate keys on the keyboard, coded signals are generated and an image of the character is reproduced on the screen. When there is a unit of information to be transmitted, the operator presses a transmit or carriage-return key, and all of the entered information is transmitted to the computer.

Through the use of special keys, editing can be accomplished on CRTs. A cursor on the screen is used to indicate the position where the operator is to enter, delete, or change data. Normally, the cursor is a blinking small bar above, below, or at a character position that corresponds to a location of buffer storage. The terminal's buffer storage is designed to hold data for batching and editing until such time as the operator desires to transmit the information to the computer. When the operator keys data onto the CRT display, the data is placed in the terminal's buffer storage. The cursor automatically steps to the next position where a character is to be displayed.

The position of the cursor can be altered rapidly by the use of cursor control keys. Once the operator has positioned the cursor, it can be homed to the first character position on the upper left-hand area of the display, or spaced down, up, to the left, or right.

Other control keys can be used to clear the display, insert information onto a previously displayed line, erase lines previously entered, delete a character or group of characters, and so on. The operating data rates of CRTs vary considerably, normally ranging from 10 to 1200 characters per second, with the operating speed usually limited by the type of communications facility used.

The composition of the coded signals required to position the cursor to specific areas on the screen, as well as to control such video attributes as highlighting, underlining, blinking, and the color of a character on a color display, can vary considerably from terminal to terminal. When a program operates on a personal computer and converts one set of coded signals that governs the operation of a terminal into a second code supported by the PC, the program is called a *terminal emulator*. Terminal emulators can be entirely software-based and loaded into the computer's memory from a file on disk, hardware-based in the form of read-only memory contained on an adapter installed in the computer's control unit, or a combination of hardware and software.

Another type of terminal rapidly being connected to computer networks is the facsimile device. These terminals are basically duplicating devices that scan a document at a sending location and reproduce the document at a receiving location. Each facsimile terminal contains a transmitter and a receiver. The transmitter converts black and white spaces on a document into electrical signals while the receiver on the device it is "talking" to takes the electrical signal and reconverts it back into a series of black-and-white spaces. Although some devices can transmit and reproduce only black and white, other terminals are capable of reproducing shades of gray, thus permitting half-tone pictures to be transmitted. Facsimile terminals currently permit one 8½-×-11-inch page to be transmitted in about 6 seconds to 2 minutes, depending on the type of machine used. When facsimile terminals are linked to a computer functioning as a message switch, documents can be routed to many users via a set of instructions. Or documents can be stored until the recipient is available to retrieve the message.

By the late 1980's, a number of vendors had introduced "fax" modem adapter cards that could be installed inside the control unit of many types of personal computers. Although these fax cards enable PCs to communicate with conventional facsimile terminals, they require the use of a scanner to first record printed information to disk prior to transmitting it to a distant device.

By the mid-1990s, fax modems had become so popular that their shipments surpassed stand-alone facsimile machines. Today, you can purchase a fax modem, fabricated on an adapter card for insertion into the PC's expansion slot, for under $100. In comparison, a similar capability in 1990, obtained through the use of a standalone facsimile machine, would have cost over $750.

Some of the typical types of terminals that may be connected to computers or other terminals via appropriate communications facilities are illustrated in Figure 3.1.2.

3.2 Forms and Uses of Data Communications Terminals

There are numerous types of data communications terminals, and there are numerous uses for such terminals.

In addition to the telegraph, one of the earliest applications of data communications terminals was to connect two or more business locations via the use of teleprinters and communications lines provided by the telephone company. One such elementary network could have utilized two teleprinters of the KSR type, for example. Here, the operator at one location would transmit a message to the terminal connected on the opposite end of the line. The receiving teleprinter would re-

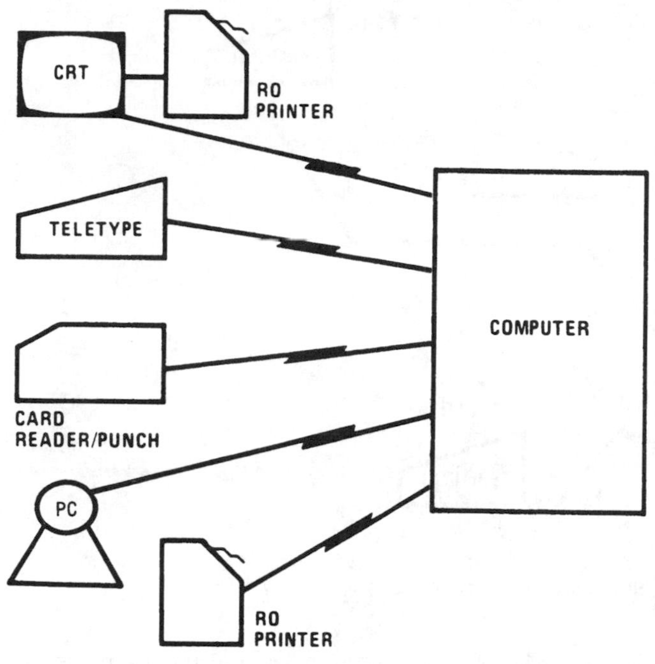

Figure 3.1.2 Typical terminals used in a computational network.

ceive and print out the message transmitted to it. After the message was completed, a response could be keyed back from the receiving end for verification.

Because of the slow operating speeds of such devices, messages were usually short and it was tedious to transmit the same message to more than one location. Later developments resulted in a party line or multipoint circuit, as illustrated in Figure 3.2.1. Through this arrangement and special equipment, any one terminal could broadcast messages to one or more other terminals at the same time, or could selectively receive messages in some predefined sequence from each terminal on the line. Although this was a considerable improvement in customer capability, it did not permit one company to send messages to an office of another company. To do so would have required the other company's office to be on the same party line. Because of the number of companies that some firms conduct business with, it was physically impossible to connect that large a number of tele-printers onto one party line.

These limitations resulted in the development of circuit and message switching by the common carriers. In circuit switching, a line physically connects each subscriber's terminal to the communications carrier's office, where circuits are physically switched to establish the required connections between the calling and called terminals. In message switching ("store and forward"), the message is first physically transmitted from a subscriber to the communications carrier, from where it is retransmitted to the intended recipient. One of the most common uses of message switching is for the transmission of Mailgrams, where a person can use a terminal to

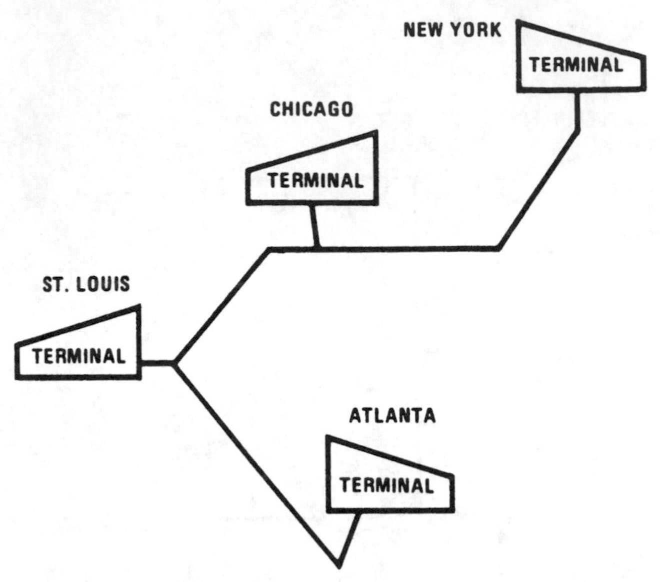

Figure 3.2.1 Multipoint or party-line network.

access the services of an electronic mail network and have a message transmitted electronically to a post office where it is printed, placed in an envelope, and hand delivered to its destination.

General business usage

Today and for the future, terminals will play an all-pervasive role in our lives. In the factory, for example, badge and card readers record the employee's time of entry and exit. Such data is then processed by the computer to prepare the worker's paycheck. Also, management is provided with an immediate status of unmanned workstations at the start of the business day so that corrective actions might be taken.

In the area of order entry and inventory control, you do not have to think about big industrial plants, but just drive to the nearest hamburger fast-food outlet to see how terminals are changing our lives. Here, when you place an order for a large hamburger, small french fries, and a medium Coke, the waitress presses several buttons on a special keyboard of a terminal to indicate the selection. Immediately, selections are transmitted to a minicomputer or microprocessor, which can be located within the store or at a central location. A table lookup then occurs as prices are assigned to the selections, the order is subtotaled, appropriate sales tax is added, and a final total is computed. This information is then transmitted back to the terminal where it is listed on the printer. As you make your selections, the inventory in the store is automatically adjusted—so many ounces of soda, one cup and lid, etc.

When inventory reaches certain critical levels, reorders can then be generated automatically by the computer. As you pay for your purchase, the order taker keys in the amount rendered. The change due, if any, is automatically printed on the terminal and the printout is furnished as your receipt. During the day, an automatic computation of sales from that terminal location and the amount received by the terminal operator can be computed, helping the owner's cash management.

Timesharing and remote computing

One of the most rapid-growth areas for terminal applications during the 1970s and through the mid-1980s was in timesharing and remote computing. In a timesharing environment, the operator normally uses a terminal to communicate with a computer, primarily by dialing the telephone number of the computer and using the switched telephone network as the communications link. After entering an appropriate identifier, which must be recognized as valid by the computer, the operator can execute, for example, an inventory analysis program. Here, the operator enters messages and data and the computer computes the number of units that should be ordered from the manufacturer to receive a discount.

When larger amounts of data must be transmitted or received, remote batch terminals (RBTs) are normally used. Some RBTs contain a small processor that enables local programs to be executed, while communications with the host or central computer occurs only when additional processing power is required. Other RBTs consist merely of a control unit and one or more peripheral units, such as a line printer, card reader, or magnetic tape unit. Such a device is mainly used to affect a batch transfer of large quantities of data to the host computer, receive large amounts of processed data from the computer, or both.

Figure 3.2.2 shows two RBT configurations. In the top portion of the figure is a simple RBT configuration, which consists of three peripheral devices: a line printer,

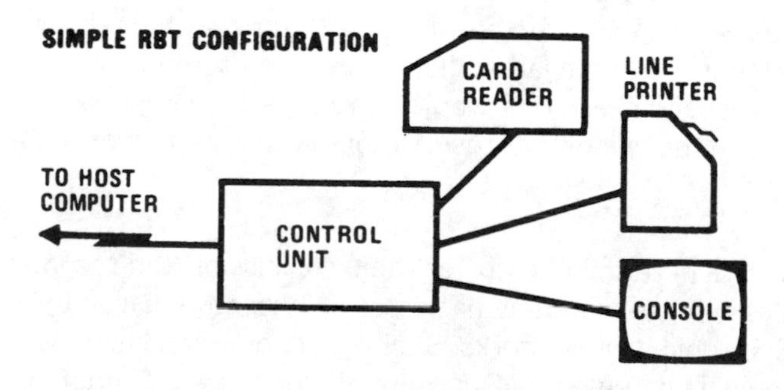

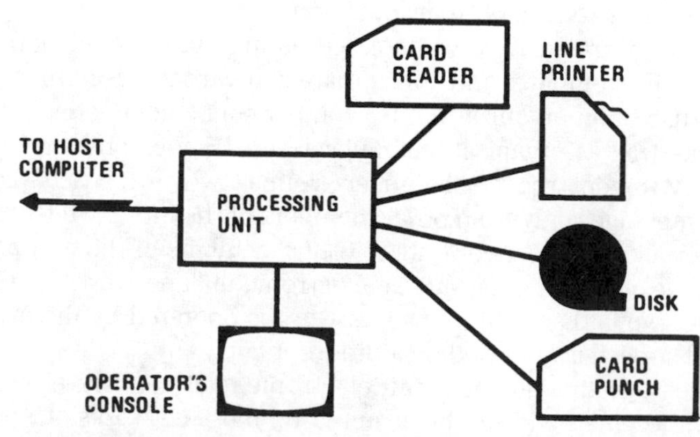

Figure 3.2.2 Remote batch terminals.

a card reader, and an operator's console. The operator's console might be used to establish communications with the computer and inform it that a job will be transmitted for processing. Once the job is transmitted to the computer, the operator waits for its execution to be completed and directs the computer's output to the RBT line printer. Or the operator could disconnect the RBT from the communications facility and reconnect it at a later time to receive the completed job.

In the lower portion of Figure 3.2.2, a more complex RBT with local processing capability is illustrated. During the day, local programs could be executed on this machine, and jobs that require the greater processing power of the host computer could be transmitted to and executed on that machine after normal working hours.

Inquiry/response applications

Two of the most common inquiry-and-response applications using terminals are airline reservation and stockbroker quotation networks. Each of these uses specially designed terminals with keyboards tailored for the particular application. In the stockbroker quotation application, keys are available to enter special symbols unique to that occupation. Thus, after the stockbroker enters the alphabetic symbol for a common stock, she can then enter one of the several symbols to determine its current price, year-to-date earnings per share, last year's earnings per share, dividend paid this year, previously paid dividends, and other information of interest to investors.

For airline reservations, a different set of symbols is used to denote unique information and changes. As an example, one key on the terminal keyboard might be used to denote a change or cancellation. Thus, once the terminal operator locates the flight a person is booked on, all that would be required is a depression of the cancellation key to remove the customer's reservation on that particular flight.

Electronic mail

One of the earliest uses of terminals was for the transmission and reception of messages. Today simple message switching has essentially been replaced by a variety of computer-based electronic mail networks. Such a network could consist of software operating on a personal computer that permits one or a few simultaneous users to access a bulletin board at the same time to store and retrieve messages and other information. Or the network could have multiple large-scale interconnected computers that provide services to tens of thousands of users.

Electronic mail networks can be categorized as in-house or commercial. An in-house one consists of a computer and appropriate software that is normally reserved for the exclusive use of an organization. By comparison, a commercial network accepts individual as well as organizational subscribers. Some commercial networks have the capability to construct "closed-user groups," which restricts group members to receiving messages only from other members of the group while transmitting messages to either all subscribers or only to other members of the group. Thus, the closed-user group is similar to an in-house electronic mail network. The key difference between the two is that an in-house network is operated by the organization, and a commercial service is operated by a different company.

Although all messages initially enter an electronic mail network electronically, a variety of distribution methods can be selected on most networks. The most common type of electronic mail is the mailbox message, in which a message is transmitted from a terminal and routed to a specific storage location, based on its destination address. Then, when the persons with the destination address access the network, they receive a "mail waiting" message, and can then scan or read their messages. Other types of electronic mail deliveries include the conversion of the message into a Mailgram for delivery by the postal service, conversion of the message onto standard 8½-×-11-inch paper with delivery by a courier, and conversion of the message into a format that permits it to be transmitted to a facsimile terminal.

The LAN workstation

As mentioned previously, adding hardware and software to a PC allows it to operate on a LAN as what is often called a *workstation* or *node*. Regardless of the terminology used, the PC's connection to a LAN enables the PC user to perform many functions beyond the capability of a conventional terminal or stand-alone computer that is not connected to a network. Some examples of these functions are [all LAN references are to the same LAN]: cooperative processing, wherein a workstation's software operates in conjunction with that of a LAN-connected file server; electronic mail, in which a PC user can send messages to other LAN-connected PCs; file transfer, wherein a scheduling program coordinates conference-room usage among employees; gateway, where the facilities of another LAN-connected PC are used to allows a common method of access (a gateway) to a mainframe computer.

Figure 3.2.3 illustrates the use of three different PC types can be used on a bus-structured LAN: workstations, file servers, and gateways. Each of these PCs can also function as a terminal device, sending information to other LAN devices. Thus, although the role and terminology of the PC might change in accordance with its networking

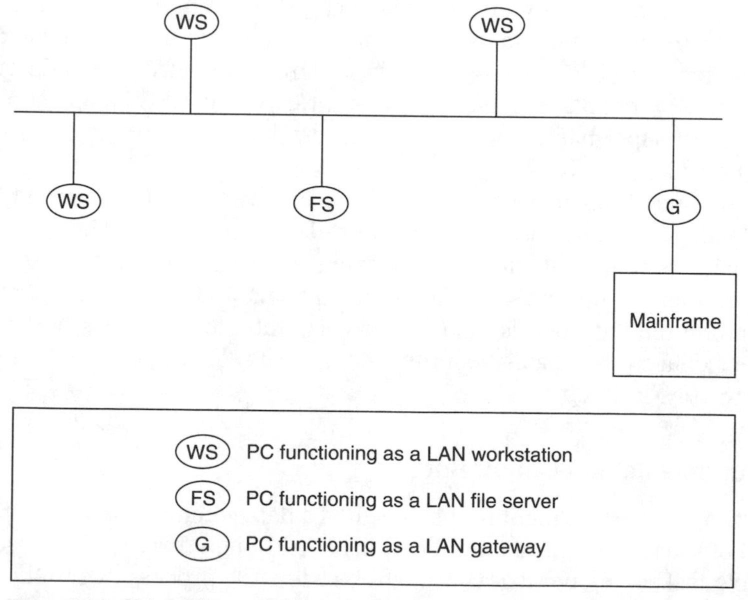

Figure 3.2.3 Using PCs on a LAN.

environment, it can still be classified as a terminal-type device. Of course, we must recognize that the PC's level of sophistication has vastly increased when compared to teleprinters and CRT-based terminal devices.

Trends

During the past decade our lives have become more and more dependent upon the utilization of terminals connected to computers.

Today, in many stores, when the customer pays by credit card, either the card is read by the terminal or the clerk enters the card number by means of a simple keyboard. Together with the amount of the purchase, this information is transmitted to the computer, which scans the customer's file to determine if the card has been reported lost or stolen, or if with this purchase the customer's allowable credit will be exceeded. In addition to performing credit verification, some terminals have additional connected devices to capture sales data in a more expedient manner than through the entry of information from a keyboard. Terminals of this type are usually referred to as *point-of-sale devices* and are rapidly replacing the conventional cash register in a variety of stores, shops, and offices where transactions between employees and customers take place. Usually one or more of these functions are performed by such devices: data capture, receipt issuance, credit verification, and local record maintenance.

One type of point-of-sale terminal—first used in supermarkets—that has great promise is an optical scanner attached to a terminal consisting of a printer, display, keyboard, and communications facilities. Many diverse operations—toll booths, department stores, laboratories, warehouses, libraries and hospitals—are now using or planning to use such devices. The use of the scanner is based on reading bar code symbols, which are a graphical representation of binary numbers.

Bar codes follow a variety of schemes, using the width, height, and distance between marks to express characters. The most commonly used bar code is the universal product code (UPC), which was adopted by the supermarket industry in 1973. Today, about 95% of all items in a supermarket carry the UPC label; and an increasing number of supermarket checkout counters have terminal devices capable of reading the code.

When the optical scanner senses the UPC label, the data represented by the bar code is read by the terminal and transmitted to a computer. The type of item is matched with its name and current price, which are then transmitted back to the terminal to illuminate the unit's display and print out on the customer's receipt. Because reading the bar code is faster, more accurate, and less costly than entering item prices on a conventional cash register, optical scanning point-of-sale terminals are here to stay.

3.3 Terminal Requirements and Constraints

Perhaps the key starting point in the design of a data communications network is the selection of the type of terminal to be utilized. Although this might at first appear to be a simple task, in many cases this might be a lengthy process due to the number of items that should be considered. In Table 3.3.1, the reader will find a list of terminal characteristics that should be reviewed prior to rendering a terminal-selection decision. Notice that the weight one gives to each of the items in the table is a function of the user's proposed or existing operating environment. Also notice that the characteristics listed in Table 3.3.1 can apply to a PC functioning as a terminal. When this occurs, the characteristics reference both the PC as a computing platform and any communications hardware and software necessary to perform a communications function.

In examining the input/output (I/O) media, the types of peripherals that can be interfaced to the terminal as well as the terminal itself should be examined. Thus, the keyboard, plastic credit card reader, and optical scanner, for example, are just three of many methods of entering data. Concerning the output media, a decision of hard-copy printout vs. the "soft-copy" image on a CRT does not have to be an all-or-nothing situation today. Some CRTs can be clustered on a party-line arrangement and a group of such terminals might share a common printer at a particular location, as shown in Figure 3.3.1.

When a PC functions as a terminal device, references to the "CRT" actually refer to its monitor. Although most PC monitors are based on the use of a built-in or at-

TABLE 3.3.1 Terminal Characteristics to be Evaluated

I/O media
Intelligent/Dumb
Operating rate(s)
Off-line capability
Data codes and character set
Operator convenience
Cost
Security
Control of user errors
Error detection and correction

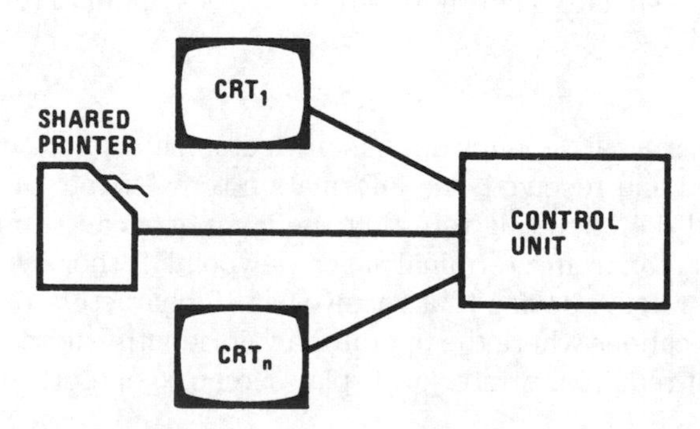

Figure 3.3.1 Eliminating all-or-nothing decisions by sharing.

tached CRT, certain other PCs—such as laptop and notebook computers that can function as a terminal—use either a light-emitting-diode (LED) or gas-plasma display. Unless otherwise stated, the information-display constraint on a CRT terminal is also applicable to the PC, regardless of the type of display it uses.

Another I/O media consideration is the type and amount of data to be displayed. If the terminal has a printing mechanism for output, there is no limit imposed on the length of print-out response. If the primary output mechanism is a display, you should limit responses to a full screen (or less) of information because it is difficult to remember parts of information that are produced on one screen display as a new display continues the information received. Most CRTs display 80 characters per line on up to 24 lines of output, for a maximum display of 1920 characters. If graphical data is to be produced, a special graphical CRT display or graph-plotting printer attached to a terminal should be considered.

If the data to be transmitted can be batched together, then a mechanism for temporary storage prior to transmission should be considered. Such devices as paper tape reader/punches, magnetic tape units, disk storage units, or data cassettes can interface with the terminal. These devices not only provide an off-line data entry capability, but they can also reduce communications costs by permitting data to be keyed at the terminal operator's typing speed and later transmitted to the computer at the higher data rate permitted by the communications facility.

When a PC is used as a terminal device, its "floppy" diskette(s) and hard drive provide both an on- and off-line data storage capability. This function facilitates the storing of received electronic messages, file transfer operations, and other activities that are normally beyond the capacity of most conventional terminals.

Intelligent/dumb

An intelligent terminal has the ability to modify its operating characteristics. This modification of operating characteristics can result from the loading of a new program into the terminal's memory from a disk, or by the insertion of a cartridge or cassette whose data the terminal can read and act upon. In comparison, a dumb terminal cannot modify its operating characteristics and always appears as the same device when used in a communications environment.

The major advantage of an intelligent terminal over a dumb terminal is the ability of the former to emulate different devices. As an example, a personal computer with appropriate software could emulate 30 or more different types of dumb terminals.

Operating rate

The operating rate of the terminal depends upon many factors, including the volume of traffic to be transmitted and received, the information transfer rate of the communications facilities, and the operating rate that the computer can support. Perhaps the most important factor, from a terminal user's viewpoint, is the psychological reaction to the operating rate. A device that receives at 10 characters per second might be too slow for applications where the operator awaits lengthy messages or reports. This situation might require a printer or display screen to operate at a higher data rate.

At the opposite extreme, you might select an operating rate too rapid for the application. Consider, as an example, a CRT whose screen can be filled with information very rapidly, with the constraining factor on its operating rate normally being the communications line connecting it to the computer. If typical operator interaction requires the entry of, for example, 20 characters and the display of 80 characters, then a data rate of 1200 characters per second would in all probability be excessive. The 80-character response would require only 67 milliseconds to be displayed, far less time than the minimum required by humans to read such output. For most situations in which a CRT display is to be filled with data, an operating speed of 600 characters per second is sufficient.

Data code and character set

When a character is transmitted from a terminal, it is represented as a set of bit patterns according to the internal code of the device. In most cases, the terminal's internal code and the transmission code are the same. Sometimes, the terminal code might differ from the transmission code or the computer's internal code. For these situations, code conversion must occur to make the terminal's code compatible with the code acceptable to the computer.

Code conversion can be accomplished by a remote computer or by certain terminals. Normally, remote batch terminals are the only ones that perform code conversion because buffer storage and the processing of instructions are required to perform the conversion. Interfacing code-conversion devices are also available commercially.

Two examples of code conversion are illustrated in Figure 3.3.2. In the top example, an older model teleprinter designed for the 5-level Baudot code communicates with a computer whose internal code is ASCII. Here, the computer converts data received from the terminal from Baudot to ASCII and data destined to the terminal

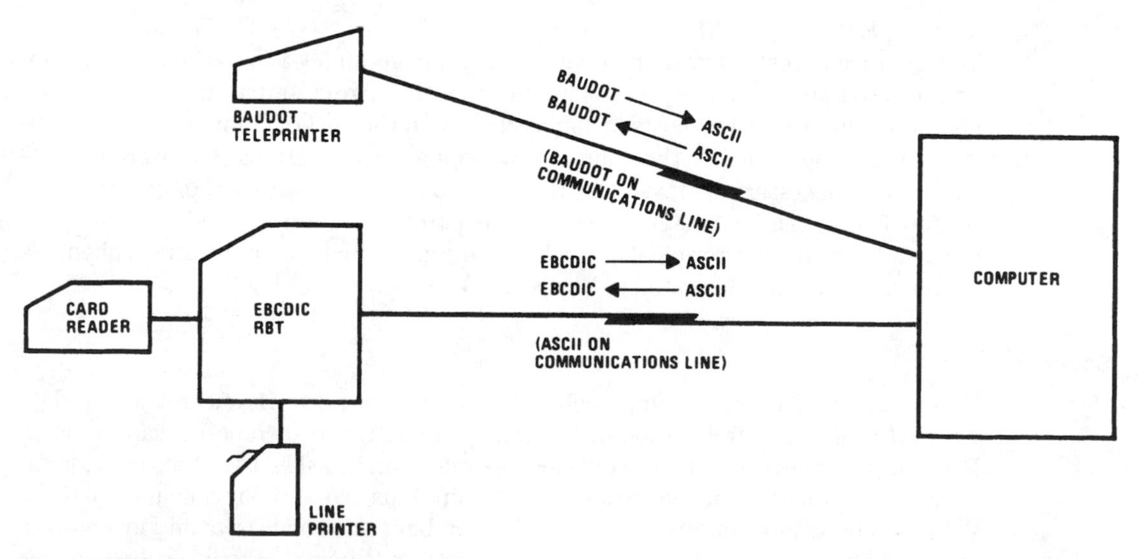

Figure 3.3.2 Code conversion.

from ASCII to Baudot. In the lower example, the RBT's processor is used to convert data in the EBCDIC code into ASCII for transmission to the computer. Data from the computer is transmitted in ASCII and converted into EBCDIC at the RBT.

When a PC is used as a terminal device, code conversion is normally performed by the PC. The conversion is accomplished through the use of software, hardware, or a combination of the two.

In addition to the data code supported by the terminal, its character set might be of equal importance. As an example, a terminal that supports the ASCII data code might only support a subset of the ASCII character set, eliminating, for example, all lower-case characters. For a text-editing application where both upper- and lower-case characters are required, this type of terminal would probably be excluded from consideration.

Operator convenience

This category can be most important, for if the terminal is difficult to operate, it might not be utilized. Items in this category include human factor considerations that ease operation of the terminal. An optical scanner or numeric keypad with no visual input display for verification is an example of inconvenience. Examples of convenience include an alert indicator, such as a bell on a teleprinter and a blinking message on a display, to attract the operator's attention to a certain situation.

One of the major reasons for PC usage growth is the PC's capability to execute programs. This capability enabled software developers to design menu-based programs as well as to use a graphical interface to support "mouse" operations. These accomplishments enabled users to move a highlight bar and press the ENTER key—or position the mouse cursor to a screen area and "click" the mouse button—to perform an operation.

Cost

Terminal lease costs vary dramatically, ranging from under $75 per month for teleprinters to many thousands of dollars per month for remote batch terminals. One should evaluate terminal costs in conjunction with the cost of communications facilities to link the device to the computer as well as any additional hardware or software costs necessary to make the terminal compatible with the computer. Thus, terminals that can be clustered to share one communications link might be less expensive than cheaper terminals, which require individual communications channels to the computer, as illustrated in Figure 3.3.3.

Security

To prevent unauthorized personnel from accessing a network and retrieving data with a terminal, a combination of hardware and software techniques can be used. From appropriate software, the terminal operator can be asked to enter a code or authorization number that is checked to ascertain if his use of the terminal is valid.

From a hardware standpoint, there are a number of methods to maintain a level of security. One method is to have the terminal operable only on the insertion of a key,

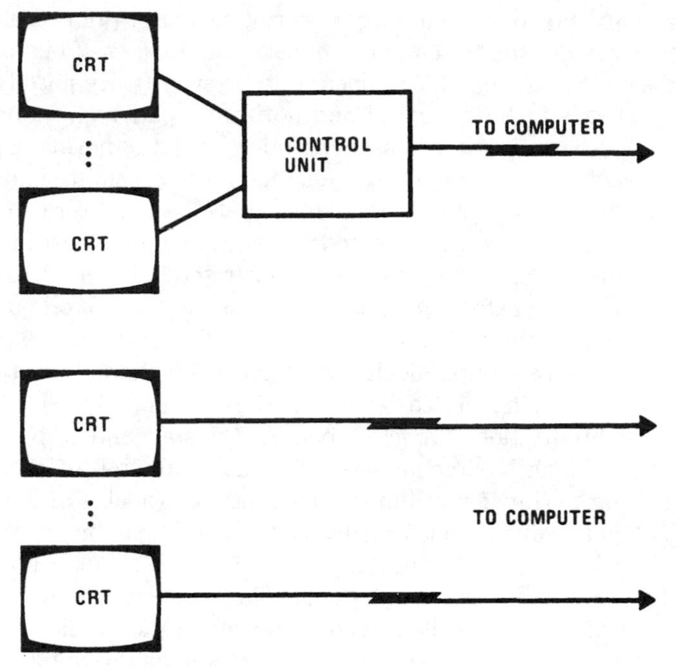

Figure 3.3.3 Evaluating configuration cost.

card, or badge issued to authorized employees. Sometimes the terminal itself can be programmed to read an employee code keyed in by the terminal operator at the start of the business day and once again after each time the terminal is shut off.

Control of user errors

Control of user errors is normally with a combination of hardware and software elements. A display screen that permits the user to move his cursor back to a previous position to change a character or characters that are in error or insert additional data is preferable to a terminal that does not permit these operations. Similarly, a keyboard without a display to let the operator verify data entered might cause operational problems.

Error detection and correction

In order for the terminal's receiver to stay in synchronization with the sender, the terminal may use one of three methods for transferring data (covered in greater detail in Section 4.4).

The first method, known as *asynchronous transmission*, is also referred to as start-stop transmission. Teleprinters and similar devices, including PCs emulating a teleprinter, use this method where characters are transmitted one at a time on a character-by-character basis. Each character is identified by one start bit and one or more stop bits. Most asynchronous terminals also generate a parity bit for each transmitted character and contain circuitry to verify the parity prior to transmitting the character. Parity is the addition of a bit to make the number of bits per character, excluding the start and stops bits, appear odd or even (as required). The two common parity methods include vertical and horizontal parity checking, as illustrated in Figure 3.3.4. The circuitry in asynchronous terminals normally provides a horizontal parity check because the bits of a character are transmitted one at a time "horizontally." The bit configuration of an 8-level ASCII-character-code terminal (which transmits and receives an 11-unit code consisting of one start bit, seven information bits, a parity bit, and two stop bits) is illustrated in Figure 3.3.5.

The second method of transferring data is known as the *block mode* or *synchronous mode*. When this method of data transfer is used, characters are blocked together for transmission as a group. Special synchronization characters are placed at the beginning and end of each block to denote the start and end of the block. The use of these special synchronization characters permits the start and stop bits used by the asynchronous terminals to be eliminated. Because a number of characters are blocked together prior to transfer, buffer storage must be available in the terminal.

Error control when terminals transfer data synchronously can be accomplished in a number of ways. One method is by the generation of a block check character. Here, as each block of characters is transferred, the terminal's internal circuitry computes a check character, which is physically placed at the end of the block. One simplistic example would be formation of a parity check character by grouping every seven characters transferred.

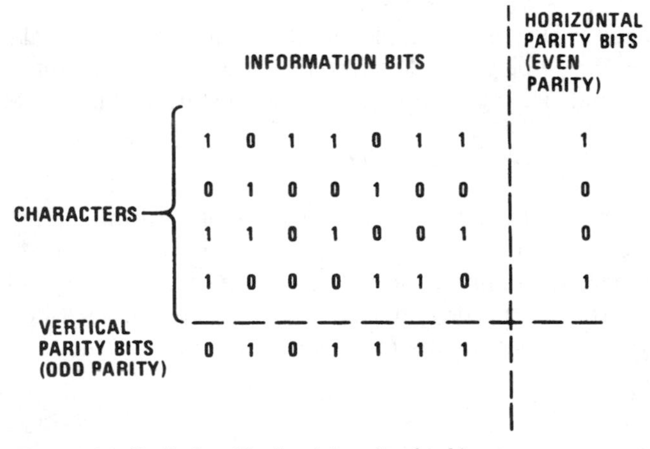

Figure 3.3.4 Vertical and horizontal parity checking.

ONE START BIT
INFORMATION BITS REPRESENT ONE CHARACTER
TWO STOP BITS
START
PARITY BIT
STOP
STOP

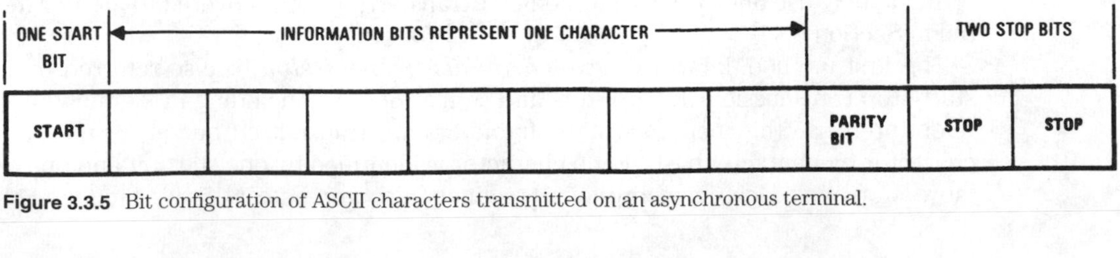

Figure 3.3.5 Bit configuration of ASCII characters transmitted on an asynchronous terminal.

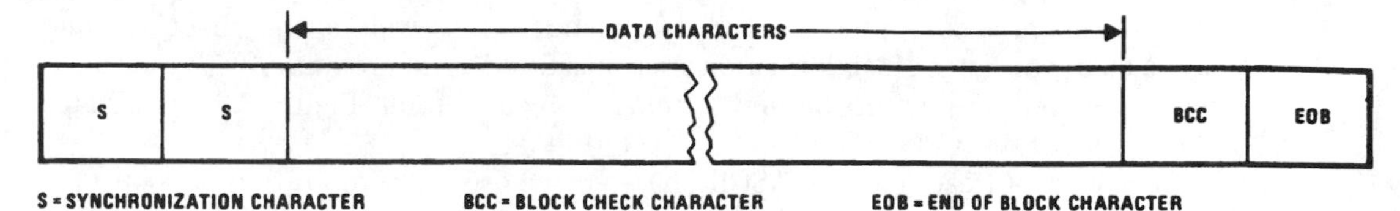

Figure 3.3.6 Typical data block format.

At the receiving end, the computer or terminal would perform an equivalent parity check character generation and compare the character it generated with the one that was transmitted. If the two match, the data is accepted; however, if they do not match, a message is sent to the terminal to retransmit the block of data.

A typical data block format for a synchronous terminal is illustrated in Figure 3.3.6. The number of characters used to form the block varies, depending on such factors as the type of input device (keyboard, card reader, etc.), the expected error rate and corresponding expected retransmission rate, and the terminal's buffer storage size.

Normally, terminals that transmit data synchronously are more expensive than asynchronous terminals because they require buffer storage and more complex circuitry. Although the use of parity was shown as one method to determine errors, it should be recognized that a double-bit error remains undetected by simple parity checking. For synchronous terminals, more advanced methods to include polynomial character generation are used as error-checking tools.

In polynomial character generation, the data to be protected in a block is treated as one long binary number. That number is divided by a fixed polynomial, which results in a quotient and a remainder. The quotient is discarded. The remainder is appended to the block as a "cyclic redundancy check" (CRC) group of characters. The actual number of characters that make up the CRC is based on the composition of the fixed polynomial.

A receiving device uses the same fixed polynomial to operate on the received block, generating a "local" CRC. The local CRC is then compared to the transmitted CRC. If they are equal, the received block is assumed to have arrived free of errors. If the local CRC does not match the transmitted CRC, one or more bit errors are assumed to have occurred. The receiver then transmits a negative acknowledgment (NAK) to the originator, which functions as a request to retransmit the errored block. Thus, error correction is accomplished by retransmission.

The third method of transferring data is actually a modified combination of the two previously discussed methods. This method, which is based on asynchronous blocking of data, was developed in response to the requirements of personal computer users for minimizing the probability of a transmission error when transferring files of information. Several ways have been developed to transfer data asynchronously in blocks, of which the XMODEM protocol and its derivatives are among the most popular.

Under the XMODEM protocol, 128 characters are grouped together and a checksum is then computed and appended to the block for error detection purposes. The checksum itself is computed by simply adding the ASCII value of each character in the block, dividing the sum by 255 and appending the remainder to the block. Once the checksum is computed, all 128 data characters as well as the checksum and three characters of prefix information are transmitted asynchronously, with each character prefixed with a start bit and terminated by a stop bit. Figure 3.3.7 illustrates the format of the XMODEM block.

The Start of Header is the ASCII character whose bit composition is 00000001, while the 1s complement of the block number is obtained by subtracting the block number from 255.

Under the XMODEM protocol, there are three ways in which a block can be rejected. If the Start of Header is not recognized, the remainder of the block will be rejected. If the block number and the 1s complement field are not the 1s complement of one another, the block will also be rejected. Finally, the receiver computes its own checksum character based on the 128 data characters in the block. Then, if the checksum computed by the receiver does not match the transmitted checksum, the block will also be rejected.

Under the XMODEM protocol, a block can be rejected up to 10 times. Once this limit is reached, the communications session is terminated and the user must initiate a new communications session.

3.4 Justification of Terminals

The decision to use terminals to communicate from remote distances to a computer is based on many of the same criteria used for the procurement of any type of equipment. Such criteria include speed of information, operational efficiency, management control, and cost savings.

Speed of information

For many organizations, the old adage, "time is money," is especially true in today's era of rapid communications. A stockbroker who cannot rapidly quote a security price is in a position similar to a salesman who cannot quote delivery data for a product—he is without a customer. For some salesmen, a portable computer terminal is as important as a display case. By using a telephone, the salesman can use the direct-distance-dialing telephone network to connect his terminal to the company computer. He can then obtain such information as the current price of an item that might be especially important for products that drastically and rapidly vary in price, quantities on hand and ready for shipment, delivery schedules for equipment to be manufactured, and other data of importance to the customer.

Operational efficiency

If an organization provides its workers with tools to reduce the time required to perform a function, the net result is to increase the operational efficiency of that firm.

START OF HEADER	BLOCK NUMBER	1s COMPLEMENT OF BLOCK NUMBER	128 DATA CHARACTERS	CHECKSUM

Figure 3.3.7 XMODEM block format.

Terminals equipped with optical scanners to read the universal product code of items in a supermarket are one example of a terminal installation designed for increasing operational efficiency. Here you can expect the clerk to "optically ring up" more items in less time, more accurately. Similarly, terminals in a fast-food hamburger store might permit more customers to have their orders totaled by fewer clerks.

Management control

Although terminals are usually first installed for applications that require speed of information or operational efficiency, management control usually results from an evolutionary maturing process. Thus, although the early terminals in fast-food stores were no more than electronic cash registers, through evolution they now are used to perform such varied tasks as inventory control and cash management, in addition to order entry. The data entered becomes part of a database for a Management Information System.

Cost savings

Although cost savings normally are a result of the previously discussed criteria, one should also consider the possibilities terminals offer to increasing company profits. Labor is a large cost for a fast-food firm, a cost that can be reduced by electronic order entry. At the same time, the data entered from the terminal can be used by a computer not only for inventory control, but also for the computation of optimum reorder levels to take advantage of supplier discounts. The result is additional cost savings to the organization.

In some firms, increased profits might be easier to obtain by the use of terminals to increase sales than by cost reductions. The previously discussed example of the salesman would fall into this category. In another example, terminals used by new car dealers query a computer to find the nearest location of a car in a specific color requested by a customer. Here, the ability of a new-car dealer to locate and arrange for the prompt delivery of the desired car results in an additional sale and increased profits for the car dealer.

3.5 Summary

The roles of terminals in data communications are wide ranging and ever increasing. Although some terminals are used only to communicate with other terminals, the majority of such devices are used for transmitting and receiving data from computers.

Many terminal types exist. In fact, a computer itself can be used as a terminal. However, terminals can generally be divided into four generic types—teleprinters, CRTs, personal computers operating as terminals, and remote batch terminals. By the addition of specialized input/output units, such as optical scanners, terminals can be utilized in a wide variety of ways.

Today, terminals are used not only for interactive timesharing applications, but also in such diverse environments as libraries, hospitals, stockbroker offices, and supermarkets. To evaluate terminal characteristics, you must investigate the I/O media, intelligent-versus-dumb operating capability, required operating rate, off-line capability, data codes and character set supported, operator convenience, cost, security, control of user errors, and the method used for error detection and correction.

Questions

3.1 In the context of the data communications field, define a *terminal*.
3.2 Describe three basic communications links and their information flow.
3.3 Discuss the differences between a keyboard send and receive (KSR), receive only (RO), and automatic send and receive (ASR) terminal. What application is best suited for each type of terminal?
3.4 What function does the cursor on the CRT perform?
3.5 What are some of the advantages and disadvantages of multipoint circuits or party lines? What technological development reduced some of these disadvantages?
3.6 Cite some examples of the use of terminals for order-entry application. What special equipment might be required?
3.7 What are the similarities of, and differences between, teleprinters and remote batch terminals?
3.8 What are some of the functions performed by a terminal emulator program?
3.9 Describe some inquiry and response applications and denote some of the special features that might be required of terminals connected to such configurations.
3.10 What advantages result from connecting an optical scanner to a terminal?
3.11 Discuss some of the terminal characteristics that should be reviewed prior to selecting a specific device.
3.12 Where can code conversion take place?
3.13 What is parity?
3.14 What is the difference between synchronous and asynchronous data transfer?
3.15 Compute the odd parity bit for the following characters:

A. 1011101 C. 1110111
B. 1110100 D. 0111001

3.16 Why does a terminal that operates in a synchronous mode normally cost more than a terminal that operates in an asynchronous mode?
3.17 What is the difference between an intelligent and a dumb terminal with respect to their communications capabilities?
3.18 Cite a parity check problem that can occur.
3.19 Discuss some of the criteria involved in a terminal justification process.

Chapter

4

Fundamental Concepts of Data Communications

4.1 Introduction

This chapter examines the basic concepts and theory of data communications. Starting at the lowest level of information, the bit, definitions and terminology is developed to show its relationship to the type and structure of information that is transmitted.

Through an examination of the transmission modes and techniques available, as well as the basic types of linkages, you should be able to develop a firm foundation and understanding of selection and connection options. Such information is of special value, not only when considering the initial establishment of a network, but also when examining the effect of changing the types of devices connected to a computer.

4.2 Terminology and Definition

Bits and bauds

The fundamental building block of electronic information transfer is the *binary digit (bit)*. The bit is the lowest level of information representation and signifies the presence or absence of a state or condition. From the standpoint of a terminal or computer, the bit can be considered as being either in the 1 condition or the 0 condition, where the 1 state usually has a higher voltage than the 0 state. Bits can be represented by square waves, as illustrated in Figure 4.2.1.

Because bits are also used to represent a quantity of data transferred per unit of time, some confusion has occurred between bits/unit of time and the term *baud*. Bits/unit of time, usually expressed as bit/second or bps, defines the amount of data transferred during a certain time interval. Thus, 110 bit/s would signify that 110 bits of information are to be transferred in one second.

The term *baud* represents a unit of signaling speed equal to the number of discrete conditions or signal events per second. When one bit is used as a signal unit,

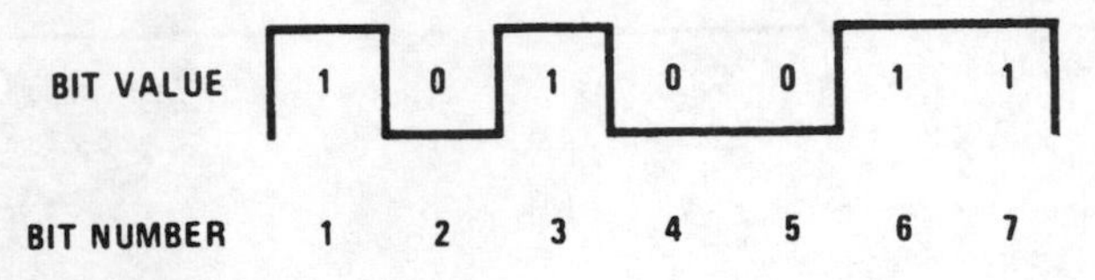

Figure 4.2.1 Bit representation.

TWO-STATE CODE, CODE UNITS 0 AND 1, BAUD EQUALS BIT/S

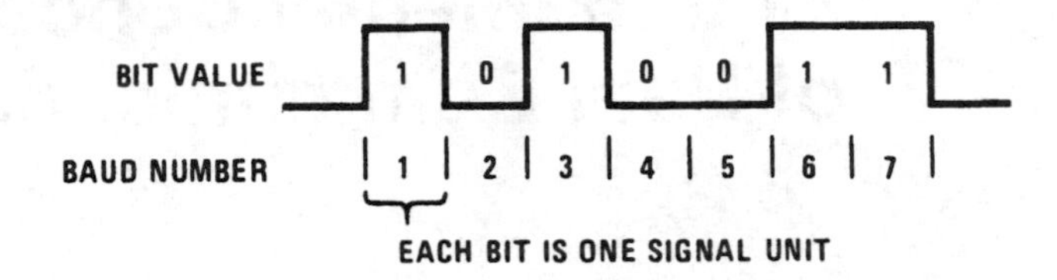

FOUR-STATE CODE, CODE UNITS 00, 01, 10, AND 11, BAUD IS HALF OF BIT/S

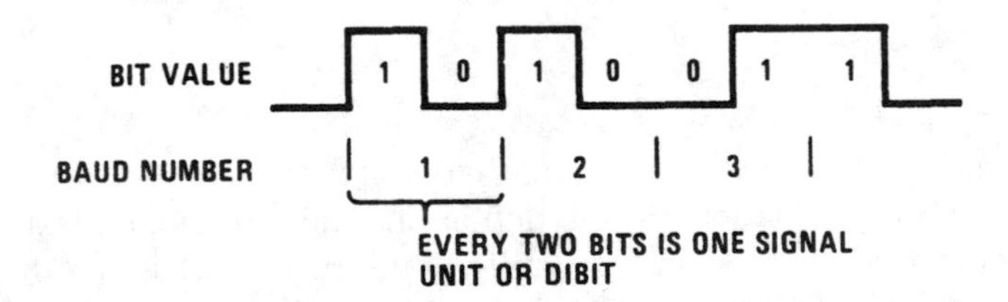

EIGHT-STATE CODE, CODE UNITS 000, 001, 010, 011, 100, 101, 110, AND 111, BAUD IS A THIRD OF BIT/S

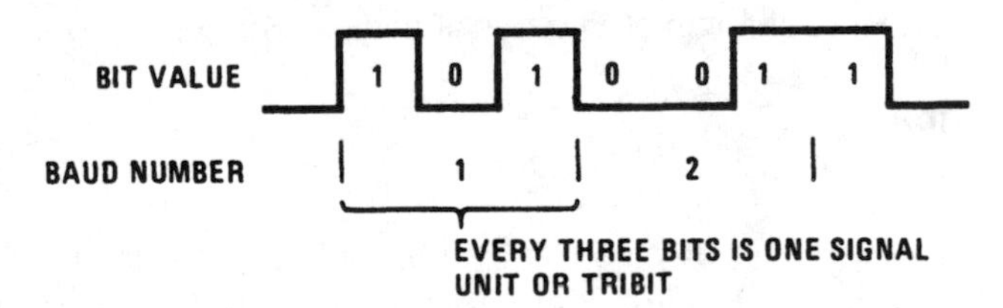

Figure 4.2.2 Correspondence between bits and baud.

baud speed and bps are equivalent, as shown in the top portion of Figure 4.2.2. When, through modulation, two bits are combined to form a signal unit, as shown in the middle portion of Figure 4.2.2., the baud rate becomes one half the bps rate. For this situation, the signal unit is called a *dibit (double bit)* and has four possible states or levels. When three bits are used to form one signal unit, as shown in the lower portion of Figure 4.2.2, eight possible states or levels can exist, and the signal unit is known as *tribit (triple bit)*. How all this happens is a function of the modulation technique used by a modem, to be discussed in Section 4.12.

Bytes, characters, and words

When a sequence of bits is operated on as a unit by computer hardware, the unit is called a *byte* of information. A byte normally consists of the number of bits required to represent one character, where a character is a letter, figure, punctuation mark, or other symbol used to represent a higher level of information than the bit. The 6-bit byte was used by most computers designed before the mid-1960s, while the 8-bit byte is the most widely used byte size in computers on the market today.

Although a computer byte is of fixed size, it can contain different size characters, as long as the number of bits grouped to represent any character does not exceed the size of the byte. This is illustrated in Figure 4.2.3, which shows the results of placing a 5-level Baudot character, a 7-level ASCII character, and an 8-level EBCDIC character into an 8-bit byte.

When referring to computers, a word is a fixed-length group of characters or bytes. In contrast to the definition of a bit or character, the definition of a word implies alignment of the data such that the word must have a certain address or it is not properly a word in the computer. In an 8-bit computer, the word size and byte size might be the same, with 8-bit bytes forming computer words of data or instructions. For this type of machine, if data is to be transmitted serially—that is, one character at a time—then one word of information is transferred at a time via the computer's I/O unit. In a 16-bit computer, the word size and byte size normally differ, with two 8-bit bytes capable of being stored in one computer word to be operated internally at one time.

Here, when data is transmitted one character at a time, the first byte is transmitted and a byte shift operation is then performed so that the second character is now positioned where the first character was previously located. Next, the second character is transmitted. Data transfer by an 8-bit and 16-bit word-size computer are illustrated in Figure 4.2.4.

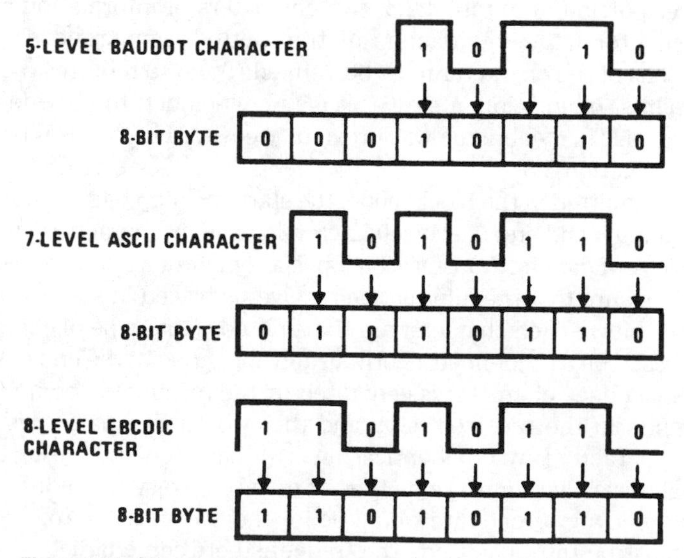

Figure 4.2.3 Many types of characters can be represented in a byte.

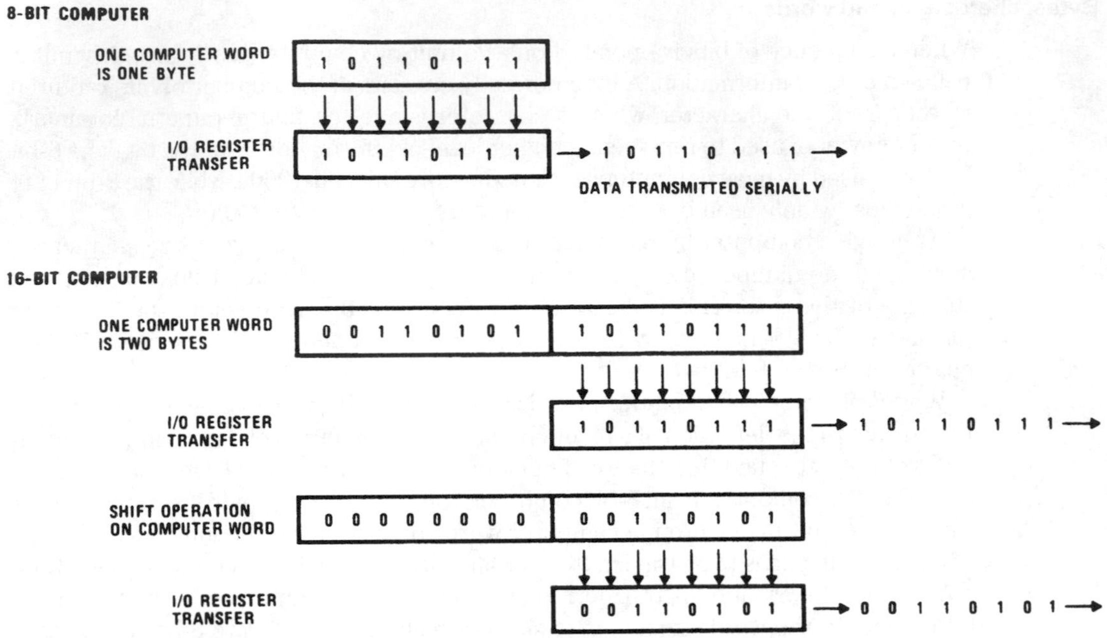

8-BIT COMPUTER
ONE COMPUTER WORD IS ONE BYTE
1 0 1 1 0 1 1 1
I/O REGISTER TRANSFER
1 0 1 1 0 1 1 1
1 0 1 1 0 1 1 1
DATA TRANSMITTED SERIALLY
16-BIT COMPUTER
ONE COMPUTER WORD IS TWO BYTES
0 0 1 1 0 1 0 1
1 0 1 1 0 1 1 1
I/O REGISTER TRANSFER
1 0 1 1 0 1 1 1
1 0 1 1 0 1 1 1
SHIFT OPERATION ON COMPUTER WORD
0 0 0 0 0 0 0 0
0 0 1 1 0 1 0 1
I/O REGISTER TRANSFER
0 0 1 1 0 1 0 1
0 0 1 1 0 1 0 1

Figure 4.2.4 Data transfer by 8- and 16-bit word-size computers.

Blocks and messages

In contrast to a message, which is an arbitrary amount of information whose beginning and end are defined, a block is a grouping of bits, transmitted as a unit over which some type of procedure is applied for error-control purposes.

From the lower portion of Figure 4.2.4, two characters of information are transmitted by the computer. If these characters are transmitted asynchronously, the start and ending positions of the character must be defined. To do so requires the addition of start and stop bits. In addition, a parity bit may be included to provide a limited form of error control. This technique was previously covered in the "Control of User Errors" portion of Section 3.3.

When data is transmitted in the block mode, the start and stop bits are not required to denote the beginning and end of each character. Instead, *N* characters are grouped together into a block of data for transmission. Special synchronization characters precede the block to ensure that the information receiver is placed in step with the data transmitter. As the bits of each character are transmitted, some type of check character is developed based on a predefined algorithm and appended to the end of the block.

At the receiver, a check character is generated by the terminal or computer using the same algorithm on the received data, and the two check characters are then compared. If the characters are not identical, an error has occurred during the transmission. This will normally require some type of error-recovery procedure to be initiated. One error-recovery procedure would be for the transmission to be repeated; that is, a block must be retransmitted. The tradeoffs between character-by-character and block-mode transmission is examined later in the section on line protocol.

In the field of data communications, the terms *block* and *message* are frequently interchanged and might result in a degree of confusion. A message, as noted earlier, is an arbitrary amount of information whose beginning and end are defined. Thus, a message could consist of one block of information. In practice, however, a message usually contains many blocks of information. The message format will vary depending on the management method of the data link used. This management, known as a *line protocol*, is covered more fully later.

4.3 Transmission Modes

One common method used to characterize terminals, communications lines, computer channels, and such communications components as acoustic couplers and modems, is by their transmission or communications mode. The three types of transmission modes available are simplex, half-duplex, and full-duplex.

Simplex

Transmission that occurs in one direction only is simplex, as illustrated in the top portion of Figure 4.3.1. In this mode of transmission, the receiver does not have a means of responding to the transmitted signal. An example of the simplex communications mode is a home radio, which can only receive signals transmitted from a radio station. In a data transmission environment, simplex transmission

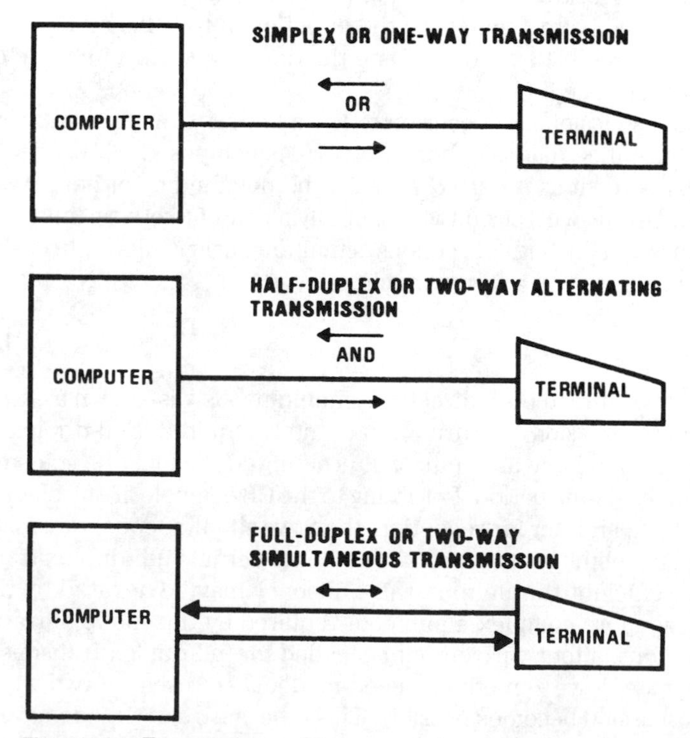

Figure 4.3.1 Transmission modes.

can be used to turn on or off specific devices at certain times of the day or when a certain event occurs.

Another example of simplex transmission is a computer-controlled environmental installation, where a furnace is turned on or off, depending on the thermostat setting and the temperature in various parts of a building. Normally, simplex transmission is not utilized where man-machine interaction is required because of the inability of the receiver to reply to the originator.

Half-duplex

As illustrated in the middle portion of Figure 4.3.1, the half-duplex mode of transmission permits data transfer to occur in two directions, but not simultaneously. This is the most common mode of transmission used in data communications today. Half-duplex transmission also is used in citizen band (CB) radio transmission, where the operator can either transmit or receive but cannot perform both operations at the same time on the same channel. After the operator has completed a transmission, the other party must be advised of this fact so that the conversation can be continued. This is accomplished by the operator saying the term "over," which tells the other operator to begin transmission.

When data is transmitted over a communications link, the transmitter and receiver of the transmission device must be appropriately turned on and off as the direction of the transmission varies. Because two wires are required to complete an electrical circuit (one for transmission and one for ground), the transmission flow on such a circuit must be halted each time the direction of travel is reversed. This halt is known as line or circuit "turnaround," and the time needed is a function of the circuit mileage on telephone lines.

Half-duplex transmission can occur on either a two-wire or four-wire circuit. The switched direct-distance-dial telephone network is composed of two-wire circuits, whereas a circuit known as a *leased line* can be obtained as either a two-wire or four-wire facility. A four-wire circuit is essentially a pair of two-wire links that can be used for transmission in both directions simultaneously (full-duplex). The turnaround time is minimized on a four-wire circuit.

Full-duplex

Transmission that occurs in both directions simultaneously is known as *full-duplex*. In this mode of transmission, no turnaround time is required. Full-duplex transmission is often used when large amounts of alternating traffic must be transmitted and received within a fixed time period. Returning to the CB example, if one channel is used for transmission and another for reception, then two simultaneous transmissions can occur. Although the elimination of turnaround time permits full-duplex transmission to provide more efficient throughput, this efficiency may be negated by the cost of two-way lines and more complex equipment required by this mode of transmission. In normal operations, a four-wire circuit is needed for full-duplex transmission.

When appropriate, low- to medium-speed modems are used on two-wire circuits, full-duplex transmission becomes possible. This is because some modems are capable of subdividing the frequency band of the two-wire circuit into two separate subchannels, with each subchannel capable of supporting transmission in a different direction.

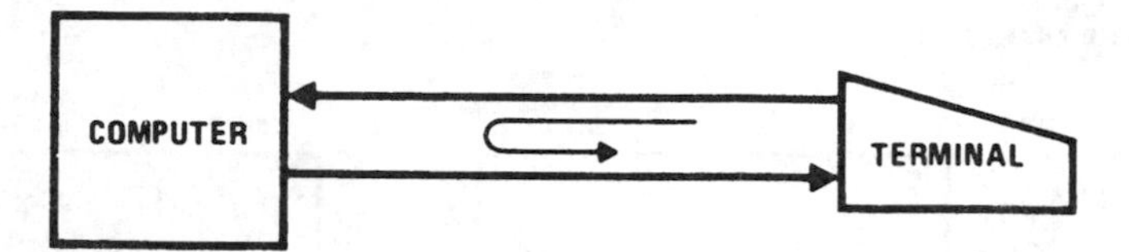

Figure 4.3.2 Echo-plex.

When referring only to terminal operations, the term *full-duplex* takes on a different meaning from the communications mode of the transmission medium. Here, the fact that the terminal is full-duplex can be used to denote that when the operator presses a key to transmit a character to the computer, it is echoed back to appear on the terminal's printer or display screen, as illustrated in Figure 4.3.2, as the operator continues to key in the data.

This mode of terminal transmission is known as *echo-plex* and is available with full-duplex operation only. When operating on half-duplex circuits, many terminals "copy" the transmitted signal back to the terminal for display to obtain a local copy or printout of characters as they are transmitted.

4.4 Transmission Techniques

Because of the necessity to synchronize a transmitting device with a receiving device, two techniques have been developed—asynchronous and synchronous transmission.

Asynchronous transmission

Asynchronous transmission is commonly referred to as *start-stop transmission*, where one character at a time is transmitted or received. Start and stop bits are used to synchronize the receiver with the transmitter, as well as to denote the beginning and end of the character transmitted. Most teleprinters, with the exception of some buffered terminals, and a large majority of personal computers used as communications terminal type devices use this method of transmission.

In asynchronous transmission, each character to be transmitted is encoded into a series of pulses. The transmission of the character is initiated by a start bit equal in width to one pulse. The encoded character (series of pulses) is followed by a parity bit and one or more stop bits that might be equal to or longer than one pulse width, depending on the transmission code used. The transmission of a 7-level ASCII character is illustrated in the top portion of Figure 4.4.1. If you assume that two stop bits are used, then 11 bits must be sent to transmit the 7 information bits that represent the encoded character. Here, the extra bits include one start bit, one parity bit, and two stop bits.

By convention, the start bit is signified by a space, logic 0, or no current, and stop bits are signified by a mark, logic 1, or by the presence of current. The start and stop elements are logically complements of each other so that when a start signal is received, the transition clearly indicates the beginning of a character. Although the parity bit can be used for error control, it is not used on many teleprinters, so bit-position 8 might have no significance. Terminals that transmit data at 10 characters per second usually use a stop element with a duration of two pulse widths. Thus, at that data rate, the transmission of an 11-unit code would require a modulation rate of 110 baud.

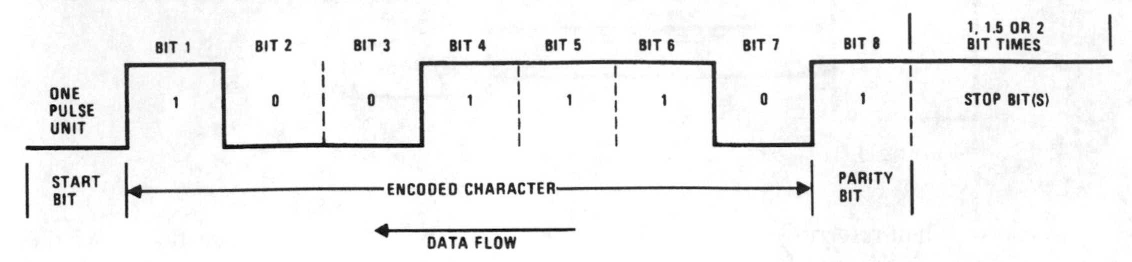

Figure 4.4.1 Asynchronous (start-stop) transmission.

For a typical 30-character-per-second teleprinter, a stop element of one pulse width is usually used. Using one signal element for the stop element would reduce the unit code per character to 10 signal elements. Thirty characters per second, each consisting of 10 signal elements or bits, would require a data rate of 300 bits per second. Because the bits are to be transmitted one at a time, a circuit with a modulation rate of 300 baud would be necessary for the transfer of information.

In the start-stop method of transmission, transmission begins anew on each character and stops after each character, as indicated in the lower portion of Figure 4.4.1. As characters are transmitted, idle time occurs between them. With synchronization starting anew with each character, any timing discrepancy is cleared at the end of the character and synchronization is maintained on a character-by-character basis.

Until 1987, asynchronous transmission was primarily used at data rates at or under 2400 bit/s. However, a few modems introduced in 1987 resulted in an increase in operation rates up to 9600 bit/s. Since 1987, there have been many design advances in universal asynchronous receiver transmitter (UART) chips and data modulation techniques. These advances have enabled modem designers to develop products that can transfer data asynchronously at operating rates up to 19,200 bps.

The UART converts parallel-formatted data, used to represent characters within a computer or terminal device, into a serial data stream. The chip prefixes each character's bit sequence with a start bit and appends appropriate parity and stop bits.

When functioning as a receiver, the UART recognizes the beginning of a character by its start bit. The chip assembles a character's serial incoming data stream, checks its parity if applicable, strips off the start and stop bits, then converts it to parallel upon which the computer or terminal device can operate.

Until 1987, most UARTs had just a one-character buffer, which made difficult the loss-less transfer of data at operating rates above 1200 or 9600 bps. Then a design change that enabled UARTs to buffer more than one character at a time permitted modem operating rates to increase beyond 9600 bps with no data loss.

A second development, new modulation techniques, resulted in a significant increase in modem operating rates together with a distinctly reduced probability of transmission errors. The techniques, collectively known as *trellis-coded modulation (TCM)*, involve the addition of one or more redundant bits to each transmitted character. This addition results in a modulation pattern that enables a receiving device to determine if a bit was misplaced because of transmission error. If so, TCM correctly repositions the bit.

Although the highest operating rate of an asynchronous modem was between 19,200 and 24,000 bps during 1993, many trade publications advertised modem performance specifications as "up to 76.8 kbps throughput." This was not the modem operating rate, but was its highest achievable throughput with built-in data compression operating at peak efficiency. For example, a modem capable of achieving a 4:1 compression ratio and having a transmission rate of 19,200 bps is capable of transferring data at $4 \times 19{,}200 = 76{,}800$ bps.

However, that elevated transfer rate does not occur continuously; it actually represents a peak rate that is only periodically obtainable. This is true because the efficiency of the modem's compression algorithm is based on the compressibility of the data being transmitted. Pure random data would not normally include redundancies. Therefore, the data would not be compressible, resulting in the modem's information transfer rate matching its operating rate.

In comparison, a long string of repeating characters would be highly compressible. The result would be a substantial increase in the modem's information transfer rate over its operating rate: the bit sequences being transmitted now represent compressed data. (Refer to Section 4.12 for specific information about the operation of compression-performing modems.) (See Table 4.4.1.)

TABLE 4.4.1 Characteristics of Transmissions Types

Asynchronous

- Each character is prefixed by a start bit and followed by one or more stop bits.
- A period of inactivity, idle time, can exist between transmitted characters.
- Bits within a character are transmitted at prescribed time intervals.
- Timing is established independently by terminals and computers.
- Transmission speeds normally do not exceed 19,200 to 24,000 bit/s.

Synchronous

- One or more synch characters prefix transmitted data.
- Synch characters are transmitted between blocks of data to maintain line synchronization.
- A number of characters are blocked together for transmission with no gaps existing between characters.
- Synchronized timing is established and maintained by the transmitting and receiving terminals, computers, or other devices.
- Terminals require storage (buffer) areas.
- Transmission speeds are normally in excess of 2,400 bit/s.

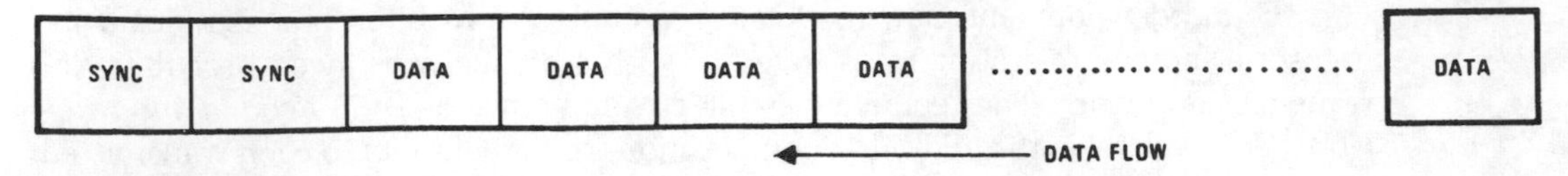

Figure 4.4.2 Synchronous transmission.

Synchronous transmission

When a high data-transfer rate is required, the asynchronous method is inefficient because of the extra time required to transmit start and stop pulses. A second type of transmission, which involves sending a group of characters in a continuous bit stream, can then be used to overcome the previously covered asynchronous limitations. This is known as *synchronous* or *bit-stream synchronous transmission.*

In the synchronous transmission method, data transfer is controlled by a timing signal (clock) at the originating device. This timing signal might originate from the terminal itself or it can be provided by a communications component, such as a modem, multiplexer, or front-end processor channel. At the receiving end, the communications component normally derives its timing from the line transitions, with a synchronized clock in the device used to control the sampling of the line conditions.

The receive clock must be kept in continuous step with the transmit clock to ensure synchronization and avoid bits being gained or lost. In order to obtain and continue this synchronization, data transmission is blocked into a group of data bits preceded by the transmission of one or more special characters. These special synchronization or sync characters are at the same code level (number of bits per character) as the coded information to be transmitted. However, each has the unique bit configuration of zeros and ones that define the sync character. The receiver recognizes and synchronizes itself onto a stream of transmitted sync characters.

Once synchronization is achieved, the actual transmission of information can proceed, as illustrated in Figure 4.4.2. Because data is grouped or blocked into groups of characters for synchronous transmission, terminals transmitting and receiving data using this mode of data transfer must have buffers for storing the character blocks.

In addition to buffers, more complex circuitry is required by synchronous terminals because the receiving terminal must remain in step with the data originator for the duration of the transmitted information block. Synchronous transmission was originally used when data rates in excess of 2400 bit/s were required. Today, the operating rate of communications equipment no longer serves as a criterion for determining if asynchronous or synchronous is used. Instead, other factors—such as the type of equipment used in a communications network and its method of use—normally govern the method of transmission. For example, an IBM Systems Network Architecture (SNA) type of network is primarily constructed using communications equipment that transmits and receives data in the form of synchronous serial bit streams. Thus, most communications devices used to construct an SNA network operate synchronously irrespective of their operating rates, which can range from 1.2 kbit/s to 1.544 Mbit/s. The major characteristics of the two types of transmission covered in this section are listed in Table 4.4.1.

4.5 Types of Transmission

The two types of data transmission that can be used are serial and parallel. In serial transmission, the bits that compose a character are transmitted in sequence over one line. In parallel transmission, characters are transmitted serially, but all n bits that are used to represent the character are transmitted in parallel—that is, simultaneously over n channels. To examine the differences between these two methods, consider the transmission of an 8-bit character. To transmit this character via parallel transmission requires a minimum of eight lines, with additional lines probably being required for control signals or for the transmission of a parity bit. Although parallel transmission is used extensively in computer-to-peripheral unit transmission, because of the high cost of the extra circuits it requires, it is not normally used other than when terminal devices are located in close proximity to a computer.

A typical use of parallel transmission is the in-plant connection of badge readers and similar devices to a computer located in that facility. One advantage of parallel transmission is that it can reduce the cost of terminal circuitry because the terminal does not have to convert the internal character representation into a serial data stream for transmission. As a result of the higher number of conductors and connectors required by parallel transmission, the cost of the transmission medium and interface will be higher, and these tradeoffs must be carefully examined. Because the total character can be transmitted at the same time using parallel transmission, higher data transfer rates than those obtainable with serial transmission facilities are possible. For this reason, most local-facility terminal-to-computer communications is accomplished by the use of parallel transmission. The two types of transmission covered in this section are illustrated in Figure 4.5.1.

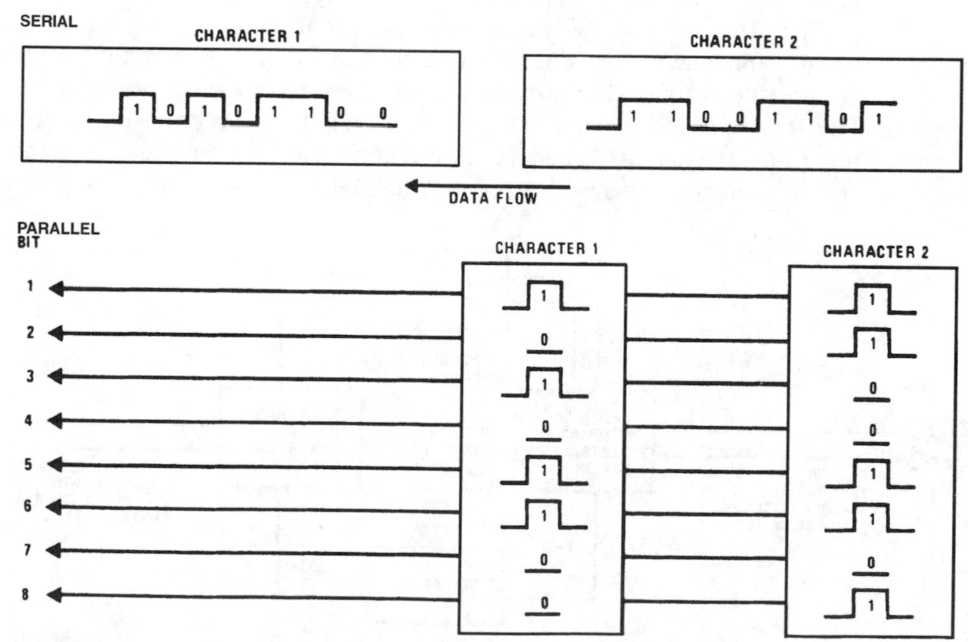

Figure 4.5.1 Types of data transmission.

4.6 Types of Circuits

The three basic types of line connections or circuits available to connect terminals to computers or to other terminals are dedicated, leased, and switched.

A dedicated line is similar to a leased line in that the terminal and computer (or other terminal) are always connected to each other on these types of circuits, transmission always occurs on the same path, and, if required, the circuit can be readily tuned (conditioned) to improve transmission performance.

The key difference between a dedicated and a leased circuit or line is that a dedicated line refers to a transmission medium internal to a user's premises, where the customer has the right of way for cable laying. A leased line, however, provides a connection between separate sites where the customer has no right of way for cable laying. Another term for a dedicated circuit is a *direct-connect line.* This type of connection normally links a terminal or business machine on a direct path through the site to another terminal or business machine located at that site. The dedicated line can be a wire conductor installed by the employees of the company or by the computer manufacturer's personnel, or it can be a local telephone line installed by telephone company personnel.

The second type of circuit, the leased line, is commonly called a *private line.* It is obtained from a communications carrier to provide an exclusive-use transmission medium between two sites that could be in separate buildings in one city or in distant cities.

The third type of circuit, known as a *switched line*, is also often referred to as a *dial-up connection*; its use permits contact with any party having access to the telephone network. The operator of a business machine accesses the computer by dialing the telephone number of a line connected to the computer. In using switched or dial-up transmission, telephone company switching centers (central offices) serve as intermediate connectors between the dialing party and the dialed party, as illustrated in Figure 4.6.1.

When the dialing party initiates the call, the first connection is made to the telephone company central office serving the calling party's telephone exchange. This connection is known as a *subscriber-loop connection* and it occurs over a two-wire line, where one wire is used for transmission in both directions while the other wire serves as a common ground. Next, based on the number dialed, the call is routed through one or more switching centers to the telephone company central office serv-

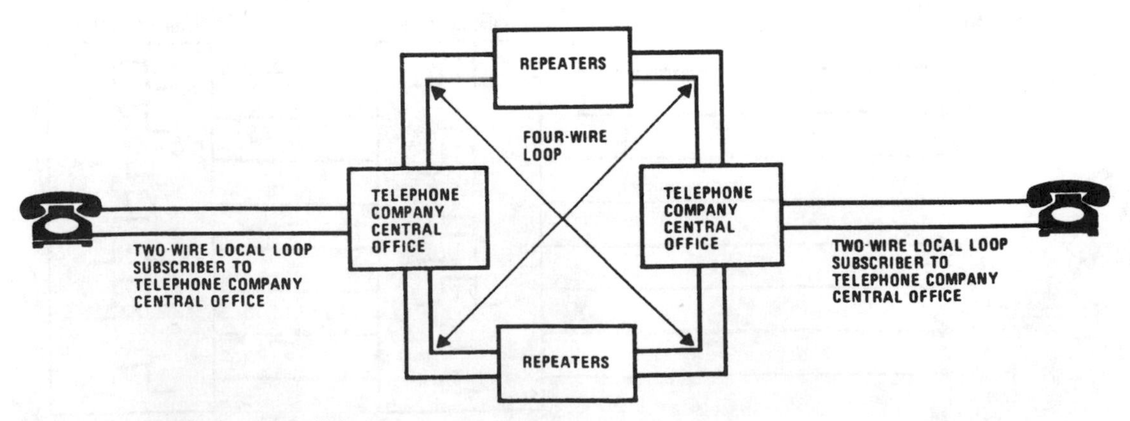

Figure 4.6.1 Switched network dialing.

ing the telephone exchange of the number dialed. The connection between telephone company central offices for long-distance calls is composed of four wires, with one pair being used for each direction of transmission. When a full-duplex leased line is required, the subscriber loops are converted to four-wire circuits.

When wire terms are used in the context of data transmission, confusion might arise because the term *four-wire* was derived from the early days of telephone, and in some cases, it does not signify four physical wires. In such cases, two wires are used to form the equivalent of a four-wire circuit. In this configuration, each terminal on the channel transmits data on one frequency and receives data on another. With each modem tuned to receive on the frequency on which the other modem transmits, simultaneous two-way transmission is achieved with a single pair of wires.

Once the telephone central office of the dialed party is reached, another local loop connection is established to finalize the connection between the dialing party and the dialed party. After the connection is established, the terminal and the computer conduct their communications. When communications is completed, the switching centers and central offices disconnect the path that was established for the connection and restore all paths used so that they become available for other connections.

When using the switched telephone network for communications, routing is accomplished according to the telephone number of the dialed party. This routing occurs according to the network switching plan used by the telephone company.

Under the network numbering plan used in the United States, which is the foundation of the network switching plan, subscriber stations are assigned 4-digit codes, such as *XXXX*, where *X* is any number 0 through 9. The telephone company central office or exchange is assigned a 3-digit code, such as *NNX*, where *N* is any number 2 through 9. One or more central offices or exchanges are used to form an area code or numbering plan area (NPA) with the assignment of a 3-digit code to prefix the central office and subscriber station numbers.

The middle digit in the area code is currently a one or zero. However, this is expected to change in the near future, since almost all available area codes are now in use. Within an area code or numbering plan area, central office codes are never duplicated. Also, one physical switching machine at a central office might serve subscribers with different central office codes. As an example, a switching machine could serve subscribers with the 444, 445, and 446 prefixes.

One variation of the area code is the international dialing access code. To dial a telephone in a foreign country from certain areas of the U.S. requires first dialing the international access code, 011, followed by a two-digit country code, a city code, and the local number.

Cost, speed of transmission, and degradation of service are the primary factors considered in choosing between leased and switched facilities. In general, if data communications requirements to a computer involve occasional random contacts from a number of terminals at different locations, and if each call is of short duration, dial-up service is used. Switched circuits are commonly used for data transmission at speeds up to 19,200 bit/s.

If a large amount of traffic is anticipated between a computer and a few terminals, leased lines are usually installed. Because a leased line is fixed as to its routing, it can be tuned or conditioned to reduce errors in transmission; this also facilitates determining the location of errors because its routing is known. Different categories of leased lines can be selected that permit transmission from under 100 bit/s to many hundreds of thousands to millions of bits per second. Additional information concerning the different categories of leased lines is covered in the section entitled "Transmission Rates." In Table 4.6.1, the reader will find a list of some of the limiting factors involved in determining the type of line to use for transmission between business machines and computers.

4.7 Line Structure

Both the geographical distribution of terminals and computers and the distance between each terminal and the computer it will transfer data to are important parameters that must be considered in developing a network configuration. The methods used to connect terminals to computers or other terminals result in one or more line structures that constitute a data communications network configuration. The two types of line structures basic to any network are point to point and multipoint, the latter also commonly referred to as *multidrop*.

Point-to-point link

When a direct connection is established between two points in a network, the line structure is known as a *point-to-point link*. Here, the two points that are connected can be two terminals, a terminal and computer, or two computers. This type of line structure can be established using a dedicated, switched, or leased circuit, as illustrated in the top portion of Figure 4.7.1. Each terminal transmits and receives information to and from the computer via an individual connection that links a specific terminal to the computer.

Multipoint link

In the early period of telephone development, and to some extent even today in some areas, a number of subscribers would be connected to a single-wire pair. This

TABLE 4.6.1 Line Selection Guide

Line type	Distance between transmission points	Speed of transmission	Use for transmission
Dedicated (Direct connect)	Local	Limited by the conductor	Short or long duration
Switched (Dial-Up)	Limited by telephone-access availability	See note below.	Short-duration transmission
Leased	Limited by communications company availability	Limited by type of facility selected	Long-duration or frequent short-duration calls

NOTE: Normally up to 19.2 kbit/s via the analog public switched telephone network (without compression) and 56 or 384 kbit/s over switched digital transmission facilities.

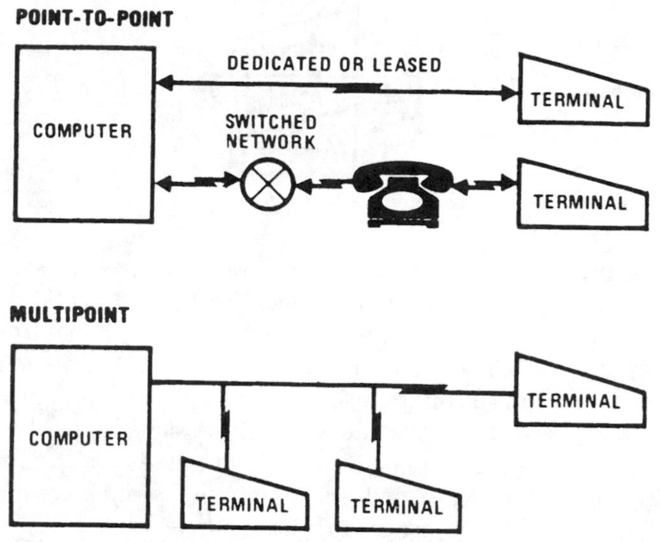

Figure 4.7.1 Network line structures.

type of connection was known as a *party line*, and communications problems would arise if two subscribers wished to place a call at the same time. The advantage of this type of circuit arrangement is that the need for individual lines from each subscriber to the telephone company office is eliminated; instead, one line links all subscribers on the party line to the telephone company central office.

Today, a line structure equivalent to the party line can be constructed, for data communications applications, which is a multipoint or multidrop circuit. The primary advantage of such a structure is to permit terminal devices to share a common line and thereby reduce the cost of the transmission medium, as illustrated in the lower portion of Figure 4.7.1.

Today, most multipoint lines are used with hierarchically structured networks based on older network architectures, such as IBM's SNA. Originally, most terminal devices serviced by multipoint lines were fixed-logic "dumb" devices. As the popularity of PCs grew, many organizations gradually replaced their dumb terminal devices with PCs containing an adapter card that could execute communications software and support multipoint communications.

Although the PC now emulates the operation of the older dumb terminal, this application does not bar the PC from performing its processing capabilities, such as spreadsheets and word processing, as well as communications with other computers. Different types of adapter cards and communications software enable a PC to function as a specific type of terminal connected to a control unit. In this case, the PC becomes the terminal illustrated in Figure 4.7.2. Thus, a PC with appropriate hardware and software can, in effect, be considered a ubiquitous terminal device.

For a multipoint line structure to be effective, one location must be the master or controlling location while the remaining locations are tributaries or slaves. Normally, each location has a unique digital address. In a data communications environment, the computer is normally the master while the connected terminal devices are the

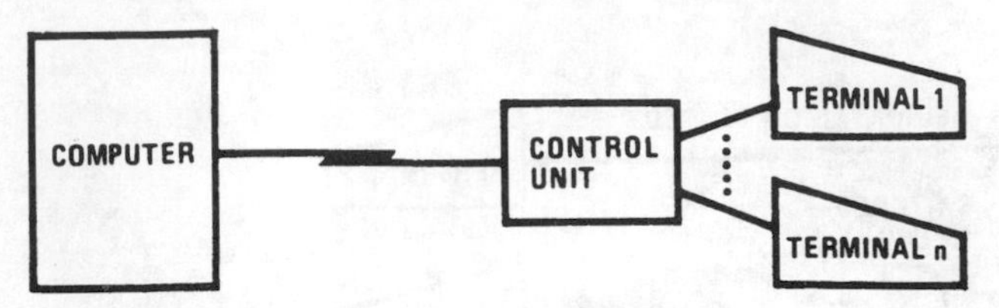

Figure 4.7.2 Point-to-point line using control units.

slaves. To regulate traffic on the circuit, the control location will use one of two methods, known as *polling* and *selecting*, detailed in the next section, 4.8. These methods are designed to prevent data transmitted from one terminal from interfering with data transmitted from another terminal, and are also known as *line discipline*. Although no two terminals on a multipoint line can transmit data at the same time, two or more terminals can receive messages at the same time. The number of terminals receiving such a message depends on the addresses assigned to the message. In some networks, a "broadcast" address permits all terminals connected to the same multidrop line to receive a message at the same time.

One variation of a point-to-point line to permit multiple terminals to communicate via a common line to a computer is obtained through the utilization of special control units, as illustrated in Figure 4.7.2. These control units are known as *modem-sharing units*, *line-sharing units*, *message interface units*, and other such terms. They permit terminals to be clustered at a common geographical location and share a common circuit for communications to a computer. Again, each terminal might have a unique address and can be a dumb, fixed-logic device or a PC containing an appropriate communications adapter card and communications software.

Both point-to-point and multidrop lines can be mixed in developing a network. Factors that must be considered in the selection of a line structure are numerous and include communications line costs, terminal support of a multipoint line discipline, computer support of that discipline, access and response times, the computer processing time, transmission delays caused by the transfer of data through various communications components, transmission speeds and distances, and the volume of traffic to be transferred.

4.8 Line Discipline

For several devices to share the use of a common, multipoint communications line without interfering with one another, a line discipline must be put into effect. The line discipline for multipoint circuits prevents transmission by more than one terminal at any time, although two or more devices can receive information simultaneously. This line discipline combines polling and selecting, and requires that each terminal on the line have a unique address of one or more characters, as well as circuitry to respond to a message sent to that address.

Polling and selecting

When the computer polls a line, in effect it samples each terminal device in a predefined sequence (polling list) to determine if the terminal has data to transmit. If the terminal has no data to transmit, the computer is informed of this fact and continues its polling sequence in accordance with its polling list until it encounters a terminal that has data to send. At this point, the computer will stop polling and permit the terminal to transmit its data. When the message transmitted by the terminal is completed, the computer will poll the next terminal on the line.

As the computer polls each terminal, the other terminals on the line must wait until they are polled before they can be serviced. Conversely, transmission of data from the computer to each terminal on a multidrop line is accomplished by the computer selecting the terminal address to which the data is to be transferred, informing the terminal that data is to be transferred to that terminal, and then transmitting data to the selected terminal.

Polling and selecting can be used to service both asynchronous and synchronous terminals that are connected to independent, multidrop lines. With polling, a considerable amount of time is utilized by the computer addressing each terminal to determine if the terminal has data to transmit. Because of the control overhead, synchronous, high-speed transmission is normally used for polling, although asynchronous transmission has been successfully used when terminals have short messages.

In the synchronous environment, terminals have a buffer. For a CRT terminal that is normally used on a multidrop circuit, the buffer area is usually equal to the number of characters that can be displayed on the screen. Thus, when the CRT operator presses the transmit key on the terminal, no data is actually transmitted to the computer until the terminal is polled. Then, the data in the buffer is transmitted and the computer polls the next terminal.

In an asynchronous environment, once the computer selects a terminal, the computer locks out all other terminals on the line until the message is completed. Asynchronous multipoint lines are usually used to connect a number of receive-only terminals, or terminals with limited transmission requirements, to a central computer. An example of this type of environment would be a state weather news distribution service that connects a number of state offices to the central prediction bureau. Here, the central location can prepare forecasts for each area and address and transmit the forecasts to each terminal via a multipoint circuit used to connect all terminals to the central computer.

By the use of signals and procedures, polling and selecting line control ensures the orderly and efficient utilization of multidrop lines. As an example of a computer polling, the third terminal on a multipoint line and then receiving data from that terminal is shown in the top portion of Figure 4.8.1. At the bottom of that illustration, the computer first selects terminal number 2 on the line and then transfers a block of data to that terminal.

When terminals transmit data on a point-to-point line to another terminal or computer, the transmission of that data is under the control of the terminal operator. This method of line control is known as *non-poll-and-select* or *free-wheeling transmission*.

Additional information concerning multipoint line efficiency and turnaround times is provided in Chapter 5.

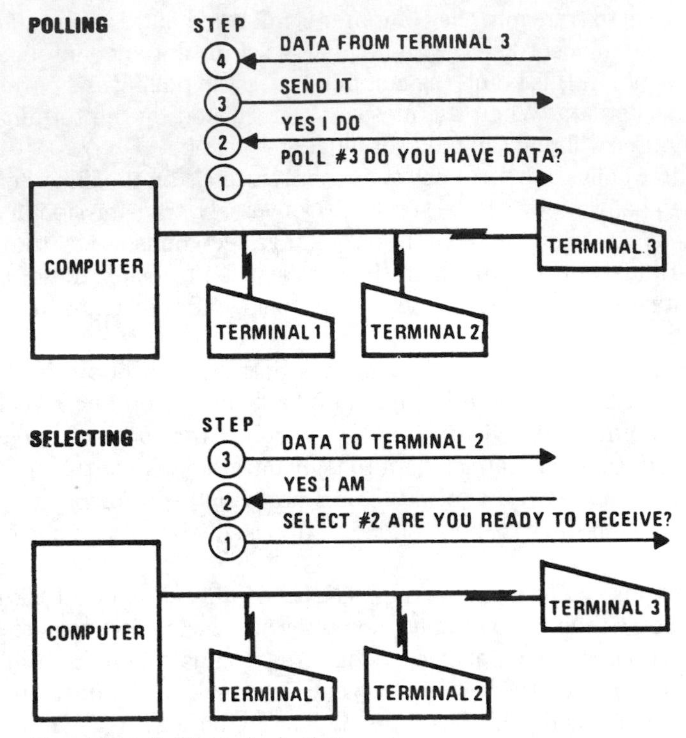

Figure 4.8.1 Polling and selecting.

4.9 Transmission Codes and Control Characters

Within a computer, data is structured according to the architecture of the machine. This internal representation of data is seldom suitable for transmission to devices other than the peripheral units attached locally to the computer. In most cases, in order to transmit data, the internal information of the computer must be reformatted or translated into a suitable transmission code. This transmission code then creates a correspondence between the bit encoding of data for transmission or internal device representation and printed symbols. Although code conversion can be performed by the computer or by terminals with internal processing capability, the transmission code used is usually dictated by the character code that the remote terminals are designed to accept. Current terminal codes include Baudot, which is a 5-level (5 bits per character) code; binary coded decimal (BCD), which is a 6-level code; the American standard code for information interchange (ASCII), which is normally a 7-level code; an extended 8-level version of ASCII used by IBM PC and compatible personal computers; and the extended binary coded decimal interchange code (EBCDIC), which is an 8-level code.

In addition to information being encoded into a certain number of bits based on the transmission code used, the unique configuration of those bits to represent certain control characters can also be considered as a code that can be used to control the line discipline used. These control characters indicate the start of header (SOH), the end of a message (EOM), and so on, with the number of permissible control characters standardized according to the code used.

In computer-to-computer data transfers, a large amount of processing time otherwise involved in converting internal data into data coded for transmission can be avoided by sending the data in the format used by the computer for internal processing. This type of transmission is known as *binary-mode transmission*, *transparent data transfer*, *code-independent transmission*, or, most commonly, as *native-mode transmission*.

Morse code

One of the most common codes, the Morse or International code, consists of a series of dots and dashes. The letter V, for example, consists of three dots and a dash. Although this code is written as a series of dots and dashes, the actual transmission consists of long and short signals with pauses between each character, as illustrated in Figure 4.9.1. By the use of long and short signals, a unique character configuration is developed that can be recognized by an experienced operator.

In the early development of mechanized telegraph equipment, Morse code was not used because the characters varied in length between pauses. Even today, this code is not practical for a computer communications environment. To overcome the variable-length problems associated with Morse code, a five-bit code, Baudot, named after its inventor, was developed.

Baudot code

The 5-level Baudot code was devised by a Frenchman, Emil Baudot, and provides a mechanism for encoding characters by an equal number of bits, in this case, five. Through the work of an American, Howard Krum, a method was developed to permit the synchronization of transmitted characters to a receiver, thus permitting the development of automatic transmission and reception equipment.

Because the number of different characters that can be derived from a code having two different (binary) states is 2^m, where m is the number of positions in the code, the 5-level Baudot code permits 32 unique character bit combinations.

Because of the necessity to transmit not only the 26 letters of the alphabet, but also digits, punctuation marks, and special symbols (the sum of which exceeds 32), it became necessary to devise a mechanism to extend the capacity of the Baudot

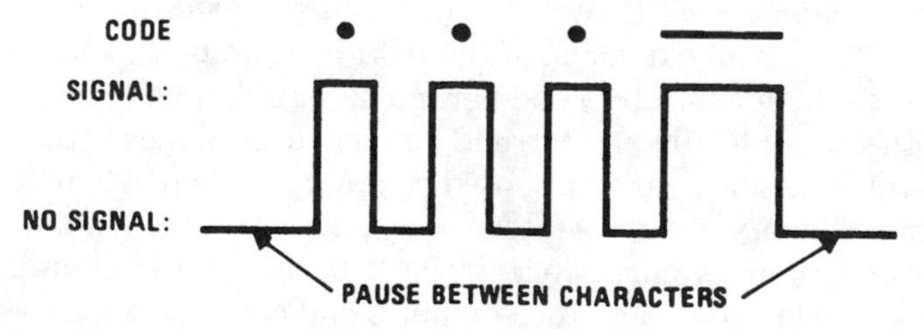

Figure 4.9.1 Morse code transmission of letter "V".

code to include additional character representations. This extension was accomplished by the incorporation of two "shift" characters into the code: "letters shift" and "figures shift." Here, the transmission of a shift character is used to inform the receiver that the characters that will follow the shift character should be interpreted from a symbol and numeric set or from the alphabetic set of characters. In Table 4.9.1, the 5-level Baudot code is listed for one particular terminal pallet arrangement. A transmission of all ones (11111) in bit positions 1 through 5 is used to indicate a letters shift, and the characters that follow the transmission of that shift character are interpreted as letters. Similarly, the transmission of ones in bit positions 1, 2, 4, and 5 (11011) is used to indicate a figure shift, and the following characters are now interpreted as numerals or symbols, based on their code structure.

BCD code

Corresponding to the development of computers was the implementation of coding to convert alphanumeric characters into binary notation and the binary notation of computers into alphanumeric characters. One of the earliest codes used to convert data from alphanumeric characters into a computer-acceptable format uses the binary coded decimal code.

The BCD coding technique permits decimal numeric information to be represented by 4 binary bits and permits an alphanumeric character set to be represented through the use of 6 bits of information, as shown in Table 4.9.2. One advantage of this code is that two decimal digits can be stored in an 8-bit computer word and manipulated with appropriate computer instructions. Although only 36 characters are listed in Table 4.9.2, a BCD code is capable of representing a set of 2^6 or 64 different characters.

In addition to the transmission of letters, numerals, and punctuation marks, a considerable number of control characters might be required to ensure line discipline is maintained. These control characters can be used to switch on and off devices that are connected to the communications line, control the actual transmission of data, manipulate message formats, and perform such additional functions as acknowledge messages received correctly and request retransmission of messages received in error.

EBCDIC format

To accommodate these control characters, an extended character set is usually required. One such character set is the extended binary coded decimal interchange code (EBCDIC), which is an extension of the BCD code and uses eight bits instead of six to represent characters. This code permits 2^8 or 256 unique characters to be represented, although only 109 are currently assigned meanings. This code is primarily used for transmission by byte-oriented computers, where a byte is a grouping of eight consecutive binary digits operated on as a unit by the computer. The use of this code in many instances might relieve the byte-oriented computer from performing code conversion when both the computer and connected terminals operate with the same character set. The EBCDIC character set is listed in Table 4.9.3.

TABLE 4.9.1 5-Level Baudot Code

Letters	Figures	Bit selection 1	2	3	4	5
Characters						
A	–	1	1			
B	?	1			1	1
C	:		1	1	1	
D	$	1			1	
E	3	1				
F	!	1		1	1	
G	&		1		1	1
H				1		1
I	8		1	1		
J	´	1	1		1	
K	(	1	1	1	1	
L	)		1			1
M	.			1	1	1
N	,			1	1	
O	9				1	1
P	0		1	1		1
Q	1	1	1	1		1
R	4		1		1	
S		1		1		
T	5				1	
U	7	1	1	1		
V	;		1	1	1	1
W	2	1	1			1
X	/	1		1	1	1
Y	6	1		1		1
Z	"	1				1
Functions						
Carriage return	<				1	
Line feed	=		1			
Space				1		
Letters shift		1	1	1	1	1
Figures shift		1	1		1	1

TABLE 4.9.2 Binary-Coded Decimal Character Set

Bit position						
b_6	b_5	b_4	b_3	b_2	b_1	Character
0	0	0	0	0	1	A
0	0	0	0	1	0	B
0	0	0	0	1	1	C
0	0	0	1	0	0	D
0	0	0	1	0	1	E
0	0	0	1	1	0	F
0	0	0	1	1	1	G
0	0	1	0	0	0	H
0	0	1	0	0	1	I
0	1	0	0	0	1	J
0	1	0	0	1	0	K
0	1	0	0	1	1	L
0	1	0	1	0	0	M
0	1	0	1	0	1	N
0	1	0	1	1	0	O
0	1	0	1	1	1	P
0	1	1	0	0	0	Q
0	1	1	0	0	1	R
1	0	0	0	1	0	S
1	0	0	0	1	1	T
1	0	0	1	0	0	U
1	0	0	1	0	1	V
1	0	0	1	1	0	W
1	0	0	1	1	1	X
1	0	1	0	0	0	Y
1	0	1	0	0	1	Z
1	1	0	0	0	0	0
1	1	0	0	0	1	1
1	1	0	0	1	0	2
1	1	0	0	1	1	3
1	1	0	1	0	0	4
1	1	0	1	0	1	5
1	1	0	1	1	0	6
1	1	0	1	1	1	7
1	1	1	0	0	0	8
1	1	1	0	0	1	9

TABLE 4.9.3 EBCDIC Character Set

Bit positions 4, 5, 6, 7	Second hexadecimal digit	00				01				10				11			
		00	01	10	11	00	01	10	11	00	01	10	11	00	01	10	11
		0	1	2	3	4	5	6	7	8	9	A	B	C	D	E	F
0000	0	NUL	DLE	DS		SP	&	-									0
0001	1	SOH	DC1	SOS				/		a	j			A	J		1
0010	2	STX	DC2	FS	SYN					b	k	s		B	K	S	2
0011	3	ETX	TM							c	l	t		C	L	T	3
0100	4	PF	RES	BYP	PN					d	m	u		D	M	U	4
0101	5	HT	NL	LF	RS					e	n	v		E	N	V	5
0110	6	LC	BS	ETB	UC					f	o	w		F	O	W	6
0111	7	DEL	IL	ESC	EOT					g	p	x		G	P	X	7
1000	8		CAN							h	q	y		H	Q	Y	8
1001	9		EM							i	r	z		I	R	Z	9
1010	A	SMM	CC	SM		¢	!										
1011	B	VT	CU1	CU2	CU3		$	,	#								
1100	C	FF	IFS		DC4	<	*	%	@								
1101	D	CR	IGS	ENQ	NAK	(	)	_	'								
1110	E	SO	IRS	ACK		+	;	>	=								
1111	F	SI	IUS	REL	SUB	|	¬	?	"								

Bit positions 0, 1 / Bit positions 2, 3 / First hexadecimal digit

CONTROL CHARACTER REPRESENTATIONS

ACK Acknowledge
BEL Bell
BS Backspace
BYP Bypass
CAN Cancel
CC Cursor control
CR Carriage return
CU1 Customer use 1
CU2 Customer use 2
CU3 Customer use 3
DC1 Device control 1
DC2 Device control 2
DC4 Device control 4
DEL Delete
DLE Data link escape
DS Digit select
EM End of medium
ENQ Enquiry
EOT End of transmission
ESC Escape
ETB End of transmission block
IGS Interchange group separator
IL Idle
IPS Interchange record separator
IUS Interchange unit separator
LC Lower case
LF Line feed
NAK Negative acknowledge
NL New line
NUL Null
PF Punch off
PN Punch on
RES Restore
RS Reader stop
SI Shift in
SM Set mode
SMM Start of manual message
SO Shift out
SOH Start of heading
SOS Start of significance
SP Space
STX Start of text
ETX End of text
FF Form feed
FS Field separator
HT Horizontal tab
IFS Interchange file separator
SUB Substitute
SYN Synchronous idle
TM Tape mark
UC Upper case
VT Vertical tab

SPECIAL GRAPHIC CHARACTERS

¢ Cent sign
. Period, decimal point
< Less-than sign
(Left parenthesis
+ Plus sign
| Logical OR
& Ampersand
! Exclamation point
$ Dollar sign
* Asterisk
) Right parenthesis
; Semicolon
¬ Logical NOT
– Minus sign, hyphen
/ slash
, Comma
% Percent
_ Underscore
> Greater-than sign
? Question mark
: Colon
Number sign
@ At sign
' Prime, apostrophe
= Equal sign
" Quotation mark

ASCII format

As a result of the proliferation of data transmission codes, attempts to develop a standardized code for data transmission were initiated. One result was the American Standard Code for Information Interchange (ASCII). This 7-level code, listed in Table 4.9.4, is based on a seven-bit code developed by the International Organization for Standardization (ISO).

ASCII characters are encoded in seven bits, while an eighth bit is available for use as a parity bit. The use of the parity bit is optional. Parity can be odd or even. Today, most terminal devices are built to conform to the ASCII code, which permits a large degree of code compatibility between different manufacturers of such devices. Although the EBCDIC character set was developed by IBM and is still used by most of that firm's computers, an ASCII character set is also available for some computers produced by that manufacturer.

Extended ASCII format

As mentioned earlier in this section, the IBM PC and compatible personal computers use an extended 8-level ASCII code. The first 128 characters are the same as the conventional ASCII character set, and the next group of 128 characters are formed by the additional bit added to the code.

Normally, the additional characters in the extended ASCII code are used to represent special graphics symbols, foreign currency symbols, and Basic "tokens," which represent an abbreviated form of a keyword in that programming language. Although many programs and data files can be transmitted using a conventional 7-level ASCII format, whenever a program or data file contains one or more extended ASCII code characters, an 8-level ASCII transmission must be used. When that situation occurs, the parity bit used in 7-level ASCII transmission is replaced by the bit used to extend the character set.

Character timings

The time required to generate a character varies according to the operating speed of the terminal. In Figure 4.9.2, the ASCII time intervals are illustrated for the character bit periods (units intervals) at terminal operating speeds ranging from 10 characters per second to 120 characters per second. Thus, at 30 characters per second each bit requires 3.33 msec for transmission. In order to determine the time required for a character to be transmitted, multiply the number of bits used to construct the character by the bit width in milliseconds. (A total of 10 bits—including parity and control bits—is assumed.)

Code conversion

One frequent problem in data communications is that of code conversion. This can be recognized when we consider what must be accomplished to enable a computer with an EBCDIC character set to communicate with a terminal that has an ASCII

TABLE 4.9.4 The ASCII Character Set

This coded character set is to be used for the general interchange of information among information processing systems, communications systems, and associated equipment.

Bits b_7 b_6 b_5 →					0 0 0	0 0 1	0 1 0	0 1 1	1 0 0	1 0 1	1 1 0	1 1 1
b_4 ↓	b_3 ↓	b_2 ↓	b_1 ↓	COLUMN → ROW ↓	0	1	2	3	4	5	6	7
0	0	0	0	0	NUL	DLE	SP	0	@	P	`	p
0	0	0	1	1	SOH	DC1	!	1	A	Q	a	q
0	0	1	0	2	STX	DC2	"	2	B	R	b	r
0	0	1	1	3	ETX	DC3	#	3	C	S	c	s
0	1	0	0	4	EOT	DC4	$	4	D	T	d	t
0	1	0	1	5	ENQ	NAK	%	5	E	U	e	u
0	1	1	0	6	ACK	SYN	&	6	F	V	f	v
0	1	1	1	7	BEL	ETB	'	7	G	W	g	w
1	0	0	0	8	BS	CAN	(	8	H	X	h	x
1	0	0	1	9	HT	EM	)	9	I	Y	i	y
1	0	1	0	10	LF	SUB	*	:	J	Z	j	z
1	0	1	1	11	VT	ESC	+	;	K	[	k	{
1	1	0	0	12	FF	FS	,	<	L	\	l	|
1	1	0	1	13	CR	GS	–	=	M	]	m	}
1	1	1	0	14	SO	RS	.	>	N	^	n	~
1	1	1	1	15	SI	US	/	?	O	____	o	DEL

Character Representation and Code Identification

The standard 7-bit character representation, with b_7 the high-order bit and b_1 the low-order bit, is shown below.

Example:

The bit representation for the character "K," positioned in column 4, row 11, is

b_7	b_6	b_5	b_4	b_3	b_2	b_1
1	0	0	1	0	1	1

The code table for the character "K" may also be represented by the notation "column 4, row 11" or alternatively as "4/11." The decimal equivalent of the binary number formed by bits b_7, b_6, and b_5, collectively, forms the column number, and the decimal equivalent of the binary number formed by bits b_4, b_3, b_2, and b_1, collectively, forms the row number.

The standard code may be identified by the use of the notation ASCII.

The notation ASCII (pronounced "as-key") should ordinarily be taken to mean the code prescribed by the latest edition of this standard. To explicitly designate a particular (perhaps prior) edition, the last two digits of the year of issue may be appended, as "ASCII 68" or "ASCII 77."

TABLE 4.9.4 Continued

Control Characters			
Col/Row	**Mnemonic and Meaning[1]**	**Col/Row**	**Mnemonic and Meaning[1]**
0/0	NUL Null	1/0	DLE Data Link Escape (CC)
0/1	SOH Start of Heading (CC)	1/1	DC1 Device Control 1
0/2	STX Start of Text (CC)	1/2	DC2 Device Control 2
0/3	ETX End of Text (CC)	1/3	DC3 Device Control 3
0/4	EOT End of Transmission (CC)	1/4	DC4 Device Control 4
0/5	ENQ Enquiry (CC)	1/5	NAK Negative Acknowledge (CC)
0/6	ACK Acknowledge (CC)	1/6	SYN Synchronous Idle (CC)
0/7	BEL Bell	1/7	ETB End of Transmission Block (CC)
0/8	BS Backspace (FE)	1/8	CAN Cancel
0/9	HT Horizontal Tabulation (FE)	1/9	EM End of Medium
0/10	LF Line Feed (FE)	1/10	SUB Substitute
0/11	VT Vertical Tabulation (FE)	1/11	ESC Escape
0/12	FF Form Feed (FE)	1/12	FS File Separator (IS)
0/13	CR Carriage Return (FE)	1/13	GS Group Separator (IS)
0/14	SO Shift Out	1/14	RS Record Separator (IS)
0/15	SI Shift In	1/15	US Unit Separator (IS)
		7/15	DEL Delete

[1](CC) Communications control; (FE) Format effector; (IS) Information separator

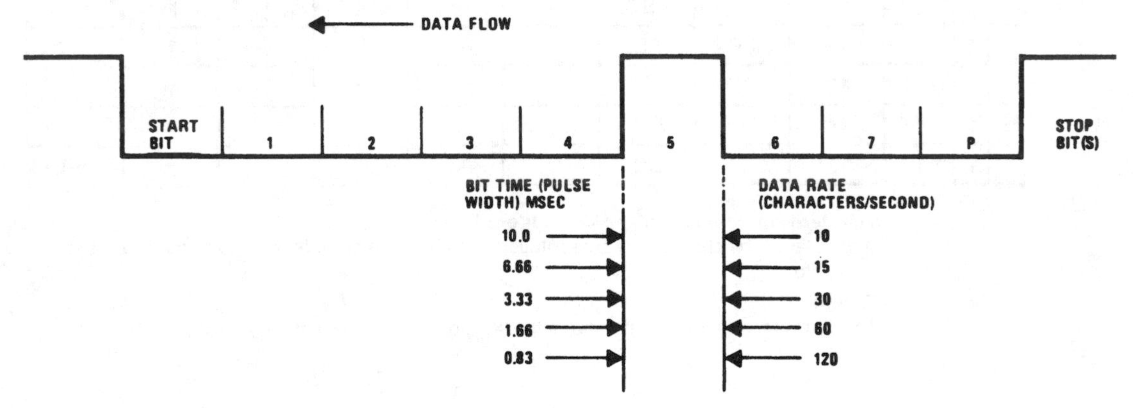

Figure 4.9.2 ASCII code bit time and data rate.

character set. When the terminal transmits a character, it is encoded according to the ASCII character code in Table 4.9.4. Thus, the letter A would be transmitted as 1000001, ignoring parity, start, and stop bits if we are using asynchronous transmission. Upon receipt of that character, the computer must convert the bits of information of the ASCII character into an equivalent EBCDIC character, in this case 11000001. Conversely, when data is to be transmitted to the terminal, it must be converted from EBCDIC to ASCII. This character translation requires computer processing time, which could be used for other activities if both the computer and terminal codes were identical.

Code efficiency

The efficiency of any two-condition (binary-based) code can be expressed by the formula:

$$E = \frac{\log_2 CS}{BC}$$

where E = the efficiency of the code
CS = number of characters or symbols required
BC = number of bits in the code

Assume that for a particular application, 32 different characters are required and a 6-bit code is to be used. By applying the formula we obtain

$$E = \frac{\log_2 CS}{BC} = \frac{5}{6} = 83.3 \text{ percent}$$

If an 8-bit code were used, then the code efficiency would be reduced to 5/8 or 62.5 percent.

Based on this methodology, the Baudot code, when using letter and figure shift to extend the character set, becomes extremely efficient. As listed in Table 4.9.1, this 5-level code can be used to represent 55 different characters. If 32 characters are required, the code has an efficiency of $\log_2 32/5$ or 5/5, which is 100 percent. If you assume that all 55 characters are required, the coding efficiency is approximately 116 percent.

Control characters

Within the transmission code, two sets of control characters may be encountered: those for terminal control and those for transmission control.

The terminal control character, as its name implies, is used to control functions on the terminal. These functions can include a line feed, ringing the bell to obtain the operator's attention, and so on.

In the area of transmission control, characters were assigned meanings to overcome the problems associated with message control, error detection and correction, and other problems associated with data transmission. A list of some of these transmission control characters and their assigned meanings will be found in Table 4.9.5.

4.10 Protocols

A protocol is a rule or management method for the data link and the attached terminals. In a data communications environment, both the terminal protocol and the data link protocol must be considered.

The terminal protocol (consisting of terminal control characters) can include such control characters as the bell, line feed, and carriage return for teleprinter terminals, cursor-positioning characters for a display terminal, and form control characters for the line printer attached to a remote batch terminal. The data link protocol is used

TABLE 4.9.5 Transmission Control Characters

SYN — Synchronous Idle	Used by synchronous transmission systems to provide message and character framing and synchronization.
SOH — Start of Header	Used at the beginning of a sequence of characters to indicate address or routing information. Such a term is referred to as a Heading. A STX character terminates a heading.
STX — Start of Text	Used at the beginning of a sequence of characters that is to be treated as an entity to reach the ultimate destination.
ETX — End of Text	Used to terminate a sequence of characters started with STX.
ETB — End of Transmission Block	Used to indicate the end of a sequence of characters started with SOH or STX.
EOT — End of Transmission	Used to indicate a termination of transmission. A transmission may include one or more records and their associated headings.
ACK — Acknowledge	A character sent by the receiving station to the transmitting station to indicate successful reception of a message.
ENQ — Enquiry	A character used to request a response from a remote station. The response that a remote station generates is predefined.

to define the control characteristics of a particular link and is a set of conventions to be followed to govern the transmission of data and control information. Further, a terminal can have a predefined control character or set of control characters that are unique to the terminal and are not interpreted by the line protocol.

Poll and select is often considered to be a type of line discipline or control, with the control character configuration to perform poll and select considered the line protocol.

In general, the line protocol permits the exchange of information, according to an order or sequence by establishing a series of rules for the interpretation of control characters that are used to govern the exchange of information. These characters control the execution of a number of tasks that are essential for the exchange of information in a data communications environment. Some of these information control tasks are listed in Table 4.10.1.

Although these tasks are important, only some of them might be required for the data transmission because the series of tasks required is a function of the total data communications environment. An example of this situation is the connection of a single terminal to a computer. Here, the establishment and verification of the connection might not be required. Conversely, when several terminals are connected to a computer via a multipoint or multidrop link, verification of the terminal's identification would be required to ensure that the data transmitted from the computer would be received by the proper terminal. In addition, once a terminal session is completed, this fact must be recognized to enable the computer's resources to be made available to other terminal users, Thus, connection disengagements of terminals other than those connected directly on a point-to-point circuit must be conducted to permit the channel on a front-end processor to become available to service other users.

Another important information control task is to establish a transmission sequence that prescribes the precedence and order of transmission of both data and control information. One example involves a number of terminals connected via a multipoint circuit to a computer. Here, this task defines the rules for when such terminals may transmit and receive information.

In addition to the transmission of control information following a sequence, the data itself can be placed into sequence. Data sequencing was first used in synchronous transmission where a long block of data could be broken into smaller blocks for transmission, with the size of the data blocks being a function of the error-control procedure used. Because of the growth in the use of PCs that support synchronous transmission, the most common data-sequencing operations at this writing are file transfers between PCs. Here, the contents of a file are divided into a series of blocks for transmission. When divided into smaller blocks, the amount of data that must be retransmitted in the event that an error in transmission is detected is reduced. Although error-checking techniques currently used are more efficient when short blocks of data are transmitted, the efficiency of the transmission will correspondingly decrease because either a positive or negative acknowledgment is returned to the transmitting device after each block is received and checked. For communications between synchronous buffered terminals and computers, block lengths of up to several thousand characters can be transmitted. However, block lengths from 80 to 512 characters are the most common. Although some protocols specify block length, most protocols permit the user to set the size of the block.

In comparison, most asynchronous file transfer protocols use a fixed block size that cannot be altered by the user. However, some file transfer protocols dynamically alter the block length to correspond to the transmission error rate, as noted by the number of negative acknowledgments received. That is, if there are no, or very few, negative acknowledgments, the protocol extends the block length. Conversely, if the number of negative acknowledgments increases (corresponding to an increase in the line error rate), the protocol decreases the block length. This action optimizes the throughput of the file transfer operation because small blocks are retransmitted when the error rate increases.

In the area of error control procedures, the most commonly used method to correct errors in transmission is to inform the transmitting device to retransmit the block of data. To correct by retransmission requires the coordination of the trans-

TABLE 4.10.1
Information Control Tasks

Information Control Tasks
Connection establishment
Connection verification
Connection disengagement
Transmission sequence
Data sequence
Error control procedures

mitting and receiving devices, with the receiving device informing the transmitting device of the status of each of the previously transmitted data blocks. If the block previously transmitted contained no detected errors, the receiving device will transmit a positive acknowledgment to the transmitting device and the sender will then transmit the next block of information.

If the receiving check character fails to match the transmitted block-check character, the receiver will transmit a negative acknowledgment and discard the block because it is in error. The transmitting station will then retransmit the previously sent block and, depending on the protocol used, several retransmissions might be attempted for recurring error conditions.

This continuous sequence of retransmissions might continue because of a bad circuit or other problems, until a default limit is reached. Once this limit to the number of retransmissions is reached, the computer might then terminate the terminal's session and the terminal operator will have to reestablish the connection.

Line control example

Consider the transmission of information contained on a reel of magnetic tape to a computer for processing. Here, the magnetic tape unit would normally be a peripheral unit of a remote batch terminal. Additionally assume that transmission is via a leased line, synchronous modems are used at both ends of the circuit, and that the transmission of data from the terminal to the computer will be in blocks of fixed size with longitudinal parity checks. To reduce this example to a manageable magnitude, assume that only two blocks of data will be transferred. If transmission is half-duplex, then the flow of data is as illustrated in Figure 4.10.1.

Prior to information on the magnetic tape being transmitted to the computer, synchronization must be established between the terminal device and the computer. With the use of synchronous modems, the bit timing for the transmitter and receiver can be provided by those devices. When the terminal begins to transmit the first block of data, two sync characters will be used as a prefix to provide block framing; thus the terminal's software will add these characters to the block while the block is in the terminal's memory being prepared for transmission. At the computer, these transmitted sync characters will be used for synchronization and then stripped from the received block.

After bit and character timing has been established, the start of the text must be indicated. This is accomplished by the use of the STX (start of text) control charac-

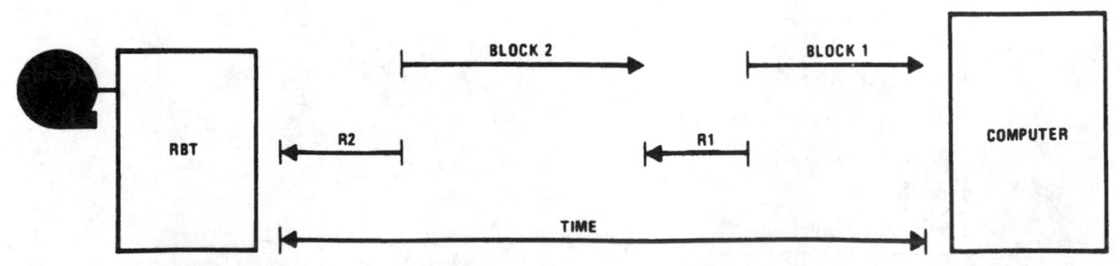

Figure 4.10.1 Typical half-duplex synchronous transmission.

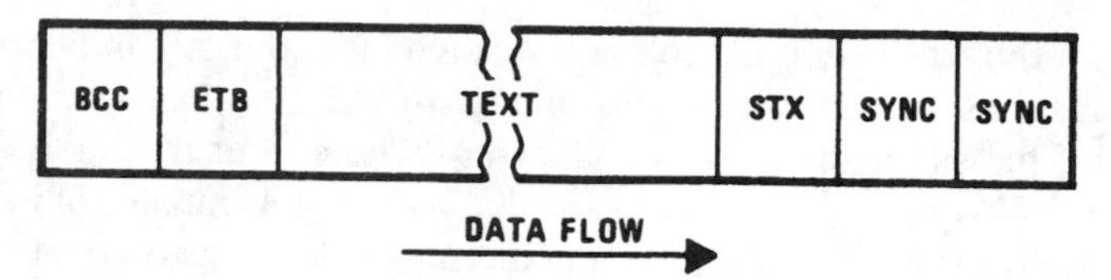

Figure 4.10.2 Primary block transmission.

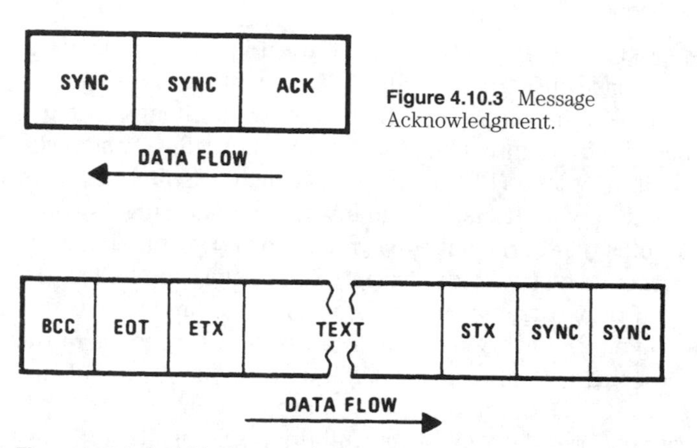

Figure 4.10.3 Message Acknowledgment.

Figure 4.10.4 Last block transmission.

ter, which again is added to the block by the software in the terminal and removed from the block by the computer's software.

Next, the actual text will follow. Here, the block size will be a function of the size of the terminal's buffer area as well as other factors, such as line conditioning and error rates. To indicate the end of the block, another control character, ETB (end of transmission block), is used. Next, for error control a block check character is added to the end of the block as shown in Figure 4.10.2. Because you can assume that a longitudinal redundancy check is used to form the block check character, this character is then made up of the sum of all the even parity bits in the text.

If the first block has been received by the computer and the computation of the BCC character matches the BCC character at the end of the block, you can assume no transmission errors have occurred. To indicate this, the computer will assemble a suitable reply to be transmitted to the terminal. If transmission was asynchronous, the reply could be simply an ACK (acknowledge) control character with suitable start and stop bits for synchronization. In the synchronous mode, you must send a reply message suitably framed with control characters, as shown in Figure 4.10.3. In Figure 4.10.1, R1 indicates the transmission of this acknowledgment.

The next and final block to be transmitted from the terminal to the computer must be framed. The starting point and end of the text must be indicated. Because the second block is the final block to be transmitted, that no additional messages will follow will be indicated by the use of the EOT (end of transmission) control character. The second block to be transmitted is shown in Figure 4.10.4.

Assuming again that the second block was received without any indicated errors, the computer's response would be as shown in Figure 4.10.3 and also indicated as R2 in Figure 4.10.1. This simplistic example illustrates the extent to which a protocol and control characters are required to ensure that a data link transfer of information is properly managed. Now examine two specific methods used to govern synchronous data transfer.

Bisync (BSC) and HDLC overview

Because of the variety of facilities that can be used for communications, as well as the number of network configurations and line structures available, many complex line protocols have been developed to effectively utilize them. Among current protocols, one of the most frequently used for synchronous transmission is IBM's binary synchronous communications (BSC or bisync), which was first introduced in 1966. This protocol is used for data transfer by many medium- and high-speed devices and provides a set of rules that carry out the synchronous transmission of binary-coded data. This protocol can be used with a variety of transmission codes. However, BSC is limited to the half-duplex transmission mode. Another limitation is that it requires receipt acknowledgment of every data block transmitted. Because of the success of BSC, of which there are several versions, a number of other protocols have been developed. Whereas BSC is a character-oriented protocol, newer protocols have been developed that are bit oriented and permit full-duplex transmission. These new protocols permit a greater volume of information to be transmitted in a given time period and thus are more efficient than BSC. One such protocol, high-level data link control (HDLC), was developed by the International Organization for Standardization (ISO) to reduce the frequent acknowledgment-negative acknowledgment series of transmissions required by the BSC protocol.

BSC examination

As mentioned previously, BSC is a procedure used to control the transmission of digital data on a half-duplex line connecting two or more devices.

The BSC protocol specifies the communications control characters used for the formatting of text, status indicators, synchronization functions, and error control. This protocol can be used on point-to-point and multipoint circuits, via dedicated, switched, or leased facilities.

Block size

The basic foundation of BSC communications is the message block illustrated in Figure 4.10.5. Each block can contain an optional header, text, and a trailer. To identify these elements, several control characters are used, including SOH (start of header), STX (start of text), and ETX (end of text). As previously covered, sync (synchronization) characters must be used to establish timing coordination between the transmitter and receiver. The number of such characters used varies with the type of application, and the two such characters shown in Figure 4.10.5 are a commonly used pattern where the message block follows the sync characters.

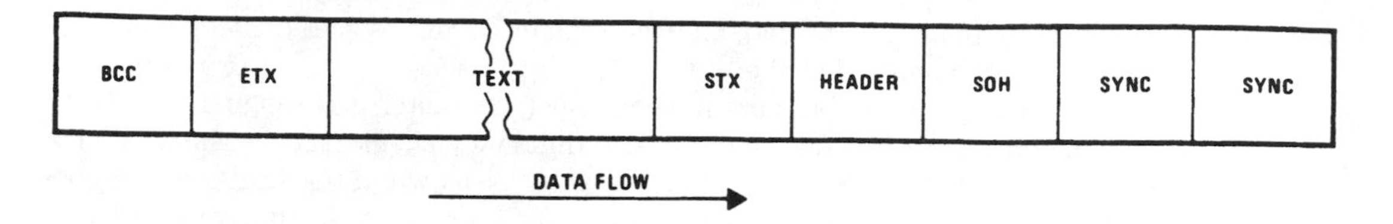

CONTROL CHARACTERS USED TO IDENTIFY HEADER, TEXT, AND TRAILER INCLUDE:

SOH = START OF HEADER, TRANSMITTED BEFORE THE HEADER CHARACTERS

STX = START OF TEXT, TRANSMITTED BEFORE THE FIRST DATA CHARACTERS

ETX = END OF TEXT, TERMINATES A MESSAGE BLOCK STARTING WITH SOH OR STX

BCC = BLOCK CHECK CHARACTER

Figure 4.10.5 Basic BSC message block.

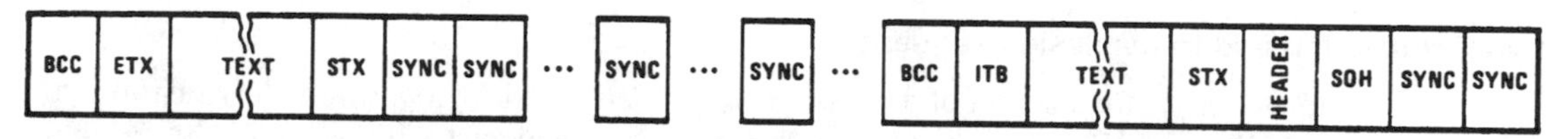

ITB = END OF INTERMEDIATE TRANSMISSION BLOCK

Figure 4.10.6 Multiblock message.

A message might consist of one or more blocks of information, where each block contains text and a trailer, but only the first block is required to contain the header. In Figure 4.10.6, a two-block message is illustrated. Here, the ETX character of a one-block message is replaced by an ITB (end of intermediate transmission block) character for all blocks, except for the last block.

The header is also known as an *address* because this information field contains a character or characters that identify the originating and/or receiving location. For a multipoint circuit, a separate control message might be used for addressing in place of the header. The SOH character indicates that the following character or characters are the header.

The next portion of the message block is the text, which is identified by a preceding STX character. Whereas a short message might be just a single block followed by an ETX, some messages consist of a multiblock, where only the last block of text is followed by an ETX character.

The last portion of the message block is the trailer, which consists of a block check character (BCC). This contains a count for error-checking in one of several ways, depending on the code used. One method is based on a cyclic redundancy check (CRC). Within BSC, two modes of CRC are used: CRC-12, which is used for 6-bit transmission codes, and CRC-16, which is used for 8-bit EBCDIC transmission codes. The CRC is a division performed by both the transmitting and receiving stations using the numeric binary value of the message as a dividend, which is divided by a constant. The quotient is discarded, and the remainder serves as the check character, which is then transmitted as the BCC.

The other method of obtaining a block check character is through a vertical redundancy check (VRC) and longitudinal redundancy check (LRC). The VRC is an odd-parity check performed on a per-character basis when the transmission code used is the 7-level ASCII code. Whereas the VRC checks characters for odd parity, the LRC checks an entire horizontal line, bit position by bit position, within a block for odd parity. The LRC becomes the BCC character and is transmitted to the receiving station.

As the block is received by the destination station, that receiver computes a BCC in the same manner the transmitting station generated its BCC. At the end of the block, the receiver compares its block count with the sender's BCC character. If the two are not equal, a negative acknowledgment (NAK) character is sent to the transmitter, requiring the block to be retransmitted. If the error continues because of some type of abnormal condition, such as line noise, and a preset number of attempts is reached in trying to transmit an error-free block, the transmitter will abort and a new connection will have to be established.

Synchronization and transmission sequence

Prior to the transmission of a message block, the sending location will transmit synchronizing (SYN) characters to synchronize the receiver with the transmitter. Next, a message exchange is initiated when one location transmits an inquiry (ENQ) control character to the other location. The ENQ character, in effect, bids for the line in both a point-to-point and multipoint circuit. If the other location can accept the message, it will acknowledge the inquiry by sending an ACK control character in response. Throughout the synchronization and transmission sequence, each acknowledgment is alternately numbered one and zero.

As illustrated in Figure 4.10.7, initially a terminal is idle. Assume that the terminal operator has entered enough data to fill the CRT display, corrected it, and wishes to transmit that data to the computer. The operator depresses the transmit key on the CRT, which in turn sends as enquiry (ENQ) control character to the computer preceded by two SYN characters. The computer responds by transmitting two SYN characters and an even, positive acknowledgment (ACK0) to the terminal.

Next, the terminal transmits two SYN characters followed by the message block, which contains the data keyed on the CRT. Now, assume that the computer receiving the message block detects an error because the computed BCC does not match the transmitted BCC character. Then, the computer will respond with two SYN characters and a negative acknowledgment (NAK). This NAK causes the terminal to retransmit the message block, preceded again by two SYN characters.

Now, assume that the block was received error free so that the computer responds with an odd positive acknowledgment (ACK1) again preceded by two SYN characters. Because one screen was transmitted, and you have assumed that only one message block was required, to conclude the transmission, the terminal will send an end of transmission (EOT) control character. This character will reset all stations (here the terminal and computer) to what is known as the control mode, in which they are neither the transmitter nor the receiver.

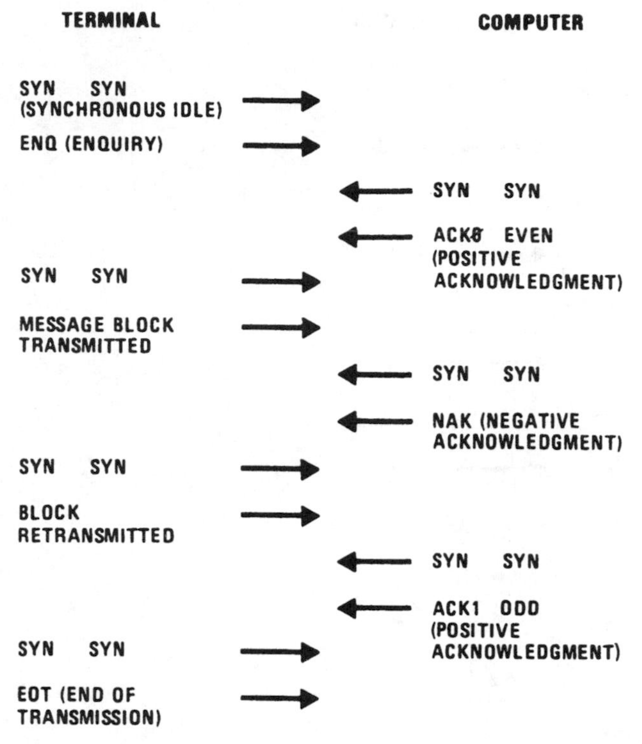

Figure 4.10.7 Synchronization and transmission sequence.

Acknowledgment sequence

Through the utilization of even and odd acknowledgments (ACK0 and ACK1), a sequential check of a series of replies to the state of transmitted message blocks can be accomplished. An ACK0 is the first affirmative reply to a polling-selection (multipoint) or a connection (point-to-point) line bid. Thereafter, the ACK0 alternates with ACK1 to inform the device transmitting message blocks that the previous block was accepted without error and that the receiver is ready to accept the transmission of the next block of information. In Figure 4.10.8, an acknowledgment sequence is illustrated that eliminates the SYN characters.

During the sequence illustrated in Figure 4.10.8, if for some reason an ACK is not received, the transmitter will send an ENQ to the receiver. An example of this type of situation is illustrated in Figure 4.10.9, where the transmitter has not received an ACK1 after the second message block was transmitted. After waiting a period of time, which is known as the *timeout period*, the sender will transmit an ENQ. If the receiver responds to the ENQ with an ACK0, then the message was lost because the receiver is still acknowledging the first message block. Conversely, if the receiver responds with an ACK1 to the ENQ, then the message was received, but the resulting acknowledgment was lost. To compensate for the lost message block, the data originator would retransmit the second message block. For the other case, where the acknowledgment was lost, when the data originator finally receives the ACK1 as a result of the transmission of an ENQ, it would then continue its transmission.

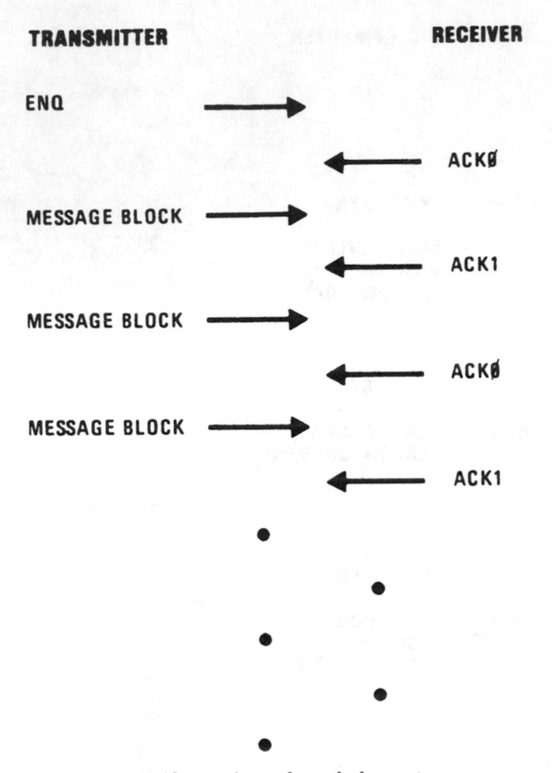

Figure 4.10.8 Alternating acknowledgment sequence.

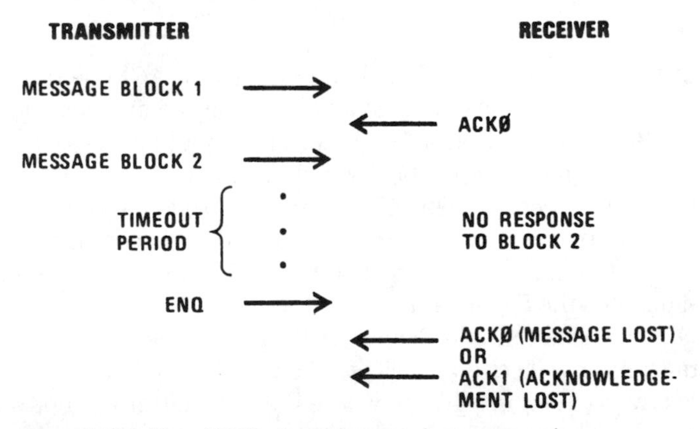

Figure 4.10.9 Use of ENQ when a message is nonresponsive.

Timeouts

In the previous example, a timeout period was mentioned as time that would elapse until the data originator would transmit an ENQ to determine the status of its previously transmitted message block. Within the BSC protocol, several timeout situations are normally used: transmit, receive, disconnect, and continue.

Transmit timeout

On a periodic basis during transmission, usually once per second, one or two SYN characters are automatically inserted into a message. These SYN insertion characters are for timing purposes, to maintain synchronization between the transmitting and receiving devices. Also, they serve as a "timefill" in the absence of a message so that the circuit remains operational and the receiver knows that transmission is still in progress. These SYN insertion characters are deleted from the message by the receiver; hence they have no effect on the message content.

Receive timeout

The receive timeout period is normally three seconds and is used to limit the waiting time for a transmitting station to receive a reply. The receive timeout also allows any receiving or monitoring station to check the circuit for SYN-idle characters. These SYN-idle characters indicate that the transmission is continuing; thus this timeout is reset and restarted each time a SYN-idle is detected.

Disconnect timeout

This timeout is an option that can be used on switched-network data links. It is usually set at 20 seconds to prevent a station from holding a connection for prolonged periods of inactivity. When a station has been inactive for that period of time, it will disconnect itself from the network.

Continue timeout

This timeout, normally set at two seconds, is used to prevent a three-second timeout, such as a receive time-out, from occurring. This timeout is used by stations where the speed of the input or output devices affect buffer availability and cause transmission delays. This timeout indicates that transmission or reception is delayed, but will continue, and that the link should be maintained. This is accomplished by sending a temporary text delay (TTD) two-character sequence within two seconds of receiving acknowledgment of the previous block. In response to the TTD, the receiving station will transmit a character sequence known as a *WACK (wait before transmitting positive acknowledgment)*. It is used to prevent further transmission until the station is ready to receive again.

Transparency

In the discussion of control characters so far, no text could contain sequences that include control characters because the receiver would detect them as control characters and perform the expected transmission control operation. BSC transparent mode allows control characters to be inserted in text; thus, it permits the transmission of many types of raw data within the standard BSC message format, without worrying about the effect that such data will have on line or message control.

Because some type of line protocol control is necessary, control characters must still be used under the transparency mode of operation. To use control characters, but reduce the probability of raw data being interpreted as such, a data link escape (DLE) character is prefixed to the control characters. Now, only when a two-character sequence of the DLE character and a normal control character is received, will the receiver recognize it as a control character.

HDLC

In order to alleviate some of the limitations of BSC, especially its design for half-duplex operations, a series of high-level data link control (HDLC) type protocols were developed. Two of these HDLC protocols include IBM's synchronous data link control (SDLC), and the standardized HDLC protocol.

The key difference between HDLC protocols and previous data link protocols is that they are bit oriented rather than character oriented. Another significant difference is that HDLCs are naturally transparent: a data bit's significance is determined by its position in the bit stream. An examination of bit patterns to determine if a control character is present is the method used in BSC. In HDLC, transparency permits unrestricted bit patterns to appear in the data because the bit patterns of control functions reside at fixed locations.

When a transmission follows a BSC protocol, large messages are broken into a series of smaller blocks of fixed size. In comparison, the HDLC protocol permits variable length messages in what is called a *frame*. Any number of frames are used in a single transmission.

Frame format

In HDLC, the same frame format is used to serve lengthy transmissions, such as remote batch, and short messages, such as inquiry and conversational transmissions.

An HDLC frame is illustrated in Figure 4.10.10. The frame itself is delimited by the flag fields, one at the beginning of the frame and one at the end of the frame. Both flags consist of the specific 8-bit sequence 01111110. Between frames, any number of flag fields can be transmitted to keep the link active and synchronized. In addition, the flag serves to alert the receiving station to the possibility that a frame is beginning. On the detection of an 8-bit nonflag field immediately after the flag, the receiving device knows that a new frame is being transmitted. To prevent a flag bit pattern from appearing between flags within a frame, zeros are inserted and deleted as required. When five 1s appear, a 0 is inserted in the bit stream after the last 1. The receiver, after detecting five 1s followed by a 0, deletes the zero.

The 8-bit field following the beginning flag is the address field. This field defines the station for which the frame is intended if the frame was originated by a primary station. The primary station is normally a computer and the secondary station a remote terminal. In a frame sent by a secondary station, the address field identifies the sending station.

Following the address field is an 8-bit control field. The functions of this field include conveying counts of frames transmitted or received, so that frames do not have to be received in sequence, but only have to be reassembled into sequence after receipt.

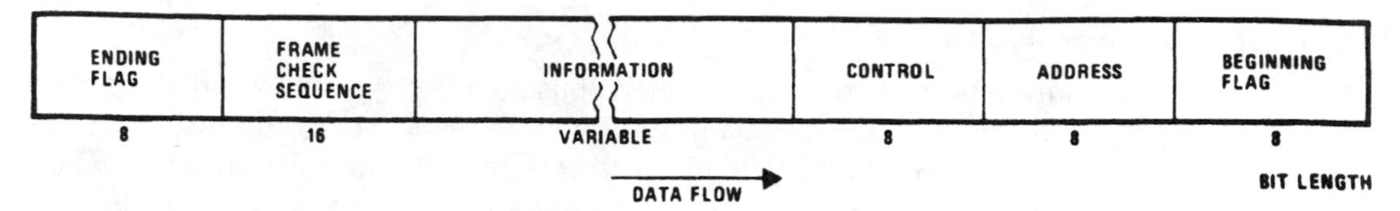

Figure 4.10.10 HDLC frame.

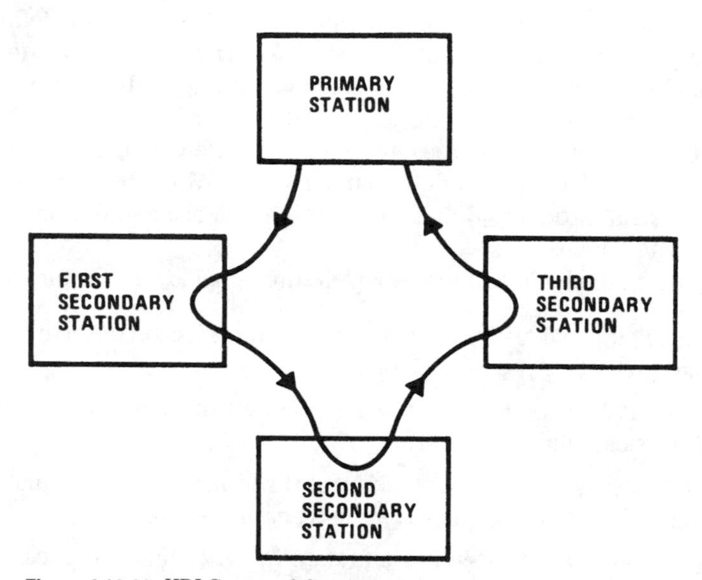

Figure 4.10.11 HDLC network looping.

After the control field, the beginning of the information field is encountered. As noted, this field is of any length and can contain text that includes control characters. Upon conclusion of the information field, a frame-check sequence field follows. This 16-bit field functions as a block check, similar to the block check character of BSC, to detect transmission errors. Finally, the frame is concluded by the 8-bit ending flag, which contains the flag bit sequence, 01111110.

HDLC operations

In addition to point-to-point and multipoint operations, HDLC can be used for network looping, as illustrated in Figure 4.10.11. When used in looping, all transmissions flow in the same direction around the loop and each frame passes through the secondary stations, where they are inspected to determine the frame's address. If the station address equals the frame address, the station acts on the data for its own use.

Secondary stations can transmit only when permitted to do so by the primary station. Once each station completes its transmission, the next station in the loop can begin transmission. The primary station can also poll individual secondary stations, in which case a response is transmitted by that secondary station to the specific poll.

Open systems interconnection

OSI is solidly in place worldwide as a major business factor in interconnecting computers. Developers of OSI are members of the International Organization for Standardization (ISO), which consists of representatives of national standards bodies. Overlapping OSI developments have been cooperatively conducted by the International Consultative Committee for Telegraphy and Telephony (CCITT), an advisory body made up of the telecommunications authorities of national governments.

The OSI reference model is the ISO's architectural framework for defining standards for linking heterogeneous computer networks. The term "open" denotes the ability of an "end-system" of one design to connect with any other end-system conforming to the reference model and the associated standard protocols. The goal is to have end-system application processes intercommunicate through the OSI environment. These processes could include a manual keyboard entry, a credit-checking program, an industrial production-line-control program, or an endless range of distributed processing applications.

The OSI reference model is defined by seven functional layers (Figure 4.10.12):

1. The *Physical* layer provides for the transparent transmission of the bit streams to and from the connecting physical media.
2. The *Data Link* layer handles the transfer of a unit of information between the ends of the physical link.
3. The *Network* layer governs the switching and routing of the information to establish a connection for the transparent delivery of the data.
4. The *Transport* layer assures end-to-end data integrity and provides for the required quality of service for exchanged information.
5. The *Session* layer coordinates the interaction between the communicating end-application processes.

END-USER APPLICATION PROCESS

FUNCTION	
SELECTS APPROPRIATE SERVICE FOR APPLICATION	7. APPLICATION
PROVIDES CODE CONVERSION, DATA REFORMATTING	6. PRESENTATION
COORDINATES INTERACTION BETWEEN END-APPLICATION PROCESSES	5. SESSION
PROVIDES FOR END-TO-END DATA INTEGRITY AND QUALITY OF SERVICE	4. TRANSPORT
SWITCHES AND ROUTES INFORMATION	3. NETWORK
TRANSFERS UNIT OF INFORMATION TO OTHER END OF PHYSICAL LINK	2. DATA LINK
TRANSMITS BIT STREAM TO MEDIUM	1. PHYSICAL

Figure 4.10.12 Seven layers.

6. The *Presentation* layer provides for any necessary translation, format conversion, or code conversion to put the information into a recognizable form.
7. The *Application* layer directly serves the communicating user-application process by providing the distributed information service appropriate to the application and its management.

Internetworking

For internetworking—communicating between end-systems (seven-layer OSI configurations or "stacks")—a layer service is defined for each of the seven layers that identifies the set of functions that the layer provides. OSI layer services are of two general types: connection-oriented, which allow the service users (entities in the next higher layer) to establish and use logical connections; and connectionless, which allow the service users to exchange information without having to establish a connection. Examples of each service type: in general, when internetworking in an environment where the majority of end systems are X.25 DTEs (data terminal equipment devices), connection-oriented is preferred; where most subnetworks are *not* X.25 packet-switched data networks, or if the application is inquiry-response, then connectionless is preferred.

Within each layer, protocols operate to provide the services defined for that layer. As a number of protocol selection options exist in some layers (for example, the Transport layer defines five distinct connection-oriented protocol classes), conformance requirements are specified by each of the layer-protocol standards. When in compliance with the required suite of standard protocols prescribed for OSI, configurations are considered to be open.

The connection-oriented and connectionless schools of thought, regarding the Network layer, originated in different parts of the data communications industry. The connection-oriented approach has adherents in telephony. There, the notion of establishing a communications channel for a fixed duration is quite natural. This style of operation carried over into the world of packet-switching data networks through the CCITT X.25 Recommendation. Another connection-oriented adherent is IBM through its Systems Network Architecture communications protocols.

Adherents of the connectionless approach are typically computer and networking vendors. The most visible connectionless technology is the TCP/IP (Transmission Control Protocol/Internet Protocol), promulgated by the U.S. Department of Defense (DOD) and its Defense Advanced Research Project Agency (DARPA). The connectionless approach was also followed by propriety protocols, such as Xerox Network Systems, and Decnet (from Digital Equipment Corp.).

TCP/IP has become the dominant solution for computer communications, both inside the DOD and in the rest of the world. For the twentieth-century data communications user, choosing between TCP/IP and OSI can be difficult. TCP/IP's five-layer protocol has been around—and unchanged—for over 20 years. But some of TCP/IP's application-level programs are years behind OSI's. And new areas, such as document interchange and transaction processing—designed to be integrated into OSI hosts—are not addressed by TCP/IP. Conclusion: TCP/IP is now, and will remain for some time, a major factor in computer communications. Eventually, as OSI and its products mature, OSI may prevail.

At this writing, the conclusion that OSI will prevail is in dispute. Based largely on the tremendous growth of the Internet—the worldwide network of networks—and its use of TCP/IP, the push to adopt OSI appears to be fading.

One further note on packet switching: the growing use of optical fiber transmission lines has measurably lowered the error rates that had been encountered with copper. This has led to error-checking packet-switched transmissions only at the sending and receiving ends. This technology is the basis for the increasingly applied "frame relay" operation.

Link access procedures

The CCITT X.25 interface standard specifies a link access procedure (LAP), the Data Link layer's link control function. LAP has been supplemented with a "balanced" version, LAPB. The Data Link layer protocol to be used on the D channel of the Integrated Services Digital Network (ISDN—see Chapter 21) is LAPD. It is similar to LAPB, except for the structure of its address field, which is used for statistically multiplexing the various services provided on the D channel. LAPD is likely to be the basis for LAPM, the proposed CCITT modem error-control standard.

4.11 Types of Service

When the distance between the terminal and the computer is relatively short, the transmission of digital information between those two devices can be accomplished simply by cabling the terminal directly to the computer. As the distance between the two devices increases, the pulses of the digital data transmitted will start to distort, until a point is reached where they become unrecognizable by the receiver. There are two methods or types of service for the transmission of data: analog and digital. For each type of service, specialized communications equipment is necessary when transmission exceeds a short distance.

Because telephone lines were originally developed to carry voice (analog) signals, the digital signals to be transmitted between a computer and a terminal over this type of medium must first be converted into a signal that is acceptable for transmission over the telephone line. To convert these digital signals into suitable analog signals, a modem is used. The term *modem* is a contraction of the terms *modulation* and *demodulation*, and it denotes the basic functions performed by that device. The modem is an electronic device that converts the digital signals generated by computers and business machines into analog tones that are suitable for transmission over the telephone network's analog facilities.

At the receiving end of the circuit, a similar device samples the transmitted tones, converts them back to the original digital signals that were transmitted, and delivers these digital signals to the connected digital device. An example of the signal conversion performed by modems is illustrated in Figure 4.11.1. This illustration shows the interrelationship between terminals, computers, and a transmission line when an analog transmission medium is used for the transfer of digital data. (Modulation techniques are discussed in Section 5.3.)

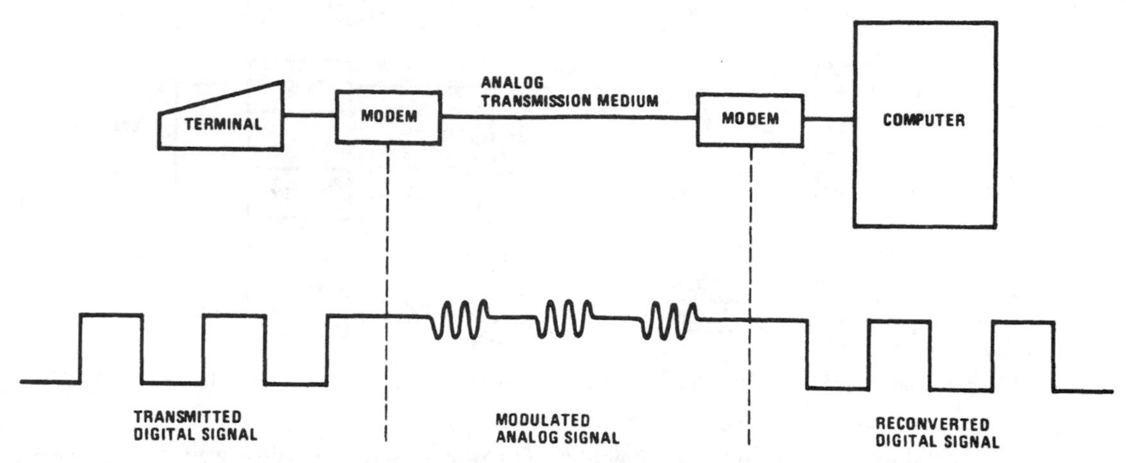

Figure 4.11.1 Digital transmission on an analog medium.

In the telephone network, both switched and leased lines are available that provide analog service. Therefore, modems or similar devices can be used for data transmission over both of these types of analog network facilities. Modems can be used on direct-connect, leased, or switched-line facilities; they are hard-wired to direct-connect and leased lines, whereas they can be either directly connected or interfaced to a switched facility. Because of this type of connection, a terminal user can only communicate with the distant location at the other end of a leased line, but can communicate with many devices when the modem is connected to a switched line.

Some low-speed terminals utilize a device called an *acoustic coupler*, which is a modem whose connection to the switched telephone line is obtained by acoustically coupling the telephone headset to the coupler. The primary advantage of this device is that no hard-wired connection to the switched telephone network is required, and a terminal interfaced to such a device can be portable and can be moved with the acoustic coupler from location to location.

Because of the growth in the use of modular telephone connectors, standard (nonacoustically coupled) modems can very often be readily connected to a telephone line. Acoustic couplers were limited, generally, to a 1200 bit/s operating rate because of their indirect connection to the telephone network. Therefore, their utilization has rapidly decreased to where they mainly serve niche markets, such as for portable computer users who transmit data from a pay telephone or hotel telephone that is not equipped with a modular connector.

The interrelationship of modems, acoustic couplers, terminals, and the analog transmission medium is illustrated in Figure 4.11.2.

Analog facilities

Currently, several categories of analog switched facilities are offered by communications carriers. Each type of facility has its own rate structure and set of operating characteristics. Normally, prior to determining which category or categories of service

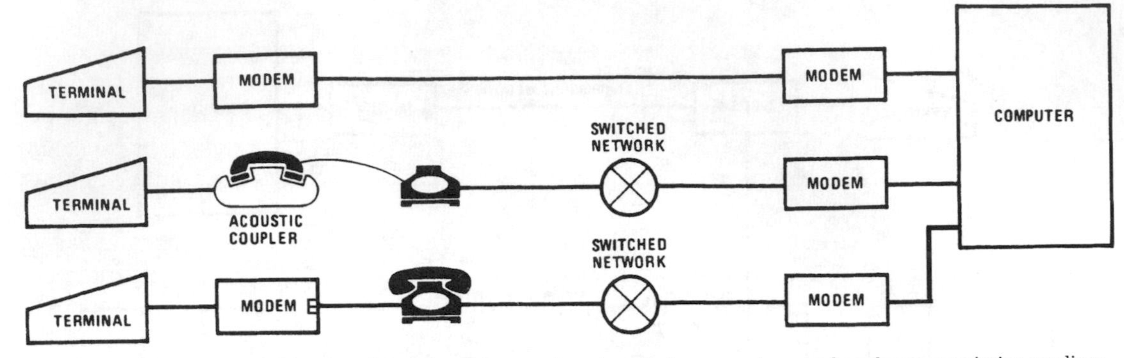

Figure 4.11.2 Interrelationship of modems, acoustic couplers, terminals, computers, and analog transmission medium.

should be used for an optimum cost-effective means of communications, an analytic study of the number of calls, quantity of transmitted data, and other parameters is conducted. The common types of analog switched facilities include direct-distance-dialing (DDD) and wide area telecommunications service (WATS). Foreign exchange (FX) can be considered a hybrid type of service because it is a combination of switched and leased analog facilities.

The first type of switched service to be covered, direct-distance-dialing, permits the user to dial directly any telephone connected to the public switched telephone network. In turn, the dialed telephone number can be connected to another terminal, a computer, or some type of business machine or data communications component. In addition to installation costs, the charge for this service can be a fixed monthly fee if no long-distance calls are made, on a message unit basis based on the number and duration of local calls, or a fixed fee plus any long-distance charges incurred. Depending on the day of the week and time of day, discounts from normal long-distance rates are available for selected calls made without operator assistance.

The second category of switched service, WATS, can be obtained in two different forms, known as *OutWATS* and *InWATS*. Each form is designed for a particular type of communications requirement common to a large number of telephone company subscribers. OutWATS can be used when a specific location has a requirement to place a large number of telephone calls to geographically distributed locations. Providing the reverse capability, InWATS permits a number of geographically distributed locations to communicate with a common facility.

Although calls initiated on WATS are similar to calls placed on the regular public switched telephone network, charges for usage can vary considerably between the two facilities. Instead of being charged on an individual call basis as on the switched network, the WATS facility user is charged a flat sum per block of hours used per month for communications. Usage in excess of that block is then billed on an overtime basis.

In order to use a WATS line, a voice-band trunk called an *access line* is installed between the telephone company central office and the subscriber's facility. Other than certain geographical calling restrictions, which are a function of the service area of the WATS line and of cost considerations, the user can place as many calls as desired on this trunk if the service is OutWATS. If the user has InWATS service, he might then receive as many calls as users dialing his telephone number.

InWATS, the well-known "800" area code, permits remotely located personnel to initiate calls to a central facilities telephone number, toll free, so long as the call originator is within the service area provided by the particular InNWATS type of service selected. The charge for both types of WATS services is a function of the service area, as well as the total call durations within a month. The service area can be intrastate, a group of states bordering the user's state where the telephone connected to the WATS access line is located, or a grouping of distant states. In 1978, WATS service was extended from the continental United States to Alaska, Hawaii, Puerto Rico, and the Virgin Islands. In the mid-1980s, AT&T introduced international 800 service, which permits calls from many European and Asian countries to be routed as INWATS calls to subscriber lines in the United States. Additional detail on WATS is included in Section 8.2.

One hybrid type of switched service that can be used advantageously under certain conditions to include geographical dispersion of terminals and period access requirements is a foreign exchange (FX) line. This circuit combines use of the public-switched network with use of a leased line. It might result in communications from one or more terminals to a computer at substantially less cost than the price of equivalent direct distance dialing.

The foreign exchange line is similar to a standard telephone in that, when the phone is placed off hook, a dial tone is received. Instead of a local subscriber area dial tone, however, the foreign-exchange-line user receives the dial tone of a distant or foreign exchange, as illustrated in Figure 4.11.3. In this manner, the user can dial long-distance calls within the foreign exchange area with the cost of such calls limited to the leased-line cost and perhaps a foreign exchange call toll. The total cost normally is considerably less than the cost of equivalent long-distance calls.

For terminal-to-computer communications, the use of the line illustrated in Figure 4.11.3 is effective when a terminal remotely located from the computer is only required to access that computer or another device within the same foreign exchange area. Here, the initiation of communications is one way, with the terminal the originator. The user of the FX line is restricted to dialing the telephone numbers of a particular exchange.

An FX line can be used to service many terminals that require communications with a computer, as illustrated in Figure 4.11.4. Here, each terminal user within a geographical area (telephone exchange) dials a local number, which is answered if the FX line is not in use.

From the foreign exchange, information is transmitted via a dedicated (leased) voice line to a permanent connection in the central (local) office of a communication carrier near the computer's location. A line from the central office is then extended to the customer's computer location to complete the connection. Because only one terminal at a time can use the FX line, normally only groups of terminals whose usage can be prescheduled or whose operational effectiveness is not reduced by contending for the line are suitable for using this method of communications.

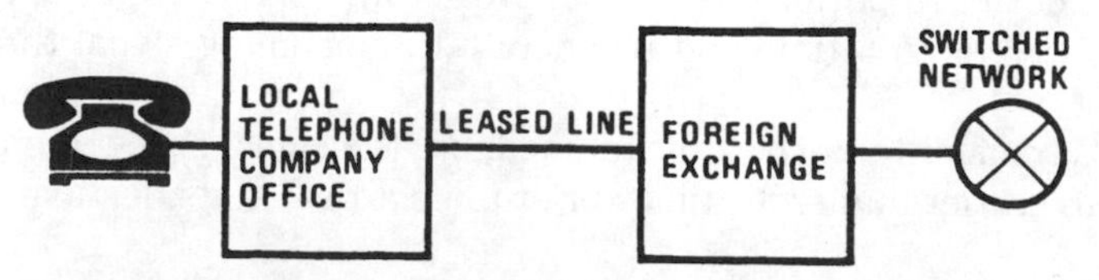

Figure 4.11.3 Foreign exchange lines.

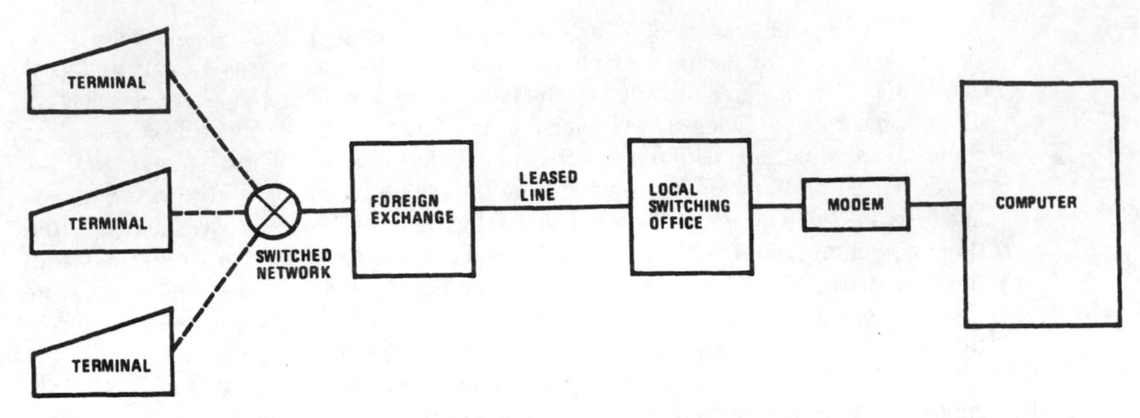

Figure 4.11.4 FX line used to support many terminals.

The major difference between using an FX and a leased line is that any terminal dialing the foreign exchange line provides the second modem required for the transmission of data over the facility; whereas a leased line used for data transfer normally has a fixed modem and terminal attached to one end of the circuit and a fixed modem and computer or terminal at the other end of the line.

Digital facilities

The second type of communications service available for user consideration, digital service has been offered comparatively recently to interconnect large geographical areas.

Using digital service, data is transmitted from its source to its destination in its original digital form. Digital service avoids the conversion of signals into an analog form for transmission and reconversion back into a digital form, as is the case when modems are used to transmit data over analog facilities.

In the place of modems, users can select one of two basic arrangements to connect their terminals or computers to a digital medium. A data service unit (DSU) provides a standard interface to a user's terminal that is compatible with modems and conducts such functions as signal translation, data regeneration, reformatting, and timing. In place of modems that are designed to operate at numerous data rates, the DSU is designed to operate at one of four speeds: 2400, 4800, 9600, and 56,000 bit/s. In operation, the transmitting portion of the DSU processes the user's signal into bipolar pulses suitable for transmission over the digital medium. At the destination, the receiving portion of the DSU extracts timing information, and regenerates mark and space data from the received bipolar signal.

The second device that can be used for communications on a digital medium is a channel service unit (CSU). It is provided by the communications carrier to those customers who desire to perform the signal processing to and from the bipolar line themselves, as well as to retime and regenerate the incoming signal through their own equipment.

When data is transmitted over a digital medium, the signal is regenerated by the communications carrier numerous times prior to its arrival at its destination. In general, digital data service gives data communications users an improved level of performance and reliability when compared to data transmission over an analog medium. (Also refer to Section 5.5.)

Hybrid digital service

Although digital service is offered at many locations throughout the United States and at several international locations, for those user locations outside the serving area of a digital facility, an analog extension is often required to connect to the service. In Figure 4.11.5, the utilization of digital service via an analog extension is illustrated. As shown, if the closest city to the terminal located at city 2 that offers digital service is city 1, then to use digital service to communicate with the computer, an analog extension must be installed between city 1 and city 2. In such cases, the performance, reliability, and possible cost advantages of using digital service can suffer. Digital service is available for the dial and leased service, where the leased digital line, like the leased analog line, is dedicated for full-time use to a particular customer.

4.12 Transmission Rate

A number of different methods exist to define the rate of data transmission. The most common method used, the data signaling rate, is usually expressed in bits per second, or bit/s. Another method is the *modulation rate*, which is a term most frequently used by a communications engineer and is expressed in terms of baud units. (Also refer to Section 4.2.)

Data signaling rate

The data signaling rate, expressed in bit/s, defines the rate at which information can be transmitted. Examine the 8-level, 11-unit coded character illustrated in Figure 4.12.1.

The data signaling rate for serial transmission is defined as:

$$R = \frac{1}{(T)} \log_2 N \text{ bit/s}$$

where T = duration of the unit signal element in seconds and N = the number of signaling conditions.

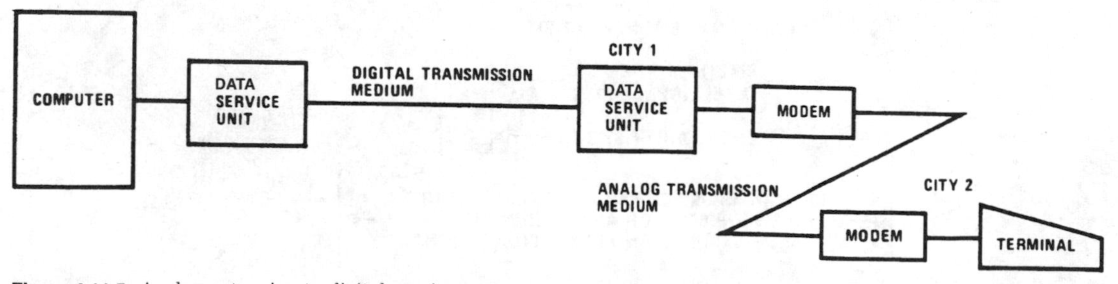

Figure 4.11.5 Analog extension to digital service.

Because the transmission of the character M requires 11 units to be transferred, with each unit time equal to 9.09 msec, and two signaling conditions can occur, with the presence of logic level 0 or 1, the data signaling rates becomes:

$$R = \frac{1}{(.00909)} \log_2 2 = 110 \text{ bit/s}$$

Sometimes the data signaling rate can be expressed in characters per second, or words per minute. In order to compare line speeds, these alternative expressions must be converted to a common denominator, such as bits per second. To do this conversion, you must know the number of bits in a character for the code used as well as the meaning of a "word." In data transmission, a word is considered to be six characters in length, normally five letters and a space. For conversion, the following formula can be used:

$$R = \frac{\text{words/min.} \times \text{chars./word} \times \text{bits/char.}}{60 \text{ seconds}}$$

Modulation rate

To describe the performance of a circuit in terms of the rate at which changes in the condition of the line can be made in a given time, the term *baud* is used. A baud is equal to one unit signal element per second. Returning to the illustration of Figure 4.12.1, the modulation rate is therefore

$$R = \frac{1}{0.00909} = 110 \text{ baud}$$

Based this equations, it is wrong to conclude that a baud is the same as one bit per second because if more than two signaling states are used, the answers would be completely different. This becomes apparent when you examine a new type of data signaling obtained by using four logic levels, as illustrated in Figure 4.12.2. Using four logic levels produces four different ways of combining two binary digits, and thus the number of signals necessary to transmit the information is only one-half

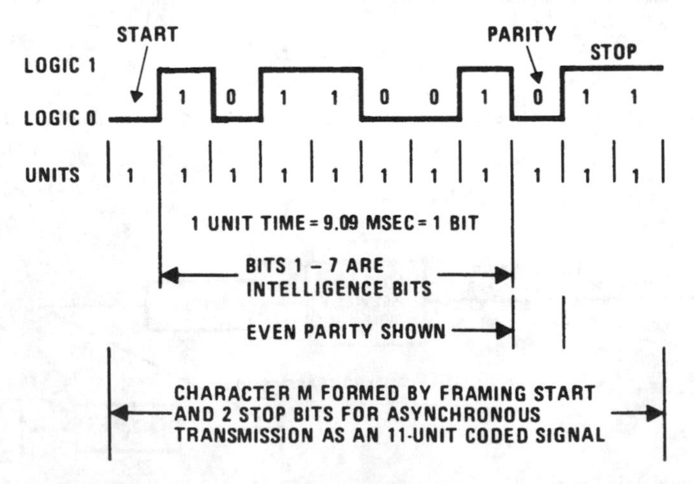

Figure 4.12.1 8-level, 11-unit coded character representation.

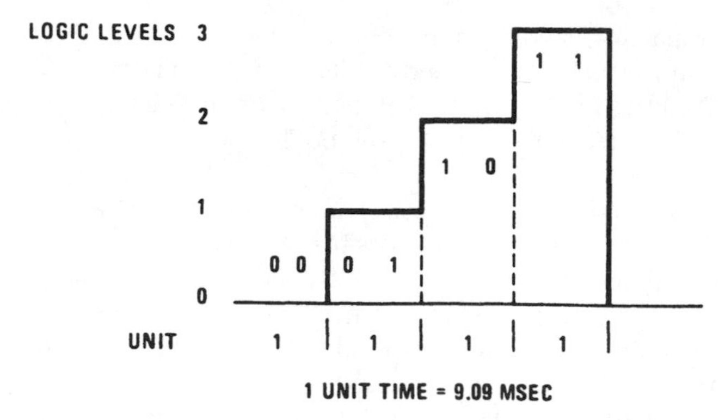

Figure 4.12.2 Data signaling using four logic levels.

that necessary with two-state signaling. For the situation illustrated in Figure 4.12.2, the data signaling rate would be

$$R = \frac{1}{0.00909} \log_2 4 = 110 \times 2 = 220 \text{ bit/s}$$

Although the data signaling rate has doubled, the modulation rate is still 1/0.00909 or 110 baud.

Factors affecting data rates

Although the types of terminals, modems, or acoustic couplers, as well as the line discipline and type of computer interfaced via a transmission medium, all play roles that affect transmission rates, the transmission medium itself is the most important factor.

The services offered by communications carriers, such as American Telephone and Telegraph Co. (AT&T), MCI, and Sprint for data transmission are based on their available facilities. Analog transmission is most readily available and can be used on in-plant dedicated lines or switched or leased telephone circuits. Available mainly in large cities and surrounding areas, digital transmission can be used from nondigital service locations by the use of an analog extension, as illustrated in Figure 4.11.5. Within analog and digital service, several grades of transmission are available for customer consideration.

In general, three grades of transmission can be obtained on analog service: narrow band, voice band, and wideband. (Ref. Sec. 8.2.) The data signaling rate that can be obtained on each depends on the bandwidth and electrical properties of each type of circuit offered within each grade of service. Summing up a very complex subject, it can be stated that transmission speed is a function of the bandwidth of the communications circuit, and the greater the bandwidth, the higher the possible speed of transmission.

When a communications carrier divides a voice-band circuit, or groups a number of transmissions from different users onto a single portion of a circuit, a narrow-band circuit is obtained. Typical transmission rates on narrow-band facilities range between 45 and 300 bit/s. One example of narrow-band usage is Teletype terminals, which are connected to message-switching networks over such facilities.

In comparison to narrow-band facilities that have bandwidths in the range of 200 to 400 Hz, voice-band facilities have bandwidths of approximately 3,000 Hz. When using voice-band facilities for data, speeds obtainable are based on the type of voice-band facility used: dial-up transmission on the public switched telephone network, or transmission via a leased line.

For the public switched telephone network, maximum data rates are between 19,200 and 24,000 bit/s, with 24,000 bit/s possible on occasion. Because leased lines can be tuned or conditioned, a data rate of up to 24,000 bit/s can be used on such a facility. (Recent modem developments, in which data compression occurs within the modem, enables an achievable throughput of up to 76,800 bit/s on the switched telephone network. Actually, the modem transmits data at 19,200 bit/s; when the maximum compression ratio of 4:1 is achieved, data throughput then reaches 76,800 bit/s. Although low data rates can be used on both narrow-band and voice-band circuits, do not confuse the two. A low data rate on a voice circuit is transmission at a speed far less than the maximum permitted by that type of circuit; a low data rate on a narrow-band facility is at or near the maximum transmission rate permitted by that type of line.)

When the bandwidths of several voice-band circuits are grouped together to provide a wider bandwidth than available on a single voice circuit, the result is known as a *wideband* or *group-band facility*. Wideband facilities can only be obtained when using leased lines and are used for transmission at rates in excess of 24,000 bit/s. Transmission rates on wideband facilities vary with the type of offerings of different communications carriers. Data rates normally available include 40.8, 56, and 230.4 kbit/s.

For the situation where terminals are directly connected to a computer, transmission rates are a function of the distance between the terminal and the computer, as well as the gauge of the conductor used to connect the two devices.

In digital transmission, present offerings by AT&T's Dataphone digital service (DDS) provide interstate, full-duplex, point-to-point, and multipoint leased line, synchronous digital transmission at data rates of 2.4, 4.8, 9.6, and 56 kbit/s, as well as data rates of 1.344 and 1.544 Mbit/s between the servicing areas of digital cities. In 1987, a high-speed digital switched communications service was introduced by AT&T. This service offers customers full-duplex, synchronous transmission over a common switched digital network at a data rate of 56 kbit/s. Other recent AT&T switched digital transmission offerings include a 384 kbit/s and a 1.544-Mbit/s (T1, proposed at this writing) service. Other communications carriers, such as MCI and Sprint, also provide and are planning a large number of digital transmission services. In Table 4.12.1, the main types of analogs and digital facilities, as well as the data rates and general use of such facilities, are listed.

Notice that one major class of emerging digital leased line service is not listed in Table 4.12.1. That service is fractional T1 (FT1), representing fractions of the T1 leased line operating rate of 1.544 Mbit/s. For most organizations, the use of FT1 service requires a communications carrier to install a T1 line between the carrier's serving central office and the organization's location. The FT1 using organization then uses a special type of communications device to transmit data on the fraction of the T1 line that was ordered. At the carrier's central office, the FT1 service is combined with other digital services via multiplexing equipment, then routed through the carrier's network.

TABLE 4.12.1 Transmission Facilities

Facility	Transmission speed	Use
Analog		
Narrowband	45–300 bit/s	Message switching
Voice band	Less than 9200 to 24,000 bit/s	Timesharing; remote job entry
Switched leased	Up to 24,000 bit/s	Remote job entry; computer-to-computer
Wideband	24,000 bit/s and up	Computer-to-computer; remote job entry; tape-to-tape transmission; high-speed terminal to high-speed terminal
Digital		
Leased line	2.4, 4.8, 9.6, 56 kbit/s, 1.544 Mbit/s	Remote job entry; computer-to-computer; high-speed facsimile
Switched	56,000 to 384,000 bit/s	Terminal-to-terminal; computer-to-computer; high-speed terminal to computer

4.13 Link Terminology [Refer to Section 3.1]

When more than one terminal transmits data over a common circuit, such as a multipoint line, this line is known as a *multiterminal-to-computer link.*

Although terminals can be configured with appropriate communications facilities to communicate directly to a computer over individual computer-to-terminal links, economics might justify the utilization of a device to combine the data from many low-to-medium-speed terminals onto one or more high-speed paths for retransmission to a computer. Many such devices that can be used to combine data transmitted from many terminals are now offered by various manufacturers.

Two of the more commonly used devices are concentrators and multiplexers. Circuits used to connect the terminals to a concentrator are known as *concentrator-to-terminal links*, and those that connect terminals to multiplexers are known as *multiplexer to terminal links*. The high-speed line that connects the concentrator to a computer or host processor is known as a *concentrator-to-host link*, and for multiplexers such connections are called *multiplexer-to-host links*. When one concentrator transmits data to another concentrator, this type of circuit is known as a *concentrator-to-concentrator link*, whereas the circuit used to connect multiplexers is known as a *multiplexer-to-multiplexer link*. Finally, the transmission path between computers is known as *host-to-host link*, to indicate the source and receiver of information. In Figure 4.13.1, some of the typical types of data links are illustrated.

Questions

4.1 What is the difference between a bit and a baud?

4.2 What is the difference between a character and a computer word? What is the relationship between a character and a computer-word size?

4.3 What are the three modes of transmission? Explain how communications operates in each mode.

4.4 Contrast asynchronous and synchronous transmission. What are the advantages and disadvantages of each transmission technique?

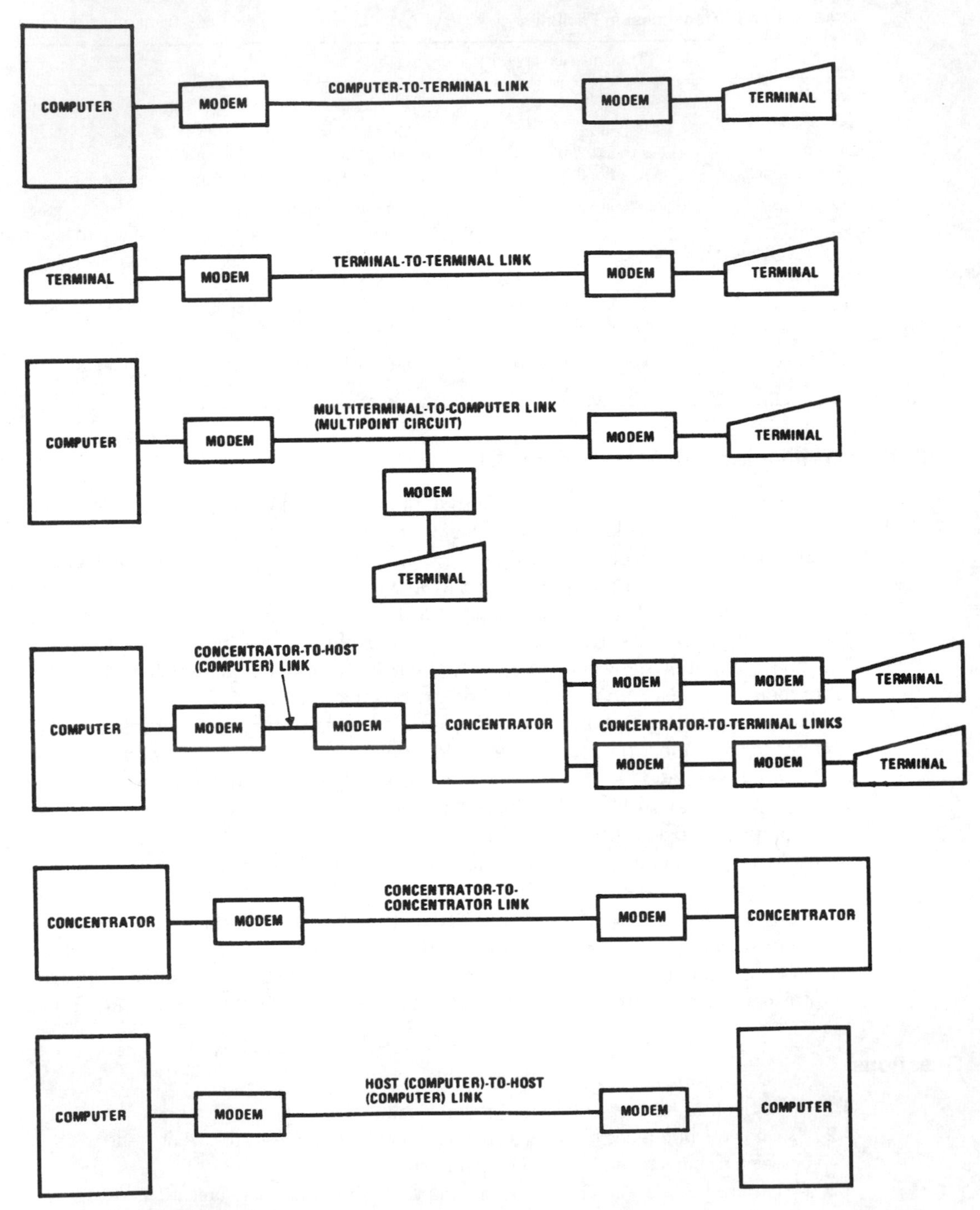

Figure 4.13.1 Link terminology.

4.5 Contrast serial and parallel transmission. Which one requires less transmission time? Explain.

4.6 Define the normal utilization of the three basic types of line connections.

4.7 Define the two basic types of line structures and the advantages and disadvantages of each.

4.8 What are some of the functions control characters perform? Cite five examples of control characters and their utilization.

4.9 Define the function of a transmission code.

4.10 Why is Morse code not suited for automatic data transmission?

4.11 How many different characters can be represented by a 10-bit code?

4.12 How long would it take to transmit one bit of a 12-bit character at 120 characters per second? How long would it take to transmit the entire character?

4.13 What does a protocol perform? Discuss several protocol control tasks.

4.13A What is the advantage of a file transfer protocol that dynamically alters the length of a block to correspond to the line error rate, versus a fixed-length block file transfer protocol?

4.14 Discuss the differences between BSC and HDLC type protocols.

4.15 Why is a transparent mode of operation for data transfer important? Contrast BSC and HDLC transparency methods.

4.16 What are the seven layers of the OSI reference model? Describe the functions of each.

4.17 Identify and discuss the two general internetworking types of OSI layer services.

4.18 What are the key differences between analog and digital service?

4.19 Discuss the differences between a leased line, a WATS line, and a foreign exchange line.

4.20 What is an analog extension and why is it used?

Chapter

5

Transmission and Equipment Basics

5.1 Introduction

The function of a data communications network is to transfer information from one location to another. This transfer is accomplished by two types of communications signals—digital and analog.

In this chapter, the basic methods of transferring information, starting with a terminal directly connected to a computer, are examined. Next, the problems, limitations, and techniques associated with transmission over the most frequently used medium, the telephone network, are covered, as well as transmission with the more recently introduced digital services.

To transmit and receive information over a medium requires a variety of devices. The basic device used for data transmission on analog media is the modem. Its transmission compatibility, frequency-band limitations, and noise problems, as well as the service unit used on digital networks, are examined in detail. In covering these components, the interrelationship of data transmission equipment, terminals, and computers is then be presented to develop an understanding of how networks can be developed, and altered.

5.2 Transmitting Digital Signals

The basic element found in many networks, a terminal directly connected to a computer, can be examined to develop an understanding of the capacity of a communications channel.

If the direct connection is of limited distance and the electrical properties of the line have no abnormalities, then the received digital pulses will be identical in shape to the digital pulses transmitted from the other end of the line—top portion of Figure 5.2.1. Unfortunately, this ideal situation seldom occurs in real life because every circuit has a degree of resistance, inductance, and capacitance that causes signal distortion.

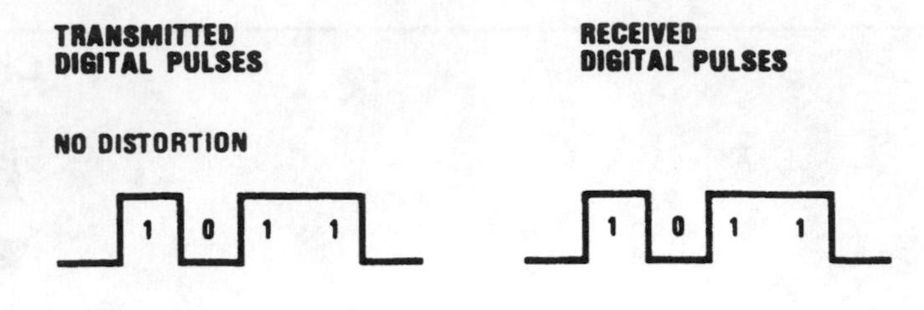

DISTORTION DUE TO ELECTRICAL PROPERTIES OF LINE

ATTENUATION EFFECT

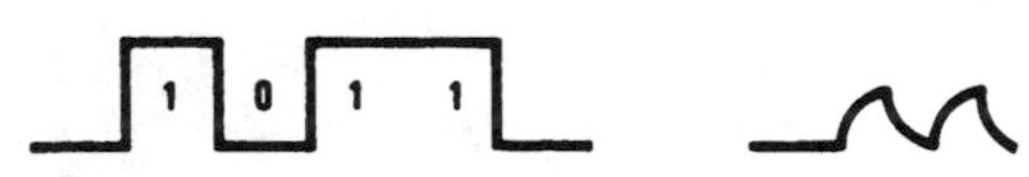

Figure 5.2.1

Because of these electrical properties, the square-edged digital pulses become distorted as they travel over the communications lines, as illustrated in the middle portion of Figure 5.2.1.

In addition to distortion, a factor known as *signal attenuation* must be considered. Attenuation results from a loss of signal strength as the distance between transmitter and receiver increases, as shown in the lower portion of Figure 5.2.1. The net effect is that the transmitted pulses might not be recognizable at the receiver.

Two additional factors can also have a bearing on the quality of the received pulses. The first factor, transmission speed, is easy to visualize since the higher the data rate, the more pulses per unit time; and the more pulses per unit time, the narrower the pulses. It is easier to distort a narrower pulse than a wide one to the point where a receiver cannot recognize it. The second factor is random noise, commonly referred to as *thermal noise*, caused by the molecular vibrations within electronic circuitry. This produces a very-low-level mixture of electromagnetic waves at different frequencies—similar to the hiss one might hear when an FM radio is tuned between stations. Thus, if the transmitted signal power falls too far, the thermal noise level can swamp the data-transmission level and thus cause errors.

Additional sources of noise include static from the atmosphere, interference from electrical equipment, and noise produced from the components of electrical circuits, including telephone company switching centers, cables, and switches.

One method used to characterize the quality of a circuit is by determining its signal-to-noise ratio. This ratio is simply the signal power divided by the noise power, and it can be increased by increasing the signal strength, decreasing noise, or both. A high signal-to-noise ratio is desirable because it maximizes the capacity of a channel. However, this does not imply that there is no limit to a signal's power placed on a communications line. In fact, a signal's power level is one of several transmission parameters whose permissible variance is specified by communications carriers.

Another method to categorize the quality of transmission over a circuit is by stating the ratio of the power transmitted to power received. In a telephone circuit environment, rather than try to determine the loss of power as a result of the transmission through such components as telephone instruments, switchboard connections, and the circuit, the total loss can be determined by end-to-end measurements. When measuring for the overall loss, a test frequency of approximately 1000 Hz is used by convention. In actuality, the test frequency is 1004 Hz in North America and 800 Hz in Europe and most other locations. The loss or gain for the entire circuit is given by the equation

$$N = \log_{10} \frac{P_1}{P_2}$$

where N = power ratio in bels
P_1 = power transmitted
P_2 = power received

Because the bel is a large unit, the decibel (dB), or one-tenth of a bel, is used so that the loss or gain becomes:

$$P = 10 \log_{10} \frac{P_1}{P_2}$$

where P = power ratio in dB.

If an amplifier produces a stronger signal at the receiver, then the received signal, P_2, will be greater than P_1 and the logarithm of P_1 over P_2 will be negative. If the power transmitted (P_1) is greater than the power received (P_2), the logarithm will be positive. As an example, consider a circuit where the power received was measured to be only a hundredth of the power transmitted. Then, the power ratio would be

$$P = 10 \log_{10} \frac{P_1}{P_2} = 10 \log_{10} \frac{100}{1} = +\ 20 \text{ dB}$$

On a telephone circuit, losses because of the line and equipment will always arise. However, one is able to obtain a zero-loss circuit when the gains from the amplifiers in the circuit are adjusted to counterbalance exactly the losses because of the circuit and component attenuation. It is important to note that the decibel is not an absolute unit but the logarithmic expression of a ratio that defines over-all loss in terms of the power sent and power received over a circuit.

Similarly, the decibel can be used to express the ratio between the signal and noise on a circuit. The correspondence between various decibel losses and power ratios is listed in Table 5.2.1. Notice that a reference point is required to define the relationships. In this case, 0 dB is equivalent to a 1:1 ratio and it serves as the reference point.

For the user who wishes to extend the physical distance between terminal and computer, the problems of distortion, attenuation, and noise might be insurmount-

TABLE 5.2.1 Decibel Related to Power Ratio

dB	Power ratio Signal-to-noise ratios
0	1:1
+3	2:1
+6	4:1
+9	8:1
+10	10:1
+13	20:1
+16	40:1
+19	80:1
+20	100:1
+23	200:1
+26	400:1
+29	800:1
+30	1000:1
+33	2000:1
+36	4000:1
+39	8000:1
+40	10,000:1

able when transmission is over an ordinary pair of telephone-type wires, unless special equipment is installed to enhance the level of transmission. One such device is a line driver, which can be used to extend the distance that a digital signal can be transmitted down a line and still be recognized properly at the receiver.

A line driver is basically a digital repeater that amplifies and reshapes signals. The line driver is normally used on direct-connect lines of in-plant facilities, where power sources and shelter are available. This device can be inserted at any point in the direct-connect line. It operates by sampling bits being transmitted, then retransmitting such bits after amplifying and regenerating them so that their structure is reformatted back into the original shape. This regeneration process is illustrated in Figure 5.2.2.

As terminal locations increase in distance from the computer, a point is reached where it becomes impractical to continue the utilization of a direct connection between the terminal and the computer. When such a distance is reached, an alternative medium must be tried. One such medium is the analog telephone network.

5.3 Signal Modulation and Demodulation

Designed for the transmission of voice signals, the analog telephone network does not use digital repeaters. Instead, analog amplifiers rebuild the level of a voice signal when it becomes faint. Unfortunately, amplifiers also amplify noise and distortion that are present on the circuit. To better understand analog transmission, first investigate the frequency spectrum and its application to both voice and data transmission.

When a person speaks, he or she is transmitting a continuous range of frequencies that travels through the air. Light waves and electromagnetic waves on a wire conductor can also be described in terms of their frequencies. That is, the amplitude of the signal at a given point in time is oscillating. The rate of oscillation is referred to as the *frequency*, which is described in terms of Hertz (cycles per second). Although the human ear can hear sounds of frequencies ranging from about 30 up to 20,000 Hz, telephone circuits transmit only between approximately 300 and 3300 Hz. This collection of frequencies, or band, that is passed is sufficient to understand speech and to recognize the speaker. When voice telephone signals travel between telephone company central offices, many of these signals can be packed together (multiplexed) electronically to enable one wideband channel to carry many conversations simultaneously. To do so requires a considerable engineering effort because this involves assigning 3-kHz bands for each voice signal, raising the signal to a higher frequency for transmission, and then separating (filtering) signals back to individual voice channels at the receiving end.

When attention was focused on telephone circuits for their use in transmitting digital data, a significant problem had to be overcome. The telephone circuit was designed for analog transmission: a continuous wave shape, such as the human voice's. If you wished to transmit digital data (discrete square-shaped pulses) over such a medium, a device would be required to convert the digital bit stream into an analog signal, and back again.

As mentioned in Section 4.11, one such device developed to perform this function is a modem, and the process it uses is called *modulation*. At the other end of the telephone line, a similar device must be used to reverse the process, converting the analog signal back into digital pulses. This process is called *demodulation*, and the device at each end of the line is built to perform modulation and demodulation. The utilization of such a device to permit a remote terminal to communicate with a computer over a telephone circuit is illustrated in Figure 5.3.1.

Modems incorporate components that protect the circuits from many signals that could cause interference with other users or with the telephone network's signals.

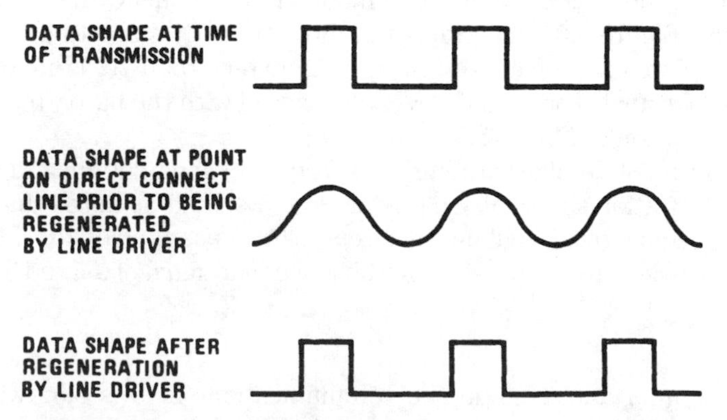

Figure 5.2.2 Restoring the bit shape.

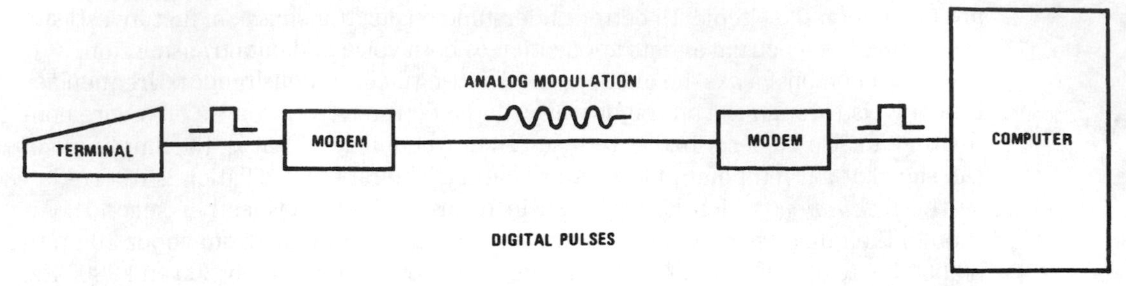

Figure 5.3.1 Modem operation.

Modulation techniques

In designing a modem, a variety of modulation techniques can be used. Each has a different set of characteristics and a different level of circuit complexity. Some designs permit transfer at a high data rate and are expensive, while other less-costly designs are used for low-speed digital signaling. The majority of modems operate by transmitting a continuous sine wave and then modulating it in accordance with the data that is to be transferred.

As illustrated in Figure 5.3.2, a sine wave is a spectral line that relates to transmission at a given frequency. Here, current or voltage is applied to produce an amplitude. In addition to the amplitude, the frequency and phase of the wave can be varied to carry information. Expressed as a function of time, the sine wave can be represented by the following equation:

$$a = A \text{ sine } (2\pi ft + \theta)$$

where a = instantaneous amplitude of voltage or current at time, t
A = maximum amplitude of voltage or current
f = frequency in Hz
θ = phase in degrees (zero in Fig. 5.3.2)

To make the sine wave carry information (that is, to modulate the sine wave), the values of A, f, or θ can be varied. The varying of these parameters produces the three basic types of modulation: amplitude modulation, frequency modulation, and phase modulation. Figure 5.3.3 illustrates tile three types. Assume that a sine wave centered in the telephone voice band at approximately 1500 Hz is modulated to transmit the digital signal. The amplitude can be varied with the bit pattern, as illustrated in the top portion of Figure 5.3.3.

The middle portion of the illustration shows frequency modulation; and the lower portion of that illustration shows phase modulation. It should be noted that the simple presence or absence of a signal on the circuit can be used to convey information that would be equivalent to the presence or absence of a signal of a fixed amplitude.

Bandwidth

One of the most important terms in the communications field is bandwidth. This term refers to the width of the range of frequencies that a channel can transmit, and not the frequencies themselves. If the lowest frequency a channel can transmit is f_1 and the highest f_2, then the bandwidth is the difference between the highest and lowest frequencies: $f_2 - f_1$. Because a telephone line can transmit frequencies from approximately 300 to 3000 Hz, its bandwidth is 3 kHz.

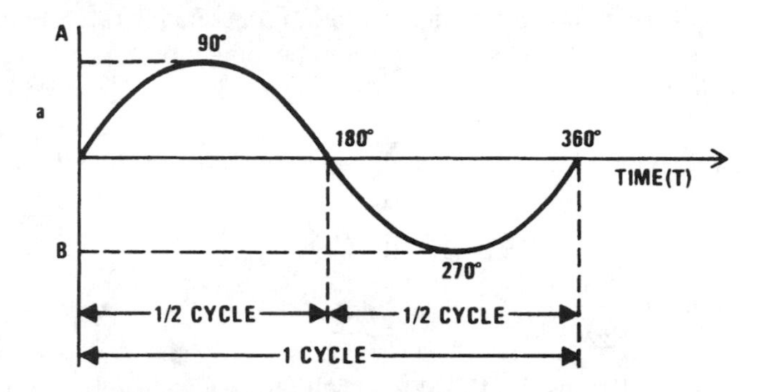

Figure 5.3.2 Sine wave.

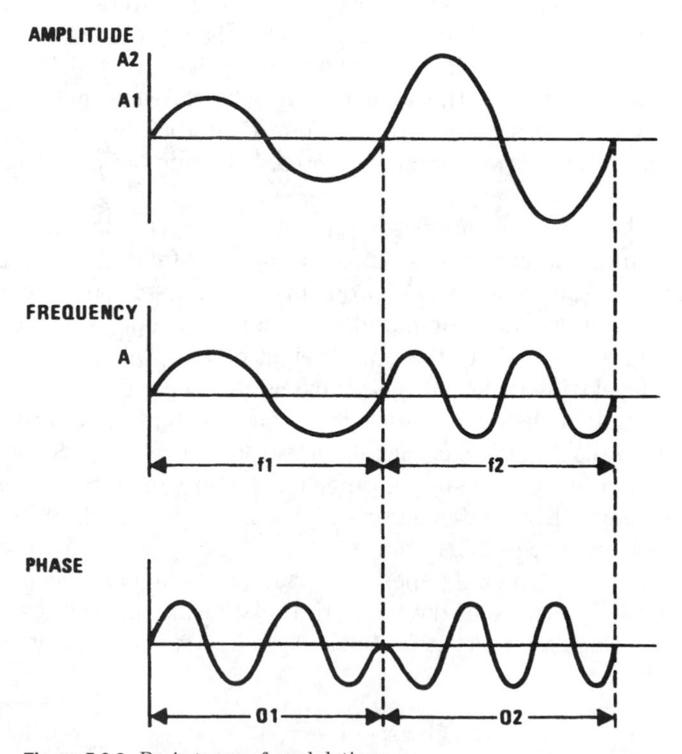

Figure 5.3.3 Basic types of modulation.

Normally, f_1 and f_2 are the frequencies at which amplitude is 3 dB down from the maximum amplitude within the band. Thus, frequencies less than f_1 or greater than f_2 can be transmitted but the amplitude will be seriously attenuated.

The capacity of a channel for information transfer is proportional to its bandwidth. A channel with a 15-kHz bandwidth can transfer five times as many bits per second as a channel with a 3-kHz bandwidth. As the maximum transfer rate of a channel is reached, the signals transmitted on that circuit become uninterpretable because they are received in too distorted a shape for the receiver to be able to recreate them.

In 1928, Nyquist developed the relationship between bandwidth and the baud rate on a circuit as

$$B = 2W$$

where B = baud rate
W = bandwidth in Hz

That is, in Nyquist's relationship, the rate at which data can be transmitted without incurring intersymbol interference must be less than or equal to twice the bandwidth in Hz. For a typical voice circuit with a bandwidth of about 3000 Hz, data transmission can only be supported at baud rates lower than 6000 symbols or signaling elements per second. Today, only the most expensive and efficiently designed modems approach the Nyquist baud rate limit of $2W$.

Because any oscillating modulation technique, such as amplitude modulation, frequency modulation, or phase modulation, immediately halves the achievable signaling rate, most modems operate at one-half to one-quarter of the Nyquist limit. Because this limit applies only to the signaling, or baud rate, and not to the actual data rate in bits per second, increasing the amount of information each pulse or signal carries (by coding) will correspondingly increase the number of bits per second transmitted.

Although this technique would appear to remove data transfer limitation, increasing the amount of information per pulse through the use of multilevel coding, as illustrated in Figure 4.2.2, introduces new problems. In a 2-level code, the difference between each level is large when compared to higher-level codes. This relationship is illustrated in Figure 5.3.4, where the pulse height of a 2-level code is eight times the height of an 8-level tribit code. Thus, as the height of the pulse is reduced, it becomes more susceptible to distortion. Another problem with multilevel coding is the circuitry required to distinguish each level transmitted. Presently, 8-level coding, representing 3 bits per baud, is the upper range used before error rates begin to rise to an unacceptable level. Error rates increase because, as the levels are reduced in size, they become more susceptible to noise.

In 1948, C. Shannon presented a paper concerning encoding and decoding methods that could be used to transmit data, and the relationship of coding to noise. In this paper, Shannon calculated the theoretical maximum bit-rate capacity of a channel of bandwidth W as

$$C = W \log_2 \left(1 + \frac{S}{N}\right)$$

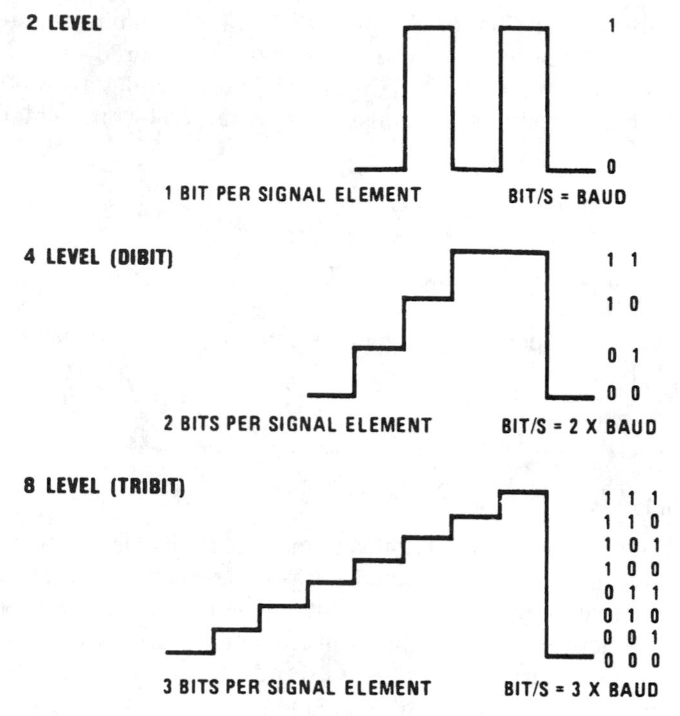

Figure 5.3.4 Multilevel coding relationship.

where W = bandwidth, in Hz
S = power of the transmitter
N = power of the thermal noise

Using a signal-to-noise ratio of 30 dB and the bandwidth of a typical telephone line of 3000 Hz, Shannon's capacity becomes:

$$C = 3000 \log_2 \left(1 + \frac{1000}{1}\right)$$

or approximately 30,000 bit/s. This value can be considered the ultimate design goal for modems on voice-grade lines. Present devices attain data rates of 19,200 to 24,000 bit/s on leased lines and normally 9600 to 19,200 bit/s or less on the switched network. Thus, the highest data transfer rates in use today are between two-thirds and three-fourths of the theoretical maximum.

Notice that Shannon's theoretical capacity limit references the information transfer or data rate capability of a modem; it does not consider the effect of data compression. This explains why you see advertisements for compression-performing modems that claim a throughput of up to 96 kbit/s. What that advertisement actually references is a device with a data rate of 24 kbit/s that can achieve a 4:1 compression ratio with data that is most susceptible to being compressed. Then, each transferred bit represents, on average, four bits of uncompressed data. Therefore, the modem's throughput becomes $24{,}000 \times 4$ or 96 kbit/s.

The relationship between Shannon's law of data transfer in bits per second and Nyquist's relationship between bandwidth and baud is covered next. Shannon's formula defines the maximum bit rate (C), which, when divided by the baud rate (B), indicates the number of bits that must be represented by one signal element. Thus,

$$N = \frac{C}{B}$$

where N = number of bits/signal element.

Once the number of bits per signal element is computed, then the number of code levels required, L, becomes

$$L = 2^N$$

Additional modulation techniques

When modems use frequency modulation, a number of problems can occur. First, as the modulation rate increases, the spread between frequencies used must be increased. This can create a problem when transmission is to occur over the public switched telephone network because different frequencies are used for such supervisory functions as circuit switching and call termination. If these frequencies are not avoided as the modem modulates the frequency in response to changing data rates, false operations could be initiated. One method used to avoid this problem is frequency-shift keying (FSK), in which just two frequencies are used: one to represent a 1 bit and another a 0 bit, as shown in Figure 5.3.5. Normally, FSK modulation is used to transfer information at low data rates.

Although multilevel signaling can be conducted using frequency modulation, such a technique is seldom used. The increase in the number of separate frequencies that have to be transmitted causes a corresponding increase in bandwidth requirements.

A second technique widely used in modem design actually consists of a number of variations of phase modulation. One method to better understand the principles of phase modulation is to think of a modem having two sine-wave generators, as illustrated in the top portion of Figure 5.3.6, with each sine wave of the same frequency, but out of phase—in this case, by 180 degrees.

If the output of the modem is switched between these sine waves as the digital pulses entering the modem vary between 0 and 1, then the phase shift can be used

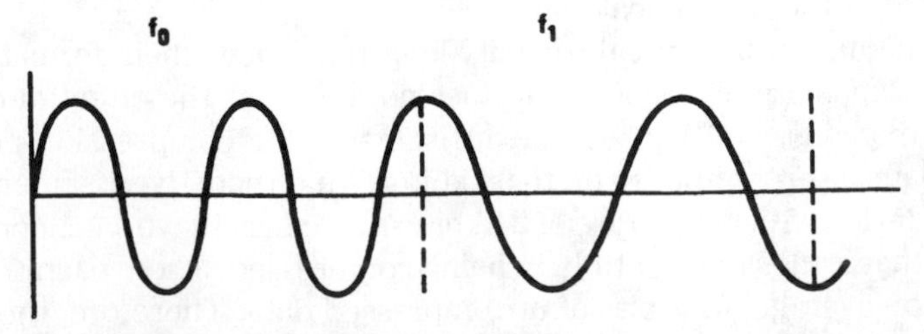

Figure 5.3.5 Frequency-shift key.

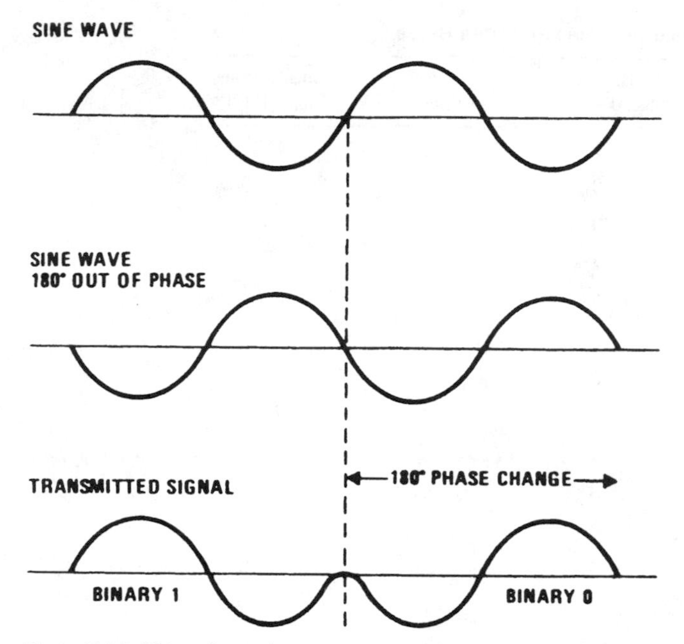

Figure 5.3.6 Using phase change to represent different information states.

to represent different information states. The result of this phase shifting is shown in the lower portion of Figure 5.3.6, where a phase change of 180 degrees is used to represent a binary 0, while the normal in-phase sine wave represents binary 1.

Currently, two types of phase modulation are used in the design of modems and the circuitry required to detect the phase information. The first method, known as *fixed-reference phase modulation*, assigns a meaning to each phase position as in the previous example. Here, a reference wave of the same frequency as the modulation signal, but of constant phase, must be generated at the demodulator to detect the phase of the incoming modulated signal.

The second method, known as *differential phase modulation*, assigns meanings to each change in phase and not to the phase conditions. This method does not require a separate reference wave and thus reduces to a degree the circuitry required in the modem. Here, one change in phase would be interpreted as a binary 0 if the preceding phase was interpreted as a binary 1, and so on. One of the primary reasons for the design of a number of modems that use phase modulation is that multiphase-generating modem circuitry is less complex than modems designed for other modulation techniques. The multiphase signals permit multiple bits of information to be represented by one phase change.

As an example, consider the multilevel coding relationships illustrated in Figure 5.3.4. By using eight different phases, three bits of information (tribits) can be represented by each discrete phase change. In Table 5.3.1, the relationships between several modulation rates and the number of bits per second of data that can be theoretically transmitted by the use of different levels of phase modulation are listed. Although the data transfer rate can substantially increase as the corresponding

TABLE 5.3.1 Modulation Rates and Data Rates

Modulation rate (Baud)	Bits per Signal element	Number of Phases	Data transfer Rate (Bit/s)
1200	1	2	1200
1200	2	4	2400
1200	3	8	3600
1200	4	16	4800
1200	5	32	6000
2400	1	2	2400
2400	2	4	4800
2400	3	8	7200
2400	4	16	9600
2400	5	32	12,000
2400	6	64	14,400
2667	6	64	16,000
2400	7	128	16,800
2400	8	256	19,200

number of phases used increases, the small changes in phase required to differentiate between states will drive up the cost of circuitry. In addition, as the number of phases increases, the degrees separating each phase decrease, increasing the probability of errors in reception.

Two of the more common methods used in the design of four-phase modulation modems are illustrated in Figure 5.3.7. The first method uses 0-, 90-, 180-, and 270-degree phase shifts, while the second method uses 45-, 135-, 225-, and 315-degree phase shifts. In the first method, a long string of repetitive dibits, such as 00 00 00 00 or 11 11 11 11, could cause synchronization problems because phase changes could be delayed for a considerable time, and most modems of this type rely on periodic phase changes to resynchronize timing. This problem can be resolved by the second method shown in Figure 5.3.7. Because no zero-degree phase is present as in the previous method, there is always a phase transmitted.

Quadrature amplitude modulation

The quadrature amplitude modulation (QAM) technique combines phase and amplitude modulation to obtain data rates in the range of 4800 to 9600 bit/s. In a QAM modem, two signals at the same frequency, but 90 degrees out of phase with each other, are used, hence the term *quadrature*. Figure 5.3.8 contains a vector diagram of the QAM technique. For each signal, four possible levels of amplitude can be applied: A1, A2, A3, and A4.

The combination of the signals P and Q, which are 90 degrees out of phase, can be used to generate 16 different conditions, each signifying four bits of information.

Proprietary and nonstandardized modems

In the quest to develop modems capable of operating at higher data transmission rates, many vendors have developed methods to more efficiently process digital data, or to modulate and demodulate such data. Essentially, these methods are either proprietary to a vendor or licensed by one vendor to others. In the latter case, while a few licensed techniques had gained widespread acceptance by the end-user community and resulted in the development of international standards, other methods were not recognized by a standards-recommending body at this writing.

Common techniques to increase the data transmission rates of modems include data compression, asymmetrical transmission, and use of a large number of carrier frequencies onto which information is packetized prior to transmission. These techniques are examined in the following paragraphs to present the reader with an understanding of modem operation based on each technique.

Data compression

Data compression is a technique that reduces or eliminates redundancies in a data stream. It is accomplished by applying one or more algorithms to the data prior to its

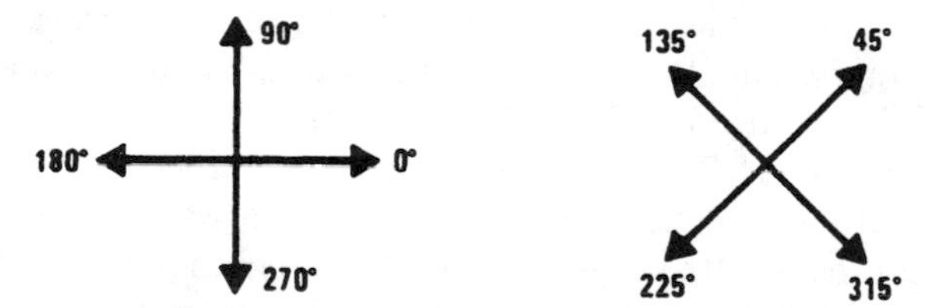

Figure 5.3.7 Different four-phase modulation techniques.

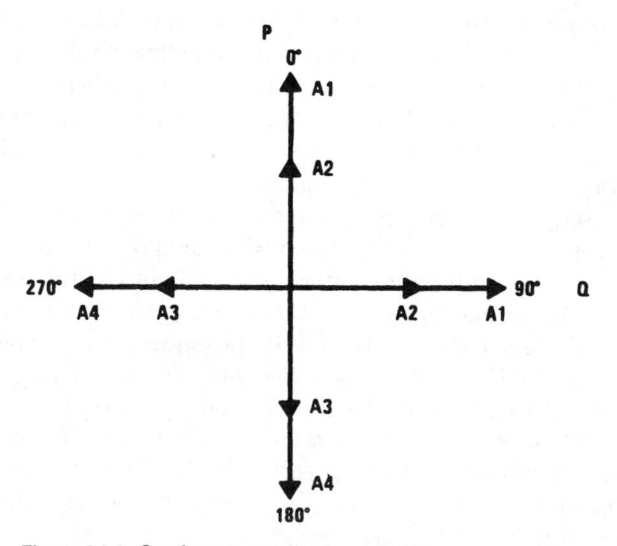

Figure 5.3.8 Quadrature amplitude modulation.

modulation, then applying a reverse algorithm or set of algorithms to demodulated data to expand it back into its original format.

A simple example of data compression is the technique known as *Run Length Encoding*. Under this technique a string of repeating characters of length greater than 4 is replaced by the three-character sequence:

$$XC_CC_I$$

where: X = Character to be compressed
C_I = Compression-indicating character
C_C = Character count

Then, as an example, a sequence of nine dashes (-) would be replaced by the three character string:

$$\text{-}9C_I$$

C_I is a special character in a character set used to indicate the occurrence of Run Length Encoding, or a bit sequence that is not commonly encountered in data transmission. In the case of the latter, an extra C_I character would be inserted into the data stream when the character naturally occurs in a bit stream, to prevent an inadvertent indication of compression. Similarly, if two C_I characters follow one another, the second would be removed from the data stream by the receiving modem.

By compressing data prior to its modulation the data transfer rate of the modem can be increased without changing the baud rate of the device. Most modems incorporating data compression use a QAM modulation technique that follows the CCITT V.29 standard—which does not assure compatibility (covered later in this chapter).

Another deficiency associated with modems that perform data compression is the variability of the data transfer rate. This variability results from data susceptibility to compression not being uniform. Thus, a modem that normally operates at 4800 bit/s and initially achieves a 2:1 compression ratio would have a data transfer rate of 9600 bit/s. Later, if the compression ratio fell to 1.5:1, the data transfer rate would fall to 7200 bit/s. In spite of these problems, data compression represents a viable method for enhancing the data transfer capability of modems.

Initially, the use of compression in modems was based on proprietary algorithms that limited interoperability. For example, a Hayes Microcomputer Products V-series modem using the Hayes data compression algorithm could not communicate with a Microcom Corp. modem using that firm's MNP Class 5 data compression algorithm. Recognizing the necessity to standardize the method used to perform compression in modems, the CCITT promulgated its V.42bis Recommendation. That recommendation specifies the use of a string-compression technique, referred to as *Lempel-Ziv*. In addition, V.42bis stipulates control techniques for one modem to "tell" another whether or not it supports this compression method. Later in this chapter, the sections on intelligent modems also cover MNP modem classes and illustrate the operation of string compression.

Asymmetrical transmission

The public switched telephone network is a 2-wire facility. Until the development of echo-cancellation technology during the late 1980s, and its widespread adoption into V.32 modems (see Section 5.4) during the early 1990s, it was both very difficult and expensive to obtain full-duplex transmission at high data rates. In place of providing full-duplex transmission, a few vendors implemented asymmetrical transmission to obtain a near-equivalent capability.

Asymmetrical transmission requires the bandwidth of the line to be divided into two segments or subchannels. The main subchannel contains a majority of the available line bandwidth, and the other subchannel contains a small portion of the line bandwidth. The division of the line bandwidth is illustrated in Figure 5.3.9.

Creating two subchannels by frequency enables data to simultaneously flow in opposite directions. However, the unequal distribution of bandwidth also results in an unequal data transfer rate. The main channel normally operates at 9600 bit/s, and the smaller-bandwidth secondary channel operates at 300 bit/s. To enhance data transfer between devices, the modem monitors the direction of data transmission and adjusts the assignment of subchannels accordingly. That is, the modem will assign the main channel to the direction in which the majority of data transmission occurs. To prevent data from being lost while transmission directions are switched, each modem contains a buffer area. This enables transmission between the modem and each connected device to continue as if the circuit were operating in a full-duplex mode.

Most asymmetrical modems are based on the use of CCITT V.29 technology for modulation and demodulation. Unfortunately, the methods used to determine when to switch subchannel directions and the bandwidths of the subchannels differ among vendors. Therefore, most asymmetrical modems manufactured by different vendors are not compatible with one another. The exception to this incompatibility occurs when one vendor builds an asymmetrical modem using technology licensed from another vendor.

One of the earliest manufacturers of asymmetrical modems was U.S. Robotics, which incorporated subchannel-switching technology into a V.29-based 9600 bit/s

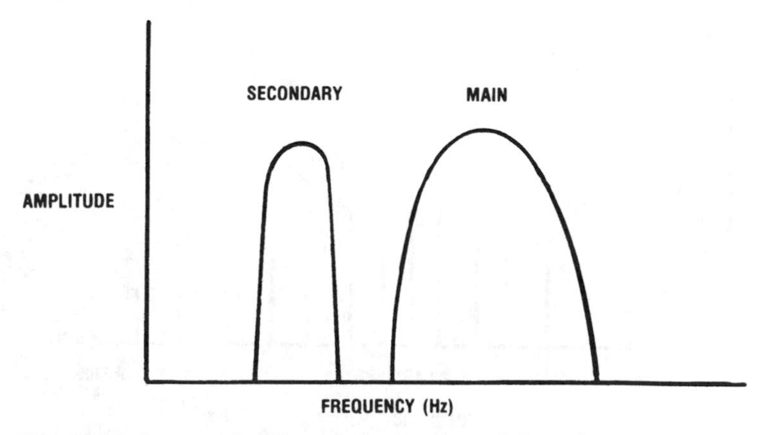

Figure 5.3.9 Asymmetrical Transmission, modem-subchannel use.

modem in 1986. In 1987, the company licensed its technology to another modem manufacturer and was attempting to interest standards bodies to accept asymmetrical transmission as a promulgated standard.

Although the U.S. Robotics asymmetrical modem technology was not standardized by the CCITT, the device has obtained a significant market share through promotions and discounts offered by the vendor to bulletin board system (BBS) operators (SYSOPS). By 1994, U.S. Robotics had introduced several asymmetrical modems under its HST (High Speed Transmission) label. The vendor also offered "dual HST" modems that could operate as either an asymmetrical modem or as a CCITT V.32 or a V.32bis device.

Packetized carrier transmission

Until 1986, modems modulated either one or two carrier signals, impressing information onto a circuit by varying the carrier amplitude. One carrier is normally required for half-duplex transmission, with the receiver turned off at a distant modem while its transmitter is turned on. Direction is reversed when the originating modem turns off its transmitter and turns on its receiver. This technique is also known as *switched-carrier transmission*. For full duplex transmission, two carriers are normally used, with each carrier providing a communications path opposite to the other.

In 1986, Telebit Corporation introduced a modem that was revolutionary in the method used to modulate data. To initiate communications with a similar device on the switched network, this modem transmits 512 tones onto a circuit. The receiving modem evaluates the tones and reports back to the originating device all unusable frequencies. Thereafter, the originating modem encodes data for transmission on each usable carrier, using 2-bit, 4-bit, or 6-bit quadrature amplitude modulation, based on the suitability of each tone.

The Telebit Trailblazer modem's modulation technique is formally called a packetized-ensemble protocol. Figure 5.3.10 illustrates the use of the bandwidth of a voice-grade circuit by a packetized-ensemble modem. Normally, the amplitude of the

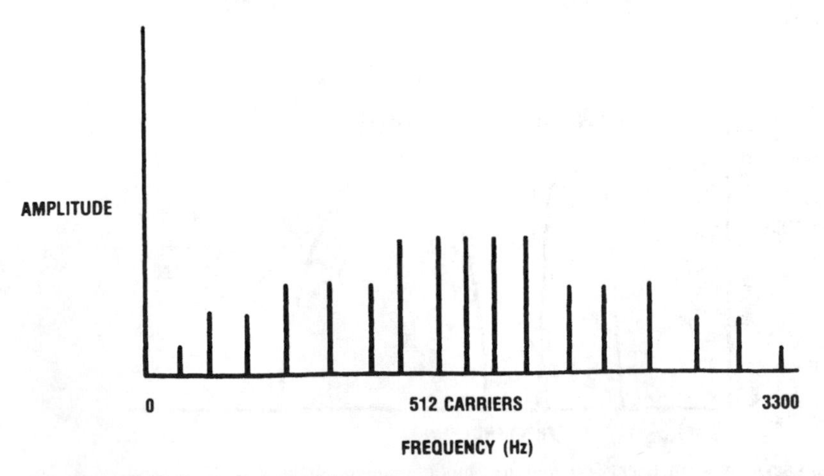

Figure 5.3.10 Packetized-ensemble modem, carrier use.

TABLE 5.3.2 Basic Pulse Modulation Techniques

Pulse code modulation (PCM)
Pulse amplitude modulation (PAM)
Pulse duration modulation (PDM)
Pulse position modulation (PPM)

carrier is higher toward the center of the frequency spectrum. This is because delay and distortion are minimal at the center of a circuit's frequency band, permitting a maximum number of bits to be packetized onto the center carriers.

The major advantages in the use of a packetized-ensemble-protocol modem are the high data rate—normally exceeding 12,000 bit/s—it can achieve over the switched telephone network and its fallback data rates when transmission impairments occur. Concerning fallback, as impairments occur that affect a carrier, the modem's throughput only decreases in increments of 50 to 100 bit/s. By comparison, a conventional modem would either become inoperative if an impairment affected its carrier frequency or would lower its operating rate by a minimum of 2400 bit/s. The latter situation would occur if the modem had a fallback data rate capability, where the lower data rate becomes achievable based on the amount of distortion or delay encountered.

The Telebit series of Trailblazer modems has achieved a significant sales success. However, like the U.S. Robotics asymmetrical modems, Telebit's packetized-ensemble protocol was never standardized by the CCITT. To provide users with the ability to communicate with CCITT-standard modems, Telebit, like U.S. Robotics, developed other modems that support both its proprietary protocol as well as CCITT-recognized protocols.

Pulse modulation techniques

Unlike the previously covered modulation techniques, pulse modulation is not used in modems. Instead, this technique is used by devices designed to transmit data over digital media. Some pulse modulation techniques are also used to multiplex, or group, a number of voice connections for transmission over a common circuit. Table 5.3.2 lists the four basic methods of pulse modulation techniques.

In pulse amplitude modulation, a signal is sampled at successive intervals in time and converted into a series of similar-width pulses from which the original analog signal can readily be reconstructed, as illustrated in Figure 5.3.11. When the pulse duration technique is used, the width of the pulse is varied in accordance with the sampled signal, while the amplitude and leading-edge position remain constant.

Using the pulse position technique, the leading edge of the pulse shifts back and forth in accordance with the sampled signal, while the width and amplitude of the pulse remain constant. In the pulse code technique, the amplitude of the sampled signal is quantized into discrete pulses at a certain amplitude level. This level is maintained until there is a significant change in the sampled signal, with the number of

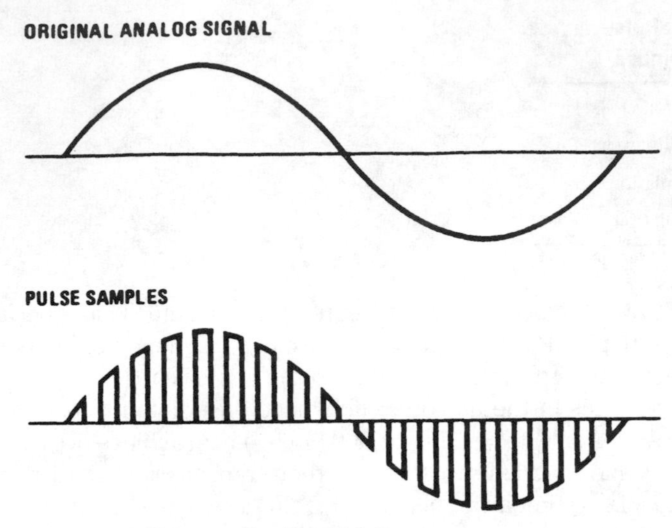

Figure 5.3.11 Pulse amplitude modulation.

levels a function of the rate of change of the signal. If the sampled signal varies rapidly, many levels will be required. A slow-changing signal would require fewer levels.

5.4 Modems and Couplers

One method that can be used to categorize data communications components is to examine the functions they are designed to perform. In this section primarily covers the characteristics, operation, and utilization of components designed primarily to affect data transmission over analog media.

Analog media employment

Today, despite the fact that several communications carriers have introduced all-digital transmission facilities that are more efficient than analog media, the analog telephone network remains the primary facility used for data transmission. Business machines, including terminals and computers, transmit digital pulses. Telephone circuits are designed to transmit analog signals, such as the human voice. Therefore, a device to convert the digital pulses into analog tones capable of being transferred on telephone circuits becomes necessary if you wish to transmit digital data over these facilities. Two such devices that can be used to convert and reconvert signals are modems and acoustic couplers.

Modem Components

In its most basic form, a modem consists of a power supply, a transmitter, and a receiver. The power supply usually takes 120 or 220 Vac, and transforms it into a dc voltage necessary to operate the modem's circuitry. In the transmitter, a modulator, amplifier, filtering, wave-shaping, and signal-level control circuitry convert the digital pulses of the business machine into a modulated, wave-shaped signal that can be transmitted over a telephone circuit.

The complexity of the modulator and its circuitry varies, depending on the modulation technique used. Normally, modems using FSK modulation at low data rates have circuitry much less complex than modems that use QAM modulation to transfer data at rates up to 9600 bit/s. The modem's receiver consists of a demodulator and associated circuitry that reverse the modulation process. They convert the analog telephone signal back into a series of digital pulses that is acceptable to the digital device at the other end of the circuit.

If the operations of the transmitter and receiver are combined to enable the device to transmit and receive data alternately, the modem is said to be capable of *half-duplex operation*. In this mode, the transmitter must be turned off at the opposite end of the line when the other modem's transmitter is turned on, as illustrated in Figure 5.4.1. Conversely, the receiver of one modem is enabled while the receiver of the second device is disabled before each change in the direction of data flow. The time interval required for these operations is called the *modem turnaround time*. This factor can have a large effect on the quantity of data transmitted (throughput) when a line is frequently turned around to transmit acknowledgments concerning the validity of previously received data blocks.

If the modem's transmitter and receiver are capable of operating simultaneously, the modem is said to operate in the *full-duplex mode*. This simultaneous transmission and reception of data can be accomplished by several techniques. One technique, echo-cancellation, enables transmit and receive signals to be carried on a two-wire circuit. Another technique involves splitting the telephone line's bandwidth into two distinct channels on two-wire circuits, or by using two two-wire pairs, such as is obtainable on a four-wire leased line.

When the bandwidth of a telephone line is split into two distinct channels, the transmitter of the modem on one side of the line operates at the same frequency as the receiver at the other end of the line, and the receiver of one modem operates at the same frequency as the transmitter of the other modem. Tying the transmitter to the receiver and vice versa establishes two transmission paths over one pair of wires, permitting full-duplex data transfer to occur. In the four-wire case, separate signal paths are formed over each of the two two-wire circuits.

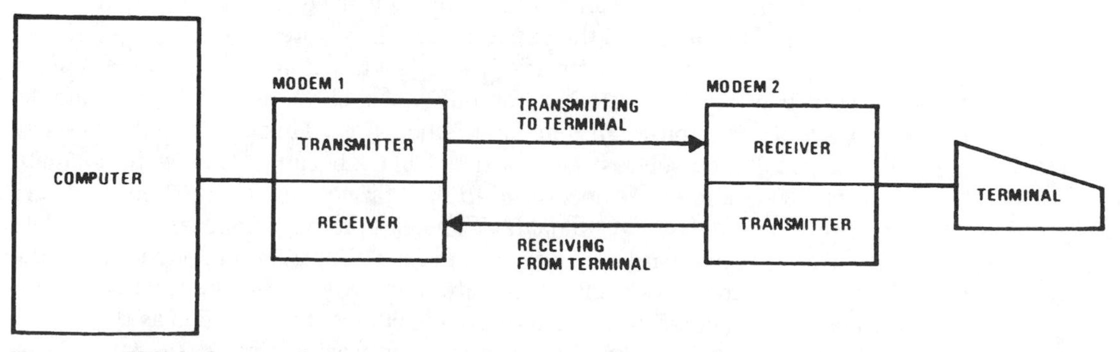

Figure 5.4.1 Half-duplex operation.

Transmission technique

Most modems are designed for either asynchronous or synchronous data transmission, although some devices can support both transmission modes. In the asynchronous mode of operation, also called *start-stop transmission*, the timing necessary for the receiving modem to synchronize itself with the transmitting modem is supplied by the transmitted character. This type of transmission is usually generated by unbuffered terminals where the time between character generation and transmission occurs randomly. Here, the character being transmitted is initialized by the character's start bit as a mark-to-space transition on the line, and is terminated by the character's stop bit (or bits), which is converted into a "space 1 marking" signal on the line. Between the start and stop bits, the digital pulses represent the encoded data that defines the character that was transmitted. (Refer to Section 4.4.)

As characters are transmitted, the asynchronous modem places the circuit in the "marking" condition between the stop bit of one character and the start bit of the next character. Upon receipt of the start bit from the next character, the modem switches the line to a mark-to-space transition and the modem at the other end of the line recognizes this transition as a signal to sample the data being sent. Both the marking and spacing conditions are audio tones produced by the modem's modulator to denote the equivalent binary data levels representing these conditions. The two tones are generated at predefined frequencies, and the transition between the two states, as each bit of the character is transmitted, defines the character. Usually used with low-speed teleprinter terminals and equivalent devices, asynchronous transmission was originally mainly used for data transfer at 2400 bit/s or less.

Today, numerous terminal devices transmit asynchronously at data rates up to 19.2 or 24 kbit/s. However, many modems supporting asynchronous high-speed data transfer actually convert the asynchronous bit stream into synchronous data for transmission between modems. Doing so permits removal of the start and start bits—in effect, reducing the 10 bits used to transfer an eight-bit asynchronously transmitted character back to eight bits.

When data is transmitted synchronously, more efficient line utilization is normally obtained because the bits of one character are followed immediately by the bits of the next transmitted character, with no start and stop bits required to delimit each character. Instead, synchronous transmission groups a number of characters into a block for transmission, with the length of the data block constrained by such factors as the terminal's buffer area and the expected line error rate.

Quite often, block length in data transmission is a function of the terminal's physical characteristics, which, in effect, is the buffer size of the device. As an example, when data is to be transmitted to represent punched-card images, it is often convenient to transmit 80 characters of one card as a block because users want compatibility with the 80-column card deck used in data processing. If punched cards are being read by a remote batch terminal for transmission to a computer, and the data is such that every three cards contain information about one employee, then the block size might be increased to 240 characters. In order to transmit data synchronously, the individual characters in each data block must be identified as they occur in time. To do so, a timing signal, which is usually provided by the modem, places each character into a unique time period, as illustrated in Figure 5.4.2.

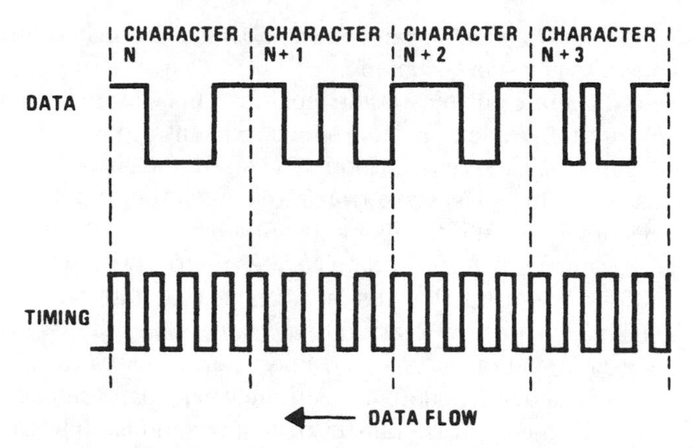

Figure 5.4.2 Synchronous timing signals.

Modem classification

Modems can be classified (Table 5.4.1) according to their modes or transmission techniques, or according to the application features they contain or the types of lines they are built to service. Generally, modems are classified in four line-servicing groups: sub-voice or narrow-band lines, voice-grade lines, wide-band lines, and dedicated lines. Subvoice-band modems require only a portion of the voice-grade channel's available bandwidth and are commonly used with equipment operating at speeds up to 300 bit/s. On narrow-band facilities, modems can operate in the full-duplex

TABLE 5.4.1 Common Modem Features

Line type / Features	Sub voice	Voice grade			Wideband	Dedicated
	Up to 300 bit/s	Low speed up to 2400 bit/s	Medium 2400 to 9600 bit/s	High speed 9600 to 19,200 bit/s	24,000 bit/s and up	Up to 1.544 megabit/s (T1)
Asynchronous	•	•	•	•		
Synchronous			•	•	•	•
Switched network	•	•	•	•		
Leased only					•	•
Half-duplex	•	•	•	•		•
Full-duplex	•	•	•	•	•	•
Fast turnaround for dial-up use				•		
Reverse/secondary channel	•	•	•	•		
Manual equalization			•			
Automatic equalization		•	•	•		
Multiport capability			•	•		
Voice/data			•	•		•

mode by using one-half of the available bandwidth for transmission in each direction, and an asynchronous transmission technique.

Modems designed to operate on voice-grade facilities can be asynchronous or synchronous, half-duplex or full-duplex. Asynchronous transmission is normally used at speeds up to and including 19,200 bit/s. Although a leased, four-wire line will permit full-duplex transmission at higher speeds, transmission via the switched telephone network normally occurs in the full-duplex mode at data rates up to 14,400 bit/s.

Voice-grade modems currently transfer data at rates up to 24,000 bit/s and usually require leased facilities for transmission at speeds above that data rate.

Wideband modems, which are also referred to as *group-band modems* because a wideband circuit is a grouping of lower-speed lines, permit users to transmit synchronous data at speeds above 24,000 bit/s. Although wideband modems are primarily used for computer-to-computer transmission applications, they are also used to service multiplexers that combine the transmissions of many low- or medium-speed terminals to produce a composite of higher transmission speed. The uses of group-band modems and multiplexers are explained later.

Dedicated or limited-distance modems, which are also known by such names as *short-haul modems* and *modem bypass units*, operate on dedicated solid conductor, twisted-pair wires, or on coaxial cables, permitting data transmission at distances ranging from 15 to 20 miles, depending on the modem's operating speed and the resistance of the conductor.

Limited-distance modems

These modems can operate at speeds ranging up to 1.5 million bits per second and are particularly well suited for in-plant usage, where the user desires to install her own communications lines between terminals and a computer located in the same facility or complex. Also, compared to voice-band and wideband modems, these modems are relatively inexpensive because they are designed to operate only for limited distances. In addition, by using this type of modem and stringing your own in-plant line, a user can eliminate a monthly telephone charge that would occur if the telephone company furnished the facilities. Limited-distance modems are explained in greater detail in Chapter 15.

In Table 5.4.1, the common applications of modems are denoted by the types of lines to which they can be connected.

Line type operations

Most modems with rated transmission speeds of up to 19,200 bit/s and some that transmit data above that speed can operate over the switched, dial-up telephone network. Since a circuit obtained from a dial-up telephone connection is a two-wire line, when this line is used to carry traffic in both directions alternately, the line and the modem operate in the half-duplex mode. The turnaround time varies by device and can become a considerable overhead factor if short bursts of data are transmitted, with each burst requiring a short acknowledgment.

To visualize some of the overhead problems associated with line turnaround, a short examination of an error-control procedure for synchronous transmission follows.

One common error-control procedure used in synchronous transmission is obtained by the use of an acknowledgment-negative acknowledgment (ACK-NAK) sequence. When this sequence is used, the terminal or computer transmits a block of data to the receiving station. Appended to the end of the block is a block check character that is computed based on a predefined algorithm. At the receiving device, the block of data is examined and a new block check character is developed using the same algorithm, which is then compared to the transmitted block check character. If the block check characters are equal, the receiving device sends a positive acknowledgment (ACK) signal.

If the block check characters do not match, then an error in transmission has occurred and the receiving device transmits a negative acknowledgment (NAK). This informs the transmitting device that the block should be retransmitted.

This procedure is also referred to as *automatic request for repeat (ARQ)* and it requires that the line on which transmission occurs be turned around twice for each block. Returning to the 80-character punched card image block, transmitting this data as a 960-bit block with control characters appended, at 9600 bit/s, would take just 100 milliseconds. If the modem turnaround time were 150 milliseconds, 300 milliseconds would be necessary to turn the line around twice. Although recently developed modem features have reduced modem turnaround time, this problem can be avoided or eliminated by using a modem with a reverse channel for acknowledgment signal, one with an echo-cancellation capability (to simultaneously transmit and receive data on a two-wire circuit), or by establishing full-duplex transmission, such as over a leased four-wire circuit.

Reverse and secondary channels

To eliminate turnaround time when transmission is over the two-wire switched network, or to relieve the primary channel of the burden of carrying acknowledgment signals on four-wire dedicated lines, modem manufacturers developed a reverse channel that is used to provide a path for the acknowledgment of transmitted data, at a slower speed than the primary channel. This reverse channel can provide a simultaneous transmission path for the acknowledgment of data blocks (transmitted over the higher-speed primary channel) at up to 150 bit/s. A second use for a reverse channel is in asymmetrical transmission. As previously covered in Section 5.3.4, in this transmission technique, the reverse channel direction is made opposite to the direction of the maximum data flow.

A secondary channel, similar to a reverse channel, can be used in a variety of applications, including providing a path for a high-speed terminal and a low-speed terminal simultaneously. If a secondary channel is used as a reverse channel, it is held at one state until an error is detected in the high-speed data transmission. It is then shifted to the other state as a signal for retransmission. Another application where a secondary channel can be utilized is when a location contains a high-speed synchronous terminal and a slow-speed asynchronous terminal. If both devices are required to communicate with a similar distant location, one way to alleviate dual line requirements as well as the cost of extra modems to service both devices is by using a pair of modems that have secondary-channel capacity, as shown in Figure 5.4.3. Although a reverse channel is usable on both two-wire and four-wire telephone lines,

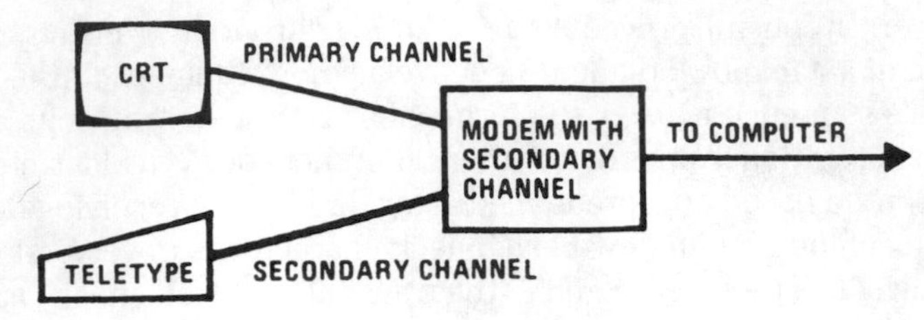

Figure 5.4.3 Secondary channel operation.

the secondary-channel technique is usable only on a four-wire circuit. A secondary-channel modem derives two channels from the same line; a wide one to carry synchronous data at speeds of 2000, 2400, 3600, 4800, or 9600 bit/s, and a narrow channel to carry slower asynchronous data. Some modems with the secondary-channel option can actually provide two slow-speed channels as well as one high-speed channel, with the two slow-speed channels being capable of transmitting asynchronous data up to a composite speed of 150 bits per second.

Error conditions and compensation

Data signals transmitted over a medium that was designed for voice conversations are often received with a degree of distortion. This distortion might be caused by several factors, including the circuit's characteristics as well as interference from other signals on other lines in the network. In general, although a large degree of distortion might not affect voice communications, a small amount of distortion can be very detrimental to data transferred by a modulated signal. As you speak on a telephone circuit, your ear and brain can effectively neutralize and compensate for any distortions present in the audible signal range because the speed of voice signal transfer is relatively slow when compared to data transmission rates.

For a modulated digital signal, even a slight degree of distortion can cause a bit position or signal level to be completely misinterpreted and thereby cause an error in the reception of the data. As data rates increase, the number of signals per unit time (baud) or the number of levels per baud, or a combination of both, must increase. Thus, reducing the signal time or reducing the differences in amplitude, frequency, or phase in a modulated signal to encode more data makes that signal more susceptible to error, to the point where even small distortions in the received signal might alter the meaning of the data.

Distortion

Signal distortion is a rather general term for an effect produced by many and various conditions. In general, as a current travels through a circuit, it encounters opposition, technically called *impedance*, which acts as a partial barrier to its flow. Some of the factors that contribute to this impedance include the circuit's resistance, capacitance, and inductance. Two of the primary causes of distortion are signal attenuation and delay.

As detailed in Section 5.2, *attenuation* is the ratio of the power of a transmitted signal to its received strength, measured in decibels as follows:

$$10 \log_{10} \frac{Power\ sent}{Power\ received}$$

To provide a valid comparison of attenuations on different circuits, a reference frequency is used for the measurements of power sent and power received. Normally, a 1004-Hz reference frequency is used for attenuation measurements in North America, while a frequency of 800 Hz is used in Europe. This reference frequency is most important because higher frequencies are subject to greater attenuation than lower frequencies. Without such a reference, comparisons would not be meaningful.

A second common cause of signal distortion is caused by the time it takes a signal to propagate to the receiver. This delay time, as expected, is known as *propagation delay*. Because different frequencies have different delay times, a term known as *envelope delay* is used as a measurements. This delay is a measurement of the differences in propagation delays of the several frequencies present in a modulated signal. As with attenuation measurements, envelope delay measurements use a reference frequency, typically between 1500 and 1900 Hz. Delay is measured in milliseconds (msec) or microseconds (μsec).

Although attenuation and delay are the two main causes of distortion, other factors, such as noise, phase jitter, amplitude jitter, phase hits, gain hits, and dropouts can cause short-term transient effects and incidental phase and amplitude modulation. *Amplitude* and *phase jitter* are terms that denote a change in amplitude and phase with respect to frequency that occurs at random. Phase hits are sudden shifts in the phase of a signal. Gain hits represent sudden increases in a received signal level. A dropout is a sudden suspension of a received signal level. Although the electromagnetic interference (EMI) resulting from the operation of electrical machinery can cause the previously mentioned problems, they are primarily caused by acts of nature, such as thunderstorms, rain, and sunspots.

Line conditioning and equalization

To increase the data-handling capacity of a telephone circuit, a reduction in amplitude attenuation and envelope delay distortion of received signals must be accomplished by techniques known as *line conditioning* and *equalization*.

Line conditioning can be obtained on leased lines by having the communications carrier add special equipment to the circuit. Such equipment includes attenuation and delay equalizers and amplifiers.

The attenuation equalizer adds a degree of signal loss to the lower frequencies of a modulated signal so that the loss throughout the transmitted band is nearly the same for all transmitted frequencies. Next, an amplifier is used to restore the signal to its original level. Because the loss is made nearly uniform by the attenuation equalizer, the amplification process becomes much easier.

To compensate for envelope delay, delay equalizers are used. These devices introduce an element of delay to some of the transmitted signals in order to adjust a uniform delay element that makes the entire signal reach the receiver at the same time. This delay equalizer is necessary because, without it, the higher frequencies would arrive at the receiver ahead of the lower frequencies.

Based on AT&T network standards, several grades of conditioning are available for selection by users of private lines. Such conditioning is only available on that type of circuit because the path is fixed and adjustments can be made to its characteristics, as previously explained. A monthly charge is assessed for line conditioning.

With the public switched network, different paths are used to establish every dialed call depending on which circuits are busy. Thus, the routing is, in effect, random and no fixed adjustment can be made to such a path.

Even when transmission is on a leased, conditioned circuit, some degree of amplitude attenuation and envelope delay distortion will be present. This is because the conditioning only ensures that these distortions do not exceed certain limits. It does not remove them entirely. Because of this fact as well as the unavailability of conditioning over the public switched network, most modems that operate at bit rates of 2400 bit/s and above are equipped with equalizers. These line equalizers can be fixed, manually adjustable, or automatic and adaptive.

An equalizer is basically an inverse filter that corrects amplitude and delay distortions, which, if uncorrected, could lead to intersymbol interference during transmission. A well-designed equalizer matches line conditions by maintaining some of the modem's electrical parameters at the widest range of marginal limits in order to take advantage of the data rate capability of the line while eliminating intersymbol interference. The design of the equalizer is critical because, if the modem operates too near or outside of these marginal limits, the transmission error rate will increase.

The faster the modem's operating speed, the greater the need for the modem to use equalizers. In addition, as the data rate of the modem increases, so does the complexity of its equalizer. Throughout the 1970s and into the 1980s, many modems with rated speeds up to 4800 bit/s designed for the dial-up network used nonadjustable, fixed equalizers that were made to match the average line conditions that have been found to occur on that type of facility. Thus, most modems with fixed or nonadjustable equalizers were designed for a normal, randomly routed call between two locations over the dial-up network. If the modem was equipped with a signal-quality light that indicated an unacceptable error rate, or if there was difficulty with the connection, the operator alleviated the problem by simply terminating the call and dialing again. This should route the connection through different points on the dial-up network.

Manually adjustable equalization is an old technique that was essentially replaced by fixed and automatic equalization. Manual adjustment is still used on a few 4800 bit/s modems used for data transfer over leased lines, with the parameters being tuned or preset at installation time, and re-equalization usually not required unless the lines are reconfigured.

At first used by 7200- and 9600-bit/s modems designed for leased-line use, automatic equalization is now used on most modems designed for operation over the public switched telephone network. With automatic equalization, a certain initialization time is required to adapt the modem to existing line conditions. This initialization time becomes important both during and after line outages because long initial equalization times can extend otherwise short dropouts unnecessarily.

Recent modem developments have shortened the initial equalization time to under 10 ms, whereas only several years ago up to 275 ms was commonly required. After the initial equalization, the modem continuously monitors and compensates for changing line conditions by an adaptive process. This process allows the equalizer to "track" the frequently occurring line variations during data transmission without interrupting the traffic flow. On one 9600-bit/s modem, this adaptive process occurs 2400 times a second, permitting the rapid recognition of variations as they occur.

Many modem manufacturers describe their equipment in terms of compatibility or equivalency with modems manufactured by Western Electric for the Bell System, prior to its breakup into independent telephone companies, or with the recommendations of the CCITT. The CCITT is based in Geneva, Switzerland, and is part of the International Telecommunications Union.

Recommendations promulgated by the CCITT at one time were primarily adopted by the postal, telephone, and telegraph (PTT) organizations that operate the telephone networks of many countries. Because of the popularity of certain CCITT recommendations, their operating characteristics have been implemented in several modems designed for use within the United States. The following examination of the operation and compatibility of the major types of Bell System and CCITT modems will be based on their primary operating rates.

300 bit/s modems (Bell System 103/113 series modems)

The Bell 103/113 series modems represent the first group of devices to gain widespread acceptance for use in transmitting data over the public-switched telephone network. Although the manufacture of stand-alone 103/113 modems ceased well over 15 years ago, their transmission capabilities are built into most modern modems designed for the North American switched telephone networks. Thus, it is important to understand the terminology associated with 103/113 modems to appreciate the capabilities of more modern modems that have a downward Bell System 103/113 compatibility.

The terminology used to describe the functions of 103/113 modems and compatible modems can lead to confusion. This confusion is a result of the terminology used in the designation of "originate" and "answer-only" operation. In this case, the terminology should not be confused with the terms to describe simplex operation because here both the originate-only and answer-only modems each have, and utilize, both a transmitter and receiver. This designation is used to describe the reversing of the transmit and receive functions between two channels at opposite ends of the telephone line. This reversal of the transmit and receive functions is accomplished within the 3000-Hz bandwidth of the telephone circuit by assigning the two channels used to different frequencies, providing filtering circuitry in the modems to separate these channels into discrete bands, and establishing design criteria. Thus, a modem that transmits in one band and receives in a second band is designated as an originate modem or an answer modem, depending on the bands used.

Bell System 103 and 113 series modems, operating at 300 bit/s, are designed so that one channel is assigned to the 1070- to 1270-Hz frequency band, while the second channel is assigned to the 2025- to 2225-Hz band. Modems that transmit in the

1070- to 1270-Hz band, but receive in the 2025- to 2225-Hz band, are designated as an originate modem, and a modem that transmits in the 2025- to 2225-Hz band, but receives in the 1070- to 1270-Hz band, is designated as an answer modem. When using such modems, their correct pairing is important because two originate modems cannot communicate with each other.

Bell System 113A modems are originate-only devices that should be normally used when calls are to be placed in one direction. This type of modem was mainly used to enable Teletype-compatible terminals to communicate with timesharing services where such terminals only originate calls. Bell System 113B modems are answer only and were primarily used at computer sites where users dial in to establish communications. Because these modems transmit and receive on a single set of frequencies, their circuitry requirements are less rigorous than other modems and their costs are thus lower. Figure 5.4.4 shows the frequency assignment for modems in this series.

Modems in the 103 series, which included the 103A, E, F, G, and J, transmit and receive in either the low or the high band. This ability to switch modes is denoted as "originate and answer" in comparison with the Bell 113A, which operates only in the originate mode, and the Bell 113B, which operates only in the answer mode.

CCITT V.21 modems

CCITT V.21 modems operate similar to Bell System 103/113 devices, using a frequency-shift keying modulation technique. Unfortunately, modems operating in accordance with the CCITT V.21 recommendation use a completely different set of frequencies for the transmission and reception of marks and spaces. Figure 5.4.5 illustrates the frequency assignment of CCITT V.21 modems. In comparing that illustration with the data contained in Figure 5.4.4, the incompatibility between Bell System 103/113 and CCITT V.21 modems becomes obvious.

Similar to Bell System 103/113 devices, V.21 modems are, for all practical purposes, no longer manufactured as stand-alone devices. Instead, a V.21 capability is built into many CCITT V-series modems that are designed to operate on European public switched telephone networks.

	ORIGINATING END	ANSWERING END
TRANSMIT	1,070 Hz SPACE 1,270 Hz MARK	2,025 Hz SPACE 2,225 Hz MARK
RECEIVE	2,025 Hz SPACE 2,225 Hz MARK	1,070 Hz SPACE 1,270 Hz MARK

Figure 5.4.4 Frequency assignment of Bell System 103/113 modems.

	ORIGINATING END	ANSWERING END
TRANSMIT	1,180 Hz SPACE 980 Hz MARK	1,850 Hz SPACE 1,650 Hz MARK
RECEIVE	1,850 Hz SPACE 1,650 Hz MARK	1,180 Hz SPACE 980 Hz MARK

Figure 5.4.5 Frequency assignment of CCITT V.21 modems.

300 to 1800 bit/s modems

Several Bell System and CCITT V Series modems operate at data rates between 300 and 1800 bit/s. A few of these modems, such as the Bell System 212A and the CCITT V.22, can operate at either of two data rates. Other modems in this speed grouping, such as the Bell System 202 and CCITT V.23, only operate at one data rate.

Bell System 212A modem. This dual-speed modem permits either asynchronous or synchronous transmission over the switched network. The 212A contains a 103-type modem for asynchronous transmission at speeds up to 300 bit/s. Frequency-shift-keyed modulation is used for 300-bit/s transmission and dibit phase-shift-keyed modulation for 1200 bit/s transmission, which permits the modem to operate either asynchronously or synchronously at this speed. The key advantage in the use of this modem is that it permits reception of two different transmission speeds. Before the terminal operator initiates a call, he selects the operating speed at the originating set. When the call is made, the answering 212A modem automatically switches to that operating speed.

During data transmission, both modems remain in the same speed mode until the call is terminated, when the answering 212A can be set to the other speed by a new call. The dual-speed 212A permits both terminals connected to Bell System 100 series modems operating at up to 300 bit/s, or terminals connected to other 212A modems operating at 1200 bit/s, to share the use of one modem at a computer site, and thus can reduce equipment requirements.

Similar to that of the 103/113-type modems, the manufacture of stand-alone 212A modems has essentially ceased. Most higher-speed modems designed for operation on North American switched telephone networks are downward compatible with the 212A modem operating characteristics. Because these operating characteristics support 300 bit/s operation, this also means that the modem is downward compatible with Bell System 100-series devices.

CCITT V.22 modems. CCITT V.22 modems operate at 1200 bit/s on the switched network or leased lines and have a fallback data rate of 600 bit/s. At 1200 bit/s, the modulation technique is the same as a 212A device. However, both the fallback data rate and the modulation techniques of V.22 and 212A modems differ at the lower data rate. The V.22 modem uses a two-phase shift keying technique when operating at 600 bit/s, whereas, the 212A uses frequency shift keying at 300 bit/s. Thus, although compatible at 1200 bit/s, these modems are totally incompatible at their lower data rates.

Like the Bell System 212A modem, the manufacture of CCITT V.22 modems as stand-alone devices has essentially ceased. Instead, modern modems designed for operation on European public switched telephone networks usually include a V.22 operating capability.

CCITT V.23 modem. The CCITT V.23 modem standard governs the transmission of data at 600 or 1200 bit/s over the switched telephone network. This standard supports both asynchronous and synchronous transmission using frequency shift keying modulation. For error control purposes, an optional 75 bit/s reverse channel is available.

At 1200 bit/s, a V.23 modem uses a frequency of 420 Hz for a mark and a frequency of 1700 Hz for a space, making it incompatible with Bell 202 devices. At the modem's lower data rate, both its frequency assignments and operating rate are incompatible with the Bell 202 series.

The CCITT V.23 modem is primarily used in Europe for Videotex operations. In this role, the 75 bit/s channel is used to transmit user selections presented as a menu on a television screen while the receiving higher-speed data channel is used to "paint" the selected picture on the screen. Because the data associated with a user's selection is only a fraction of the data required to paint a screen, the reversal of primary and reverse channels makes sense in this type of operation.

2400 bit/s modems

Modems that operate at 2400 bit/s include the Bell System 201, CCITT V.26 series, and the V.22bis modem.

Bell System 201 Series modems. The Bell 201 series modems are designed for synchronous bit-serial transmission at data rates of 2000 and 2400 bit/s. The 201A modem is an obsolete device and was phased out of use. The 201A was designed to operate over the switched network at 2000 bit/s. The 201B modem is designed for 2400 bit/s transmission over leased lines. The 201C modem is designed to operate at 2400 bit/s over the switched network or leased lines and its introduction made both the 201A and 201B obsolete. It provides increased data transfer rates over the 201A and can operate on either switched or leased lines, whereas the prior models did not have this flexibility. A more modern version of the 201C is AT&T's 2024A modem, which is compatible with the 201C.

CCITT V.26 modems. The CCITT has defined three V.26 modem recommendations, with the second and third standards known as *V.26bis* and *V.26ter*, respectively. Here, the terms *bis* and *ter* are from Latin to mean second and third. Each of the modems in the V.26 series uses a dibit phase-shift-keying modulation technique.

The V.26 modem is designed for 2400 bit/s synchronous communications on a four-wire leased line. The similar V.26bis modem is designed for operation on the switched network and includes an optional reverse channel that can be used for data transfer at 75 bit/s. The V.26ter incorporates an echo-canceling technique, which permits full-duplex 2400 bit/s operations on the switched network. Echo canceling will be described later in this section when the operation of the CCITT V.32 modem is covered.

CCITT V.22bis modem. The V.22bis recommendation doubled the data rate of V.22 modems by adding amplitude modulation on top of phase shifts, resulting in a quadrature amplitude modulation technique. Under this technique, the data stream is divided into 4-bit groupings—with 2 bits used to specify the same phase—as a V.22 or Bell System 212A, while the other 2 bits are used to change the amplitude of the carrier signal. This technique permits a full-duplex data transmission rate of 2400 bit/s over either the switched network or a leased line and supports both asynchronous and synchronous transmission.

Included in the V.22bis recommendation is a requirement for the modem to operate at a 1200 bit/s fallback data rate, which makes the modem compatible with CCITT V.22 and Bell System 212A. Unfortunately, some V.22bis modems manufactured in Europe follow the V.22 design by falling back to both 1200 bit/s and 600 bit/s. Similarly, some V.22bis modems manufactured in the United States fall back to data rates of 1200 and 300 bit/s to emulate a Bell 212A device. Because of this, care should be taken to use the 1200 bit/s fallback rate when a lower data rate becomes necessary. Otherwise, if modems are set to operate at their lowest data rate, they will be incompatible with one another.

4800 bit/s modems

Two common modems that operate at 4800 bit/s are the Bell System 208 series and the CCITT V.27 series.

Bell System 208 series modem

The Bell System 208 series modems use a quadrature amplitude modulation technique. The 208A modem is designed for either half-duplex or full-duplex operation at 4800 bit/s over leased lines. The 208B modem is designed for half-duplex operation at 4800 bit/s on the switched network. Newer versions of the 208A are marketed by AT&T as the 2048A and 2048C. Both versions are designed for use on 4-wire leased lines, with the 2048C having a start-up time less than one-half of the 2048A, which makes it more suitable for use on multidrop lines.

CCITT V.27 modems. The three members of the V.27 series operate similar to Bell System 208 modems, using a differential phase-shift-keying modulation technique. The V.27 modem is designed for full-duplex transmission at 4800 bit/s on 4-wire circuits. The V.27bis recommendation uses automatic equalization and governs half-duplex 4800 bit/s transmission with a fallback speed of 2400 bit/s on the switched network, while the V.27ter recommendation eliminates equalization.

Both Bell System 208 and CCITT V.27 modems encode three bits of data into one of eight phase angles. Unfortunately, each series of modems uses different phase angles to represent a tribit value, resulting in the modems in one series being incompatible with modems in the other series.

9600 bit/s modems

Three modems that are representative of devices that operate at 9600 bit/s are the Bell System 209A and the CCITT V.29 and V.32 modems.

Bell System 209 Series modem. The 209A modem is designed for single-channel transmission at 9600 bit/s over leased lines or for selective data rates, depending on the number of channels in increments of 2400 bit/s. The 209A modem has a built-in synchronous multiplexer that combines up to four data rate combinations for transmission at 9600 bit/s. The multiplexer combinations are shown in Figure 5.4.6.

2,400-2,400-2,400-2,400 BIT/S
4,800-2,400-2,400
4,800-4,800
7,200-2,400
9,600

Figure 5.4.6 Bell 209 A multiplexer combinations.

CCITT V.29 modem. The CCITT V.29 modem recommendation is similar to the Bell System 209 because both use a quadrature amplitude modulation technique. In this technique, the serial data stream is subdivided into groups of four consecutive bits. The first bit in each group specifies the amplitude to be transmitted, while the remaining three bits select one of eight phase changes.

The amplitude and phase angles in a V.29 modem differ from those in a Bell 209 device. Therefore, these modems are incompatible with one another.

Although the CCITT V.29 modem was originally developed as a full-duplex device for leased lines, a modified version of this modem has achieved greater utilization. The modified version, designed for half-duplex operation on switched telephone networks, is manufactured as a chip set commonly called a *data pump*. This chip set is used in asymmetrical applications as well as in facsimile. The chip set is well suited for fax operation because, after an initial "handshake" between the calling and called modems, the fax transmission is in one direction, which eliminates the effect of turnaround delays on throughput.

CCITT V.32 modem. The CCITT V.32 modem uses a modified quadrature amplitude modulation technique, which permits full-duplex 9600 bit/s transmission on the switched network. The key to the ability of the V.32 modem to obtain full-duplex transmission is an echo-canceling technique. This technique enables the modem to establish two high-speed channels in directions opposite to one another. This is accomplished by the design of the modem's receiver, which allows it to cancel out the effect of its transmitted signal. This in turn enables the modem to simultaneously distinguish its sending signal from the signal being received.

When the V.32 modem was first marketed, the cost associated with implementing echo-canceling resulted in several vendors modifying conventional V.29 modems to achieve near full-duplex transmission capability over the switched network. As covered in the previous section in this chapter, the use of asymmetrical transmission using a V.29 modem "engine" provides a near-equivalent full-duplex transmission capability.

A reduction in the cost of implementing echo-cancellation was achieved with the use of very large scale integration (VLSI) chip sets. This, in turn, significantly reduced the cost of manufacturing V.32 modems. By 1994, many V.32 modems could be obtained for under $200.

V.32 modems support operating rates of 9.6, 7.2, 4.8, 2.4, and 1.2 kbit/s. When operating at 2.4 kbit/s, a V.32 modem is V.22bis-compatible and can be V.22-compatible. When operating at 1.2 kbit/s, a V.32 modem can be compatible with a Bell System 212A, CCITT V.21, or both modems, depending on its manufacture.

14,400 bit/s modem. Until the early 1980s, the maximum practical transmission obtainable on voice grade lines was limited to 9600 bit/s. Although a few vendors introduced 14,400 bit/s modems, the throughput attainable on average quality lines was in many cases lower than that using 9600 bit/s modems. In 1984, a new generation of modems using Trellis Coded Modulation (TCM) reached the market, resulting in the availability of reliable transmission at data rates of up to 19,200 bit/s—in some instances, 24,000 bit/s—on leased lines. The CCITT developed the V.33 recommendation, which standardizes the operation method of TCM at 14,400 bit/s on leased lines, and the V.32bis Recommendation, which standardizes the TCM operation method at 14.4 kbit/s over the public switched telephone network. Because the V.33 Recommendation preceded the V.32bis Recommendation, V.33 is covered first.

CCITT V.33 modem. The CCITT V. 33 modem is designed for synchronous operations at 14,400 bit/s on leased lines. This modem incorporates Trellis Coded Modulation to overcome the limitations of conventional QAM by reducing the probability of a signal point being erroneously interpreted.

To understand the principle behind TCM, it is first necessary to examine the operation of conventional high-speed modems.

AV.29 modem operating at 9600 bit/s packs 4 bits into an amplitude and phase change. The plotting of the position of the modulated signals on the basis of all possible bit combinations results in the modem's signal constellation pattern. Figure 5.4.7 illustrates the V.29 modem constellation pattern, which has 16 distinct signal points.

If 5 bits were packed into 1 baud, 32 distinct signal points would result. The signal constellation pattern of a modem that packed 6 bits into each baud would consist of 64 points. The resulting signal constellation patterns of these modems would be 2 or 4 times as dense as the V.29 pattern, with an inversely proportional reduction in the distance between the points in the constellation. Thus, the distance between signal

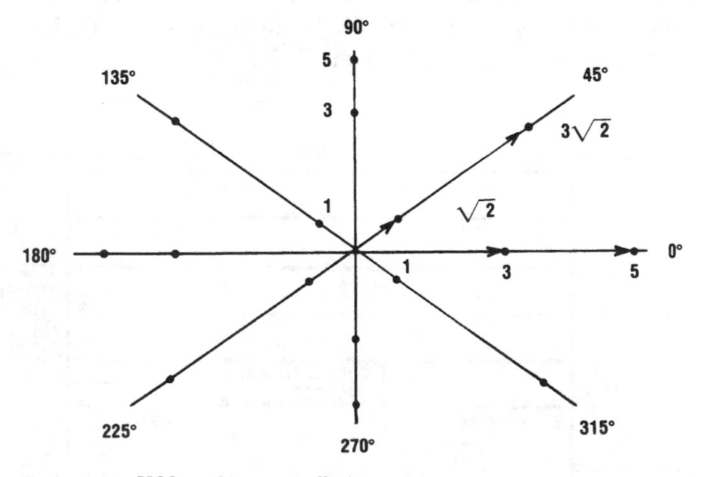

Figure 5.4.7 V.29 modem constellation pattern.

points of a QAM modem operating at 12,000 bit/s would be one-half that of a 9600-bit/s QAM modem. The distance between signal points of a 14,400-bit/s modem would be one-quarter that of a 9600-bit/s QAM modem.

The decrease in the distance between possible signal points in the modem's constellation pattern would make higher speed modems more likely to demodulate a distorted signal incorrectly. This incorrect demodulation would result from the modem matching the received signal to the nearest point in the signal constellation, then converting the point into its predefined sequence of bits. If a transmission impairment caused a change in amplitude or phase, a modem with a dense constellation pattern would be more likely to select a point in the constellation pattern incorrectly than a modem with a constellation pattern that was not as dense.

For error correction, the TCM scheme adds a redundant bit of information to the data bits. Although the additional bit results in a denser constellation pattern that would normally be more susceptible to transmission impairments, an encoder in the modem establishes redundancy and dependency between successive signal points. The encoder causes only certain signal points to be valid, enabling the decoder to detect an invalid signal point and select the closest valid point to the point received.

Figure 5.4.8. shows a block diagram of the TCM process for a modem operating at 14,400 bit/s. The serial data stream is grouped into a set of 6 bits by the serial-to-parallel converter and passed in parallel toward the signal point mapper. The first 4 bits are passed to the signal point mapper as is; bits 5 and 6 are first passed to a convolutional encoder. The encoder adds another bit of information to establish the redundancy and dependency between successive signal points. This redundant bit ensures that only certain sequences signal points are valid.

CCITT V.32bis modem. The CCITT V.32bis modem reflects the addition of echo-cancellation to the V.33 modem. This addition enables data rates of up to 14.4 kbit/s to be achieved on the public switched telephone network. The V.32bis modem is downward compatible with the V.32 modem. Also, the V.32bis modem adds support for a 12-kbit/s rate. This means that a V.32bis modem can operate at either 14.4 or 12 kbit/s when communicating with another V.32bis modem, or at 9.6 kbit/s and lower rates when communicating with a V.32 modem.

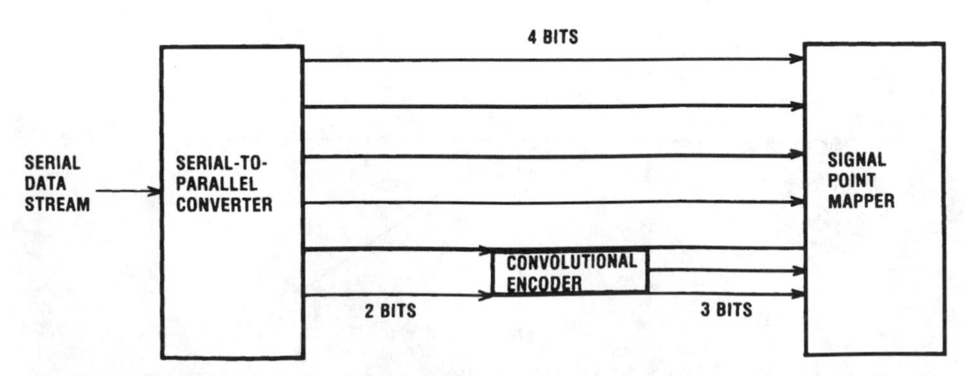

Figure 5.4.8 The Trellis-Coded modulation process.

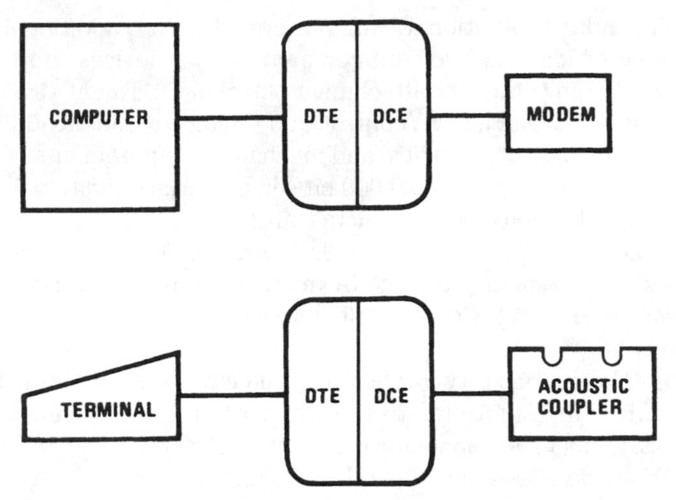

Figure 5.4.9 Typical DTE-DCE interfaces.

New modem standards

Previously proposed as the V.fast modem standard, it was renamed and adopted in 1994 as V.34. Another new modem standard is V.32terbo. Both represent extensions of V.32bis technology. Here, TCM and echo-cancellation provide a foundation for near error-free full-duplex transmission on the switched telephone network.

The V.32terbo proposed standard identifies new data signaling rates of 16.8 and 19.2 kbit/s. At those rates, the data stream is divided into groups of 7 or 8 consecutive data bits, respectively, and are encoded to generate three trellis-encoded bits.

The V.34 standard extends trellis encoding to three dimensions. It is designed to support a data rate of up to 24 kbit/s over the switched telephone network. At this writing, several vendors had delivered V.terbo-compatible modems, while V.34 modem shipments were imminent.

Modem interface and handshaking

In data communications, the term *interface* is often used to denote the type of electrical connection between the business machine and its associated data communications equipment. The business machine, which might be a computer or terminal, is called *data terminal equipment (DTE)*, and its associated data circuit-terminating equipment (DCE) can include such devices as modems and acoustic couplers. Two typical DTE-DCE interfaces are illustrated in Figure 5.4.9. Currently, three types of interfaces are used: voltage, contact closure, and current loop.

Voltage interfaces are the type most frequently used in DTE-DCE connections. Most voltage interfaces conform to the Electronic Industries Association (EIA) standard EIA-232-C/D. Here, the C/D references the EIA-232-C standard and Revision D of the standard, which occurred in January 1987. EIA-232-D is very similar to EIA-232-C, with the newer standard now including a specification for the familiar D-shaped 25-pin interface connector, whereas, EIA-232-C simply referenced the connector in an appendix. In addition, EIA-232-D provides a few additional functions that support testing of local and remote communications devices. Both interface standards specify a 25-pin connector attachment between DTE and DCE, with specified pin assignments for ground, data, control, and timing circuits. In addition, each interface standard specifies the electrical and mechanical requirements of the interface and has an operating range up to 20,000 bit/s in bit-serial operation for both the asynchronous and synchronous modes of data transfer.

Through the utilization of the EIA-232-C/D interface, standard interaction between many types of equipment produced by many different vendors becomes possible, which permits users a high degree of flexibility in selecting equipment for their specific requirements.

In the EIA-232-C/D interface, two types of connectors are required to effect a DTE-DCE connection. In conforming to the standard, a female connector is connected to the DCE, and a male connector is used with the DTE.

This standard also specifies a cable length of 50 feet or less between devices, regardless of the data transfer rate. When using this interface, the pin assignments listed in Table 5.4.2 must be used. However, unassigned pins can be used for additional functions, if they are required for operation.

Another voltage interface that closely resembles the EIA-232-C/D interface and is used as a European standard is the CCITT V.24 interface. Table 5.4.2 also shows numbered CCITT V.24 circuits. Another standard, EIA RS-449 (and its associated standards), provides capabilities beyond those of EIA-232-C. These include higher data rates and longer cable lengths. Although EIA-449 was introduced with great expectations, its utilization has been considerably below expectation. Part of the problem was probably the requirement for two connectors (37-pin and 9-pin), which added complexity and cost in designing equipment to operate in conformance with the standard.

The contact closure interface is a mechanical relay type that is used with bit-parallel data transmission. The interface, because of its electromechanical nature, operates at a low speed, usually under 100 characters per second. Presently, no standard exists that covers this type of interface. However, voltage and current limits are specified by the communications carrier when this interface is used. Within these limits, the user can determine his own electrical requirements.

A current loop interface permits the direct connection of two DTE devices at extended distances within a facility. As an example, a terminal can be connected to a computer at a distance in excess of the 50-foot EIA-232-C standard when a current loop interface is used. Presently, military interface standard MIL 188C defines one type of current loop connection.

Modem handshaking

Modem handshaking is the exchange of control signals necessary to establish a connection between a modem and a business machine at one end of a line and another modem and business machine at the other end of the circuit. The signals required to set up and terminate calls, as well as the signals used for the transmission of information, are predetermined according to the standard that the devices are designed to follow and the circuits that the devices are designed to operate with.

TABLE 5.4.2 EIA-232-C/D and CCITT V.24 Equivalent Interface Pin Assignment

Pin number	Signal direction DTE — DCE	EIA RS-232-C/D circuit	CCITT V.24 equivalent	Description
1		AA	101	Protective ground
2	→	BA	103	Transmitted data
3	←	BB	104	Received data
4	→	CA	105	Request to send
5	←	CB	106	Clear to send
6	←	CC	107	Data set ready
7		AB	102	Signal ground (Common return)
8	←	CF	109	Received line signal (detector)
9		—		Reserved for data set testing
10		—		Reserved for data set testing
11		—		Unassigned
12	←	SCF	122	Secondary received line signal detector
13	←	SCB	121	Secondary received line signal detector
14	→	SBA	118	Secondary transmitted data
15	←	DB	114	Transmission signal element timing (DCE)
16	←	SBB	119	Secondary received data
17	←	DD	115	Receiver signal element timing (DCE)
18		LL		Local loopback*
19	→	SCA	120	Secondary request to send
20	→	CD	108.2	Data terminal ready
21	←	CG/RL*		Signal quality detector/Remote loopback*
22	←	CE		Ring indicator
23	→	CH/CI	111/112	Data signal rate selector (DTE/DCE)
24	→	DA	113	Transmit signal element timing (DTE)
25		TM*		Test mode*

*Unique to EIA-232-D

As an example, examine the operation of a Bell 113-type modem as illustrated in Figure 5.4.10. The handshaking routine commences when an operator at a remote terminal dials the telephone number of a similar modem connected to a computer. At the computer site, the telephone rings and this activates a ring indicator signal (circuit CE), which is set on by the answering modem and passed to the computer. This signal informs the computer that the associated data set has received a ringing signal.

In response, the computer will send a data terminal ready (circuit CD) to its modem to enter and remain in the data mode. When set, it permits the modem to automatically answer the incoming call; when reset, it commands the modem to

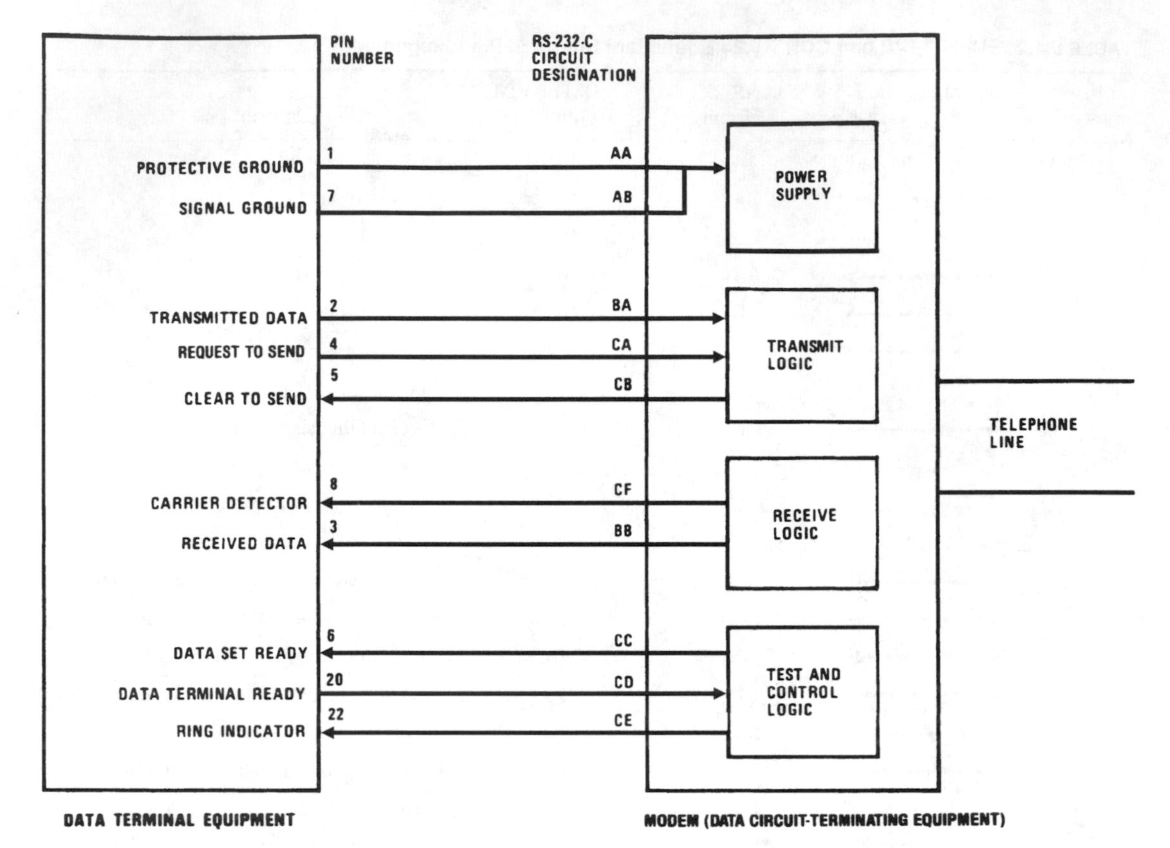

Figure 5.4.10 Bell type 113 modem interface.

disconnect the line at the end of the call. When the computer sends the data-terminal-ready signal to its modem, the modem transmits a tone signal, which to the human ear sounds like a high-pitched beep. The operator at the other end, on hearing this tone, will press the data button on her modem.

Once this data button is depressed, the originating modem will transmit a dataset-ready (circuit CC) signal to the terminal and the answering modem will send the same signal to the computer that it is interfaced to. At this point in time, both modems are placed in the data mode of operation.

In a timesharing environment, the computer normally transmits a request for user identification to the terminal operator. To do this, the computer sets request-to-send (circuit CA), which informs the terminal's modem that it wishes to transmit data. The terminal's modem will then respond with a clear-to-send (circuit CB) signal and will transmit a carrier signal. The computer's modem detects the clear-to-send and carrier-on signals and begins its data transmission to the terminal. When the computer completes its transmission, it drops the request-to-send signal and the terminal's modem then terminates its carrier signal.

Depending on the type of circuit that transmission occurs over, some of these signals might not be required. For example, on a switched-network two-wire line, the request-to-send signal determines whether a terminal is to send or receive data, whereas on a leased four-wire circuit this signal can be permanently raised. For further information, refer to specific vendor literature or appropriate Bell System technical reference publications. A list of some of the modem-handshaking control signals and their functions is presented in Table 5.4.3.

Modem features

Over the last several years a number of features have been incorporated into modems that have increased their operational flexibility. Among these features are multiport capability, multiple-speed selection, voice/data capability and intelligent operations. Multiport modems are primarily designed for operation on leased lines because the sharing of a line's capacity is difficult to accomplish on a dynamic dial-up basis. A modem with a multiport capability offers a function similar to that provided by a multiplexer, which is a device that combines several data streams into one composite (higher-speed) data stream for economies in transmission. At the receiving end, another multiplexer then breaks out the composite data stream into its original parts. In fact, multiport modems contain limited-function multiplexers that provide users with the capability to transmit more than one synchronous data stream over a single transmission line, as illustrated in Figure 5.4.11.

In comparison with conventional multiplexers, the limited-function multiplexer used in a multiport modem combines only a few high-speed synchronous data streams, whereas multiplexers can normally concentrate a mixture of asynchronous and synchronous, high- and low-speed data streams.

Multiple speed selection. For data communications networks that require the full-time service of dedicated lines, but also require access to the switched network if the dedicated circuits should fail or degrade to the point where they cannot be used, dial backup capability for the modems becomes necessary. Because transmission over dedicated lines usually occurs at a higher data rate than one can obtain over the

TABLE 5.4.3 Modem Handshaking Signals and Their Functions

Control signal	Function
Transmit data	Serial data sent from device to modem
Receive data	Serial data received by device
Request to send	Set by device when user program wishes to transmit
Clear to send	Set by modem when transmission may commence
Data set ready	Set by modem when it is powered on and ready to transfer data. Set in response to data terminal ready
Carrier detect	Set by modem when signal present
Data terminal ready	Set by device to enable modem to answer an incoming call on a switched line
	Reset by device to disconnect call
Ring indicator	Set by modem when telephone rings

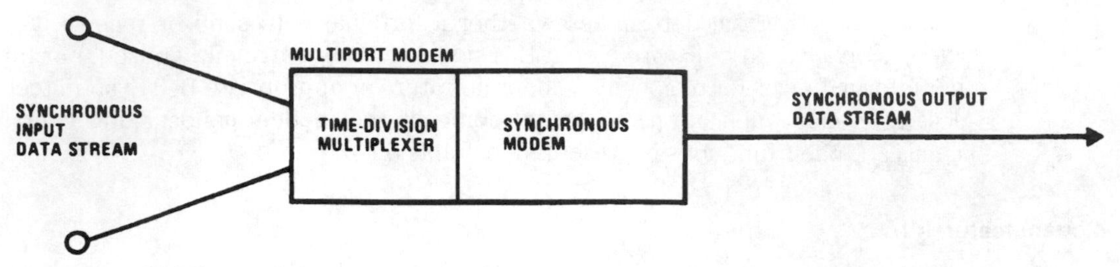

Figure 5.4.11 Multiport modem.

switched network, one method to facilitate dial backup is through switching down the speed of the modem. Thus, a multiple-speed modem that is designed to operate at 9600 bit/s over dedicated lines can be switched down to 7200 or 4800 bit/s for operation over the dial-up network until the dedicated lines are restored.

Voice/data capability. Some high-speed modems can be obtained with a voice/data option that permits a specially designed telephone set, commonly called a voice adapter, to provide the user with a voice communications capability over the same line that is used for data transmission. Depending on the modem, this voice capability can be either alternate voice/data or simultaneous *voice/data*. Thus, the user can communicate with a distant location at the same time data transmission is occurring or he can transmit data during certain times of the day and use the line for voice communications at other times. Voice/data capability can also be used to minimize normal telephone charges when data transmission sequences require voice coordination.

Self-testing features. Many low-speed and most high-speed modems have a series of test switches that may be used for local and remote testing of the modem and line.

In the local, or analog, test mode, the transmitter output of the modem is connected to the receiver input, disconnecting the customer interface from the modem. A built-in word generator is used to produce a stream of bits, which is checked for accuracy by a word comparator circuit, and errors are displayed on an error lamp as they occur. The local test is illustrated in Figure 5.4.12.

To check the modems at both ends, as well as the transmission medium, a digital loop-back self-test can be used. To conduct this test, personnel must normally be at each modem to push the appropriate test buttons, although a number of vendors have introduced modems that can be automatically placed into the test mode at the distant end when the central-site modem is switched.

In the digital loop-back test, the modem at the distant end has its receiver connected to its transmitter, as shown in Figure 5.4.13. At the other end, the local modem transmits a test bit stream from its word generator and this bit stream is looped back from the distant end to the receiver of the central-site modem, where it is checked by the comparator circuitry. Error lamps indicate if either the modems or the line is at fault.

The analog loop-back self-test should normally be used to verify the internal operation of the modem; the digital loopback test will check both modems and the carrier. Although analog and digital tests are the main self-tests built into modems, several vendors offer additional diagnostic capabilities that might warrant attention.

Intelligent operations. Although Hayes Microcomputer Products is generally attributed with having developed the concept of incorporating intelligence into a modem, in actuality Bizcom was the first vendor to obtain a patent on the technology. The technology has been licensed to many other vendors, of which Hayes Microcomputer Products is most commonly known, because of the wide acceptance of its modems, which represent de facto intelligent-operation standards.

Intelligent modems can analyze and execute commands sent to them and, with the appropriate software, permit users to develop predefined dialing directories that automatically set the communications parameters. This enables the modems to access various information utilities and computers and dial distant devices. Because of the importance of Hayes compatibility in the intelligent modem marketplace, the operation of Hayes-compatible modems is closely examined.

The Hayes Command set

Almost all current personal computer communications programs are written to operate with the Hayes command set; therefore, the degree of Hayes compatibility supported by a non-Hayes modem determines which communications software can be used with that modem. In some cases, non-Hayes modems can work as well as or

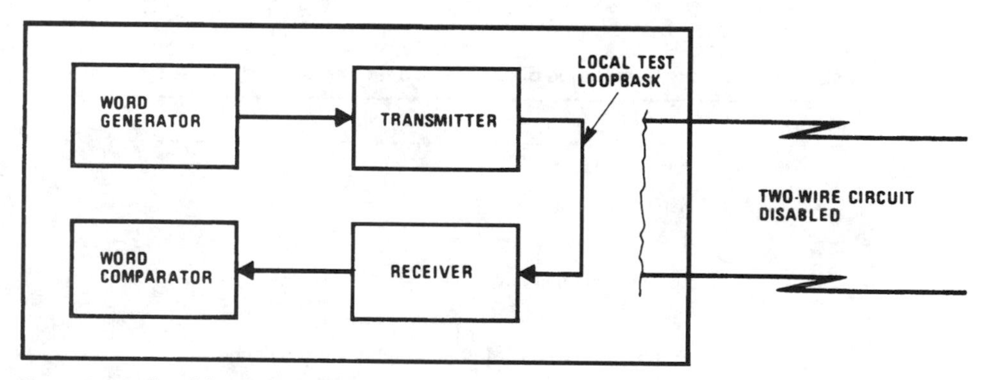

Figure 5.4.12 Local (analog) testing.

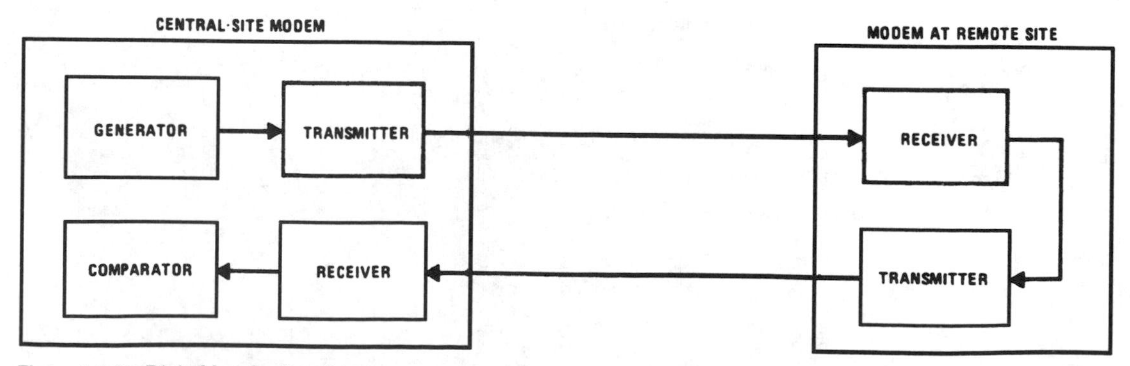

Figure 5.4.13 Digital loopback self-test.

even better than a Hayes modem if the software supports the non-Hayes features of that device. In other cases, if one or more Hayes Smartmodem features are omitted, users might have to reconfigure their communications software to work with a non-Hayes modem; this, however, usually results in degraded performance.

The Hayes command set is a basic set of commands and command extensions. The basic commands, such as placing the modem off hook, dialing a number, and similar operations, are common to all Hayes modems. The command extensions, such as placing a modem into a specific operating speed, are applicable only to modems built to transmit and receive data at that speed.

Commands are initiated by transmitting an attention code to the modem, followed by the appropriate command or set of commands. The attention code is the character sequence AT, which must be specified as all uppercase or all lowercase letters. Because all command lines must include the prefix code AT, many modem manufacturers have denoted their modems as Hayes AT-compatible.

The command buffer in a Hayes Smartmodem holds 40 characters, which permits transmission of a sequence of commands to the modem on one command line. This 40-character limit does not include the attention code, nor does it include spaces in a command line that make the line more readable. Table 5.4.4 lists the major commands included in the basic Hayes command set.

TABLE 5.4.4 Major Commands of a Hayes Command Set

Command	Description
A	Answer call
A/	Repeat last command
C	Turn modem's carrier on or off
D	Dial a telephone number
E	Enable or inhibit echo of characters to the screen
F	Switch between half- and full-duplex modem operation
H	Hang up telephone (on hook) or pick up telephone (off hook)
I	Request identification code or request checksum
M	Turn speaker off or on
O	Place modem online
P	Pulse dial
Q	Request modem to send or inhibit sending of result code
R	Change modem mode to originate-only
S	Set modem register values
T	Tone dial
V	Send result codes as digits or words
X	Use basic or extended result code set
Z	Reset the modem

The basic format for transmitting commands to a Hayes-compatible intelligent modem is as follows:

```
AT Command [Parameter(s)] Command [Parameter(s)]..Return
```

Each command line includes the prefix AT, followed by the appropriate command and the command's parameters. The command parameters, which are usually the digits 0 or 1, define a specific command state. For example, an HO command tells the modem to hang up or disconnect a call, and HI tells the modem to go off hook (i.e., the action that occurs when the telephone handset is lifted). Because many commands do not have parameters, those terms are enclosed in brackets to illustrate that they are optional. A number of commands can be included in one command line as long as the number of characters does not exceed 40, which is the size of the modem's command buffer. Finally, each command line must be terminated by a carriage return character.

The use of the Hayes command set is illustrated when, for example, someone automatically dials. New York City information. First, the modem must be told to go off hook, which is similar to manually picking up the telephone handset. Then, the modem is notified of the type of telephone system being used (i.e., pulse or tone) and the telephone number to dial. Thus, if a terminal or personal computer is connected to a Hayes-compatible modem, the following commands are sent to the modem:

```
AT H1
AT DT1, 212 555 1212
```

In the first command, the 1 parameter used with the H command places the modem off hook. In the second command, DT tells the modem to dial (D) a telephone number using tone (T) dialing. The digit 1 is included in the telephone number when the call is long distance, and the comma between the long-distance access number (i.e., 1) and the area code (i.e., 212) creates a two-second pause before the modem dials the area code. This two-second pause ensures that the long-distance dial tone is received before the area code number is dialed.

Because a Smartmodem automatically goes off hook when a number is dialed, the first command line is not actually required and is typically used for receiving calls. Moreover, the type of dialing need not be specified in the second command line if a previous call was made, because the modem uses the last type specified. Although users with only pulse dialing availability must specify P in the dialing command when using a Hayes Smartmodem, several vendors now offer modems that can automatically determine the type of dialing facility to which the modem is connected and select the appropriate dialing method. Other non-Hayes modems automatically attempt a tone dial and, if unsuccessful, redial using pulse dialing.

Result codes. The Smartmodem's response to commands is known as *result codes.* The Q command with a parameter of 1 enables the modem to send result codes when command lines are executed, while a parameter of 0 inhibits the modem from responding to the execution of each command line.

If the result codes are enabled, the V command can be used to determine the format of the result codes. When the V command is used with a 0 parameter, the modem transmits result codes as digits, and a 1 parameter causes the result codes to be transmitted as words. Table 5.4.5 lists the basic result codes set of the Hayes Smartmodem 1200. The commands for the use of these result codes are written as follows:

```
AT Q0
AT V1
```

The first command, AT Q0, causes the modem to respond to commands by transmitting result codes after each command line is executed. The second command, AT V1, causes the modem to transmit each result code as a word code. As shown in Exhibit 2, this causes the modem to generate the word code `CONNECT` when a carrier signal is detected. If the command `AT V0` was sent to the modem, a result code of 1 would be transmitted by the modem.

By examining both the result codes issued by a Smartmodem and the commands that generated them, software can be developed to perform such operations as redialing a previously dialed telephone number to resume transmission in the event of interruption and automatically answering incoming calls when a ring signal is detected.

Modem registers. A third key to the degree of compatibility between non-Hayes devices and Hayes Smartmodems is the number, use, and programmability of registers contained in the modem. A Hayes Smartmodem contains a series of programmable registers that govern the function of the modem and the operation of some of the commands in the modem's command set. Table 5.4.6 lists the functions of the first 12 registers built into the Hayes Smartmodem 1200 and includes the default value of each register and the range of settings permitted. These registers are known as *S registers* because they are set with the S command in the Hayes command set. In addition, the current values of each register can be read under program control, thus enabling software developers to market communications programs that permit the user to easily modify the default values of the modem's S registers.

Almost all intelligent modems are fully compatible with the functions of the first 12 S registers listed in Table 5.4.6. However, many vendors extended the use of S registers to define other modem operations. The result: some third-party products have 100 or more S registers, many of which are used to perform functions not covered by the original series of basic AT commands.

To illustrate the advantages of reading and resetting the values of the modem's S registers, the length of time a Smartmodem must wait for a dial tone before going off hook and dialing a telephone number is provided as an example. Because the dial

TABLE 5.4.5 Smartmodem 1200 Basic Result Codes Set

Digit Code	Word Code	Meaning
0	OK	Command line executed without errors
1	CONNECT	Carrier detected
2	RING	Ring signal detected
3	NO CARRIER	Carrier signal lost or never heard
4	ERROR	Error detected in the command line

TABLE 5.4.6 S Register Control Parameters

Register	Function	Default value	Range
S0	Ring to answer on		0..255
S1	Counts number of rings	0	0..255
S2	Escape code character	ASCII 43	ASCII 0..127
S3	Carriage return character	ASCII 13	ASCII 0..127
S4	Line feed character	ASCII 10	ASCII 0..127
S5	Backspace character	ASCII1 8	ASCII 0..127
S6	Dial tone wait time (seconds)	2	2..255
S7	Carrier wait time (seconds)	30	1..255
S8	Pause time caused by comma (seconds)	2	0..255
S9	Carrier detect response time (1/10 second)	6	1..255
S10	Time delay between loss of carrier and hang up (1/10 second)	7	1..255
S11	Tone duration and spacing time (milliseconds)	70	50..255

tone wait time is controlled by the S6 register, a program that allows users to change this wait time might first read and display the setting of this register during the program's initialization. The modem can read the S6 register when the program sends the following command to the modem:

```
AT S6?
```

The modem's response to this command would be a value between 2 and 255; that is, the number of seconds that the modem will wait for dial tone. Assuming that the user desires to change the waiting period, the communications program would then transmit the following command to the modem, where *n* would be a value between 2 and 255:

```
AT S6 = n.
```

Compatibility. Full compatibility with a Hayes modem requires command set, result codes, and modem register compatibility. Of the three, modem register compatibility is usually the least important. In fact, many users prefer to consider only command set and response codes compatibility when acquiring intelligent modems.

Register compatibility can be omitted from consideration because many non-Hayes modem vendors manufacture compatible modems using the default values of the Hayes Smartmodem registers. In this manner, manufacturers can avoid building the S registers into their modems and can reduce their size, complexity, and cost. Thus, if the default values of the S registers are sufficient for the user, and the modem under consideration is both command-set and result-code compatible, the issue of register compatibility can usually be eliminated as an acquisition issue.

Notice that the commands listed in Table 5.4.4 are applicable to all modern modems. In addition to these basic commands, Hayes developed a series of extended commands, represented by letters prefixed by an ampersand (&). Some of these extended commands are applicable to most Hayes-manufactured modems. Many other of these commands are applicable only to specific modems or to modems with a specific built-in operational capability.

Examples of extended commands applicable to most Hayes modems include: `&Zn = x`, used to store up to *n* strings in a modem's nonvolatile memory for later dialing; `&T1`, which places a modem into its analog loopback mode of operation to permit the echo of characters keyed at a terminal connected to the modem; and `&W`, which stores the active modem configuration in memory. Extended commands that differ between modems include commands to: place a modem into a specific type of error detection and correction mode of operation; initiate a specific type of data compression; initiate another feature applicable only to certain modems. In addition to differences in the support of extended commands among Hayes modems, there are also differences between the manner in which the operations of extended commands are defined by Hayes and by other vendors. For example, some vendors require setting an S register to a specific value to initiate a certain type of error detection and correction. By comparison, Hayes modems would use an extended command prefixed by an ampersand. Also, other vendors use a different extended command code. Thus, it is important to obtain communications software that supports the extended commands of the modem you intend to use.

The MNP protocol. Among the first modems to offer an error detection and correction feature were those manufactured by Microcom, Inc. This company created a revolution in modem technology by developing a modem protocol known as *MNP (Microcom Networking Protocol)*.

MNP is a communications protocol built into MNP-compatible modems that support interactive and file transfer applications. In developing MNP, Microcom structured it to accommodate ongoing changes in its implementation. To accomplish this, the protocol's major functions are divided into classes. When one MNP modem communicates with another MNP modem, the two devices negotiate with each other to operate at the highest mutually supported class of MNP service.

Table 5.4.7 summarizes the features associated with available MNP classes. Until 1990, Microcom licensed MNP only through Class 5 to other modem manufacturers. In that year, the company began to offer full MNP licenses. Thus, an MNP-compatible modem, while compatible with all other MNP modems, might actually be compatible only with a subset of available MNP classes, unless a third-party vendor obtained a full license and incorporated all MNP classes into its product.

V.42 recommendation. Despite the impressive popularity of MNP error detection and correction, Microcom's scheme remains a de facto standard. Instead, in 1990, the CCITT promulgated its V.42 Recommendation.

This standard does not address modulation techniques, as other CCITT V series do. Instead, V.42 defines a protocol where modems form data into blocks for transmission. For error detection, the modems generate and add a CRC to each block. The data-block flow is in accordance with a "link-access procedure" (LAP)—different from the data flow under the MNP protocol. But, in deference to the MNP modems' large installed base, V.42 supports MNP error detection and correction as a secondary stan-

TABLE 5.4.7 MNP Classes

Class	Description of functions performed
1	Asynchronous half-duplex byte-oriented transmission providing an approximate 70 percent efficiency. A 2.4-kbit/s modem using this MNP class will attain a throughput of 1.69 kbit/s.
2	Asynchronous byte-oriented full-duplex transmission providing an approximate 84 percent efficiency. An MNP Class 2, 2.4-kbit/s modem will attain a 2.0 kbit/s throughput.
3	Asynchronous start/start bits are stripped. This enables, between modems, synchronous bit-oriented full-duplex transmission, providing about 108 percent efficiency. This permits a 2.4-kbit/s modem to attain a throughput of about 2.6 kbit/s.
4	Adds "adaptive packet assembly" (based on the number of retransmission requests, packet sizes are dynamically adjusted). Also adds "data phase optimization" (providing a protocol-overhead-reducing mechanism). The resulting efficiency is about 120 percent, enabling a 2.4-kbit/s modem to attain a throughput of 2.9 kbit/s.
5	Adds data compression—average compression ratio of 1.6 to 1—to Class 4 service. This means that every 16 characters are compressed into 10 characters for transmission, increasing the protocol efficiency to about 200 percent, which enables a 2.4-kbit/s modem to attain a throughput of about 4.8 kbit/s.
6	Adds "universal link negotiation" and "statistical duplexing" to Class 5. Universal link negotiation permits MNP modems to start operation with a common low-speed modulation method, then negotiate the use of an alternative higher-speed modulation method. At the end of a successful Class 6 link negotiation, the modem pair operates at 9.6 kbit/s with V.29 technology. With statistical duplexing, user traffic patterns are monitored, enabling the dynamic allocation of V.29 half-duplex transmission. Under Class 6's 9.6-kbit/s operation, MNP renders an average throughput approaching 19.2 kbit/s.
7	Adds a data compression capability to MNP, based on a Huffman statistical encoding technique. Under this class, a 2.0:1 to 3.0:1 compression ratio can be reached, increasing throughput to two to three times a modem's operating rate.
8	No longer available.
9	Adds V.32 modulation support to Class 7. This enables a throughput of up to three times a V.32 modem's full-duplex 9.6-kbit/s operating rate.

dard. This works as follows: A V.42-compatible modem first tries to use LAP to communicate in its error-free mode to another modem. If the other modem does not support V.42, the first (V.42) modem then tries to communicate under MNP error control.

Data compression. The CCITT's V.42bis Recommendation, a data compression scheme, entails V.42 protocol use. However, V.42bis does not support any other compression method. Theoretically, this means that, for a V.42bis modem to function in a compressed fashion, it must communicate only with another V.42bis modem. However, most V.42bis modems support MNP through Class 5, therefore permitting both data-compressed and error-controlled communications with another MNP modem.

The data compression technique is the main difference between V.42bis and MNP Class 5 or 7. V.42bis uses "Lempel-Ziv," which is capable of operating on either single characters or strings. In contrast, the MNP compression method operates primarily on single characters, while being somewhat less efficient. With data highly susceptible to compression, a V.42bis modem with V.32 modulation operating at 9.6 kbit/s can attain a throughput of up to 28.4 kbit/s.

Acoustic couplers

Unlike a conventional modem, which requires a permanent or semipermanent connection to a telephone line, an acoustic coupler permits data transmission to occur through the handset of an ordinary telephone. Similar in function to a modem, an acoustic coupler is a device that accepts a serial asynchronous data stream from data terminals, modulates that data stream into the audio spectrum, and then transmits the audio tones over a switched or dial-up telephone connection.

Acoustic couplers are equipped with built-in cups into which a conventional telephone handset is placed. Through the process of acoustic coupling, the modulated tones produced by the acoustic coupler are directly picked up by the attached telephone handset. Likewise, the audible tones transmitted over a telephone line are picked up by the telephone earpiece and demodulated by the acoustic coupler into a serial data stream that is acceptable to the attached data terminal.

Acoustic couplers use two distinct frequencies to transmit information, while two other frequencies are used for data reception. One of the frequencies from each pair is used to create a mark tone, which represents an encoded binary one from the digital data stream; another from each pair of frequencies generates a space tone, which represents a binary zero. This utilization of frequencies permits full-duplex transmission to occur over the two-wire switched telephone network.

Because acoustic couplers enable any conventional telephone to be used for data transmission purposes, the coupler does not have to be physically wired to the line. Thus, it permits considerable flexibility in choosing a terminal working area, which can be anywhere a telephone handset and standard electrical outlet are located. Acoustic couplers are manufactured as both separate units and as built-in units to data terminals, as shown in Figure 5.4.14.

Acoustic couplers are normally used to permit portable terminals and personal computers to communicate with electronic mail networks, information utilities, and data processing facilities. Although couplers have to a large degree been replaced by the use of modems that contain jacks into which modular telephone plugs can be inserted, they are still used by many persons who travel. This is because most hotels and public telephones do not have modular telephone connectors, which excludes the use of a modem with a modular jack.

Manufacturers of acoustic couplers design them to be either Bell System or CCITT V Series compatible. In the United States, most couplers are either Bell System 103 or 212A compatible; in Europe, couplers are designed to be V.21 or V.22 compatible.

Operations. When a terminal is attached to or has a built-in acoustic coupler, and the operator wishes to send data to a computer, she merely dials the computer's tele-

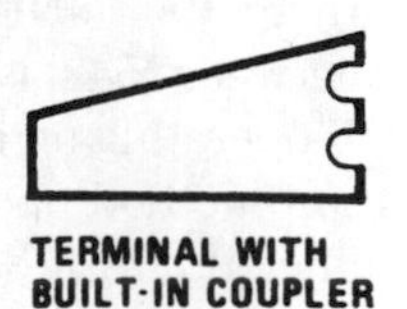

Figure 5.4.14

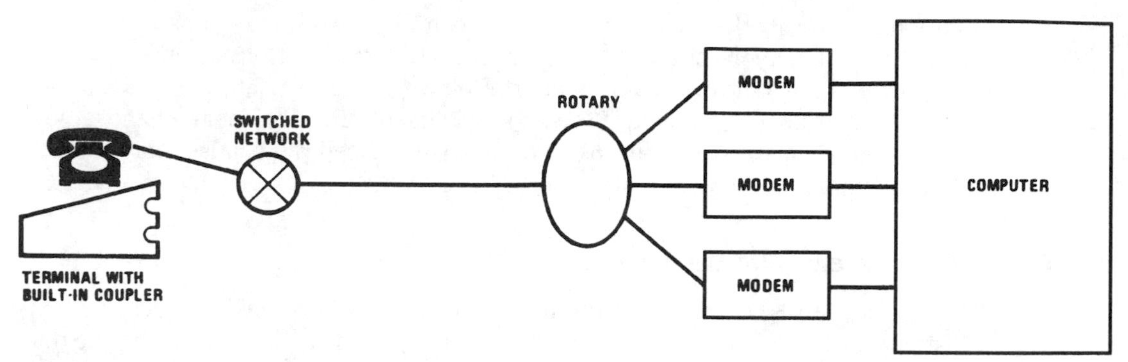

Figure 5.4.15 Network access in a timesharing environment.

phone access number and, on establishing the proper connection (hearing a high-pitched tone), places the telephone headset into the coupler.

Although terminal usage varies widely because applications are numerous, the prevalent utilization of acoustic couplers is for obtaining access to timesharing networks. In a timesharing network, a group of dial-in computer telephone access numbers may be interfaced to a rotary switch that enables users to dial the lowest telephone number of the group and automatically "step" past or bypass busy numbers. Each telephone line is then connected to a modem on a permanent basis, and the modem in turn is connected to a computer port or channel. An answering device in each modem automatically answers the incoming call, establishing a connection from the user who dialed the number to the computer port, as shown in Figure 5.4.15.

In contrast to when modems are permanently connected to telephone lines, the telephone can be used for conventional voice communications when not connected via the acoustic coupler. To obtain the use of a line for voice communications when that line is connected to a modem, a device known as a *voice adapter* must be installed.

One disadvantage associated with the use of acoustic couplers is a deficiency of transmission rates, compared to rates that can be obtained by using regular modems. Because of the properties of carbon microphones in telephone headsets, the frequency band that can be passed is not as wide as the band that modems can pass. Although typical data rates of acoustic couplers vary between 110 and 300 bit/s, some units do permit transmission at 450, 600, and even 1200 bit/s. For usage with low-speed terminals, the acoustic coupler can be viewed as a low-cost alternative to a modem, while increasing user-transmission-location flexibility.

Problems in usage. One possible cause of errors in the transmission of data can be ambient noise leaking into the acoustic coupler. The coupler should be kept far away from the terminals to reduce noise levels. Similarly, if the terminal is not in use, remove the telephone from the coupler because the continuous placement of the headset in that device can cause crystallization of the speaker and receiver elements of the telephone. This will act to reduce the level of signal strength. Another item that might warrant user attention is the placement of a piece of cotton inside the earpiece, behind the receiver of the telephone. Although the placement of cotton at this location is normally done by most telephone companies, this should be checked because the cotton keeps speaker and receiver noise from interfering with each other and prevents transmitted data from interfering with received data.

One easily resolved problem is the poor placement of the telephone handset into the coupler. On many occasions, users have hastily placed the handset only partially into the coupler, a situation that reduces signal level.

5.5 Digital Transmission and Service Units

In the early 1970s, carriers began offering communications networks designed exclusively for the transmission of digital data. Specialized carriers, including the now defunct Datran, performed a considerable service to the information processing community through their pioneering efforts in developing digital networks. Without their advancements, major communications carriers might have delayed the introduction of all-digital service media.

In December 1974, the FCC approved the Bell System's Dataphone digital service (DDS), which was shortly thereafter established between five major cities. Since then, the service has been rapidly expanded to the point where more than 100 cities have been added to the DDS network. Western Union International set another milestone in February 1975 by applying to the FCC for authority to offer its international digital data service (IDDS) from New York to Austria, France, Italy, and Spain, resulting in digital data transmission by major carriers becoming a reality.

Since the introduction of DDS, AT&T has considerably expanded its offerings of digital transmission services. The new offerings include a variety of leased-line services under the label, Accunet Spectrum of Digital Services (ASDS). These services range from fractional T1 lines operating at $n \times 56$ or $n \times 64$ kbit/s ($n = 1$ to 24) to T1 service operating at 1.544 Mbit/s, to T3 service operating at approximately 45 Mbit/s (actually, 44.736), and fractional T3. In addition, AT&T offers switched 56 and 384 kbit/s digital transmission.

The other major communications vendors, such as MCI and Sprint, also offer a full range of digital transmission services.

Comparison of facilities

When analog, or voice-grade, transmission facilities are utilized, the data stream can be modulated into two distinct amplitudes, frequencies, or phases, each of which represents marks and spaces, or binary ones and zeros, respectively. In so-called voice-grade telephone circuits, the usable bandwidth is approximately 3000 Hz and the power transmitted at the highest frequency is significantly lower than the power transmitted at the lowest one. This bandwidth limitation not only causes a loss of distinction between the vocal "s" and "f" sounds, but also limits the amount of information that can be transmitted via modulated analog signals.

In the switched telephone network, the characteristics of a data path cannot be exactly determined because each new call may take a different set of links. Over long distances, multiple voice-grade lines were often combined into 3600 channels of 4000 Hz each and sent by microwave transmission. In this combining, or multiplexing process, an original 2225-Hz signal could be shifted to 19,225 Hz for transmission, and end up as a 2220-Hz or 2230-Hz signal at the receiver. This transmission over the switched network normally occurs at data rates up to 19,200 bit/s. By obtaining a leased line, using automatic equalization, and conditioning the line, data rates of 24,000 bit/s or greater can be readily achieved.

With voice-grade type of analog transmission, the data is easily amplified, although any noise or distortion along the path is also amplified. In addition, the data signals become highly attenuated, or weakened, by the telephone characteristics originally geared to voice transmission. For the analog transmission of data, expensive and complex modems must be used at both ends of the link to shape (modulate) and reconstruct (demodulate) the digital signals.

When digital transmission facilities are used, the data travels from end to end in its near-original form, with the digital pulses regenerated at regular intervals as simple values of one and zero. Data and/or channel service units are used at both ends of the link to condition the digital signals for digital transmission. (Ref. Section 4.11.)

DDS is strictly a synchronous facility providing full-duplex, point-to-point and multipoint service limited to speeds of 2, 4, 4.8, 9.6, 19.2, and 56 kbit/s.

Terminal access to the DDS network is accomplished by means of a data service unit, which alters serial unipolar signals into forms of modified bipolar signals for transmission, and returns them to serial unipolar signals at the receiving end. The various types of service units will are covered in detail later in this section.

Digital signaling

It is important to understand what modified bipolar signaling is and why it is necessary because this form of signaling is the foundation of digital data transmission. Using the top portion of Figure 5.5.1, first examine a serial, unipolar signal commonly produced by such devices as teleprinters and other data terminals. In normal return-to-zero bipolar signaling, a binary zero is transmitted as zero volts and a binary one as either a positive or negative pulse, opposite in polarity to the previous binary one. This alternation of positive and negative pulses produces an alternating polarity, which returns the voltage sum to zero and avoids any undesirable direct current buildup, as illustrated in the middle portion of Figure 5.5.1.

Because DDS incorporates its own network codes to include such information as zero suppression, idle, and out-of-service data, the original bipolar format will be violated when such network control information is transmitted. Such a bipolar violation occurs when the alternate polarity rule (previously covered) is violated. An example of this violation is when the last pulse transmitted is sent as a positive pulse and the next pulse is transmitted in a similar manner. In the lower portion of Figure 5.5.1, a bipolar sequence containing bipolar violations is illustrated. Here, the letter B in a pulse indicates that the polarity was determined by the bipolar rule while the letter V indicates that the polarity was formed in violation of the bipolar rule.

One of the key aspects of the insertion of DDS network control codes that modify and violate the bipolar signal is the use of a zero suppression code. Because a long succession of binary zeros would not provide the necessary transitions to maintain proper timing recovery, strings of more than six zeros are replaced with zero sup-

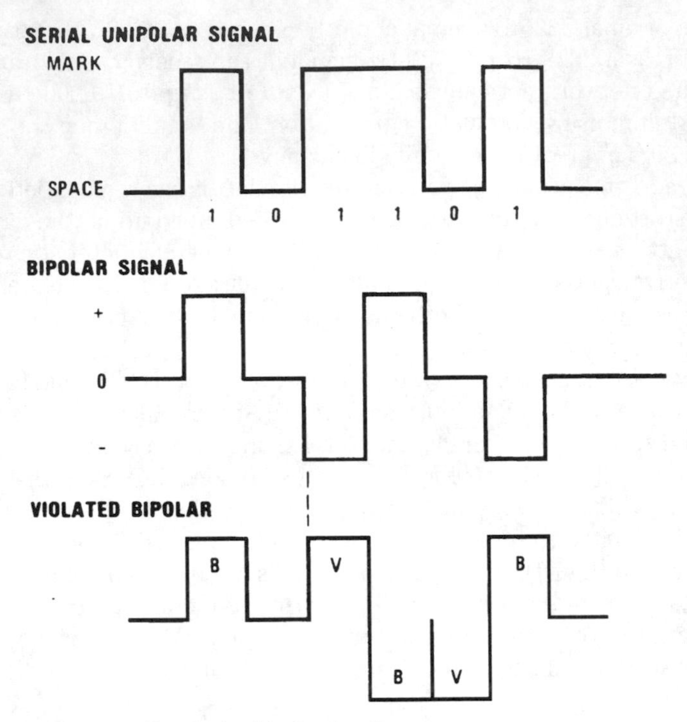

Figure 5.5.1 Developing bipolar signaling.

pression codes to maintain synchronization. Because the unrestricted insertion of violations in the DDS pulse stream would be meaningless, time slots are reserved prior to a violation for application of a binary pulse or no-pulse in such a way that successive violations alternate in polarity.

Figure 5.5.2 shows how a bipolar signal undergoes violation insertion. In this example, a zero suppression sequence is inserted into the binary channel signal. The resultant signal, as shown, returns the voltage sum to zero.

For digital transmission, precise synchronization is the key to success of an all-digital network. It is essential that the data bits be generated at precise intervals, interleaved in time, and read out at the receiving end at the same interval to prevent loss or garbling of data sequences. To accomplish the necessary clock synchronization on the Bell digital network, a master reference clock is used to supply a hierarchy of timing in the network. Should a link to the master clock fail, the nodal timing can operate independently and retain synchronization for up to two weeks without excessive slippage during outages.

Service units and network integration

In covering the characteristics of service units that interface terminals to digital networks, it is important to understand the functional differences between channel service units (CSU) service units (DSU).

The Data Service Unit (DSU) operates from 2.4 through 56 kbit/s for DDS transmission. It can function at 64 kbit/s to access a fractional T1 line. DDS and FT1 operation require both a DSU and a Channel Service Unit (CSU). The CSU is positioned between the DSU and the digital line, with the DSU between the user's terminal equipment and the CSU. The CSU's functions are: physically terminate the line, amplify the signal, and initiate remote loopback in response to codes that it receives over the line. The DSU's functions are: convert the unipolar signals that are received from terminal devices into bipolar (for transmitting over the digital network); supply timing recovery; control signaling; and provide synchronous sampling.

When DDS service was first offered, a DSU could be obtained only from AT&T under tariff. In contrast, a CSU was operated and owned by the carrier. Following FCC deregulation in 1983, users could buy both CSUs and DSUs from AT&T or other vendors. Even though some vendors offer separate DSUs and CSUs, most of the units are now supplied combined, integrating both of them into one common housing. This combined product is now commonly referred to as a *DSU*, to differentiate it from the CSU that is designed to support the termination of a T1 line.

Examine the operation of the combined DSU/CSU used on DDS and 64 kbit/s digital lines. A good way is to explore the operation of each device by using the Bell System 550A-type CSU and 500A-type DSU as models. Although they are no longer manufactured as separate devices, an overview of their operation will explain their functions as what are the two modules in the self-contained housing now called a *DSU*.

Figure 5.5.3 contains a simplified schematic diagram of the Bell System 50OA-type DSU and the 550Atype CSU. When a channel service unit is installed, the customer must supply all of the transmit logic, receive logic, and timing recovery in order to use that device, whereas the DSU performs these functions.

The CSU is devoid of circuitry necessary to provide timing recovery and detect, or generate, DDS network control codes, which becomes the customer's responsibility when this device is used. Nominal 50% duty cycle bipolar pulses are accepted from the customer on the transmit and receive data leads. The pulses, synchronized with

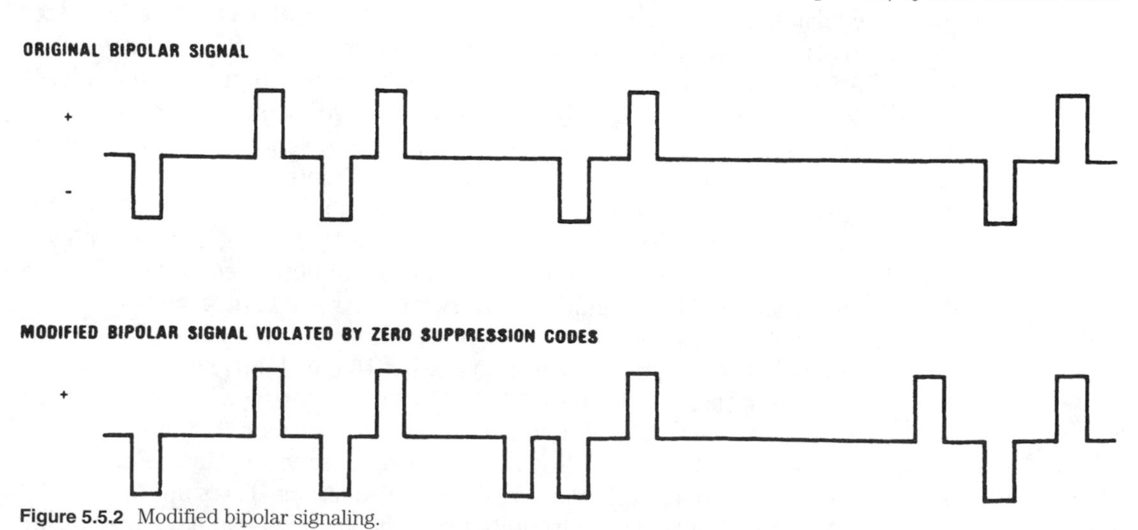

Figure 5.5.2 Modified bipolar signaling.

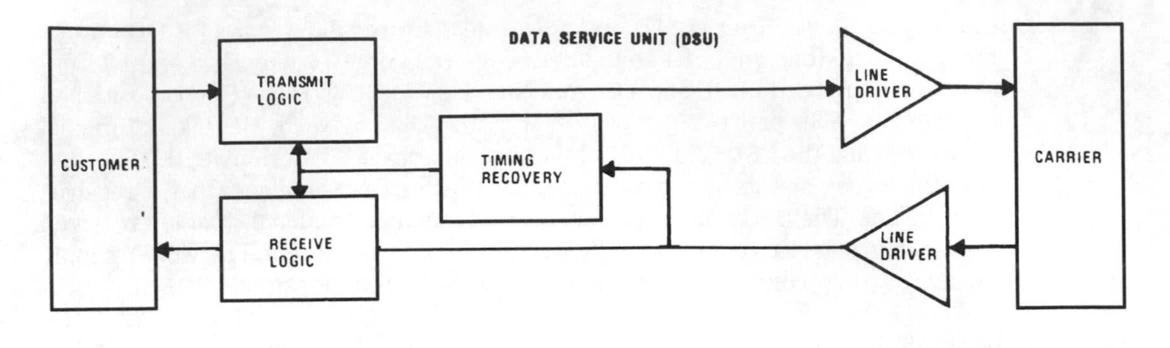

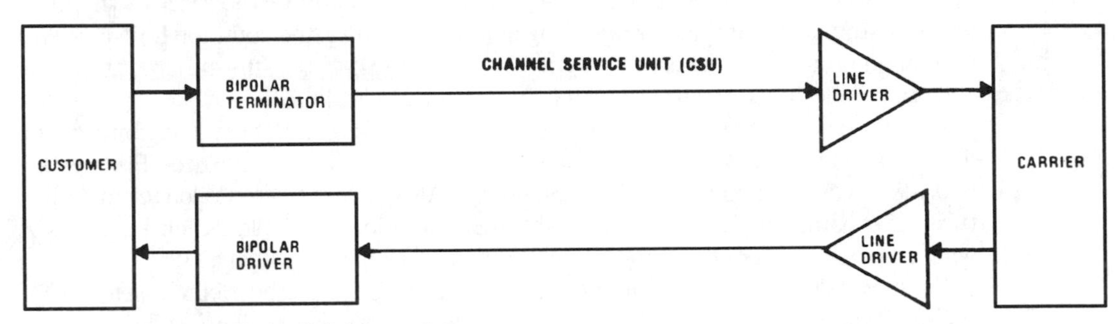

Figure 5.5.3 Service units for digital transmission.

the DDS, are amplified, filtered, and passed on to the 2-wire metallic-pair telephone company cable. The signals on the receive pair are amplified and equalized by the line receiver. The resultant bipolar pulses are then passed to the customer, who must recover the synchronous clock used for timing the transmitted data and sampling the received data. The customer must further detect the DDS network codes, enter appropriate control states, and remove bipolar "violations" from the data stream.

CSU interfacing is accomplished by use of a 15-pin female connector that utilizes the first six pins: the four previously described plus a status indicator and ground lead. In addition to the communications carriers, several vendors offer compatible channel service units for customer connection to digital networks.

In comparison to channel service units, data service units incorporate all the circuitry necessary to make the devices plug compatible with existing modems and terminals. The unit includes an analog circuit, similar to that described in the CSU, plus a digital circuit that handles all timing-recovery and network-control codes.

DSU interfacing is accomplished by use of a standard 25-pin EIA-232-C female connector on the 2.4- through 9.6-kbit/s units, using 10 pins for signaling. The wideband 56-kbit/s unit utilizes a 34-pin CCITT, V.35 (Winchester-type) connector using 14 pins for signaling. Several independent suppliers also manufacture DSU-type units, which offer even more flexibility in the form of multiport options. Bell System 550A channel service units and 500A data service units are listed in Table 5.5.1. Both the DSU and CSU devices incorporate properly balanced and equalized terminations for the 4-wire loop, as well as the circuitry to permit rapid remote testing of the channel. The signals on the 4-wire loop are the same for both devices and are terminated in the serving central office of the communications carrier into a complementary unit called an *office channel unit (OCU)*. From here, the time-division multiplexing hierarchy begins, as illustrated in Figure 5.5.4.

TABLE 5.5.1 Dataphone Digital Service Interface Units

	Speed	List code
Bell 500A-Type DSUs	2.4 kbit/s	500A-L1/2
	4.8 kbit/s	500A-L1/3
	9.6 kbit/s	500A-L1/4
	56 kbit/s	500A-L1/5
Bell 550A-Type CSUs	2.4 kbit/s	550A-L1/2
	4.8 kbit/s	550A-L1/3
	9.6 kbit/s	550A-L1/4
	56 kbit/s	550A-L1/5

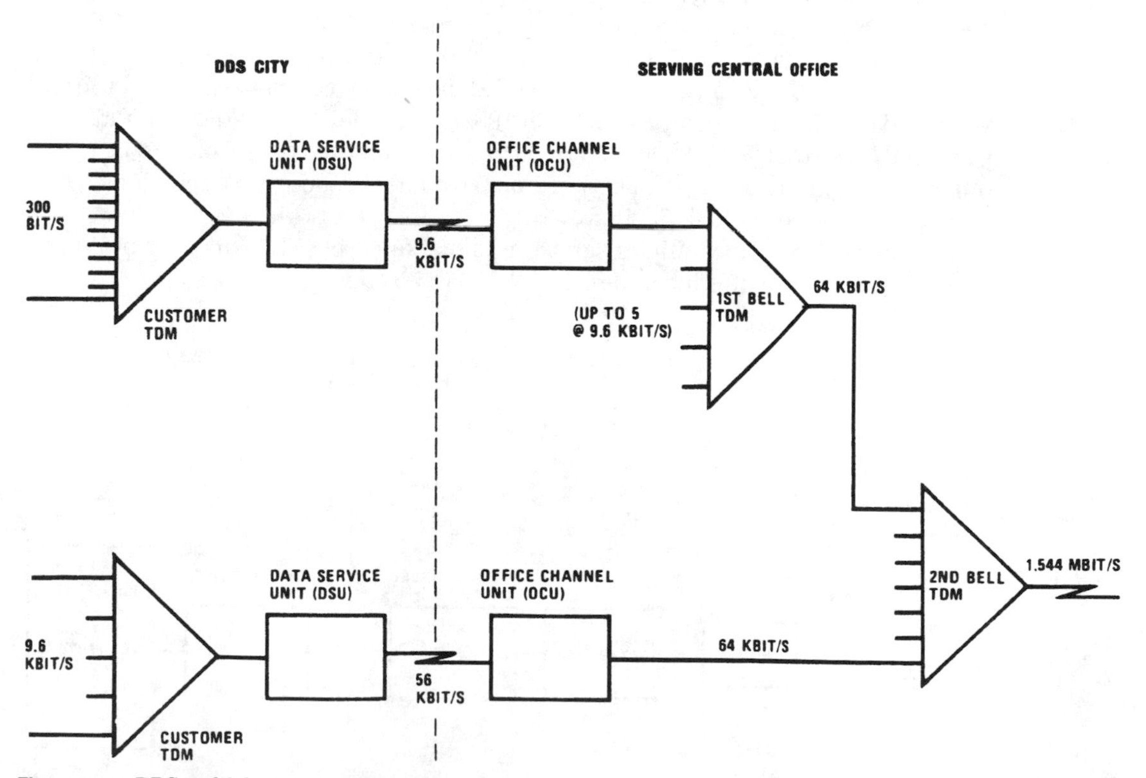

Figure 5.5.4 DDS multiplexing arrangement.

Signals from the OCUs are fed into the first stage of multiplexing, which combines up to twenty 2.4-, ten 4.8-, or five 9.6-kbit/s signals into a single 64-kbit/s channel, which is the digital capacity of a voice channel in T1 digital transmission. A second stage of multiplexing takes the 64-kbit/s streams and efficiently packs them into a T1 bit stream operating at 1.544 Mbit/s, which can carry voice as well as data signals over existing long-line facilities. Using this scheme, future expansion of DDS can be accomplished at a very rapid pace. This could, at a later date, relegate analog transmission of data to history.

Analog extensions to DDS

The Bell System provides an 831A data auxiliary set that allows analog access to DDS for customers located outside the DDS servicing areas. The 831A connects the EIA-232-C interfaces between a data service unit (500A type) and a voiceband data set. The 831A contains an eight-bit store, control, timing, and test circuits that allow loopback tests toward the digital network. Figure 5.5.5 illustrates a typical analog extension to a DDS servicing area.

The T1 carrier

The 1.544 Mbit/s line interface at the second TDM illustrated in Figure 5.5.4 is more commonly known as a *T1 carrier*. It has provided the basis for spawning the burgeoning T1 multiplexer market, which is covered in Chapter 11.

Until 1982, the use of the T1 carrier was restricted to telephone companies; it was used internally by such organizations to send digitized voice signals or a mixture of voice and data between central offices. In 1982, AT&T filed Tariff No. 270 for its High Capacity Terrestrial Digital Service (HCTDS), which allowed digital data to be transmitted between customer premises, or from customer premises to a telephone company central office, at 1.544 Mbit/s.

In January 1984, 1.544 Mbit/s service was marketed by AT&T as part of its Accunet digital service offering under the T1.5 classification.

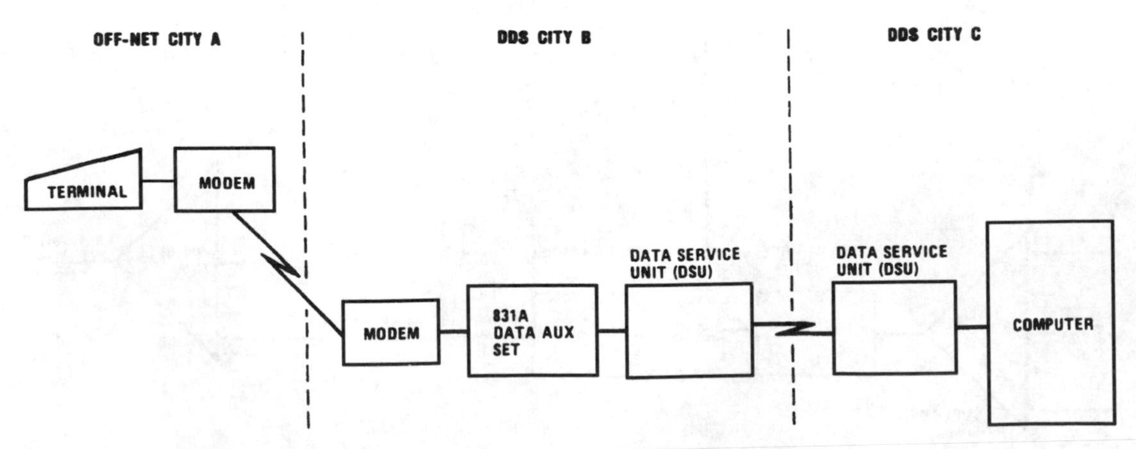

Figure 5.5.5 Analog extension to DDS.

The T1.5 service could not have arrived at a more opportune time. With interest rising in implementing distributed data processing and automated offices, a more economical method was needed to transmit large amounts of data. With 1.544 Mbit/s service available, users can more readily fit a wide range and volume of diverse data types, including voice, digital, and video onto one T1.5 circuit.

A cornerstone of the digital revolution in telephony and T1 carrier use is the *channel bank*, a device that transmits 24 independent channels over copper cable that had previously carried only one voice call. The early channel banks, which acted as multiplexers, were also called *T-carriers*. The T-carrier is built around two or more channel banks, or D banks, which convert analog signals into a digital format and multiplex them into one digital signal for transmission over a digital trunk. The multiplexed digital signal is called the *T1 bit stream* or the *DS-1 signal*.

Over the years, specifications for the performance of a number of channel banks have evolved. They are generally referred to by the designations first used at Western Electric Co.: D1, D2, D3, and D4. Although D1 and D2 channel banks are no longer being manufactured, many are still used in the telephone network. The D3 and D4 banks are state of the art.

D3 and D4 banks are very similar. The D4 bank is basically two D3 banks in a single chassis; the two share common equipment. The D3 and D4 banks continue to use the T1 bit stream for each set of 24 channels. The D3 bank has a 24-channel capacity using one T1 bit stream, and the D4 bank uses two T1 bit streams for a total capacity of 48 channels.

The T1, or DS-1, bit stream encoded by the D banks in today's time-division multiplexer (TDM) equipment is a digital stream of ones and zeros that runs at a rate of 1.544 Mbit/s. PCM (pulse code modulation) is the technique used to transmit analog (voice) signals on D banks. Bipolar [also called alternate mark inversion (AMI)], is the digital form that the T1 signal takes once it is passed to the digital trunk facility.

Digitizing voice

An analog signal is first sampled at preselected, equally distributed time intervals. The resulting analog sampling is referred to as *pulse amplitude modulation (PAM)*. Each analog PAM sample is quantized and coded as a digital eight-bit PCM byte. The eight-bit byte can be transmitted on a digital facility to another location where the PCM process is reversed and the information is reconstructed or decoded into its original analog form. Figure 5.5.6 illustrates the sampling and coding of a voice (analog) signal.

The T1 rate involves from the sampling rate applied to the analog signal. The Nyquist theorem of information theory states that to encode an analog signal, it must be sampled at twice its bandwidth. The telephone industry's voice-grade band runs at about 300 Hz to about 3300 Hz. The upper limit of frequencies on a voicegrade facility is normally taken as 4000 Hz. Hence, twice the bandwidth (2×4000) is 8000 samples per second. This sampling rate allows reconstruction of the analog signal.

An example of a bipolar signal used in a T1 bit stream (DS-1) is illustrated in Figure 5.5.7.

The 0s are at zero voltage, and 1s are produced by a nominal 3.0-volt peak positive- or negative-going pulse. The 1s pulses have an alternating polarity. If the first 1

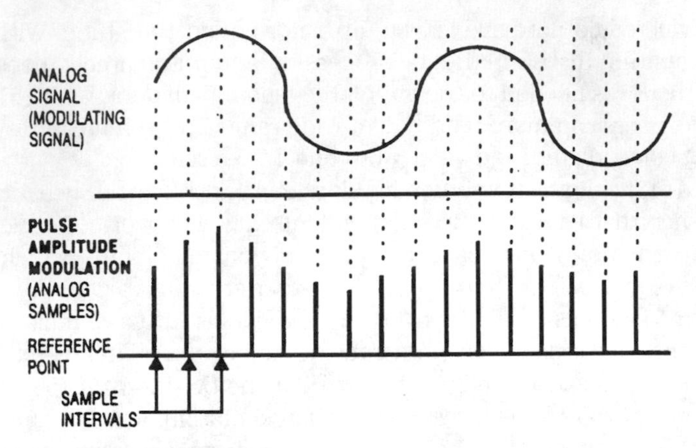

Figure 5.5.6 Digitizing voice.

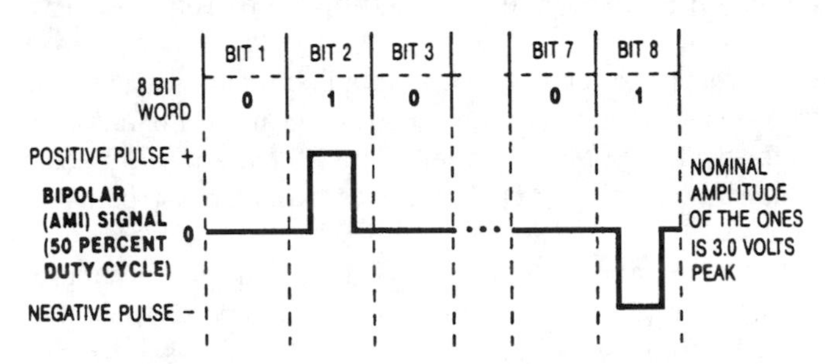

Figure 5.5.7 Bipolar signaling in a T1 bit stream.

is represented by a positive-going pulse, the next 1 will be a negative pulse, and vice versa. These alternating pulses are called *bipolar* or *alternate mark inversion (AMI)*. A violation of the bipolar signaling scheme would have occurred if bit 4 had been a positive pulse in relation to bit 2 or if bit 5 had been a negative-going pulse, in relation to a negative bit 4.

As a fundamental representation of the T1 or DS-1, bit stream transmitted from a D bank or TDM equipment, the 1.544 Mbit/s stream must meet certain basic requirements:

- It is a bipolar AMI, return-to-zero signal.
- Each pulse has a 50 percent duty cycle with a nominal voltage of 3.0 volts.
- There can be no more than 15 consecutive 0s present in the stream.

A D4 framing pattern has been added to the T1 specification. D4 framing has traditionally been used in D banks, although it did not become a requirement in network TDM equipment until early 1985. Prior to that time, suppliers of TDM equipment were required only to comply with the DS-1 specifications.

D4 framing in a T1 bit stream begins with a single frame bit (F1) followed by 192 data bits. Then comes frame bit 2 (F2) and another 192 data bits. Next comes frame bit 3 (F3) followed by its 192 data bits. This sequence continues up through frame bit 12 (F12); each frame bit is followed by 192 data bits. Then, the precise 12-bit frame pattern repeats itself, with each frame bit followed by 192 data bits. This repetition enables the transmission equipment to keep the bit stream in synchronization.

A sequence of 12 frames in D4 framing is known as a *superframe*. Each superframe presents a predefined framing pattern based on the frame bit value in each of the 12 frames. The top portion of Figure 5.5.8 illustrates the construction of a superframe while the lower portion of that illustration shows the resulting D4 framing pattern.

The D4 format consists of 24 consecutive 8-bit words, or "channels," following a framing bit, as illustrated in Figure 5.5.9. Much as the D bank transmits and receives 24 voice channels, the D4 framing and formatting will enable customer premises TDM equipment to mimic the separation requirement of the central office equipment.

To summarize, 8000 samples per second are required to reproduce an analog signal using PCM, with each sample represented by an 8-bit byte (channel). There are 24 eight-bit channels following each frame bit in a T1 stream that uses D4 framing and D4 formatting. The frame bit is followed by 24 eight-bit channels yielding 193 bits per frame. These frames are being produced at a rate of 8000 per second. Thus, a total of 1.544 Mbit/s is derived from the 193 bits at a sampling rate of 8000 bit/s.

SUPERFRAME CONSTRUCTION

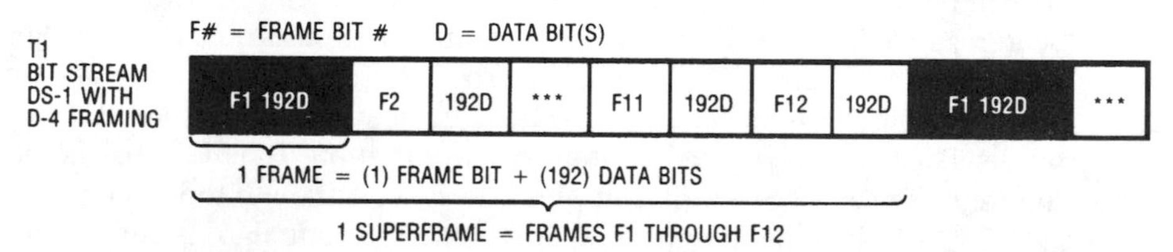

D4 FRAMING PATTERN

	F1	F2	F3	F4	F5	F6	F7	F8	F9	F10	F11	F12
D4 FRAMING PATTERN (FRAME BITS)	1	0	0	0	1	1	0	1	1	1	0	0

1 FRAME = (1) FRAME BIT + (192) DATA BITS
1 SUPERFRAME = FRAMES F1 THROUGH F12

Figure 5.5.8 Superframe construction and D4 framing pattern.

1 FRAME OF A T1 STREAM	FRAME BIT (1) BIT	CHANNEL 1 (8) BITS	CHANNEL 2 (8) BITS	CHANNEL 3 (8) BITS	. . .	CHANNEL 24 (8) BITS

(1) FRAME

1 FRAME = (1) FRAME BIT + [8 BITS PER CHANNEL × 24 CHANNELS]
1 FRAME = (1) FRAME BIT + [192 CHANNEL BITS OR DATA BITS]
1 FRAME = (193) BITS [FRAME AND DATA]

Figure 5.5.9 The D4 format.

Some restrictions on transmission apply to the T1 bit stream. In voice transmissions, there is a need to transmit not only the voice signal, but also the mechanical signaling information. The D bank at each end must know occurrences, such as when the line is "on-hook" or "off-hook," or if there is a battery reversal. To accomplish this, every 6th and 12th frame is used to transmit and receive voice-channel signaling information. This is accomplished by using the 8th bit in frames 6 and 12 in a voice channel (DS-0) for signaling. Figure 5.5.10 illustrates the positioning of voice-channel signaling information by frame (top) and by bit in the modified frame (bottom).

If all 24 DS-0s are used for voice transmission, then there will be 24 signaling bits in every 6th and 12th frame (one bit per channel per signaling frame). This technique is known as *bit robbing*. The theft of the signaling bits does not interfere with the PCM voice signal.

Those bits taken from the DS-0s during the 6th frame are referred to as the "A" bits, and those taken from the channels during the 12th frame are referred to as the "B" bits. From every superframe there will be one A and B bit pair per DS-0 (voice channel). Each pair of A and B bits is used to relay the signaling information for its respective channel between D banks: for example, loop closures, battery reversals, on-hook, off-hook. Not surprisingly, bit robbing affects data transmissions in the T1 bit stream.

Another T1 requirement that affects the volume of data transmitted in the T1 bit stream is the "Is density" requirement. To keep the telephone company T1 line, repeaters, and channel service units (CSUs) in synchronization, the DS-1 specification calls for no more than 15 consecutive zeros in the bit stream at any one time. This is referred to as *1s density*. The telephone company repeaters and CSUs use the 1s pulses like a clock signal to maintain synchronization. Too many 0s (no pulses), and the repeaters and CSUs will drift and lose sync. Although today's equipment might actually be able to handle more than 15 zeros, the standard is set. Private facilities using microwave and fiber optic might not have a 1s density requirement.

To ensure 1s density in the T1 Bit Stream, the telephone company uses B7 zero code suppression. This technique ensures that if all eight bit positions in a DS-0 are

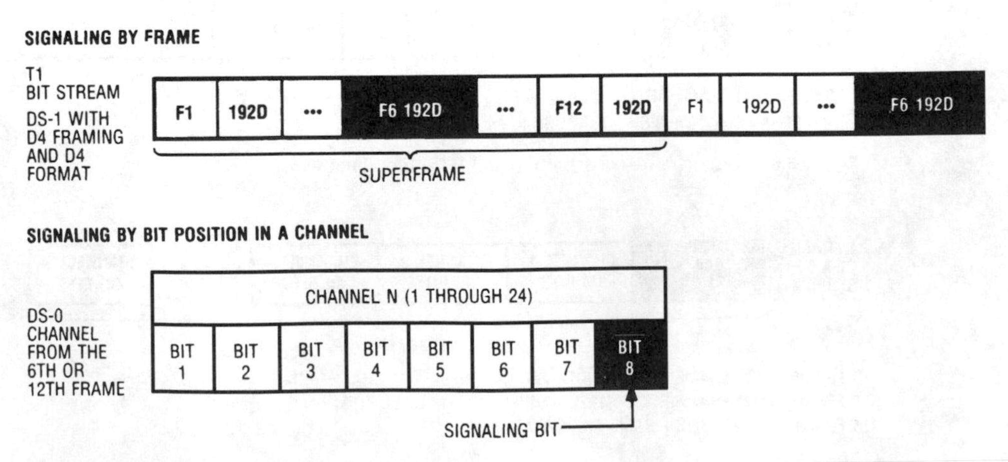

Figure 5.5.10 Inserting voice signaling information.

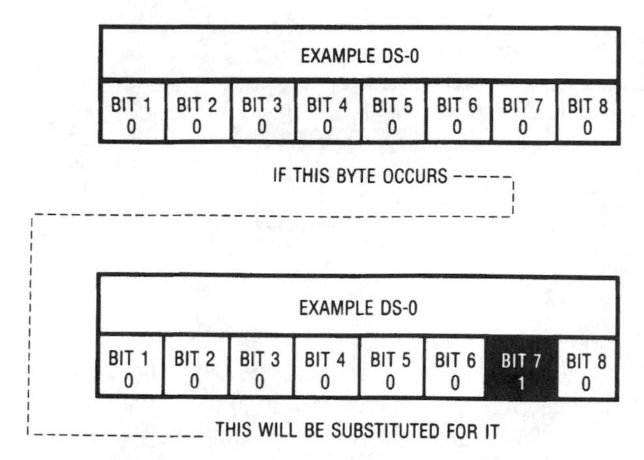

Figure 5.5.11 B7 zero code suppression.

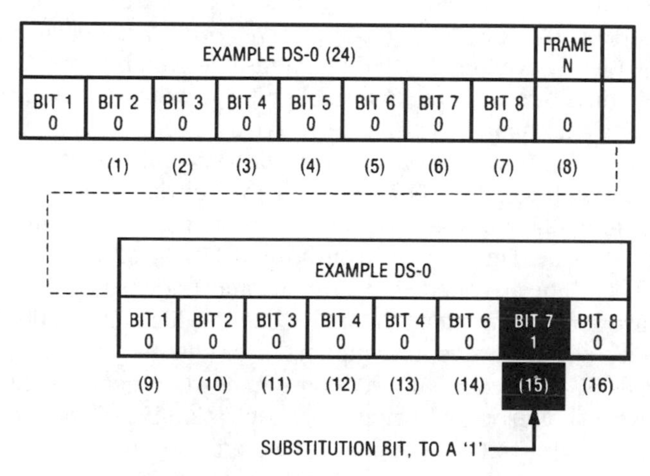

Figure 5.5.12 Worst-case B7.

0s, a 1 bit will be substituted in the Bit 7 position. Figure 5.5.11 illustrates the use of B7 zero code suppression.

Now, consider a worst-case scenario, illustrated in Figure 5.5.12. This illustration shows the longest possible string of 0s with B7.

Within the T1 bit stream, if Channel 24 were followed by a frame bit "0," and then Channel 1 were all 0s, there would be a total of 16 consecutive 0s. B7 cuts the number of 0s to 14, maintaining the integrity of the DS-1 specification.

How does B7 and bit robbing affect both a voice and data channel at the DS-0 level? Figure 5.5.13 shows an example of a voice channel (DS-0) in which either the 6th or the 12th frame contains all 0s.

In this example, the original voice channel is losing its Bit 8 to a 0A or 0B signaling bit. A PCM channel can tolerate the loss of the 8th bit to signaling, as well as the loss of Bit 7 to the B7 zero code suppression.

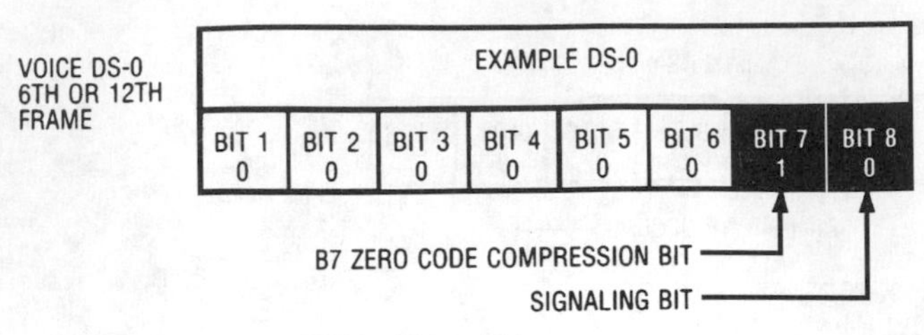

Figure 5.5.13 Effect of B7 and bit robbing.

There is no need to rob bits in a data channel (DS-0) because no signaling is required. However, having all 0s in the data channel is a possibility. If this occurs, data is corrupted by the B7 zero code suppression. Although a voice channel can tolerate this, a data channel cannot. To compensate for this situation, a data channel generally contains only seven usable data bits. One of the eight data bits in the data channel (DS-0) is made a 1. This prevents the data channel from being corrupted by B7 zero code suppression.

This type of arrangement is called a *nonclear channel*. A nonclear channel is a data channel (DS-0) with seven usable and one unusable data bits.

The seven usable data bits yield 56,000 usable bit/s and 8000 unusable bit/s. The 56,000 bit/s on a nonclear channel is also referred to as a DS-A.

The mathematics shows that in a T1 bit stream in which all channels are considered nonclear, there will be only 1.344 Mbit/s available for data transmission. There are 8000 frame bits and 192,000 usable data bits that hold to the 1s density requirement in the T1 bit stream. Thus, in a T1 bit stream (1.544 Mbit/s) with nonclear channels, fully 200,000 bit/s are needed for framing and 1s density.

Clear channels are available. A clear channel is one in which all 64,000 bits are usable. A medium, such as microwave, can support clear channels; private microwave links do not require B7 zero code suppression nor do they need 1s insertion. Although there is always a question of losing synchronization if too many consecutive 0s appear on the link, with today's technology the quantity of consecutive 0s can generally run in excess of 15. Just because eight consecutive 0s are in a particular DS-0 does not mean that will be 15 consecutive 0s in the T1 bit stream.

Clear-channel capability can also be obtained on DS-1 telephone company equipment by using bipolar transmission with Binary 8 zero substitution (B8ZS) coding in the T1 bit stream (DS-1). With B8ZS coding, each eight consecutive 0s in a byte are removed and the B8ZS code is substituted.

Extended superframe format

Another change in T1 carrier network applications is the Extended Superframe format (ESF), also called the F_e *format*. ESF redefines the D4 framing pattern. Instead of looking for the resident 12 consecutive bits in the D4 framing, 24 consecutive frame bits are being processed. Figure 5.5.14 illustrates the construction of an ESF frame.

A single ESF frame contains 24 frame bits. Unlike D4 framing, in which the 12 framing bits follow a specific pattern, the ESF is not entirely a specific repeating pattern. The ESF framing pattern always keeps the same format, yet the actual frame bits are broken down into three types of frame bits.

In an ESF frame, the "d" bits appear in frame bit positions 1, 3, 5, 7, 9, 11, 13, 15, 17, 19, 21, and 23. Of the 24 bit positions, the "d" bits use 12 positions or half of the available frame bits. Even though there are now 24 frame bits, as opposed to 12, there are still only 8000 frame bits per second in the T1 stream. Because the "d" bits are using half of the available frame bits, the "d" bits total 4000 bit/s. The "d" bits will be used by the telephone company to perform network monitoring, alarms, and reconfiguration.

The remaining bits are split. Frame bit 2, 6, 10, 14, 18, and 22 will be used for a cyclic redundancy check code consuming 2000 bit/s. The remaining 2000 bit/s constitute the basic frame pattern. This part of the F_e format is very specific, with a pattern of 0 0 1 0 1 1. The pattern will always appear in frame bits 4, 8, 12, 16, 20, and 24. The ESF, like B8ZS, will be applicable only when the equipment to recognize it is deployed in the network. Figure 5.5.15 illustrates the ESF framing pattern or F_e format.

Although D4 framing is dominant, wide implementation of ESF is gaining and is expected to bring about several improvements in overall network performance. Not the least of these will be the use of the ESF to monitor network performance. With the ESF it is possible to provide continuous performance checking without inhibiting traffic performance. Overall network control and performance measurement will also benefit from the ESF. The ESF makes it possible to use a 4 kbit/s channel to control and report on performance. Another important advantage that will flow from network-wide implementation of the ESF is the elimination of false framing patterns. False framing can lead to serious error conditions going unreported, such as instances when a NAK is mistaken for a network crash.

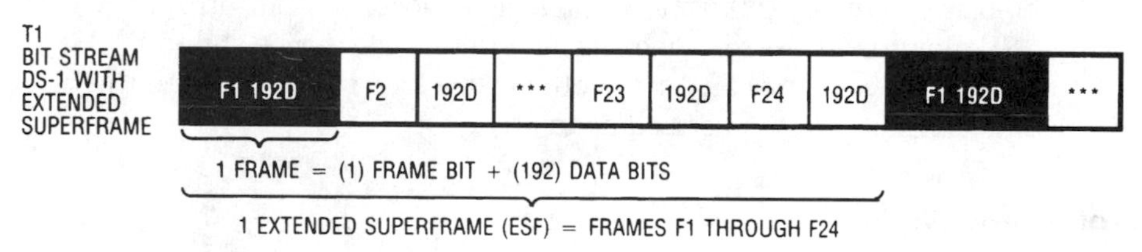

Figure 5.5.14 ESF frame construction.

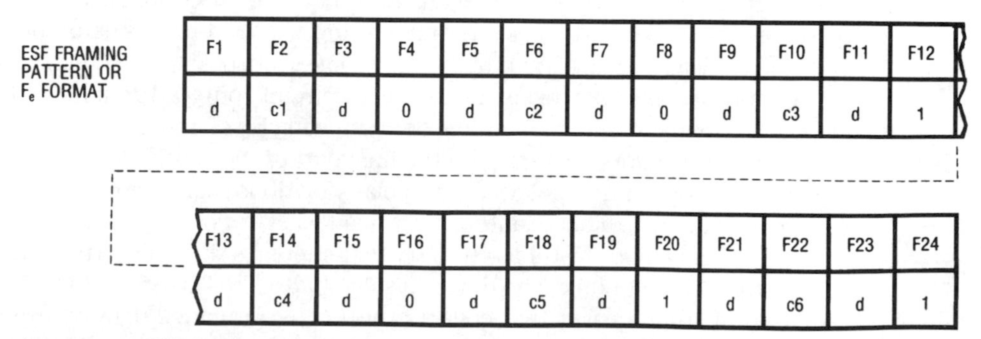

Figure 5.5.15 ESF framing pattern.

Channel service unit

DSU functions are incorporated into most types of data terminal equipment (DTE) intended for operation on T1 lines. This differs with DDS transmission, where the DSU started as a separate device.

Besides terminating a line, a CSU operating on a T1 facility amplifies signals and initiates remote loopback. These functions are similar to those of the CSU half of the combined DSU/CSU that operates on DDS facilities. A CSU used on a T1 facility also formats frames and computes performance-measurement statistics. To format a frame, the CSU encodes every 193rd bit, to supply synchronization framing. In some instances, this operation also generates and stores performance-measurement statistics. This latter function is defined by industry standards, which oblige CSUs to compute and store various performance data. When the CSU receives a predefined code (implanted in the T1 framing), it transmits the stored performance statistics.

The T-carrier CSU physically terminates the transmission facility. It is also the final signal-regeneration point before data arrives at the terminal equipment. Because T1 multiplexers and channel banks already have the DSU functions incorporated, T-carrier CSUs are normally supplied as separate devices, not as part of a combined DSU/CSU. However, most subrate CSUs are supplied in combination with DSUs.

Examine the T-carrier CSU functions—especially those of its transmitter and receiver. The CSU's transmitter scans the received user-equipment signal for excess zeros, a low 1s density, and bipolar violations (if Binary 7 Zero Code Suppression is used). Then, the CSU encodes the data (based on the signaling procedures of the T1 circuit) and transmits a regenerated digital signal.

The CSU's receiver examines the received digital signal for several parameters, depending on the CSU's frame format and line coding. These parameters include: bipolar violations, B8Zs line coding, signal loss, and out-of-frame signals. Also, the CSU monitors the error-checking algorithm if the device supports ESF framing. The CSU's receiver can scan for remote loopback codes, regenerate the received signal, and transmit it to the user equipment.

Types of T-carrier CSUs

There are two types of T-carrier CSUs: basic and ESF-compliant. A basic CSU, which incorporates front-panel controls and displays, meets FCC circuit-interface specifications. Front-panel features enable a user to activate far-end-CSU loopbacks, attach test equipment to insert signals for monitoring a looped-back signal's bit error rate, and display diagnostic information, such as a line's loss of signal and bipolar violations.

A high-order multiplexer might have a T1 circuit routed through it, while linking several T1s onto an optical fiber line and removing bipolar violations. Then, the CSU senses only the errors occurring on the final part of the circuit. Therefore, the CSU's bipolar-violation count displays only bipolar violations that occur between the multiplexer and the CSU. This result might give a false view of the T1 circuit's quality.

The ESF-compliant CSU has circuitry that supports the ESF format, including access to the ESF data link. The device also includes CRC error testing. This CSU has a built-in microprocessor that collects various performance data and sends it over the ESF data link using a special communications protocol called *BX.25*.

The collected performance data includes CRC-6 errors, out-of-frame events, error events, errored and severely errored seconds, and unavailable seconds. A CRC-6 error occurs when the ESF's computed CRC code does not jive with the received CRC code. An out-of-frame event is defined as occurring when any two of four consecutive frame-synchronizing bits are erroneous. An error event is either a CRC-6 or an out-of-frame error. An errored second is a second with one or more error events. If 320 or more error events occur in a second, then you have a severely errored second. The unavailable second is one with 10 or more consecutive severely errored seconds preceding it.

Originally, most ESF-compliant CSUs were based on AT&T Technical Publication 54016. Today, ESF compliance is defined by both the AT&T publication and the ANSI T1E1 committee's standard T1.403. Both documents define the same ESF framing format, with most of their error terminology similar. The major difference between the two is the technique by which data passes over the ESF data link between communications carrier monitor equipment and the CSU.

Under the AT&T document, a CSU can store up to 24 hours of performance data. Then, responding to a command over the ESF link, the CSU sends it to monitoring equipment. Under ANSI T1.403, the CSU sends performance statistics every second to carrier equipment using a performance report message (PRM) scheme. The PRM has the last four seconds of performance statistics.

Questions

5.1 What factors affect the shape of digital signals transmitted in their digital form?
5.2 What is a signal-to-noise ratio and how is it used?
5.3 What power ratios result from the following circuit measurements?

A. Power transmitted 1 watt; Power received 0.1 watt.
B. Power transmitted 0.2 watt; Power received 0.005 watt.
C. Power transmitted 0.05 watt; Power received 0.00025 watt.

5.4 Where would a line driver be used and how does it function?
5.5 Discuss the differences between amplitude, frequency, and phase modulation.
5.6 What is the relationship between a channel's bandwidth and its signaling capacity?
5.7 How can an increase in the amount of information a signal pulse represents lead to an increase in the number of bits per second transmitted'?
5.8 Discuss some of the problems associated with multilevel coding.
5.9 What is the relationship between Nyquist's baud limitation law and Shannon's formula that defines the maximum bit rate that can be transmitted over a channel?
5.10 What basic functions does a modem perform?
5.11 What timing differences exist between asynchronous and synchronous modems?
5.12 What are the four line-servicing groups modems can be classified into? What are the differences in operating speeds between these groups?
5.13 Why would a modem designed to operate as a Bell System 103/113 device be unsuitable for use in most foreign locations?
5.14 Discuss how V.22bis modems can be incompatible with one another.
5.15 What are the two key causes of circuit distortion and in what units are they measured?
5.16 Why is a reference frequency important for attenuation measurement?
5.17 What devices can be used by a communications carrier to condition a leased line? How do these devices operate?

5.18 Why is line conditioning available only on leased lines? What devices do modems designed to operate on the switched network use as a result of the unavailability of conditioning on that facility?
5.19 Why is the EIA-232-C interface standard an important consideration when obtaining communications equipment?
5.20 What function does modem handshaking perform?
5.21 What key advantages are obtained from using acoustic couplers instead of modems?
5.22 What is the difference between serial unipolar signals, bipolar signals, and modified bipolar signals such as the violated bipolar signal used for digital transmission of the Bell System's Dataphone digital service (DDS)?
5.22A Why is a modified V.29 modem designed for half-duplex operation on the switched telephone network very suitable for facsimile transmission?
5.23 Convert the following serial unipolar signal into a bipolar signal:

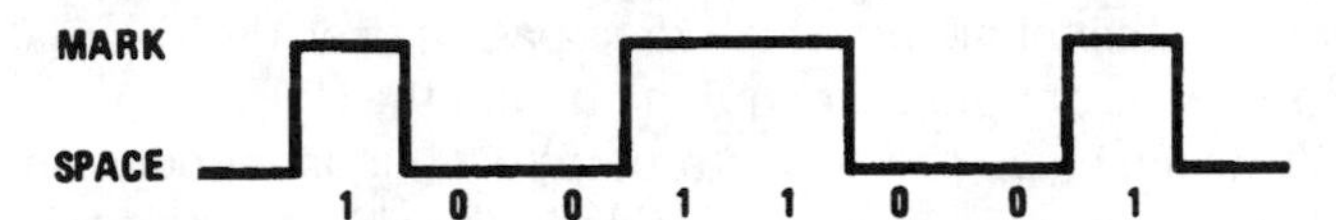

Figure 5.23 Convert the following serial unipolar signal into a bipolar signal.

5.24 What is the difference between a data service unit and a channel service unit?
5.25 What devices and facilities are required to obtain an analog extension to the Bell System's DDS?
5.26 Discuss the relationship between PAM, PCM, and the maximum number of voice channels that can be digitized using PCM on a T1 circuit.
5.27 If the maximum number of voice channels on a T1 carrier is multiplied by the data rate required to digitize each voice conversation, an operating rate of 1.536 Mbit/s is obtained. Discuss the difference between this data rate and the T1 transmission rate of 1.544 Mbit/s.

Chapter

6

Regulation Agencies and Communications Vendors

6.1 Introduction

Any company has the right to develop its own data transmission facilities as long as such facilities are for its own exclusive use and do not interfere with other existing or potential public or private transmission services. As an example, lines connecting several terminals to a computer in an organization's building may be strung if those lines do not use portions of existing telephone company circuits. The communications carrier may agree to the use of its facilities in return for a prescribed payment. When a terminal and a computer are located in different buildings, it might possible for the company to run a very long cable between the two locations. However, the company not only has to construct the transmission medium, but it might also have to obtain property rights from numerous companies and individuals to install its lines over their property. Because of both the high cost of constructing facilities for individual use and to the complexity and costs involved in obtaining property rights and then maintaining such constructed facilities, private transmission facilities are mostly impractical and not commonly encountered. Because of the preceding problems, most data transmission networks, therefore, use transmission facilities provided by communications common carriers.

In the United States, thousands of companies are currently recognized as communications carriers. They provide a wide range of services, including facilities for voice, data, facsimile, printed messages, and packet switching, as well as appropriate communications channels for television transmission, over such media as switched and leased lines and satellite channels.

Communications common carriers furnish communications services to the public. These carriers are regulated by one or more federal, state, and international agencies and organizations. In the United States, the early growth of communications was by telegraph, which generally followed railroad company rights-of-way. Until 1934, both telephone and telegraph services were regulated by the Interstate Commerce Commission (ICC). Because of congressional dissatisfaction with the ICC's inability to regulate communications effectively, the Communications Act of 1934 was passed. This legislation created the Federal Communications Commission (FCC).

6.2 Regulation Agencies and Tariffs

The FCC is an independent Federal agency whose role is to regulate interstate and international communications originating in the United States. This regulation applies to communications by radio, telephone, telegraph, facsimile, and other transmissions by wire, cable, or radio.

Under the provisions of the 1934 Communications Act, all communications common carriers are required to furnish services at reasonable charges upon reasonable request. No carrier may construct, acquire, or operate interstate or foreign facilities originating in the United States without the approval of the FCC.

Tariffs

Under the provisions of the Communications Act, every common carrier must file schedules with the FCC that show all charges, practices, classifications, regulations, and other pertinent data for interstate communications services offered to the public. These schedules are known as *tariffs* and are normally filed several months before their terms are to become effective. Once a tariff is filed, it will automatically become effective, having the force of law, unless it is suspended or disapproved by the FCC.

In addition to a tariff being a schedule of rates, practices, and regulations, it should also be viewed as a contract between the communications common carrier and the service subscriber. Thus, tariffs form a significant portion of the machinery by which the FCC enforces the duties and prohibitions imposed on common carriers. Not all communications common carriers have to file tariffs of their own for a particular type of service. If these carriers concur with the tariffs filed by other common carriers, they may apply to use that tariff. However, because (in many cases) jurisdiction over common carriers is divided between FCC and state or municipal regulatory agencies, there is frequently a variation in tariff rates for the same service in different locations.

One example of a tariff with a high degree of concurrence among a large number of communications common carriers is tariff 245, which was filed with the FCC by the American Telephone and Telegraph Company. This tariff covers the V-H measuring plan that establishes the basis for determining the cost of a telephone call by providing a uniform method for calculating the distance between an originating station and a receiving station. In this tariff, areas in the United States and Canada are assigned a mathematical coordinate on a vertical (V) and a horizontal (H) basis. By dividing each area into a series of small squares, each with equivalent latitude and longitude, the distances between any two points in airline miles can be calculated from the coordinates by the Pythagorean theorem. Because the cost of telephone circuits and calls is a function of distance, this tariff provides a foundation for determining the cost of telephone calls on the switched network and on private leased lines.

FCC and State Regulation

Interstate services cross state lines; intrastate services do not cross state lines and are normally regulated by state or municipal agencies. Because of this division of jurisdiction over communications common carriers by Federal, state, and perhaps municipal regulatory agencies, there is frequently a difference in tariff rates for the same service in different locations. In addition, carriers may operate under separate tariff regulations. This can result in minor variations of a generally available basic service.

International regulations

As many companies in different nations began to construct telegraph and telephone facilities, network incompatibilities resulted in subscribers of one country not being able to communicate with subscribers of communications common carriers in a second country. These incompatibilities and the difficulties they caused made the development of an international engineering and standards organization a necessity. In 1865, the International Telegraph Union (ITU) was formed to promote cooperation and compatibility. In 1934, the ITU became the International Telecommunication Union.

The ITU is an administrative international organization that is responsible for the allocation, registration, and utilization of the radio frequency spectrum; it studies and makes recommendations on technical, operating, and tariff questions in regard to international telephone and telegraph communications; performs such other functions as the coordination and publication of telecommunications service data required for the international operation of such services; and plans and manages technical cooperation programs for developing countries.

In 1992, based on a reorganization, important ITU functions previously carried out by the Consultative Committee on International Telephony and Telegraphy (CCITT) were assigned to its replacement, the Telecommunication Standardization Sector (ITU-T). Similarly, the Consultative Committee on International Radio (CCIR) was replaced by the Radiocommunication Sector.

Within the ITU-T are specialized study groups that are periodically formed to conduct technical and administrative conferences in order to develop international operating practices, pricing, policy arrangements, and technical specifications. A few of the ITU-T study groups are listed in Table 6.2.1.

The ITU-T holds a Plenary Assembly every four years. Among other procedures, the Study Groups are reviewed; their functions are subject to change.

Intelsat

During the early years of the space age, the feasibility of using satellites for communications was proven to be economically and technically practical. In the early 1960s,

TABLE 6.2.1 ITU-T Study Groups

Group no.	Study group
2	Network operation
3	Tariff and accounting principles
4	Network maintenance
5	Protection against electromagnetic environment effects
7	Data networks and open system communications
10	Languages for telecommunication applications
11	Switching and signalling
12	End-to-end transmission performance of networks and terminals
13	General network aspects
14	Modems and transmission techniques
15	Transmission systems and equipment

SOURCE: U.S. State Dept.

the U.S. Communications Satellite Corporation (Comsat) was established by an Act of Congress to construct and operate a global public communications satellite network.

Because of the growth in communications between countries via satellite, an international regulatory agency, the International Telecommunications Satellite Consortium (Intelsat), was established in 1964. From an initial membership of 11 countries, Intelsat has grown to about 100 members today. Representing the United States in Intelsat, Comsat is responsible for the design of the Intelsat series satellites that were built by Hughes Aircraft and launched by the National Aeronautics and Space Administration (NASA). In the United States, Comsat owns and operates a number of earth stations, but most such stations overseas are owned and operated by the local, government-owned Postal Telephone and Telegraph (PTT) organizations.

The Intelsat network consists of a number of satellites positioned over the Atlantic, Pacific, and Indian oceans, as well as earth stations located throughout the world. This satellite network permits direct connections between any two operating members, and is designed to permit voice, data, and television transmission.

Changes in U.S. regulatory environment

Until the early 1950s, telephone companies in the United States in general prohibited their subscribers from attaching subscriber-provided equipment to a telephone company circuit. In 1956, a U.S. Circuit Court of Appeals overturned the Federal Communications Commission in a legal case, where the FCC had ruled that the Bell Telephone System could prevent a subscriber from attaching a foreign device to his telephone. The device was called a *Hush-A-Phone* and was a mechanical attachment that snapped on to the handset of a telephone.

Following the guidelines set by the Circuit Court of Appeals, the FCC issued its now-famous Carterfone decision in June 1968. The FCC ruled that the Bell System could not prevent subscribers from using their telephones in conjunction with a device known as a *Carterfone*, which acoustically coupled the telephone handset to a two-way radio. The Carterfone was designed to permit conversations between private mobile radio units and telephone subscribers without the intervention and cost of special telephone equipment.

One of the key impacts of the Carterfone decision was to liberate the growth of data communications applications from tariff restraints imposed by the telephone industry that had previously prohibited the attachment of foreign equipment to telephone company circuits.

The Carterfone decision was the most significant of several moves made by the FCC that spurred the utilization of data communications. Others were the specialized carrier decision, the "Open Skies" decision that fostered communications satellites and the approval of packet-switching networks as value-added carriers. As a result of these and other decisions, the number of vendors manufacturing data communications devices of offering data communications services, and the general use of data communications for information transfer, has grown in leaps and bounds.

FCC registration program

In 1976, the Federal Communications Commission created an equipment registration program to permit devices produced by nontelephone company manufacturers to access the public switched telephone network. This access was to be accomplished without the requirement of installing a data access arrangement, on the condition that such equipment was built to conform to certain interconnection specifications established by the FCC.

On October 1, 1976, the FCC equipment registration program rules were finalized and resulted in the elimination of access arrangements when the user obtained certified equipment. This enabled subscribers to reduce costs and ease the installation of nontelephone company equipment to telephone company lines.

Under the FCC interconnect registration program, customer-provided equipment that is registered—as a result of meeting a series of operational characteristics—can be interfaced via a plug to a telephone-company-supplied jack for connection to the public switched telephone network. Equipment manufactured prior to the program or equipment that does not meet the series of FCC operational characteristics can still be used. But such equipment must interface the telephone company jack through a registered data access arrangement provided either by the telephone company or by an independent manufacturer.

In Figure 6.3.1, the effect of the FCC registration program is illustrated for the interconnection of various types of transmitting and receiving equipment to telephone company facilities.

If the subscriber desires to transmit data over a leased telephone company circuit (Fig. 6.3.1A), no data-access arrangement is required, regardless of the type of modem used. This is because the leased line is for the exclusive use of that subscriber and any interference caused by nontelephone company equipment on that circuit will only affect that user. For the subscriber who obtains telephone company equipment (Fig. 6.3.1B), no access arrangement is required because the telephone company equipment is registered. A similar interface arrangement occurs when any

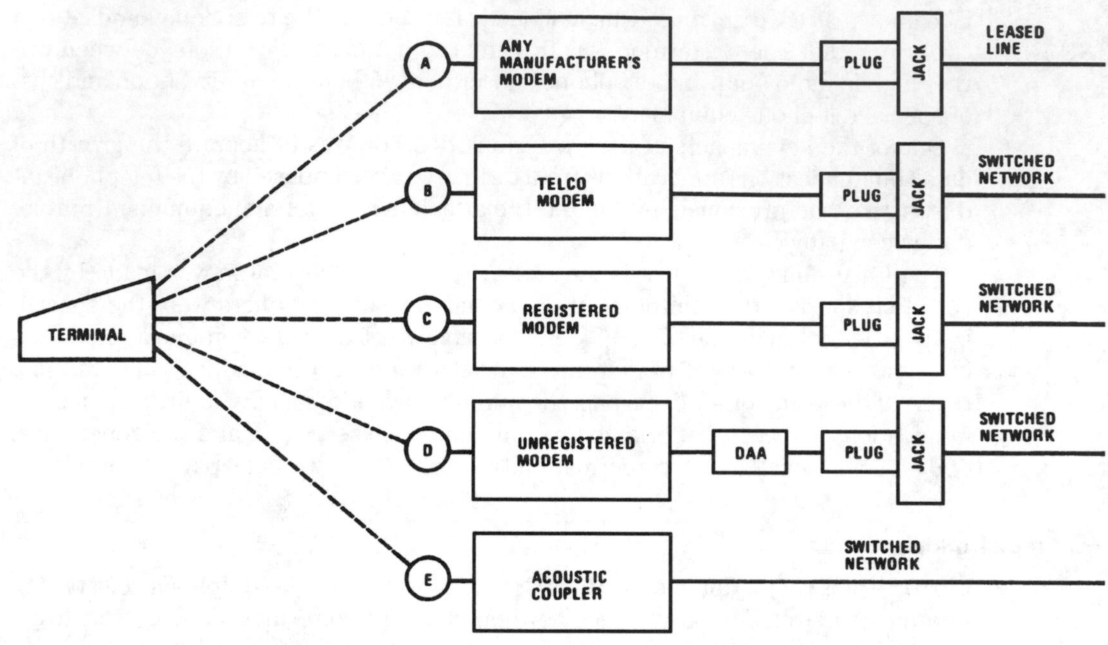

Figure 6.3.1 Interconnection under the FCC registration program.

vendor-manufactured modem that is certified under the FCC registration program (Fig. 6.3.1C) is connected to the switched network. When an unregistered modem (Fig. 6.3.1D) is connected, a data access arrangement must be obtained to interface the telephone company jack. For users who install devices that are acoustically coupled to a telephone circuit, such as a terminal with a built-in acoustic coupler (Fig. 6.3.1E), neither an access arrangement nor a plug-jack connection is required because the interface to the telephone line is via the telephone headset and the coupler's power supply is thus segregated from the telephone company line (no hard-wire connection).

Computer II inquiry, then divestiture

In 1980, the FCC's Computer II (C II) inquiry decision virtually negated the 1956 Consent Decree, which had restricted AT&T to providing only regulated common-carrier services. In effect, C II allowed AT&T to enter the computer field. Starting in 1982, under a series of decisions by the Justice Department, both AT&T and IBM became free to pursue just about any data communications endeavors they desired. With the Justice Department's divestiture decision in 1982, the Bell operating companies (BOCs) became independent of AT&T. (Refer to Chapter 7 for a discussion of the consequences of divestiture.) Many vendor announcements—both from AT&T and IBM and from the independents—have been keyed to these regulatory decisions.

Questions

6.1 What is a tariff and why is it important?

6.2 Discuss the division of jurisdiction in the United States with respect to communications common carriers.

6.3 What is the purpose of the International Telecommunications Union?

6.4 How has the Federal Communications Commission's equipment registration program affected the connection of equipment to telephone company facilities?

6.5 What type of subscriber equipment can be connected to the public switched telephone network via a plug-jack connection?

Chapter

7

The Communications Industry

7.1 Introduction

To place in perspective a discussion of carrier offerings, operation, and cost, which will be covered in Chapter 8, first consider the organization and structure of the communications common-carrier industry in the United States and several foreign countries.

As covered in Chapter 6, communications common carriers are companies that are licensed by one or more international, federal, state, or municipal agencies to provide service to the public at reasonable rates upon reasonable request. In the United States, the communications industry consists of the telephone companies, specialized carriers, satellite carriers, and international record carriers. Each of these communications groups consists of a number of companies and subsidiaries, ranging from thousands of telephone companies to several international record carriers. Each of these groups is licensed to provide a variety of services to include the transmission of voice, data, printed messages, video, facsimile, and telemetry.

7.2 The U.S. Telephone Industry

Although there are thousands of telephone companies in the United States, this industry is dominated by five large companies that generate over 95 percent of all telephone company revenues. The largest and most important of these large companies is AT&T, which is the undisputed leader in the field.

AT&T

AT&T was incorporated in 1885. At its peak, the company provided telephone service to over 70 percent of the U.S. population and handled over 90 percent of the long-distance calls placed in the nation. Through various degrees of control, ranging from ownership of a majority to all of the stock, AT&T controlled 22 associated telephone companies—called *Bell operating companies (BOCs)*—together with its manufacturing arm, the Western Electric Co. With Western Electric, AT&T owned the Bell Telephone Laboratories, which, up to 1925, had been Western Electric's Engineering Dept. Bell Labs performed—and still performs—centralized research. Among its many inventions are the transistor and the solar battery.

AT&T's role has changed extensively, starting with the Computer II (CII) inquiry in 1980 (Chapter 6). CII established a distinction between basic and enhanced transmission services. Basic services would continue to be regulated; enhanced services—those involving computer processing—and customer premises equipment (CPE) were deregulated and offered on a competitive basis.

In 1982, AT&T and the U.S. Justice Dept. settled their long-standing antitrust case. Under the divestiture decision—later called the *Modified Final Judgment (MFJ)*—AT&T relinquished ownership and control of the 22 BOCs. It retained ownership of Western Electric, which it renamed *AT&T Technologies (AT&T-T)*. A retained subdivision of AT&T-T is the renowned Bell Labs. Also retained was Long Lines, renamed *AT&T Communications*, which is the company's long-distance operation. AT&T's unregulated affiliate, AT&T Information Systems or ATTIS (originally called *American Bell*), was formed to market CPE (although in 1985 the FCC responded affirmatively to an AT&T petition to sell CPE without going through a subsidiary). The divested BOCs were permitted to market CPE via their own unregulated affiliates. (In 1986, the FCC also dropped this requirement.)

The divested BOCs were grouped into seven regional companies (RBOCs). They are Ameritech, Bell Atlantic, BellSouth, NYNEX, Pacific Telesis (PACTEL), Southwestern Bell, and US West. See Figure 7.2.1 for the approximate geographic boundaries of the regions and for each RBOC's component BOCs.

Under AT&T's approved divestiture plan, the nation was divided into local access and transport areas (LATAs), ranging from metropolitan areas to entire states. Within each LATA, the BOC provides local and "long-distance" (local-exchange) services. Inter-LATA services are reserved for interexchange (long-distance) carriers, such as AT&T, MCI, and US Sprint.

Strengthening its competitive posture, AT&T has undertaken a series of acquisitions and mergers. Notable among them is the acquisition of the NCR Corp., a computer manufacturer, and McCaw Cellular Communications Inc., a leading wireless-network operator.

Independent telephone companies

The independent telephone companies, in general, offer the same facilities as the BOCs and are compatible with and also interconnect to them. These independent telephone companies range in size from small local companies that provide service to a few hundred subscribers to General Telephone and Electronics (GTE), which is usually considered the leader of the independents.

GTE is a highly diversified communications, research, and manufacturing company with operations throughout the United States and abroad. GTE operates a net-

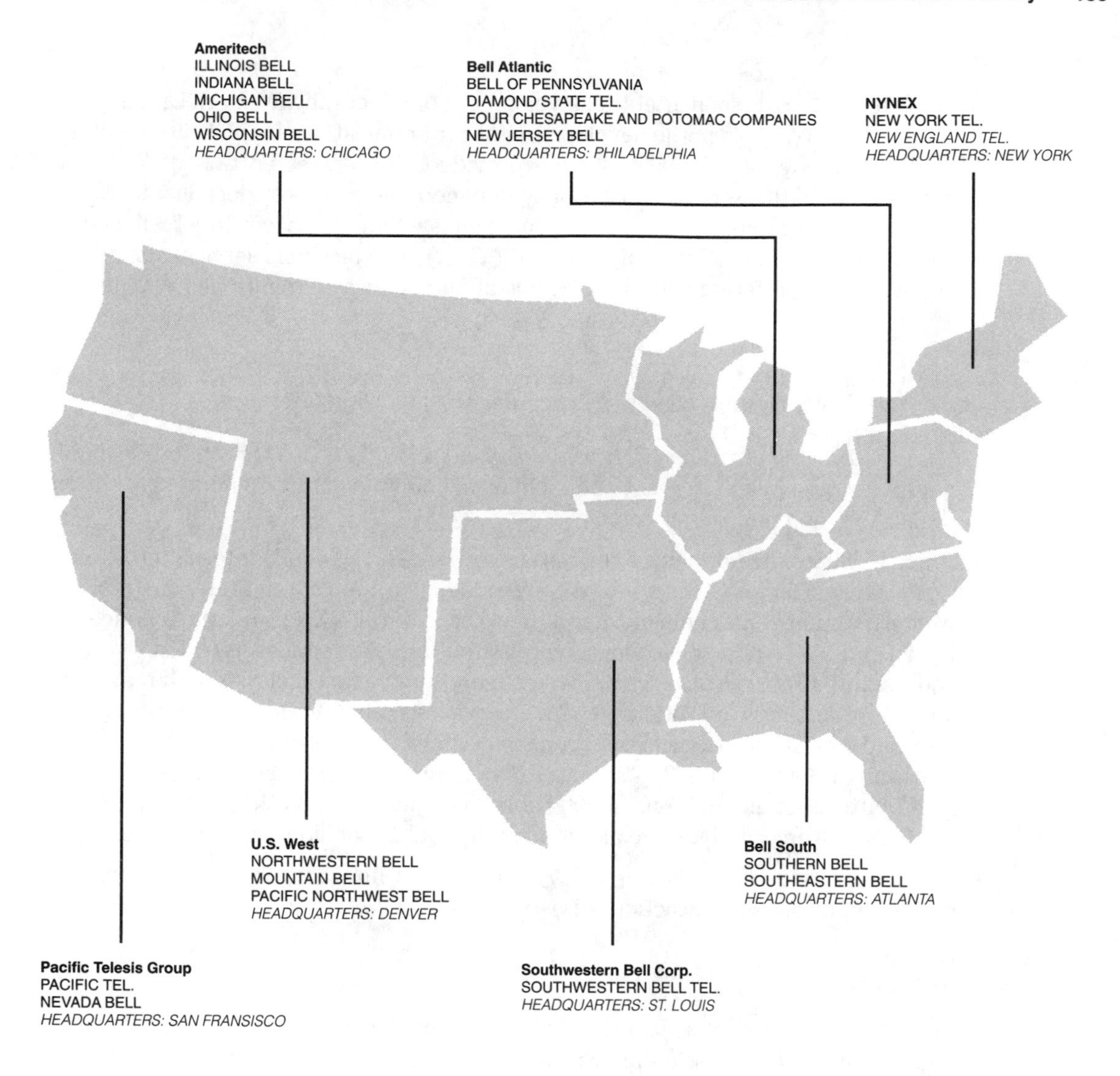

Figure 7.2.1 The seven regional phone companies. Courtesy of North American Telecommunications Association.

work quite similar to AT&T's. Its structure includes a number of operating companies in the United States, and several international subsidiaries.

Today, there are hundreds of independent telephone companies that have thousands of subscribers. To represent these independent companies, the United States Independent Telephone Association (USITA) was formed, with headquarters in Washington, D.C. USITA coordinates their operations through committees dealing with such subjects as technical practices, accounting standards, and legal matters.

7.3 Western Union

Incorporated in 1851, soon after the development of telegraphy, Western Union's history includes a vital role in the development of communications in the United States. Although the telegram for a long time provided the largest portion of Western Union's revenue, the service substantially declined over the past decades. In 1991, Western Union reorganized and sold its message switching services to AT&T. These services were EasyLink, Telex, and Telex II (TWX). The facsimile service, Mailgram, which was a joint offering with the U.S. Postal Service, was discontinued.

EasyLink

Under the EasyLink umbrella, AT&T includes the following offerings:

- *International electronic mail and messaging services.* Delivery options include X.400, fax, and Telex. Access is possible from communicating computers, such as PCs, LANs, and mobile devices.
- *Telex.* This service remains a ubiquitous low-speed (50 baud) international teletypewriter facility. Telex service was introduced to the United States in 1958 by Western Union. The company purchased the older TWX (110 baud) service in 1971 from AT&T and subsequently integrated the two services. AT&T now owns and operates Telex as an electronic messaging service that includes both real-time and store-and-forward offerings. The service is able to accept higher-speed buffered inputs, but transmission remains at 50 baud; the 110-baud TWX operation was discontinued. AT&T's domestic network comprises about 60,000 terminals (with about an additional 40,000 able to be linked from other networks, such as MCI's). Overseas, there are about an additional 1.5 million Telex terminals.
- *Electronic Data Interchange (EDI).* This is an intercorporate, computer-to-computer, business-transaction service.

7.4 Specialized Carriers

The birth of the specialized carrier industry occurred in 1969 when the Federal Communications Commission approved the application of Microwave Communications Inc. (MCI) to offer shared private microwave service as a common carrier.

Initially, such service was offered between Chicago and St. Louis and nine intermediate locations. This service permitted up to five subscribers to share a single channel of 2000-Hz bandwidth for the transmission of voice, facsimile, and data. Because the bandwidth per channel was well below customary telephone company channel bandwidths, MCI's rates were substantially lower than rates charged by existing communications carriers. During the early years of MCI's development, its growth was restricted because of problems it had in linking its microwave network with subscribers through local facilities leased from telephone companies—which were not anxious to assist a competitor in the removal of revenue from their organizations. In the early 1970s, a series of legal rulings required the telephone companies to provide interconnections with specialized common carriers. Thus, several such carriers were able to compete with AT&T Long Lines Department for a small, but growing percentage of the interstate private-line market.

MCI has widened its data communications offerings, and has acquired WUI, one of the major original international record carriers, and SBS (see 7.6), and BT North America. Today, MCI's offerings closely match AT&T's. Based in Washington, DC, MCI is the second-largest U.S. long-distance carrier.

Sprint

Headquartered in Kansas City, MO, Sprint is one of the survivors that challenged the American Telephone and Telegraph Company in the long-distance communications market.

This company is a specialized long-distance common carrier that provides coast-to-coast private-line communications services via microwave, cable, and satellite facilities for business, institutional, and governmental customers. Founded in 1970 during a diversification effort of its former parent, Southern Pacific Co., Sprint initially built its network along the railroad right-of-way, linking major California cities and the Southwest. It has since rapidly expanded to the East and North, purchasing the voice and data portions of United Video in Missouri, Oklahoma, and Texas in 1974, and the assets of Data Transmission Company (Datran) in 1976.

Sprint acquired the packet-switching carrier Telenet (based in Reston, Va.) from GTE. The bulk of usage on Sprint, advertised as a nationwide optical fiber network, is for long-distance calls. With Telenet, now called SPRINTNet, providing most of Sprint's data services, the carrier's offerings are very similar to the competing services from MCI and AT&T. Sprint has increased its global competitive efforts by recently announcing a "communications partnership" with Deutsche Telekom and France Telecom. One of the partnership's stated goals is to provide integrated voice, data, and video communications services to businesses around the world.

7.5 Value-Added Carriers

A *value-added carrier* uses transmission facilities leased from the communications common carriers, such as AT&T and the independent telephone companies. However, these facilities are connected through computers (switches) and other specialized equipment that produce a new method of transferring data, by adding (compared to raw transmission) such functions as error control and code conversion.

Traditionally, charges for the use of the public switched telephone network have been based on the duration of the connection and the distance between the originator and the receiver. When data transmission requirements increased to the point that a leased line connecting a remote terminal to a computer at a fixed monthly rate was more economical than the use of the switched network, most users still only used the leased-line facility a fraction of the available time for data transmission.

The Federal Communications Commission's rulings over the last few years have sought to encourage competition in the transmission of information. Several approaches to user requirements have emerged. One particular type of value-added network designed to meet the needs of certain types of data communications is based on packet-switching technology. Packet switching is a form of store-and-forward message switching in which the information to be transmitted is grouped into packets. Each packet receives a destination address, so that packets from many subscribers can be routed over common facilities so long as they can be interrogated at certain locations in the network and then routed to their appropriate destinations.

In general, public packet-switching networks offer subscribers an in-place network that one cannot usually cost-effectively develop for oneself by establishing a private network. This is because a public network spreads the cost of the service and resources among many subscribers through the use of the technology. In addition to providing packet-switching service at hundreds of locations throughout the United States, these companies provide transmission service to Europe, Asia, and the Middle East. For example, one carrier, MCI, provides value-added shared-access services via its Tymnet Global Network (acquired in 1993 from BT North America). Coverage is to over 850 U.S. cities and to over 100 foreign countries. Another prominent public packet-switching network is ADP Autonet.

7.6 Satellite Carriers

Satellite communications has become a reality. The growth in satellite communications has been especially large for long-distance communications, especially transoceanic communications, where laying cable is a long and tedious process when compared to a satellite launching.

In 1963, the Communications Satellite Corp. (Comsat) was incorporated to design, launch, and operate communications satellites, as well as to sell circuit capacity to other communications common carriers. Comsat is a privately owned U.S. corporation formed by Congress under the Communications Satellite Act of 1962. Under the provisions of the Act, Comsat was directed to establish a global commercial communications satellite system in cooperation with other countries as quickly as possible. Currently, Comsat derives most of its revenues from the assorted satellite services it provides to a number of U.S. communications common carriers serving the public between the U.S. and foreign locations. These services are offered through the satellites of the International Telecommunications Satellite Organization (Intelsat), which was covered in Chapter 6. Because the cost of transmission via a satellite is independent of distance, satellite transmission has distinctively different cost characteristics when compared to terrestrial communications circuits.

In 1974, Western Union became the first company in the United States to operate domestic satellite communications using its own satellite. In that year, its Westar I and II satellites were launched. Each of these satellites contains 12 transponders, each capable of relaying 1200 one-way voice channels, or 24,000 telegraph channels, or 50 million bits per second of data—or some combination of these. By 1976, Western Union obtained revenues of almost $20 million from Westar services, including almost $9.5 million for facilities leased for fixed terms to three other communications common carriers that were providing communications services in competition with Western Union. In 1976, RCA discontinued its use of Westar service when it launched its own satellite. Subsequently, Westar was acquired by Hughes Network Systems and the satellites were "retired." Western Union has since gone out of the data network business.

In 1974, RCA inaugurated its Phase I domestic satellite service known as *RCA Satcom*, using Telesat Canada's Anik II Satellite and four RCA-operated earth stations near New York City, San Francisco, and Juneau and Anchorage, Alaska. RCA's Phase II satellite service was established in 1976 when RCA-owned satellites were launched into orbit, and several additional earth stations were established. Subsequently, RCA was absorbed by General Electric. In 1993, Martin-Marietta acquired RCA's satellite services from GE.

One of the pioneers in satellite-based data communications, the all-digital Satellite Business Systems (SBS) was formed by IBM, Comsat, and Aetna Insurance. In 1985, IBM became the sole owner. Subsequently, SBS was sold to MCI, which now offers satellite-based transmission with its Communications Network Service (CNS).

Another FCC-authorized domestic-communications satellite common carrier was Contel ASC, since acquired by GTE Spacenet (IDB Worldcom acquired Contel's international services). The circuits transmit voice, data, facsimile, and video (see Chapter 8 for more detail).

Whereas traditional satellite communications, such as the Intelsat network, rely on service interconnections with terrestrial telephone company facilities, Contel ASC, which commenced operation in July 1974 as American Satellite Communications Corp., offered subscribers a service that removed many of the constraints of telephone company interconnections. Formerly, subscribers were limited not so much by the physical capacity of the satellite channel but mainly by the limitations of the interconnecting telephone company facilities, as illustrated in the top portion of Figure 7.6.1. In addition to this public-access type of service, Contel ASC also provided its subscribers with a direct-access (private) service by installing an antenna and assorted equipment necessary to comprise an earth station at the customer's site. This station is usually mounted on the customer's roof or parking lot and connected by a short cable to the subscriber's equipment in the building (see Fig. 7.6.1, bottom.)

7.7 International Record Carriers (IRCs)

Prior to 1983, communications between the United States mainland and international overseas locations were normally provided within the U.S. by six communications common carrier organizations. These organizations were: International Telephone and Telegraph (ITT) World Communications, Western Union International, Inc. (not related to Western Union), TRT Telecommunications Corp., RCA Global Communications, Inc., FTC Communications Inc., and AT&T Long Lines.

United States value-added carriers, such as Telenet and Tymnet, extended their U.S. facilities overseas through contractual arrangements with various international record carriers. Communications from the United States to Canadian locations were provided by AT&T and Western Union.

Part of the FCC's Computer II decision, in effect, removed the monopoly of the IRCs. Following the FCC's August 1982 ruling, just about any domestic communica-

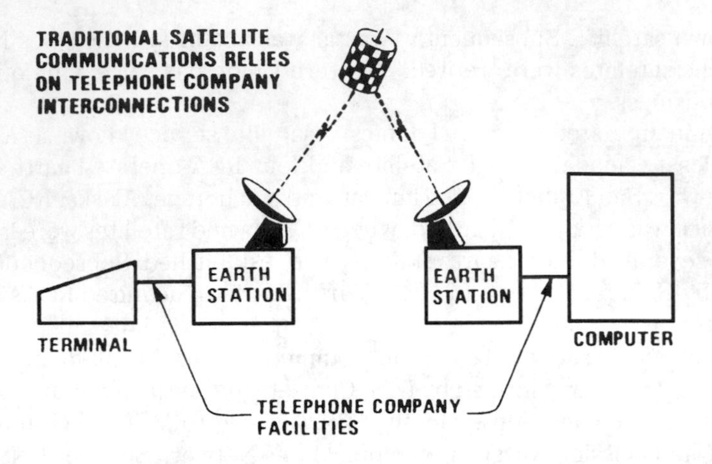

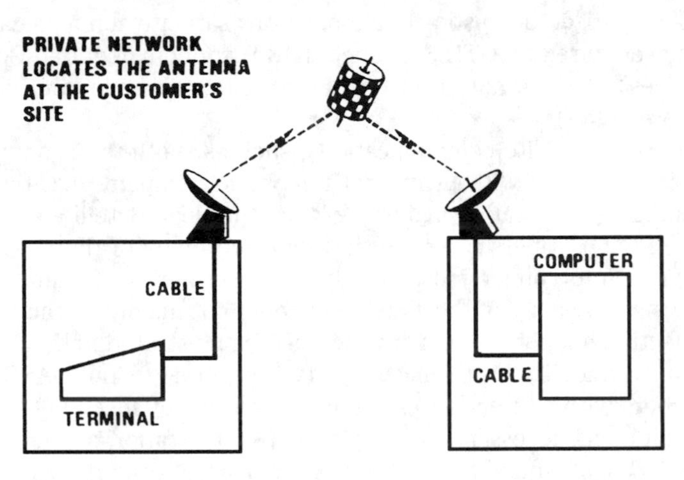

Figure 7.6.1

tions company that could establish an agreement with a foreign government's postal telegraph and telephone (PTT) department was no longer prevented from doing so. Concurrently, the previous restrictions on "gateway" cities for overseas transmission were generally eliminated.

Subsequent to—and in some respects, as a consequence of—the IRC ruling, most of the "old order" changed. ITT Worldcom was acquired by Western Union. WUI became a division of MCI. When RCA was purchased by General Electric, RCA Globcom was sold to MCI. AT&T Long Lines has become AT&T Communications. AT&T acquired the data communications services of Western Union. And IDB WorldCom acquired ITT WorldCom and TRT/FTC. At this writing, LDDS Communications has announced its intention to acquire Wiltel and the IDB Communications Group, IDB WorldCom's parent company. These acquisitions reputedly make LDDS the fourth largest long-distance carrier in the U.S. (behind AT&T, MCI, and Sprint).

FCC rulings have blurred the distinction between domestic carriers and IRCs. Domestic carriers are no longer restricted to domestic operations. Correspondingly, the

IRCs now offer their services within the continental United States. Competition appears to be the keynote.

7.8 Common Carriers in Foreign Countries

In place of independent companies, public telecommunications facilities in most foreign countries are provided by national organizations that are a part of the government. Most such organizations fall under the cognizance of a Minister for Posts and Telecommunications.

In the United Kingdom, until recently, public telecommunications facilities were provided by British Telecom. Although this organization maintained all the equipment that it installed and had some manufacturing facilities, it placed orders for new equipment with several private telephone equipment manufacturers in Britain, as well as a number of overseas firms. The U.K. has undergone its version of divestiture. British Telecom no longer has a monopoly position, and the company is free to compete in the open market.

As an element of the U.K. deregulation process, the British government formed OFTEL as part of its Department of Trade and Industry. OFTEL licensees include British Telecom, Cable and Wireless, Sprint, MCI, IDB, and AT&T as international carriers operating in the U.K.

The European community is continuing to undergo a deregulation process. Lowered barriers are facilitating acceptance of foreign devices.

In France, all public telecommunications facilities are provided by the Telecommunications Division of the State Secretariat for Posts and Telecommunications (PTT). Unlike its former British counterpart, the French PTT owns no manufacturing facilities and orders necessary customer equipment primarily from four French telephone equipment manufacturers.

In most western European countries, a large variety of services are offered for data transmission. As an example, the German Federal Post Office offers switched data communications services at 50 bit/s on the Telex network and at up to 28,800 bit/s (with V.34 modems) on the public switched network, with private line channels available for subscribers to transfer data at E1 (2.048 Mbps).

Questions

7.1 Discuss the new structure of AT&T and the regional companies.

7.2 What function does the Bell Telephone Laboratories perform for AT&T?

7.3 What effect did the 1982 antitrust settlement have on AT&T's business activities?

7.4 What interconnect problems restricted the growth of MCI and other specialized carriers? How were these problems resolved?

7.5 Explain the origin of the term *value-added*.

7.6 What data rates are available to European users?

Chapter

8

Common Carrier Offerings: Cost and Selection Considerations

8.1 Introduction

Today, many communications carriers have facilities designed for data transmission. These facilities range from the public-switched telephone network, which was designed for analog-voice-type transmission, to more recently introduced digital facilities designed expressly for the transmission of digital data. In between these extremes are a number of hybrid services that, through the utilization of computers and the combination of various transmission media, provide specialized offerings. Examples of these offerings include specialized common carriers, value-added carriers, and satellite carriers.

In this chapter, carrier offerings will first be examined as they apply to data communications. Starting with the public switched telephone network, a number of telephone company facilities will be examined to include their operational limitations, cost considerations, and flexibility and ease of utilization. Next, the concepts, interface requirements, and financial considerations of a few specialized offerings will be examined.

Note that all costs and means of establishing them, such as those detailed in the rate tables included herein, should be considered as typical, not necessarily current, and to be used primarily for the examples and exercise questions. If you have actual current applications, refer to the latest figures, obtainable from the appropriate vendors or government organizations.

Finally, a number of general guidelines will be developed based on the comparison of several facilities, assuming predetermined or expected traffic volumes.

8.2 Telephone Company Analog Service

Today, one of the most common methods of data communications occurs over the public switched telephone network. Because this network was designed for the transmission of analog signals, modems (data sets) are used at the transmitting stations to modulate the digital pulses of computers and business machines. The signals are then suitable for transmission over the analog medium. At the receiving station, the modem reconverts (demodulates) that signal back into its original digital format.

Within the category of switched network transmission, several types of services are available for subscriber consideration. The most common service is direct distance dialing (DDD) over the switched network, in which the originator dials the number of the called party. This technique is known as *station-to-station dialing*, in contrast to operator station-to-station and person-to-person switched network calls, where operator intervention is required.

A second category of service is wide-area telecommunications service (WATS), in which calls are made over the switched network. But the economics and geographical areas that can be called might differ from general public switched telephone network usage.

A third type of switched network facility is a foreign exchange (FX) line, really a hybrid combination of a leased line and the switched telephone network. WATS and FX service is covered later in this chapter.

Switched network utilization

Although the past few years have witnessed an increase in the number of specialized communications carriers and their data transmission offerings, today much of data transmission traffic is still carried by the switched telephone network. In this network, a telephone call is switched through to its destination once dialing has been completed.

Once a telephone number is dialed, the number wanted is passed from the subscriber's telephone to the local telephone company central office serving that subscriber, as illustrated in Figure 8.2.1. At that central office, the requested number is examined and circuits are switched to provide the subscriber with a connection to a distant central office, where the number is examined again, and more circuits switched, until the dialing party is connected to the dialed party. Although only two central offices are shown in Figure 8.2.1, a call placed on the switched network can travel through many such offices, with the routing being a function of existing traffic on the network. Thus, a call from New York to Atlanta could conceivably be routed through Boston, Los Angeles, and Kansas City.

Because a different path is likely to be selected each time a call is placed, telephone connections established on the switched network cannot be conditioned for

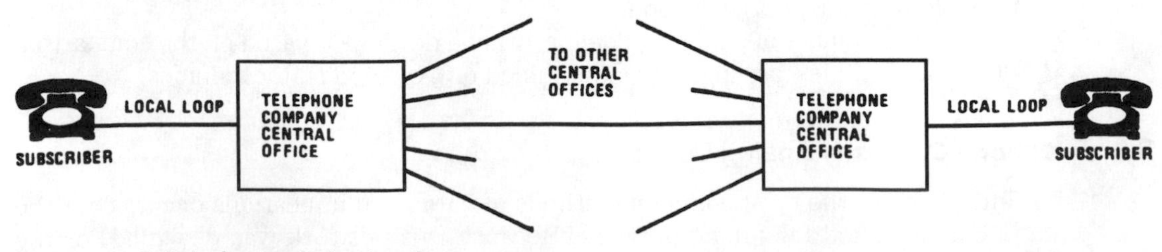

Figure 8.2.1 Switched network circuit path.

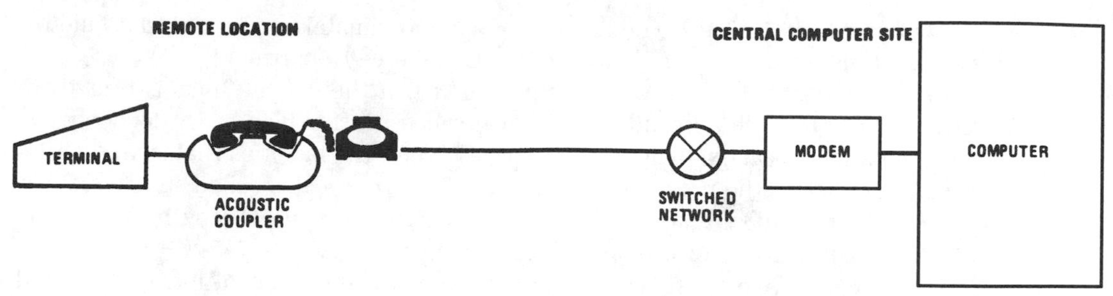

Figure 8.2.2 The switched network can be used for computer timesharing.

the transmission of data. Data rates on analog facilities depend on the modems used. For example, as mentioned previously, the V.34 modem enables operation as high as 28.8 kbps.

The key advantage to using the switched telephone network is that it is universally available. Also, it is less expensive than other types of service when usage is low and the number of users in a geographical area is not large. The primary data use of the switched network today is for timesharing applications at rates ranging from 110 to 14,400 bit/s. Here, a terminal operator dials the telephone number of a modem (connected to a computer) that has an automatic answer feature. At the terminal, the operator has either an acoustic coupler or another modem, as illustrated in Figure 8.2.2 (see Chapter 12 for operation with a PC and its modem). Although only one modem is shown at the computer site, in general, that location has many such devices connected to a rotary switch. The rotary permits a user to dial any one of a block of telephone numbers assigned to the modems and have the call automatically "stepped" to the next available line if the dialed number is busy. One advantage to using the switched network is that, once a subscriber completes the transmission, the modem at the computer site becomes available to service another user. If, for some reason, the quality of the switched connection leaves something to be desired (for example, it is too noisy), the terminal operator has only to terminate the call and redial the number to obtain a probable new routing over the switched network.

Although timesharing applications over the switched network mainly use asynchronous terminals, synchronous devices can also use that network for data transmission. One example of the latter is remote batch terminals, which frequently use the switched network to transmit and receive large volumes of data.

Post-divestiture

Until 1984, when AT&T's divestiture took effect, there were only two types of tariffs: interstate and intrastate. The former were filed by AT&T and the other common carriers (OCCs) with the Federal Communications Commission (FCC); the latter, by the local operating companies with a state's public utility commission (PUC).

In the post-divestiture environment, the distinction between interstate and intrastate survives and determines the appropriate regulatory authority: federal or state. But there is a new criterion—whether a service is within the local areas served by the divested local operating companies. These areas are called *local access and transport areas (LATAs)*. They correspond, approximately, to the standard metropolitan statistical areas defined by the U.S. Commerce Department.

If a service is within a LATA, it is intra-LATA. If the service links two or more LATAs, it is inter-LATA. In addition, there are access tariffs filed by the local operating companies (local exchange carriers or LECs) for the facilities that span their local areas to access the network of an inter-LATA carrier—also called an *intercity* or *interexchange carrier (IEC)*. These access tariffs are the means by which the LECs charge the IECs for use of the local facilities.

So, instead of two tariff types, there are now six: inter-LATA, intra-LATA, and LATA access, for both federal (interstate) and intrastate. The federal inter-LATA tariffs are filed by the IECS. The federal intra-LATA tariffs apply to LATAs like Washington, D.C., which includes substantial sections of Maryland and Virginia. Similarly, the LATAs for Philadelphia, Chicago, St. Louis, Kansas City, Louisville, and Cincinnati include sizable out-of-state sections. These tariffs are filed by the LEC with the FCC.

The intrastate, inter-LATA tariffs apply to the intrastate business transferred to AT&T from the LECs by the divestiture agreement. The intrastate, intra-LATA tariffs cover the intrastate business still remaining with the LECS. The LATA access tariffs, both federal and intrastate, govern local access to an inter-LATA carrier's network.

Within each LATA are interface points—called *points-of-presence (POPs)*—to the IECs. Each IEC has its own list of POPs. AT&T calls its POPs "serving offices." These points are the only places within a LATA where the IEC can receive and deliver traffic. Therefore, a customer's premises must use facilities of the LEC to connect to an IEC at the latter's POP. However, not every IEC service—such as DDS (Dataphone digital service), and T1—is available at every POP.

Switched network cost

The cost of a telephone call on the switched network depends on such factors as the time of day that the call is originated, the day the call is originated, whether the [1] call is interstate or intrastate (inter- or intra-LATA), the distance between the called and calling parties, and the duration of the call.

When calls are placed within a LATA, a number of different types of rates might be available. Some telephone companies offer a category of service that permits the subscriber to dial an unlimited number of local calls for one monthly fixed rate. Or the subscriber can select a lower monthly fixed rate that permits less calls, and then must pay an additional amount for each call dialed in excess of the base number permitted.

Because switched telephone network utilization is normally economical compared with leased-line service (which is charged on a flat-fee monthly basis), one common procedure is to estimate the monthly communications cost of a terminal via the switched network. This can be done in several ways. First, if usage varies from day to day, a log can be kept for the month of total transmission minutes per

[1] Much of the following material in this chapter is adapted from reports of The Aries Group and of CCMI ("Guide to Networking Services"), both of Rockville, MD.

day, and a monthly cost developed. If an estimate of daily average transmission length is available, then simply multiplying that number by 22 working days per month might suffice.

Note that in a switched network, several seconds might elapse for call completion, during which period no data can be transmitted.

Private leased lines

A private leased line is a permanent circuit dedicated to the exclusive use of a subscriber. Such a line bypasses the public switched telephone networks and their signaling equipment. Because the circuit is permanent, the same circuit is always used for each transmission, so conditioning can be applied to the path to improve its transmission quality.

There are now two possible implementations of private-line service: either the transport circuit (also called *baseline service*) or the total-service circuit. A transport circuit connects just two central offices, leaving the user responsible for dealing with the LEC for the local connections at each end (to be acquired under Federal access tariffs). Under FCC Tariff 9, users also have the option to construct their own connections to the nearest CO and bypass the LEC.

If the private line is ordered as a transport circuit, the user is responsible for ordering, engineering, maintaining, and testing the total end-to-end connection, besides overseeing the actual installation. The user receives three bills:

- One for special access from the LEC at the local end—at its rates;
- One from the IEC for the circuit between COs;
- One for special access from the LEC at the remote end—at its rates, which could be quite different and be billed on a different cycle.

Total-service circuits are similar to the pre-divestiture private lines installed under the Multischedule Private Line (MPL) tariff. The IEC has end-to-end responsibility for ordering, installing, servicing, maintaining, and billing the circuit as a premises-to-premises service. The IEC deals directly with the LEC on both ends to ensure proper line operation. In certain cases, the user might have to pay a surcharge for this extra service (the Access Coordination Function in FCC Tariff 11). At the user's option, there would be just one bill from the IEC, or two (or more) bills—one from the IEC and one for the Special Access facilities from each LEC involved. This latter option is called *Coordinated Access*.

One advantage of a leased line for many applications is that higher data rates are possible than can be obtained on the switched network. Another advantage of a leased line is that no setup time is required to dial the other party, have the call routed, and have a device on the other end answer the call. Because this line is used exclusively by the subscriber, there is no possibility that a busyline signal will be encountered, as when calls are placed over the switched network. Also, the monthly cost of a leased line is fixed, regardless of its usage, whereas the cost of transmitting data on the switched network normally increases as the transmission time increases.

Types of circuits

Presently, communications common carriers offer seven types or classes of private circuits:

- Telegraph grade
- Voice grade
- Wideband
- Audio and TV
- Dataphone Digital Service (DDS)
- T1 (1.544 Mbit/s)
- T3 (44.736 Mbit/s)

The more specialized the type of private line, the fewer the number of central offices available for that service.

A telegraph-grade or subvoice-band channel is a private line, for the exclusive use of a subscriber that has a bandwidth of less than 300 Hz. This type of circuit permits transmission at data rates up to about 150 bit/s, has the narrowest bandwidth, and permits the lowest rate of data transfer of the six types of circuits offered.

A voice circuit has a bandwidth of approximately 3000 Hz. This is the most-frequently used leased line and permits data transfer at speeds up to 28,800 bit/s (refer to Sec. 4.12).

The bandwidth of a number of voice-grade lines can be combined into a wideband (broadband) circuit. Such a circuit has a bandwidth well over 3000 Hz, which supports data at about 500,000 bit/s or more.

The most frequently used leased line for data transmission is a voice-grade circuit with a bandwidth of about 3000 Hz. AT&T refers to this type of line used for data transmission as a *Type 3002 circuit*. The user should contact the telephone company representative to obtain current pricing information because tariffs change quite frequently. In addition, you can also ask for specific pricing information from telephone company representatives for the other types of circuits listed above.

When the quantity of transmitted data increases, faster terminals and higher-speed modems can be used to maintain or reduce the transmission time over the switched network. At a certain volume of transmission, the cost of telephone calls made over the switched network will equal or exceed the cost of a private leased line. This crossover point between leased and switched-network facilities varies because of a number of leased-line factors. These factors can best be explained by examining the structure of FCC Tariffs 9 and 11 (which AT&T filed, but of which the other IECs have equivalents).

Review of FCC Tariff 9

Under FCC Tariff 9, the portions of interstate private lines connecting AT&T's points-of-presence have three rate elements:

- *Interoffice Channel.* A distance-sensitive charge for the channel between two AT&T COs, based on the V&H coordinate method. The list of AT&T COs and their coordinates is defined in FCC Tariff 10. Each link in a multipoint circuit is priced separately, using a distance table (Table 8.2.1). At each step in the distance table, there is a fixed charge plus a per-mile charge that is applied against *all* mileage.
- *Central Office Connections.* Per-service-termination charges for the functions at the AT&T CO that interconnect interoffice channels with the local channels and "other access" (special access and bypass facilities). They are assessed as a monthly charge (Table 8.2.2).
- *Channel Options.* A per-channel charge for the features added to a channel to change or augment its transmission characteristics. These features include signaling, data conditioning, and special routing. The charges are assessed on each length of interoffice channel (Table 8.2.3).

TABLE 8.2.1 Interoffice Channel

(Monthly rates as of July–Nov. 1993 and March 1994)

Mileage band	Telegraph/voice grade	DDS (to 9.6 kbit/s)
All	$281 + 0.33/mi	$316 + 0.37/mi

Mileage band	19.2 kbit/s DDS	56 kbit/s DDS	T1*
All	$364 + 0.43/mi		$2800 + 3.95/mi
1 to 50		$365 + 6.29/mi	
51 to 100		$477 + 4.05/mi	
101 to 500		$627 + 2.54/mi	
501 to 5750		$1183 + 1.43/mi	

* 3-yr. discount is 22%; 5-yr. discount is 31%.
SOURCE: CCMI

TABLE 8.2.2 Central Office Connections

(Rates as of November 1993)

	Monthly	Installation
Metallic	$95.45	$82.70
Telegraph	159.00	82.70
Voice grade data	22.10	215.00
DDS (all but 56 kbit/s)	21.30	167.00
DDS (56 kbit/s)	53.25	252.00
T1	270.00	340.00

Note: In counting central office connections, there is one charge for each physical connection at the AT&T central office.

TABLE 8.2.3 Channel Options

(Rates as of Nov. 1993 & March 1994)	Monthly	Installation
Signalling	N/C	NC
C-1 Conditioning	N/C	N/C
C-2 Conditioning	N/C	N/C
C-3 Conditioning	N/C	N/C
C-4 Conditioning	N/C	N/C
C-5 Conditioning	N/C	N/C
C-7 Conditioning	N/C	N/C
C-8 Conditioning	N/C	N/C
D-1 Conditioning	N/C	N/C
D-3 Conditioning (data)	N/C	N/C
D-5 Conditioning	N/C	N/C
D-6 Conditioning	N/C	766.00
M-24 Multiplexing (T1)	208.00	1095.00
M-44 Multiplexing (T1)	675.00	N/C

Review of FCC Tariff 11

AT&T's rates for local channels are filed in FCC Tariff 11. Local channels represent the local facilities that AT&T acquires from the local operating companies and that are needed to connect a customer's premises to its point-of-presence. These rates reflect the rates in the Access Tariffs filed by the local operating companies. These Access Tariffs govern the facilities and services provided by the local operating companies to AT&T and the other intercity carriers. Because each local operating company has its own cost structure and its own revenue requirements, the rates for local channels in FCC Tariff 11 have been averaged by LATA. So, there are 192 different sets of rate tables for each type of local channel, one for each of the 192 LATAs that AT&T recognizes.

Local channels are measured between wire centers [the wire center definitions are in NECA's (National Exchange Carrier Association) FCC Tariff 4] and connect two points located in the same LATA (local channels cannot cross a LATA boundary). These channels are priced by the same "fixed plus per mile" method used for interoffice channels in FCC Tariff 9.

AT&T can provide service to multiple customer premises within a LATA with bridges at local exchange carrier (LEC) central offices (COs), unless AT&T cannot meet design and technical performance criteria.

Voice-grade and telegraph-grade local channels are also described as being either one-way (half-duplex) or two-way (full-duplex). For voice-grade channels, the two-way ones are about $10 to 15 more expensive than one-way. For typical voice-grade local channels in the New York Metro LATA effective with the latest major change on March 1994, see Table 8.2.4.

Other charges—FCC Tariff 11

Under FCC Tariff 11, there are also charges assessed for signaling and conditioning, again different charges for each LATA. For example, the charges in the New York Metro, Washington DC, Los Angeles, and Chicago LATAs are shown in Table 8.2.5.

Access Coordination Function

Under the Access Coordination Function of FCC Tariff 11, customers can have AT&T engineers acquire and maintain the local connections. This feature is required for each termination on a local channel ordered under FCC Tariff 11. The cost of this service varies by private line type (Table 8.2.6).

The specific services that AT&T provides as part of the Access Coordination Function are:

- Access compatibility review
- Access-service-request preparation and design-layout-report review
- Coordination of access installation
- Installation of a physical connection
- Preservice testing and turn-on

TABLE 8.2.4 Local Channel Rates—Voice Grade—NY Metro LATA 132

(Rates as of March 1994)

	Local		Bridged		Interbridged	
Mileage	Fixed	Per mile	Fixed	Per mile	Fixed	Per mile
0	$78.71	—	$48.21	—	—	—
Over 0	111.79	$6.03	81.28	$6.03	$50.79	$6.03

Local installation: $751.36
Bridged installation: $455.55
Interbridged installation: $47.48

Local Channels—Other Grades—NY Metro LATA 132

Line speed (bps)	Mileage	Monthly Fixed	Monthly Per Mile	Install
2.4K	0	$166.56	—	$428.15
	Over 0	192.45	0.64	
4.8K	0	173.96	—	428.15
	Over 0	204.11	0.76	
9.6K	0	185.74	—	444.47
	Over 0	245.87	1.38	
56K	0	243.11	—	479.53
	Over 0	322.68	2.16	

SOURCE: CCMI

TABLE 8.2.5 Rates for Channel Options—FCC Tariff 11

(Rates as of August and September 1993)

	New York	D.C.	L.A.	Chicago
Signaling	$10.84	$ 7.87	$ 1.09	$ 6.89
C-Conditioning	5.11	1.59	9.78	7.10
Installation ("C")	—	—	186.95	85.80
D-Conditioning	1.16	0.00	1.41	7.29
Installation ("D")	32.78	211.12	294.86	343.18
DDS Bridging (all sites)				
2.4–9.6K	$21.30 (installation: $167.00)			
56K	53.25 (installation: 252.00)			

NOTES: Conditioning and signaling charges on local channels must be applied twice per drop: once at the customer's premises and once at AT&T's central office.

SOURCE: CCMI

TABLE 8.2.6 Access Coordination Function

(Rates as of August 1993 & February 1994)

	Monthly	Installation
Telegraph (30 baud)	$ 9.55	$149.00
Telegraph (75 and 150 baud)	15.90	149.00
Voice Grade	29.75	174.00
DDS (2.4 to 9.6)	29.75	232.00
DDS (56)	29.75	287.00
T1	78.00	215.00

SOURCE: CCMI

Foreign exchange service (refer to Section 4.11)

A foreign exchange (FX) line is a hybrid combination of a leased line and public switched telephone network access. A foreign exchange line can be used in a variety of ways for economical data transmission. When the subscriber in city A picks up the telephone, a connection is made via a leased line to the switched network at city B. Therefore, the subscriber can have "toll-free" access to other telephone numbers in that city. In a data communications environment, city B usually has an automatically answering modem and thereby permits an unlimited number of terminals within that area to communicate, one at a time, with a computer located in city A. In addition to an interexchange-channel monthly bill, which is computed similarly to the previously covered private leased-line charge, several additional costs can be incurred when using foreign exchange service. First, a monthly business exchange line charge, which varies by city, is provided and billed by the local telephone company in city B.

Next, the connection from the customer's location in city A to the local telephone company central office is provided and billed by the telephone company serving that area.

Sample point-to-point (voice) circuits

To help understand the tariffs better, a few of examples follow:

Example 1: Tie Line (4-wire) Between Bethesda, MD and Los Angeles, CA

The tie line, if acquired from AT&T as an end-to-end connection, requires local channels at each end—one from Bethesda to AT&T's central office in Silver Spring, MD and the other within the city of Los Angeles. The Bethesda end is rated from the CHCHMDBE wire center and is four miles away from AT&T's central office. The Los Angeles end is rated from the LSANCA03 wire center, the same one that serves AT&T's central office in that city. These two intraLATA connections are governed by FCC Tariff 11.

The connection provided directly by AT&T goes from Los Angeles to Silver Spring (2292 miles), and is governed by FCC Tariff 9.

AT&T portion:		
Interoffice channel:	$281.00 + (2292 × $0.33)	$1037.36
Central office connections 2 @ $22.10		44.20
Signaling:		—
Total:		$1081.56
Local channels:		
Los Angeles end (LATA 730)		
Local channel:		75.85
Surcharge:		—
Signaling: (2 @ $1.04)		3.18
Total:		$ 79.03
Bethesda end (LATA 236)		
Local channel:		107.50
Surcharge:		—
Signaling: (2 @ $11.87)		23.74
Total:		$ 131.24
Access coordination: 2 @ $29.75		$ 59.50
Local channel total:		$ 269.77
Grand total:		$1351.33

Example 2: FX line (2-wire) from Bethesda to a Los Angeles open end

AT&T portion:	$1081.56
Local channels:	
Bethesda end (LATA 236)	
Local channel:	107.50
Surcharge:	—

Signaling: (2 @ $11.87)	23.74
Total:	$ 131.24
Access coordination: 2 @ $29.75	$ 59.50
Switched access (Premium FX):	
Los Angeles open end:	
(Pacific Bell Tariff FCC 128)	0.1347 to 7.72 cents/min.

Notice that there is no charge for signaling on the open end.

A sample data circuit

In FCC Tariff 9 (AT&T-provided facilities), multipoint connections can be specified among the AT&T central offices, following the minimum distance "spanning tree" required to connect all points. FCC Tariff 11 is not so clear about the methodology to be used in connecting multiple drops by a single local channel to the appropriate AT&T central office. The rules for multidrop local channels are (unless the customer wishes to use the much more complicated rules for bridged, multipoint local channels):

- Each drop must have its own local channel to its AT&T central office
- Each drop pays for one central office connection
- Each drop pays the necessary mileage to reach its AT&T central office

As an example, consider a circuit from a host in Philadelphia (PHLAPALO wire center) connecting five drops in Rockville, MD (RKVLMDRV wire center) and three in Gaithersburg, MD (GTBGMDGB wire center). The closest AT&T central office to the two Maryland cities is Silver Spring, MD. The connection between the AT&T central office in Philadelphia and Silver Spring is governed by FCC Tariff 9. The local portions on each end are governed by FCC Tariff 11. The appropriate charges are, assuming D-5 conditioning (9600-bit/s multipoint) and surcharge exempt:

AT&T Portion:	
Interoffice channel (120 mi.): $281 + 120 × 0.33=	320.60
Central Office connections: 9 at $22.10	198.90
D-5 conditioning:	1.94
Local Channels:	
Philadelphia end: (LATA 228)	
Local Channel (4-wire, 1 mile)	87.46
D-5 conditioning (1 link at 2 × $1.94)	3.88
Access Coordination (1 link at $29.75)	29.75
Silver Spring End: (LATA 236) IR	
D-5 conditioning (8 links at 2 × $0.00)	0.00
Access Coordination (8 links at $29.75)	238.00
5 Local Channels to Rockville (at $124.56)	622.80
3 Local Channels to Gaithersburg (at $124.56)	373.68
Total:	$1877.01

Using special access with FCC Tariff 9

AT&T itself pointed out that its multidrop customers do have another choice in linking their remote drops to its points-of-presence—either special-access services acquired directly from the local operating companies or privately acquired facilities (facility bypass). The transport circuit option frightened many people because it left them totally responsible for ensuring that the AT&T and the locally provided connections were ordered, installed, serviced, maintained and billed properly.

If a user has AT&T acquire the necessary special access facilities on the user's behalf (called *coordinated access*), AT&T does not have the responsibility for paying the bills for special access. These bills would now come directly to the user.

An example of special access with FCC Tariff 9

If the last example were configured as a transport circuit under the special access rates, it would be priced as:

- A special access connection in Philadelphia
- Intercity line under FCC Tariff 9 (AT&T-provided facilities)
- Special access connections in Maryland

The transport design costs less than the pricing using FCC Tariff 11. However, it is a more complicated design requiring detailed knowledge of the various access tariffs—and a greater responsibility for the performance of the line. Using coordinated access will reduce the total service cost—but still more expensive than the Transport cost, possibly a reasonable compromise between cost-savings and end-to-end maintenance. Remember that under coordinated access, AT&T still orders, installs, services, and maintains an end-to-end connection.

Bridged local channels

When AT&T bridges multidrop lines at its point-of-presence, it is billed by the local operating companies for two-channel terminations *for each customer drop* under the access tariffs. AT&T now allows local channels to be bridged at the wire centers of the local operating companies as well.

In changes to FCC Tariff 11, there are just three types of local channel involved in a bridged local channel—an option AT&T calls "LEC Bridging":

- *Local channel.* Connects a point to an AT&T central office.
- *Bridged local channel.* Connects a point to a customer-designated bridging hub of the local operating company.
- *Interbridge local channel.* Connects two bridging hubs.

A sample bridged local channel

In the previous sample data circuit, the points in the Washington LATA can be placed on a bridged local channel. The portion in the Philadelphia LATA and the intercity portion remain the same.

In using bridged local channels, you must specify to AT&T the bridging hubs to be used. From the optimized transport design in this section, you know that Rockville is the best choice from among the three candidate bridging hubs. With a single hub in the Washington LATA, all nine terminations, including the one to AT&T's central office in Silver Spring, will count as bridged local channels. The five in Rockville involve no mileage. The three in Gaithersburg are six miles away from the bridging hub, and the AT&T central office is nine miles away.

The use of bridged local channels in the Maryland portion of this circuit reduces the cost from the unbridged case. Coordinated access is a good compromise.

Rationalizing the pricing of local channels

In pricing the sample data circuit, there are very different ways to price essentially the same service from Philadelphia to eight terminals in the Maryland suburbs of Washington:

- Total service (FCC Tariff 11)
- Transport service (special access)
- Total service (coordinated access)
- Total service (LEC Bridging)

AT&T has been quite aggressive since FCC Tariff 11 went into effect in closing off potential opportunities to lose money in pricing its local channel service.

In computing its rates for local channels, AT&T takes the special access channel rates of the Bell operating company and multiplies them by 1.101. This 10.1% surcharge is intended to recover the nonchannel special-access expenses that AT&T incurs in maintaining a local channel for a customer. These incidental expenses include additional labor, engineering, testing, and on-site visits that are billed to AT&T. Then, AT&T adds a $4 per month charge to this surcharged amount to recover its own labor expenses for "direct billing inquiry." This item compensates AT&T for about $45 million of expenses that it incurs in printing and handling private-line bills and in supporting the inquiries of its private line prospects for local-channel service that do not lead to circuit orders.

The 10.1% surcharge alone is applied against the local operating company's special access rates for channel options, such as signaling and conditioning. It is not imposed against the special access surcharge.

The nonrecurring charges for local channels in FCC Tariff 11 are also not a pure pass-through of the special access charges. AT&T adds an extra amount to compensate it for "lead/lag" delays—for the time it must accept delivery of a special access line, but cannot bill it because the end-to-end connection has not been established yet. So, the nonrecurring charges factor in an extra seven days of lead time in establishing a local channel, and an extra day in disconnecting one.

To determine the approximate per-drop difference between the three major private line choices—local channels, coordinated access and transport, use the following:

$$\textit{Total service} = \textit{ACF} + \$4 + (1.101 \times \textit{special access})$$

$$\textit{Coordinated access} = \textit{ACF} + \textit{special access}$$

$$\textit{Transport} = \textit{Actual special access expense}$$

(ACF is the charge for the Access Coordination Function.)

Remember that under the rate-making methodology used by the FCC, AT&T cannot earn any return on its expenses—it can only earn a return on its investment. So, these changes to its local channel tariff do not earn any revenue for AT&T, just enable it to recover its incidental expenses.

Steps for pricing private lines

Under FCC Tariffs 9 and 11, private lines are no longer priced by airline mileage between rate centers, but through AT&T's central offices as intermediate homing points. Tariff 9 describes the algorithm for pricing "total service" private lines:

1. Refer to the pricing point tariff, FCC Tariff 10, Part 3.0 to find the rate center assignment for the NPA-NXX code of each drop on the line.
2. Turn to Part 4.0 and find the LATA assignment of each rate center.
3. Go to Part 5.0 and look up each LATA found in Step 2. Note the central offices available. Remove the central offices that do not have the desired private line service available. You will find nearly 500 central offices that are only usable for radio and television program transmission services.
4. Refer to NECA's (National Exchange Carrier Association) pricing point tariff, FCC Tariff 4 to find the wire center assignment for the NPA-NXX code of each drop on the line. NECA's FCC Tariff 2 is sorted by locality (rate center, almost) so your work in Step 1 will give you at least a chance to find your NPA-NXX codes.
5. Note the LATA assignment in NECA's tariff. Do not be surprised if it is different than the one you found in Step 2. AT&T and NECA still disagree on LATA assignments, occasionally.
6. While in NECA's FCC Tariff 2, look up the wire center V&H coordinates shown for each wire center on your line.
7. Assign each wire center to the closest central office, within its LATA, that has the desired private line service available. Remember that central offices are now specified as wire centers with their own wire center V&H coordinates.
8. Connect the central offices with a minimal spanning tree. Use the interoffice channel charges to price the links between these central offices.
9. Connect each wire center by a point-to-point link to the local operating company's wire center that serves the assigned central office. Use the local channel charges to price each of these links. Apply the special access surcharge, if applicable.

10. Make the decision to use bridged local channels, if there are more than two drops served by the same central office. If you use bridging, you will have to decide which bridging hub(s) you will use. Special codes mark bridging hubs in NECA's FCC Tariff 2. For example, voice-grade data bridging is flagged with a code of "H".
11. List the appropriate signaling and conditioning charges. Remember to apply these charges to both the interoffice and local channel segments of the line. Also remember that signaling and conditioning are imposed twice per drop, once at the central office and once at the premises.
12. Add in the charges for the inside wire recovery charge (telegraph, voice and DDS only).

Overall structure

AT&T's private lines that come under Federal jurisdiction (note: the major issue here is jurisdiction (Federal vs. State), not physical routing, (inter-state vs. intra-state) are covered by three tariffs:

- *FCC Tariff 9—Private Line Services.* Covers those channels and services that AT&T provides directly from its own facilities and the labor of its own people.
- *FCC Tariff 11—Private Line Local Channels.* Covers those local channels and services that AT&T acquires on behalf of its customers from the local operating companies and resells in order to assemble an end-to-end connection.
- *FCC Tariff 10—Mileage and Administrative Information.* The rating point tariff used with the other two. Also contains the list of AT&T's points-of-presence (AT&T central offices) with their V&H coordinates and services available (such as T1 or CSSA).

References:

AT&T's FCC Tariff 9		
Terrestrial Digital Circuits	Sect. 5	pp. 80–91
Dataphone Digital Circuits	Sect. 6	pp. 92–102
Wideband/Group/Supergroup	Sect. 7	pp. 103–115
Voice Grade Circuits	Sect. 8	pp. 116–183
Telegraph Grade Circuits	Sect. 9	pp. 184–190
Terrestrial Audio Circuits	Sect. 10	pp. 191–217
Terrestrial Television	Sect. 11	pp. 218–236
AT&T's FCC Tariff 11		
Special Access Surcharge	Sect. 5.2.4	pp. 68–68.4
Inside Wiring	Sect. 6.2.7.B	pp. 74.1–74.5
Terrestrial Digital Circuits	Sect. 5	pp. 66–69
Dataphone Digital Circuits	Sect. 6	pp. 70–78.12

Wideband/Group/Supergroup	Sect. 7	pp. 79–79.1
Voice Grade Circuits	Sect. 8	pp. 80–95.5
Telegraph Grade Circuits	Sect. 9	pp. 96–105
Terrestrial Audio Circuits	Sect. 10	pp. 106–114
Terrestrial Television	Sect. 11	pp. 115–123

Wide Area Telecommunications Service (WATS)

WATS is a packaged discount telecommunications plan arranged for subscribers who either make many outgoing long-distance calls to many geographical areas or who receive many calls from certain geographical areas. WATS provides the subscriber with dial-switched network access between a station connected to a WATS access line and other stations in designated service areas.

Under the WATS arrangement, customers have a choice of five service areas (Zones 1 through 5) for service within the contiguous United States and one service area (Zone 6) for service to Alaska and Hawaii. Zone 1 consists of all of the subscriber's adjacent states while the higher zones progressively expand the area of service. For example, a user in Georgia with a band 1 WATS can receive toll-free calls from the states of Alabama, Florida, Kentucky, Mississippi, North Carolina, South Carolina, and Tennessee. Service to a higher number zone includes service to all lower numbered zones.

Two categories of WATS service can be selected by subscribers—outward service known as *Outwats* and inward service, *Inwats*. Outwats service provides for the origination of calls from a subscriber with an outward WATS access line to all telephones in a designated zone of service. Similarly, Inwats service provides for the termination of calls to an Inwats access line from all telephones in a designated zone of service. Inwats is the commonly known toll-free "800" number.

Both Inwats and Outwats, when obtained as interstate offerings, do not permit calls to or from points within the same state in which the access line terminates. Such services are considered intrastate and are provided by local operating companies in the subscriber's state.

WATS cost

Both Inwats and Outwats charges are based on mileage and the total monthly communications time. There are also several miscellaneous charges, including access line, installation, and so on. Additional rate elements include volume discounts and optional features. Tables 8.2.7, 8.2.8, and 8.2.9 detail typical AT&T Outwats and Inwats (800) rate information.

8.3 Telephone Company Digital Service

In comparison to leased private voice-grade lines, in which digital data must first be modulated into an analog signal for transmission, Dataphone digital service (DDS) is designed exclusively for the transmission of digital data in digital format, without the

TABLE 8.2.7 Rate Period Definitions

Day	Monday to Friday	8:00 am to 4:59 pm
Evening	Monday to Friday	5:00 pm to 10:59 pm
	Sunday	5:00 pm to 10:59 pm
Night	Sunday to Saturday	11:00 pm to 7:59 am
	Saturday	8:00 am to 10:59 pm
	Sunday	8:00 am to 4:59 pm

NOTE: AT&T no longer recognizes holiday discounts for both WATS services.
SOURCE: CCMI, March 1994

TABLE 8.2.8 Outward WATS/800 Rate Table (Mainland USA)

	Day		Evening		Night	
Miles	First 30 sec.	Add'l 6 sec.	First 30 sec.	Add'l 6 sec.	First 30 sec.	Add'l 6 sec.
55	$0.1215	$0.0243	$0.0835	$0.0167	$0.0700	$0.0140
292	0.1295	0.0259	0.0890	0.0178	0.0730	0.0146
925	0.1360	0.0272	0.0945	0.0189	0.0780	0.0156

SOURCE: CCMI, March 1994

TABLE 8.2.9 Inward WATS/800 Rate Table (USA) per Hour

Day	Evening	Night
$14.80–15.79	$11.84–12.63	$11.84–12.63

SOURCE: CCMI, December 1993

need for modems or other modulation equipment. DDS offers users full-duplex transmission of digital signals at synchronous speeds of 2.4, 4.8, 9.6, 19.2 and 56 kbit/s, using end-to-end digital technology. Both point-to-point and multipoint digital service are available.

The primary advantage of DDS is that noise and distortion are not amplified on that facility as they are in analog transmissions. Instead of amplifying the digital signal, the signal is regenerated at regular intervals as it travels through the DDS network, thus reconstructing new, clean pulses that minimize the possibility of bits being received in error. Because of all-digital technology, DDS is guaranteed to provide the subscriber with an average performance in excess of 99.5-percent error-free seconds for operation at all data rates. If the error performance level should drop below that figure, the period of substandard performance is considered by the telephone company as a circuit interruption, and the subscriber receives a credit allowance for that period of time.

In place of modems that are required on analog circuits, users of DDS can select one of two types of interface devices (refer Section 4.11). The first device, known as a *data service unit (DSU)*, performs such functions as signal regeneration, timing recovery, signal reformatting, proper coding, and decoding of signals, formatting and the generation and recognition of control signals. If the subscribers decide to provide their own equipment to perform the functions of the DSU, the telephone company can furnish the user a device known as a *channel service unit (CSU)* as part of a digital access line to provide network protection and remote loopback testing capability. CSUs are also available from other vendors.

DDS is available as a leased, point-to-point service, or as a switched service. DDS point-to-point channels are for transmission of synchronous sequential binary data signals at speeds of 2.4, 4.8, 9.6, 19.2, or 56 kbit/s. DDS is available for two-point or multipoint applications in full- or half-duplex modes on a full-period basis. The 9.6- or 56-kbit/s service can be used to connect a station on a customer's premises to a basic packet switching service (BPSS) arrangement or to connect two BPSS arrangements.

DDS is configured by combining components to connect two or more central offices or between central offices and points of connection. These offices may be in the same or different LATAs, although DDS is not available in every LATA, nor to all rate centers within a LATA.

Interoffice channels are covered in AT&T's FCC 9 tariff, which offers interLATA, interstate private line service between AT&T central offices. The local portion of a channel, including all coordination elements, is covered in FCC 11.

Rates

Rates for an interoffice channel are fixed monthly amounts based on the following elements:

1. Mileage between AT&T central offices, either two-point or multipoint.
2. A fixed-mileage band charge based on the selected service and the total distance.
3. A flat-rate central-office connection charge.
4. Channel options and office functions.

How to Rate the Interoffice Channel

1. Select the DDS rate table.
2. Determine the CO connection charge (Table 8.3.1) and the following charges.
3. Use the V&H coordinates of the COs found in FCC 10, and determine the mileage of the circuit.
4. Determine the mileage charges from the rate tables by adding the per-mile charges (Table 8.3.2).
5. Add any channel option charges. The result is the applicable charge.

TABLE 8.3.1 DDS CO Connections

Monthly		Install.
2.4 kbit/s–19.2 kbit/s	$21.30	$167.00 + 167.00
56 kbit/s	N/A	252.00 + 252.00

SOURCE: CCMI, September 1993

TABLE 8.3.2 DDS Rates By Transmission Speed—Monthly

Speed (kbit/s)	Mileage limit	Fixed cost	Per mile cost
2.4	All	$ 316.00	$ 0.37
4.8			
9.6			
19.2	All	364.00	0.43
56	50	365.00	6.29
	100	477.00	4.05
	500	627.00	2.54
	5750	1183.00	1.43

SOURCE: CCMI, March 1994

Switched digital service (SDS) is furnished for the switching and transmission of simultaneous two-way 56-kbit/s digital signals. It consists of a common-user digital network furnished between designated AT&T central offices.

A digital access line—56 kbit/s or 1.544 Mbit/s (one channel of T1.5 line)—and terminal interface unit (CSU/DSU) are required to connect a customer's premises to a 56 kbit/s SDS central office, except when local exchange carriers provide a switched digital access service.

Service is available 24 hours a day, 7 days a week. Credit allowances are given for interruptions in service because of the failure of a component furnished under tariff.

Rates

Rates are a composite of 1) usage charges (Table 8.3.3) and 2) the charge for a digital or switched digital access line between the customer's premises and an AT&T central office where 56-kbit/s switched digital service is available, and 3) a monthly minimum-usage charge of $20.00 per switched access line, or $75.00 per special access surcharge.

To price the access line, refer to tariffs FCC 9 and 11. Rates for 56-kbit/s switched digital service as furnished in FCC Tariff 4 are for usage only.

Usage charges depend on whether a call is between 56-kbit/s SDS central offices or between access lines (two customer premises served by the same SDS CO): to price a call between access lines, use the zero mileage rates in Table 8.3.3; to price a call between 56 kbit/s SDS central offices, use the mileage and duration shown.

Accunet Reserved 1.5 (T1) is an interstate service furnished on a two-point or multipoint basis (Table 8.3.4). Two-point service provides for the simultaneous two-way transmission of digital signals at 1.544 Mbit/s. For multipoint service, one customer premises is designated as the originating customer premises for the duration of the call. Digital signals are transmitted from the originating premises to two or more receiving customer premises, which have no return capability.

TABLE 8.3.3 DDS Usage Rates (Dollars)

	Day		Evening		Night	
Miles	Initial 30 sec.	Add'l 6 sec.	Initial 30 sec.	Add'l 6 sec.	Initial 30 sec.	Add'l 6 sec.
0	0.1375	0.0115	0.1275	0.0095	0.1275	0.0095
55	0.1375	0.0115	0.1275	0.0095	0.1275	0.0095
124	0.1515	0.0143	0.1450	0.0130	0.1450	0.0130
292	0.1625	0.0165	0.1495	0.0139	0.1495	0.0139
430	0.1730	0.0186	0.1635	0.0167	0.1635	0.0167
925	0.1855	0.0211	0.1750	0.0190	0.1750	0.0190
1910	0.1895	0.0219	0.1785	0.0197	0.1785	0.0197
3000	0.1955	0.0231	0.1845	0.0209	0.1845	0.0209
5750	0.5550	0.0950	0.5115	0.0863	0.5115	0.0863

SOURCE: CCMI, March 1994

TABLE 8.3.4 Two-Point/ Multipoint 1.544 Mbit/s Rates ($)

	(Day, Evening, or Night)	
Miles	Initial 30 sec.	Add'l 6 sec.
0	3.2250	0.2610
55	3.2250	0.2610
124	3.6425	0.3445
292	3.8310	0.3822
430	4.1480	0.4456
925	4.4465	0.5053
1910	4.5495	0.5259
3000	4.6920	0.5544

SOURCE: CCMI, March 1994

Accunet Reserved Service is provided on a common-user network that is available, on a reservation basis, 24 hours a day, seven days a week.

A call can access the Accunet Reserved network through a DS1 Switch Port at designated serving offices via an access line. DS1 ports and access lines operate at 1.544 Mbit/s and each is dedicated to a single customer.

- *Rates.* Rates are a composite of the following elements: 1) usage charges, 2) call set-up charges, and 3) extension channel charges.
- *Usage charges.* Usage charges depend on whether a call is a two-point or multipoint call and if service is between cities or between access lines. Minimum payment period is one half hour.
- *Service between high-speed switched digital serving (HSSDS) cities.* Service between two or more customer stations that are located in different cities and crossing a state line.
- *Call set-up charges.* Set-up charges apply to each Accunet Reserved call.

Two-point and multipoint service is available to Canada via Buffalo Peace Bridge, New York. For the portion of service within the U.S., domestic rates apply. For the portion of service in Canada, the Canadian company rates apply. Rates to International Points:

Location	Rate/30 Min.
New York to Federal Republic of Germany	$468.00
New York to France	$468.00
New York to Italy	$468.00
New York to Sweden	$468.00
New York to Switzerland	$468.00
New York to United Kingdom (London)	$468.00

Analog/digital connections

If a digital access line from the customer premises is not available, an analog/digital adapter can be used to join two separate services, analog leased line and a DDS circuit. This device provides interface compatibility for control signals and retiming for data signals. It is furnished by the telephone company at its central office for the connection of private-line service and a Dataphone digital service operating at 2.4, 4.8, or 9.6 kbit/s. This device permits an analog extension with modems on both ends of the circuit to be used to link a subscriber outside of a digital serving area to the DDS network.

8.4 Value-Added Carriers

Value-added carriers began operation in the United States during 1975, interconnecting a handful of cities via a new technological approach. These carriers link hundreds of cities in the United States as well as worldwide.

The value-added carriers are different from previously covered data transmission services and private data networks in both the enhanced and extensive offerings to users and the sophisticated technology they use. The technology, called *packet switching*, makes it possible for the value-added carrier to provide any user, large or small, with the kind of fast-response, error-free, low-cost-per-transaction data transmissions previously available only to companies that invested in their own private networks.

In essence, the value-added carrier takes advantage of the substantial economies of scale resulting from one very large network—fully utilizing such expensive resources as transmission lines and concentration equipment by sharing the network among the subscribers. These carriers pass on a portion of the consequent savings to the individual subscriber through tariff charges based mainly on connect time and characters transmitted.

Beyond the economics, leasing existing communications facilities allows the carrier to obtain just as much transmission capacity for each location as is required by the traffic load. This provides the flexibility to adapt quickly to subscriber traffic and geographical demands, and permits the incorporation of other transmission offerings (such as satellite services, as they become available).

In a value-added network, data is transmitted in units called *packets*, with each packet containing a specified number of characters. These packets are then routed through the carrier's network by minicomputers placed in the carrier's switching centers. Other minicomputers in the carrier's network are used as terminal interface processors, connecting computers and terminals to the network so that such devices become compatible with one another for transmissions into packets, and transmit the packets to the carrier's switching centers via common carrier facilities, where they are retransmitted to the packets' destinations.

History

Conceptually and technologically, value-added carriers had their origin in Arpanet (Advanced Research Projects Agency Network), a nationwide consortium of computers at numerous research centers tied together over a packet-switching network. However, Arpanet is operated on behalf of the U.S. Government to support research activities of various Federal agencies and not as a communications common carrier.

In 1973, the Federal Communications Commission approved the concept of value-added networks and determined that they should be regulated as common carriers, permitting potential public network operators to propose a value-added network as long as they applied for FCC approval. On April 16, 1974, Telenet Communications Corp. (now part of the Sprint Communications Co. as the SprintNet service) received FCC authorization as a value-added carrier. Beginning service in 1975 between 16 cities, by 1976 Telenet's service was extended to 30 cities and by 1988 over 18,000 local exchange areas were linked.

Carrier use

Value-added networks (VANs) enable customers with dissimilar data communication devices to communicate with one another. VANs accomplish this by performing speed and protocol conversion to customer data transmissions. Because they supply enhanced services, VAN carriers are not regulated.

VANs usually lease facilities from interexchange or local exchange carriers and supply the switches that tie the network together. These switches perform the conversion functions that would otherwise be done by customer computers.

Most of the carriers use packet-switching technology that assembles data into packets and sends the packets from several channels over one path on high-speed lines. In addition to the data that is to be transmitted, each packet includes control and destination information, such as the header that is used by the packet switch to sort packets as they come in on one channel to be switched to another channel. Packets are checked for transmission errors and are retransmitted if errors are detected.

Figure 8.4.1 shows a typical network. The method of accessing the network is either by private lines or dial-up (switched) services.

Carrier office interface and routing

The criteria for choosing either a dial-up line via the switched network or a leased line to access the carrier's central office are substantially the same as those used in configuring private data communications networks. In short, such factors as transmission speed, acceptable busy-signal incidence, response time, volume of traffic, length of individual transmissions or transactions, and whether line use is substantially continuous or mostly occasional must be considered. In addition, multipoint circuits can be used to connect many terminals on a common line, or individual terminals can be connected on point-to-point leased lines.

As an example of the packet-switching technology, examine a subscriber located in Seattle who wants to transmit a message via a value-added carrier to his computer

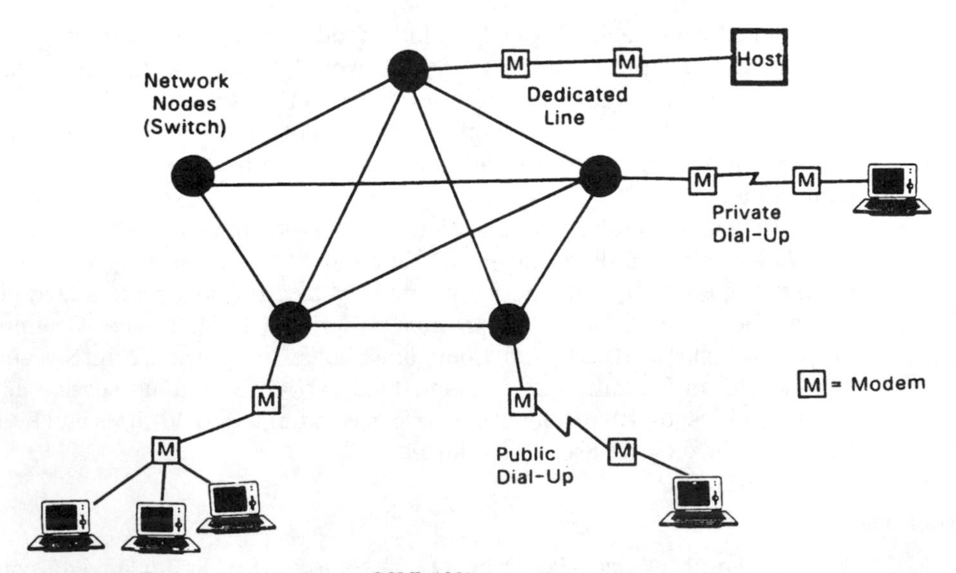

Figure 8.4.1 Typical network. Source CCMI, 1985

center in Houston. To initiate a connection to the carrier, the subscriber dials a Seattle access number or she might have a leased line to the carrier's central office if the volume of transmission is high. Once the subscriber connects to the carrier, she enters the computer center's code to inform the carrier where she desires to be routed to. The carrier then makes it virtual connection for the subscriber to be routed to the computer.

As the subscriber transmits messages to the computer, the carrier's equipment formats the messages into prescribed data packets. After each packet is formed, it is transmitted to and through a number of carrier central offices to Houston. Although the first packet might travel via San Francisco, Los Angeles, and Dallas to Houston, other packets could travel different routes, with all packets being error checked over each hop of the journey and buffered at Houston to await the arrival of the other packets.

As packets arrive at the carrier's Houston central office without error, they are released from the central office and passed to the subscriber's computer via a leased line connecting the computer facility to the carrier's central office. The computer receives the packets in the same order as they were sent because the carrier's equipment rearranges the packets into proper sequence. Although the original message might have started out at a low data rate (for example, at 300 bit/s), once it enters the carrier's network, it will be transmitted at 9600, 50,000, or 56,000 bit/s. Even including electrical propagation, queuing, and acknowledgment delays, a packet will proceed from any source office to any destination office in, on average, one-half second.

Carrier differences

SprintNet and MCI/Tymnet, two of the leading value-added carriers, use markedly different approaches to packet switching. These differences significantly influence the performance and cost of the services offered. Although one carrier is less expensive in certain cases, the other provides better response times in different situations. When both provide equivalent performance, the costs become more competitive. Both carriers offer similar benefits for new users, including immediate nationwide service, high reliability, flexibility, network management, and reduced costs—especially in start-up situations.

SprintNet offers its users support for dial-in asynchronous, dedicated, and TWX connections to the network. Asynchronous connections support transmissions at 300 bit/s (30 characters per second), 1.2 and 2.4 kbit/s (120 and 240 characters per second). Figure 8.4.2 shows one of the ways a remote user can be connected to his host computer via Telenet. Here, the interfaces to the terminal and to the host computer are the same. Telenet also supports the CCITT X.25 host-computer packet-switching interface standard.

In addition to providing packet switching, SprintNet provides dynamic or adaptive routing of packetized data between its major network nodes, as illustrated in Figure 8.4.3. As shown, consecutive packets from a source might follow different routes to the destination node. This dynamic routing permits the network to distribute the intra-network data transmission workload evenly over all the nodes, thereby attaining high line utilization.

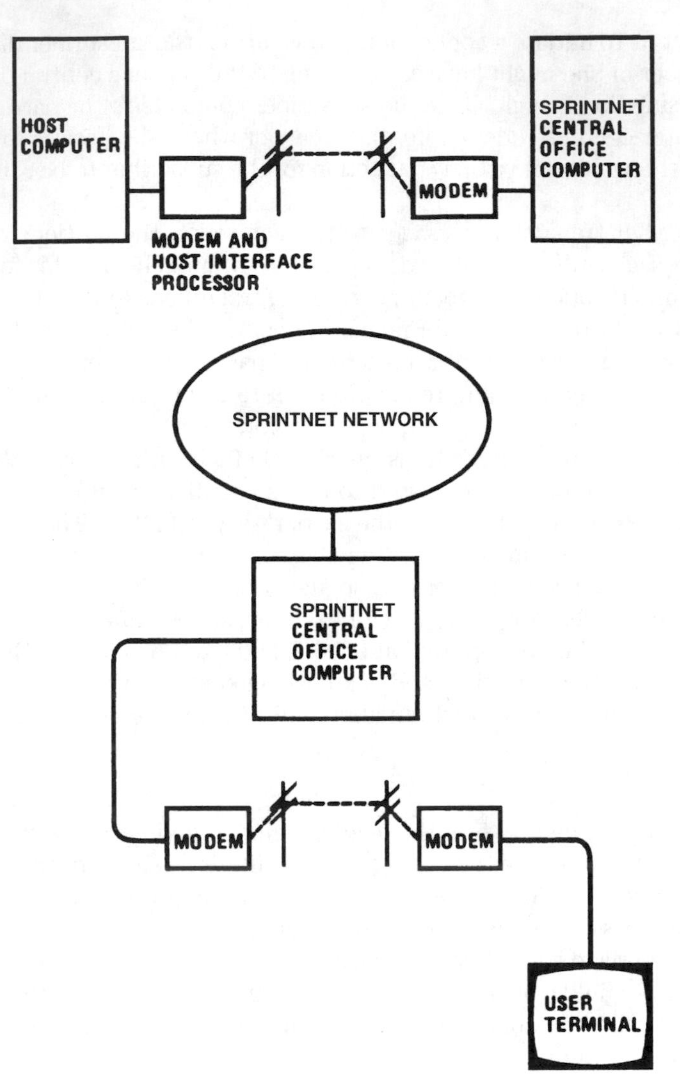

Figure 8.4.2 Connecting to user's host.

Also offered is automatic network recovery from failure or congestion on any node or link. The dynamic alternate routing feature is the main reason for SprintNet's high data transmission reliability. Also, in every transmission between a terminal and a host computer, there are two nodes that must operate—the node directly connected to the terminal and the node directly connected to the host computer. Because most SprintNet nodes are equipped with "hot standby" hardware, total node failures are extremely rare.

Proximity to SprintNet's service cities is important for dial-in users because such users must pay a distance-dependent telephone charge to connect terminals to SprintNet. For subscribers that anticipate large volumes of traffic, SprintNet provides Inwats service to connect a subscriber's terminal or computer to a SprintNet central office. For the connection to a customer's computer site, leased access charges vary with the type of computer interface, the number of computer ports to be accessed, and the transmission rate. These leased-line fees can vary from a few hundred dollars per month to thousands of dollars per month. In addition, SprintNet charges subscribers a monthly account fee.

Interfacing the host

Two types of host interfaces are available from SprintNet. With the first type of interface, SprintNet connects a network access processor at the customer's computer, through multiple low- and high-speed asynchronous ports. It appears to the customer computer as a rotary of dial-in modems. No hardware or software changes to the customer computer are required.

The second type of host-to-network interface available from SprintNet is the packet-mode interface: X.25. This interface is available to SprintNet customers throughout the United States at a flat monthly charge (detailed later) independent

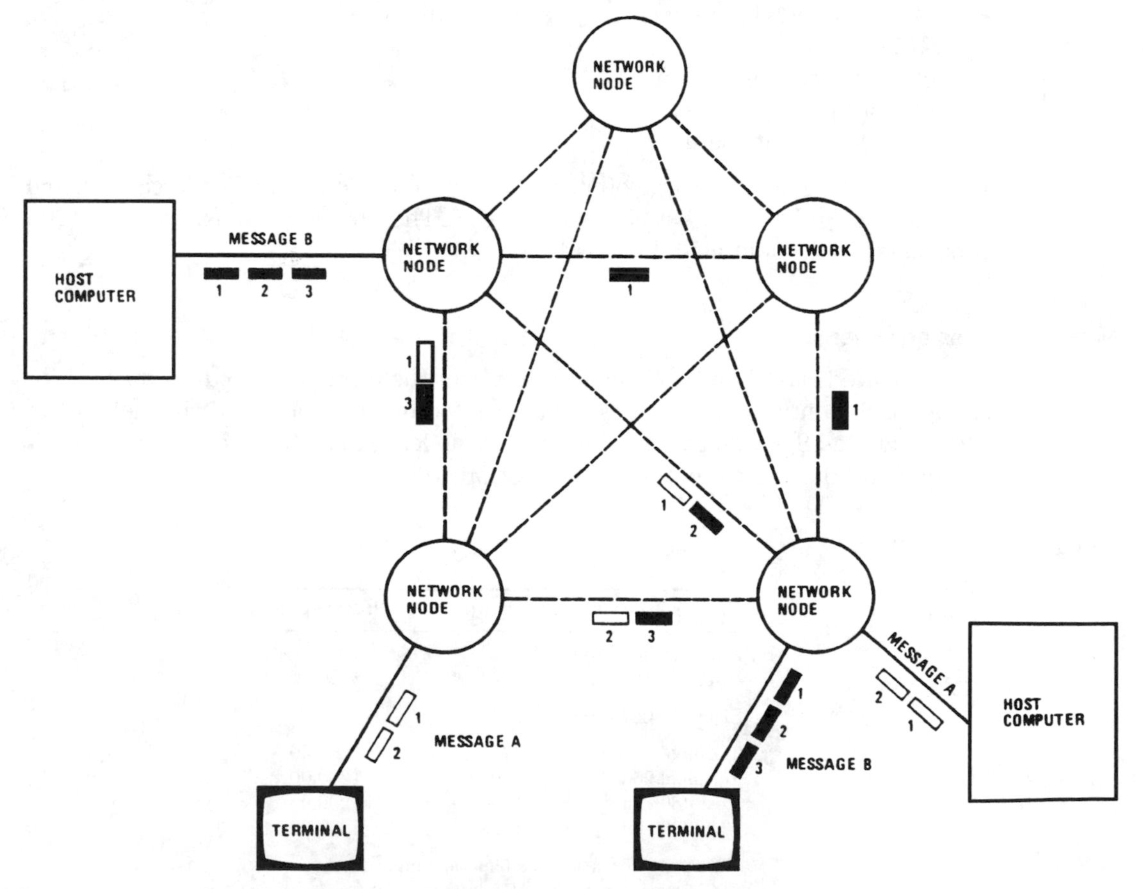

Figure 8.4.3 Dynamic routing.

of customer location. The X.25 interface is also available on public packet networks in other countries, including Bell Canada's Datapac network, the French Transpac network, and the British International Packet Switching Service.

SprintNet services & rates (typical)

The SprintNet public network consists of an optical-fiber backbone transmission facility, interconnecting switching centers located in major U.S. cities. These switching centers are called *access centers*. The switches handle terminal and host connections, tie in remote access localities, and route data to its destination. SprintNet's network complies with the X.25 standards established by the international CCITT (now ITU-T) governing board. Public dial-up access is presently available at 300, 1200, and 2400 bit/s with optional local error protection. Terminal and host leased line service is available from 300 bit/s to 56 kbit/s dependent on the host application. Data is segmented into character blocks called *packets*. Automatic error detection and correction of transmitted packets is a standard feature of the service.

- *Rates*. The total monthly charge is a composite of:
  ~traffic
  ~network access arrangements
  ~optional services and features
  ~monthly account charges.
- *Traffic*. A minimum charge of 50 packets per connection applies to each virtual connection originating within mainland U.S. The charge per kilosegment (1000 packets) in the continental U.S. is $1.40.

Network access arrangements

For dedicated leased-line service SprintNet uses the term *dedicated access facility (DAF)*, the monthly cost includes the associated modems, a leased channel port in the designated SprintNet access center, and the leased line between the customer's location and the access center. (Source: CCMI, 1991)

DAFs

Terminal speeds	Monthly	Installation
Asynchronous 300 to 1200 bit/s	$ 600.00	$1200.00
SDLC* & 3270 BSC*		
Terminal 4800 bit/s	425.00	1200.00
Terminal 9600 bit/s	650.00	1200.00
Host 4800 bit/s	1200.00	1200.00
Host 9600 bit/s	1525.00	1200.00

*SDLC: Synchronous Data Link Communications;
BSC: Binary Synchronous Communications

X.25#		
2400 bit/s	1000.00	1200.00
4800 bit/s	1200.00	1200.00
9600 bit/s	1525.00	1200.00
14,400 bit/s	2000.00	1500.00
56,000 bit/s	Contact co. rep	Contact co. rep.

#X.25 network interface must be used in conjunction with an X.25 DAF.

For public dial-in access, SprintNet offers asynchronous dial service at 300 to 9600 bit/s. Minimum call duration is two minutes.

Dial location	Hourly rate
Class A access center	$ 7.50
Class B access center	9.00
Class C access center	9.00
Inwats	15.50

Async-to-3270 service is also available. Public dial-in rates apply. Other services include: X.25 dial, 1200 to 2400 bit/s; SDLC and 3270 BSC, 2400 and 4800 bit/s; X.25 dial, 1200, 2400, and 4800 bit/s; async to 5250.

Transactions

A SprintNet transaction occurs when a message is sent from a terminal to a host computer, or vice versa, and a response message is returned.

Because the charges are based on 64-character packets, sending a 65-character message costs twice as much as sending a 64-character message. Clearly, the closer each of the two messages in a transaction is to a multiple of 64, the more efficient the data transmission over the SprintNet network. As indicated later, the efficiency is also dependent on the network packetizing algorithm when it is applied.

The packet charges on the SprintNet network can be calculated for an application using the transaction concept. The fixed monthly data transmission load (in millions of characters) for any application can be expressed in packets by assuming that data transmission occurs as transactions—a message is sent from a source to a destination, and, at a later time, a response message is received at the original source from that destination; that all message sizes can be expressed as ratios of sent to received—1:1, 1:2, 1:3, 1:4, 1:5, etc.; and that protocol handshaking characters are not transmitted over the network. Based on these assumptions, the minimum number of packets in a transaction are determined:

$$\textit{Packets per transaction} = \frac{\textit{smaller message size}}{64} + \frac{\textit{larger message size}}{64}$$

Notice that each message-size division must be rounded to the next higher whole number—there are no fractional packets.

The number of characters in a transaction is determined by simply summing the message sizes:

Characters per transaction = *(smaller message size)* + *(larger message size)*

The number of transactions in a fixed monthly load is calculated by dividing the characters per transaction into the total characters per month:

Transactions per month = (*Total characters per month*)/(*Characters per transaction*)

Finally, the monthly packet transmission costs for SprintNet can be calculated using the formula:

Monthly packet costs = ($1.40/1000 packets) × (*packets/transaction*) × (*transactions/month*)

To illustrate this formula, consider the following two cases. In the first, the smaller outgoing message is 12 characters and the larger return message is 60 characters (a 1:5 ratio), yielding a minimum of two packets per transaction. In the second case, the smaller message is increased by one character (13 characters) and the 1:5 ratio maintained, making the larger message 65 characters. The larger message now exceeds 64 characters. As a result, a minimum of three packets is now required for each complete transaction. The packet transmission costs for a 10-million-character monthly load are $389.20 in the first case and $537.60 in the second. This is a 38-percent difference for the six extra characters per transaction.

The smaller message-size ratios (such as 1:1) are almost always more expensive, and a sharp increase in cost occurs as the messages get smaller and packet usage becomes more inefficient. The network packetizing algorithm might also introduce more inefficiency in order to decrease response time (covered later).

In typical interactive transaction applications, average output messages, which are computer generated, are greater in length than input messages, which are people generated. Thus, message-size ratios are generally 1:5 or larger.

Response time

The *response time* is defined as the time between the end of the operator input and the beginning of the receipt of computer output—that is, the time waiting for a reply to begin. The response-time length has three components: access-line transmission time, network transition time, and host computer time.

Delays introduced by SprintNet's network, as well as the overall message transmission time, can be estimated. Host time depends on the user application.

Tymnet's character

In contrast to Telenet's message orientation, MCI/Tymnet is character oriented. Although it uses packet switching within the network, the packets might contain data from multiple users. The packets (64 characters each) transfer data between nodes. Because they can contain data from many users, they are almost always full.

In this way, the transmission bandwidth between nodes is efficiently used. Tymnet does not use a dynamic routing algorithm for each packet, but rather dynamically establishes a path that remains fixed for the duration of the logical connection.

To understand how the shared-packets/fixed-routing technology works, let's walk through a terminal transaction of 25 characters to the host, 125 characters back to the terminal, noting the time required for each step to execute:

1. Terminal logs onto the network. As soon as the network validates the user account, a path to the host computer is established. This path is based on the most cost-effective routing at the time of the log-in (not included in response-time calculation).

2. Terminal enters characters. The network passes each character along to the host, assembling and disassembling packets along the way (see Figure 8.4.4) until the message reaches the host.

3. The terminal enters the last character. (1 character at 30 characters per second = 0.03 sec)

4. The network inserts this character into the first available packet, and passes it along to the host computer. The packets consist of 512 bits (8 bits/character), and are assumed to pass through three intermediate nodes. (512 bits @ 9.6 kbit/s = 0.053 sec × 4 = 0.21 sec)

5. The host computer processes the message and then returns the first character of the response message to its network node. (1 character @ 30 characters per second = 0.03 sec)

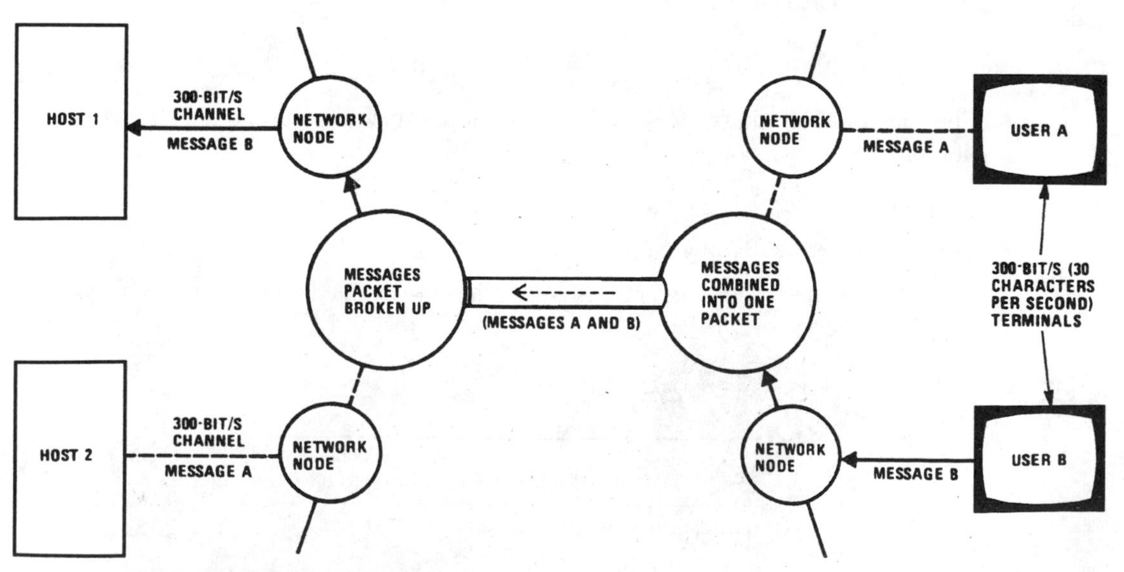

Figure 8.4.4 MCI Tymnet's routing with shared packets.

6. The first character passes through the network to the terminal's network node as soon as the net receives it—MCI/Tymnet does not wait for the rest of the message. (512 bits at 9.6 kbit/s = 0.053 sec × 4 = 0.21 sec)
7. The terminal's network node sends the first character to the terminal. (1 character at 30 characters per second = 0.03 sec)
8. The host sends more characters. Each is passed to the terminal as the network receives it (not included in response-time calculation).
9. Steps 2 to 8 continue, each character being transmitted over the same established path, one at a time, until the logical connection is terminated (not included in response-time calculation). Total response time equals 0.51 sec.

This character-at-a-time method yields a similar response time to that of SprintNet. The time required for transmission is kept to about 0.51 seconds, assuming a 30-character-per-second line into the host computer. MCI/Tymnet's actual response time includes additional delays within the network.

Tymnet's charges are not based on the number of packets (because these are shared by many users), but rather on characters transmitted. This algorithm is more representative of actual transmission volumes.

Tymnet's services and rates (typical)

In 1988, MCI/Tymnet had nodes in over 780 cities (interconnecting over 12,000 local exchange areas), divided into high- and low-density and overseas categories. Because of economies of scale, the higher-density cities have reduced rates. MCI/Tymnet's character charges are a substantial portion of the monthly costs.

Access is via public telephone network (including WATS) or customer-dedicated access line. The service accommodates most transmission codes and protocols for speeds up to 4800 bit/s.

Rates

The total monthly charge for MCI/Tymnet's data transmission service is the combination of:

1. usage
2. network access arrangement
3. interface equipment. Several discount plans are available for extended period (12 month) service.

Usage:

	Connect hour	$/kilocharacter (kch) block
U.S. mainland	See network access	$.06

Dial access charges per connect hour for cumulative hours per month are shown for transmission speeds of 110 to 2400 bit/s.

Business day 7 A.M.—6 P.M. Weekdays	First 1000 hrs.	Next 1000 hrs.	Next 3000 hrs.	Over 5000 hrs.	Connect time min.
High density city	$ 4.45	$ 3.45	$ 2.70	$ 2.20	2.5 min.
Medium density	6.95	5.95	5.20	4.70	2.0 min.
Low density	11.45	10.45	9.70	9.20	1.5 min.
WATS	23.20	22.20	21.45	20.95	1.5 min.

Off-Peak Hrs.	High density	Med. density	Low density	WATS
Per connect hour	$2.30	$3.30	$4.55	$17.30
Connect time min. minimum	2.5 min.	2.0 min.	1.5 min.	1.5

Public 3270 multidrop service is available for terminal clusters in selected locations called *primary access areas*. Clusters outside the primary area will be assessed a mileage charge. Rates depend on service level selected. Check MCI for rates.

Dedicated access (monthly per port)

Speed (bit/s)	Access	High density	Med. density	Install.
110 to 1200	Dial-up	$ 250.00	$ 400.00	$ 500.00
110 to 1200	Leased line	200.00	350.00	500.00
2400	Dial-up	325.00	450.00	500.00
2400	Leased line	250.00	375.00	500.00

Asynchronous out-dial service (110-1200 bit/s monthly)

	High density	Med. density	Installation
First Port	$225.00	$275.00	$500.00
Ea. Add'l Port	125.00	125.00	250.00

Comparing vendors

For a direct comparison between packet-switched network vendors, potential customers must make several assumptions and pick specific network configurations. As an example, consider terminal access at 30 characters per second from five geographical areas for four hours of transmission per day via public dial-in service provided by each vendor. Next, you can estimate the total monthly data transmission character load and obtain the interconnection cost for the subscriber's central computer to be connected to the vendor's nearest central office facility.

When such a specific configuration is examined and costed out, the differences in tariffs for interfacing a host computer can offer significant savings between vendors. A second factor that becomes readily visible from an actual configuration is the terminal access cost. Although one vendor currently offers service in more cities, the cost of accessing that network might be more expensive than that of the other vendor at certain locations. In the final analysis, network users must evaluate network charges using characteristics of their own applications.

Other value-added carriers (Refer to Sec. 7.5) compete with SprintNet and MCI/Tymnet, including AT&T, IBM, ADP, Compuserve, and CSC. Functions and characteristics of the SprintNet and MCI/Tymnet operations are shown in Table 8.4.1. The key point that all data communications users must remember is that if they want a variety of cost-effective communications service offerings, they must—on a periodic basis—evaluate their networks and the latest offerings. This evaluation should include the consideration of nontelephone company communications services where cost effective, or where unique and highly desired user benefits are provided.

8.5 Satellite Services

Over the last three decades, satellite communications has captured the interest of diverse users in business, government, and international agencies. The interconnections provided by satellite networks permit the economical and reliable transmission of voice, television, and data.

Because of technological advances in developing both launch vehicles and satellites, it has become possible to place into orbit larger satellites with expanded capacities for data transfer to and from earth stations. In addition, the size and costs of earth stations have decreased to the point where user-owned stations are now a practical consideration for a number of organizations.

Compared to conventional communications methods, satellite transmission offers a number of unique advantages. First, the cost of transmission is distance insensitive: Conventional communications methods have a cost proportional to distance. Next, satellites have a low-cost point-to-multipoint (broadcast) capability that is most expensive to duplicate with conventional techniques. Broadcasting means transmission from an earth station to a satellite will be relayed back to earth, addressed so that many earth stations can receive the transmitted message at the same time. On a terrestrial link, a message-switching service, or other computer-based services, interfacing to many land lines would be required to duplicate this feature.

United States Services

After numerous delays to resolve complicated legal and technical problems, the Federal Communications Commission authorized U.S. domestic satellite services in 1972. After this landmark ruling, a rapid development of U.S. domestic satellite networks took place.

TABLE 8.4.1 Functional Characteristics of SprintNet and MCI/Tymnet

Feature
Virtual Call Service
■ High priority (interactive users, minimum transit delay)
■ Low priority (noninteractive, longer packets transferred)
■ Past select (inquiry transactions)
■ Receive collect calls
Network Message Switching Services
■ Message delivery A. Upon demand B. Automatic delivery
■ Message storage
■ Multidestination (broadcast) messages
■ Delivery acknowledgment
■ Message text editing
Private Network Systems
■ Under tariff
■ Lease/purchase
■ Custom protocols by request
Network Management
■ Network control center
■ Customer network control console
■ Detailed call accounting
■ Logical subnetworks
■ User authorization IDs/passwords
■ Network command language
Host Interfaces
■ Emulation A. TTY B. IBM 3270 BSC C. IBM 2780/3780 BSC D. IBM HASP multileaving
■ Message level X.400
■ X.25 packet mode
Terminal Interfaces
■ Asynchronous contention (unbuffered terminals)
■ Asynchronous contention (buffered terminals)
■ Asynchronous polled
■ Synchronous polled
■ Synchronous contention A. IBM 2780 BSC B. IBM HASP multileaving

The first domestic satellite, Westar 1, was launched for Western Union Telegraph Co. in 1974 (see Sec. 7.6). This satellite was built by Hughes Aircraft and launched by the National Aeronautics and Space Administration. Westar 2 followed in the same year. With excess capacity, Western Union leased bulk channel capacity to other communications common carriers, as well as to military and commercial customers.

Typical satellite charges

AT&T's Skynet is one of several domestic communications satellite common carrier services authorized by the FCC. As stated in FCC Tariff 7 (August 25, 1993), Skynet Digital Service provides for the transmission via satellite of digital signals at speeds ranging from 56 kbit/s to 1.544 Mbit/s between two or more premises. These premises can be on the mainland, Hawaii, Puerto Rico, or the U.S. Virgin Islands. Typical rates are listed in Table 8.5.1.

Rates are for a full-time space segment. FCC Tariff 7 defines a space segment as "a partial transponder providing unidirectional satellite transponder capability to a customer." [2]

8.6 Service Selection Considerations

Although this section primarily examines service or facility selection from an economic stand-point, the reader should note that other factors may govern a final decision. These factors include installation cost, which should be amortized over the expected life of the transmission method selected; the time between placing an order and having service operational; the anticipated error rate on each medium; and the expansion potential of both the medium and the subscriber's requirements.

TABLE 8.5.1 AT&T's Skynet Digital Satellite Service

Per space segment*	Monthly
56 kbit/s	$992.00
64	1102.00
128	1983.00
192	2756.00
256	3527.00
384	5291.00
768 kbit/s	6613.00
1.544 Mbit/s	9920.00

*Rates are for a full-time space segment. FCC Tariff 7 defines a space segment as "a partial transponder providing unidirectional satellite transponder capability to a customer."

SOURCE: FCC Tariff 7, August 25, 1993, provided by CCMI.

Leased line vs. dial-up

In comparing leased line with dial-up calls made over the public switched telephone network, several factors must be considered, including flexibility, operating data rates, reliability, expansion, and economy of usage.

With a leased line, the subscriber has a business machine at one end of the line permanently connected to the business machine at the other end of the circuit. Thus, unless special devices, such as fallback switches are used, the terminal and its associated modem cannot be used for communications to another business machine. When the public switched telephone network is used, business machine operators merely dial the telephone numbers of the modems connected to the business machines with which they wish to communicate. When a transmission is completed, the operator hangs up the telephone and then can receive a new dial tone and initiate a call to another business machine. Thus, in a terminal-computer communications environment, the terminal and its associated data communications equipment (such as a modem or acoustic coupler) become multifunctional.

Operating data rates on both leased lines and the public switched network generally permit transmissions at minimal error (1 bit in error per 100,000 transmitted) at data rates of 9600 bit/s or more. Because a leased line has a fixed routing, it can be conditioned to permit higher data transfer rates than does the switched network. Thus, data transfer rates over 9600 or 14,400 bit/s for relatively long periods of time (an hour or more per day) normally require a leased line as the transmission medium. This is because, economics aside, transmission on the switched network at such data rates would usually cause too many errors in received data blocks, resulting in many retransmissions that would extend the transmission time. On the switched network, under extremely noisy conditions, there would be so many retransmissions that the session would, for all practical purposes, never be accepted at the receiving terminal.

As for reliability, when a leased line experiences an outage, the subscriber must wait until the circuit is repaired or replaced before transmission can restart. For transmission over the public switched network, however, the telephone company equipment will automatically switch the subscriber around faulty equipment and lines. Thus, unless the local loop from the subscriber to the telephone company central office is affected, the user can normally redial when the switched circuit becomes inoperative and be routed around the impairment.

Considerations

To examine the economic tradeoffs between using the public switched telephone network and leased lines, review Section 8.2. The factors to consider are listed in Table 8.6.1.

For dial-up, use the formula:

$$\textit{monthly cost} = \textit{cost/call} \times \textit{calls/day} \times \textit{days/month}$$

The computed monthly cost is an average. Some months, the cost will be over that amount and some months less. However, if the subscriber's computer utilization should increase, perhaps as a result of new applications or a growth in business that

TABLE 8.6.1 Dial-Up and Leasing Cost Factors

Average number of calls to computer per day
Average call duration
Number of characters transmitted and received per call
Working days/month
Distance between operating cities

For dial-up, use the formula: (cost per call) × (calls/day) × (days/month)

requires additional processing power, then the cost of utilizing the public switched telephone network will similarly increase.

For the leased-line, costs are simply those of mileage and station terminals' access. The cost of modems is excluded. And no conditioning would be required if the terminal were to operate at 300 bit/s.

Because the leased line has a fixed monthly cost, in comparing use of leased lines with use of the switched network, you must assure that call estimates are fairly accurate or the analysis could result in a wrong decision. Note that in such comparisons, the number of characters per call (Table 8.6.1) is not necessarily a factor. That information is relevant only for costing transmission via a value-added carrier where one part of the charge element is based on the number of characters or packets transmitted.

Leased vs. foreign exchange

The leased vs. foreign exchange comparison compares the use of two leased lines, one of which is connected to the public switched network. The normal use of a foreign exchange line is to place the "open end" of the circuit in a city where several terminals requiring contention or scheduled access to a computer are located. The "closed end" of the circuit is located at the computer site. Thus, terminal operators in the open-end city dial the telephone number of that line and are then connected to the modem at the computer site, as shown in the top portion of Figure 8.6.1.

One alternative configuration to service several terminals located in a distant city (from where the computer is located) can be accomplished by the installation of a multipoint leased line, as shown in the lower portion of Figure 8.6.1.

For this type of operating environment, poll and select software must be available for use by the computer, and buffered terminals capable of recognition of their addresses must be installed.

If you assume all terminals are near each other, then the leased-line cost for each circuit is approximately equivalent. Other cost differences between the two circuits can result from differences in the modems used and in local telephone company costs. For the FX line, terminals could use low-cost acoustic couplers or asynchronous low-speed, low-cost modems.

For a multipoint environment, synchronous transmission, normally used, results in more expensive terminals and modems. For access to the FX line, each dialed call might incur a local toll charge, depending on the local operating telephone company.

For the multipoint leased line, no such toll charge will be incurred. Although the terminals that access the computer via the FX line can also access other business machines through the use of the switched network, terminals on the multipoint circuit can access only the computer interfaced to that line.

Although poll-and-select permits all terminals to communicate with the computer at the same time, terminals using the FX line can do so only one at a time and must therefore either contend for access to the FX line or have their usage scheduled.

Switched vs. WATS

Normally, an Inwats access is used to link a number of terminals scattered throughout a geographical area (or areas) to a central computer via the public switched telephone network. Usually, the connect time per terminal is such that, on an individual basis, transmission via the switched network is less expensive than installing leased lines from each terminal location to the computer center. When this situation arises, the subscriber might then want to determine the cost of Inwats service in order to see if savings can be obtained by the use of such a facility.

Review the applicable sections of Section 8.2 to make the comparison of switched network vs. WATS costs. For switched considerations, include average cost per call,

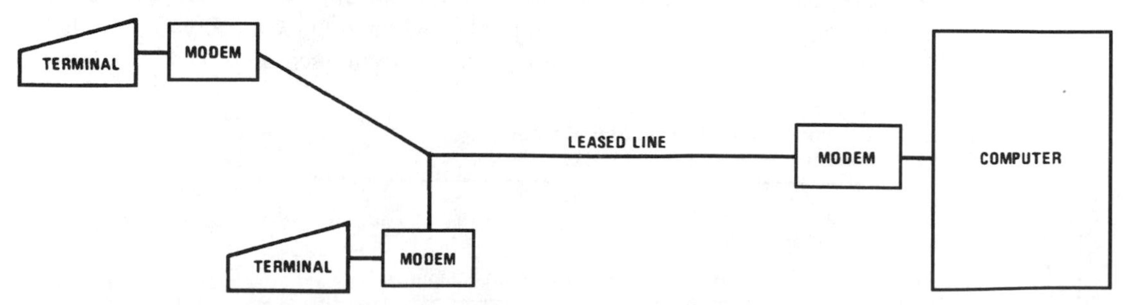

Figure 8.6.1 Foreign exchange vs. multipoint line.

total calls per day, terminals per site, and working days per month. Come up with an equivalent cost per hour.

Compare this with the cost of an Inwats line plus the Inwats per-hour charge. Based on these figures, it might be more economical for a subscriber to select a measured-time Inwats service.

Switched vs. WATS vs. value-added carrier

When the services of a value-added carrier are to be considered, one of the first items to be determined is the service area of that carrier. Today, value-added carriers service several hundred cities in the United States, providing subscribers with dial-in connections via the public switched telephone network at data rates up to at least 4800 bit/s. In comparison, switched network service is available at every location throughout the United States.

There are three basic components in the cost of using a value-added carrier that are billed to subscribers. These components include a terminal-connect-time charge, a character or packet transmission charge, and a host computer interconnection charge. The latter is the charge to link the central office of the value-added carrier to the subscriber's computer and to permit a certain number of terminals to simultaneously dial and connect to the value-added carrier's network and then be connected via the network to the computer.

For the switched cost, consider the transmission time per terminal per day, convert to hours per month, then to cost per month.

In trying to estimate the cost of a value-added carrier, you must estimate the number of characters or packets to be transmitted and received, depending on which method the carrier uses for billing purposes. Assume that the carrier under consideration bills by characters, and its cost components are as listed in Table 8.6.2. Assume the characters per hour transmitted and received by each terminal. Convert to characters per day, then characters per month. Figure the monthly cost.

Next, figure the connect-hours per month per terminal, then the connect-time charge component.

Last, figure the carrier's fixed charge per month for providing the subscriber with a leased line from the carrier's central office to the user's computer, and with equipment to permit terminals to simultaneously access the computer from any entry point in the network.

Adding the three components produces a total monthly charge that you can compare to the cost of transmission via the public switched network. If you then divide the computed cost by the total monthly terminal connect time, cost per hour is

TABLE 8.6.2 Hypothetical Value-Added-Carrier Cost Components

A. Character charge
B. Connect time charge
C. Central office to subscriber host computer line and equipment

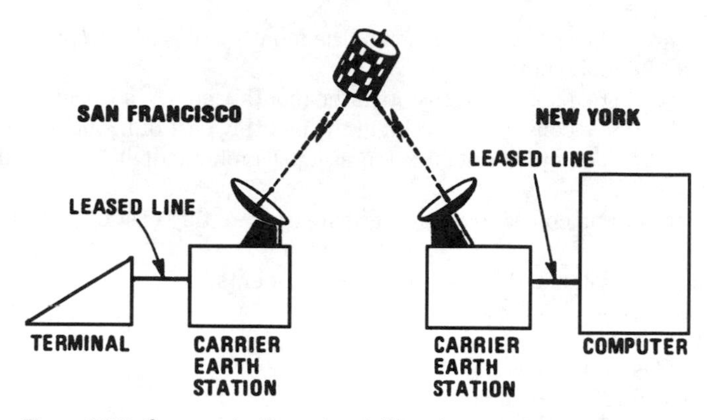

Figure 8.6.2 Communications via satellite: charge components.

reached. To that figure, compare WATS service costs from Tables 8.2.8 or 8.2.9, thus completing comparison of the costs of the three services.

Terrestrial vs. satellite

Examine the transmission of data both via satellite and over conventional land facilities. For the latter, numerous possibilities exist. To narrow our choices, assume the requirement is to transmit data from a remote batch terminal located in San Francisco to a computer located in New York City at 9600 bit/s, 10 hours per day, seven days per week. Because of these operating times as well as the high operating data rate, use of the public switched telephone network is excluded.

The next section estimates the cost of a leased analog circuit or 9.6-kbit/s Dataphone digital service and compares those rates to the cost of transmitting by satellite.

Satellite considerations

Because data will be transmitted at 9.6 bit/s, the physical installation of a satellite ground station at the San Francisco terminal location and the New York City computer site would not be cost-effective because these stations are designed to transmit data at 56 kbit/s and rent for upwards of $10,000 per month.

From Table 8.5.1, the cost of a New York City to San Francisco channel is $992 per month plus termination and special equipment charges. To this cost, add the interconnect charges to link the facilities in each city to the earth station. These cost components are illustrated in Figure 8.6.2.

Add up the point-to-point terrestrial costs, following the steps of Example 1 in Sec. 8.2. Compare the result to that of the satellite case.

Questions

8.1 Discuss the differences between using the public switched telephone network, leased lines, foreign exchange lines, and WATS service. Under what conditions might each service be used for data transmission?

8.2 What factors should be considered to obtain the monthly cost of data transmission over the public switched telephone network'?
8.3 What factors should be considered to obtain the monthly cost of a leased line?
8.4 What factors should be considered to obtain the monthly cost of Inwats service?
8.5 Discuss some of the advantages obtained in using Dataphone digital service in comparison to leased lines.
8.6 What factors should be considered to determine the monthly cost of Dataphone Digital Service?
8.7 Given the following value-added carrier cost components:

- Connect-time cost,
- Character-transmission cost,
- Monthly central-office-to-computer cost.

What are the factors to consider in comparing the costs of using a value-added carrier, the public switched telephone network, or leased lines for data transmission?

Chapter

9

Interfacing Data Transmission Devices

Chapter 5 covered the methods and problems associated with modulating analog signals. Next, two basic transmission devices, acoustic couplers and modems, were examined with respect to their transmission characteristics and compatibility, frequency-band limitations, and noise and distortion problems. In that chapter, elementary modem interfacing and handshaking via data terminal equipment (DTE) to data circuit-terminating equipment (DCE) interface was covered. Table 5.4.2 listed EIA-232-C/D and CCITT[1] V.24 equivalent interface pin assignments, and some of the control signals for the operation of a Bell Type 113 modem were explained.

Using the information presented in Chapter 5 as a foundation, this chapter concentrates on the interfacing of DCE to DTE over different types of transmission media. The basic circuits of the EIA-232-C/D interface are examined in detail, and several problems and limitations of the interface and transmission method employed are covered. Also, the Electronic Industries Association (EIA) recommended standard EIA-449, which was expected to replace the widely used EIA-232-C interface, is covered.

9.1 EIA-232-C Modem Interface

As examined in Chapter 5, the EIA-232-C specification was revised with the issuance of the EIA-232-D standard in January, 1987. The differences between these standards lie primarily in the specification of the D-shaped 25-pin interface connector and the inclusion of new functions by the revised specification.

The EIA-232-D standard includes the D-shaped 25-pin interface connector as part of the standard. In comparison, EIA-232-C merely made reference to the connector in an appendix, explicitly stating that the connector was not part of the standard.

[1]CCITT and its replacement ITU-T, are used as interchangeable designations.

As to functions, EIA-232-D can be considered as a superset of the EIA-232-C standard, since it adds functions to support the testing of both local and remote data communications equipment. In addition, EIA-232-D changed the names of some signals. As a result of the basic similarity of the two standards, however, they are collectively referred to as *EIA-232-C/D* in this chapter, differentiating only when it becomes necessary to denote a specific characteristic of a particular standard.

The 25-pin interface listed in Table 5.4.2 can be rearranged into five basic groupings, based on the function the circuits are designed to perform. The rearrangement is shown in Table 9.1.1, in which the circuits have been categorized into ground, data, control and timing, secondary operation, and testing.

Ground circuits

The two ground circuits included on the EIA-232-C/D standard are protective ground and signal or common ground. Protective ground, circuit AA, pin 1, must be electrically bonded to the device or the frame of the equipment, or to an external ground. The purpose of this circuit is to ground any voltage leak and thus protect the user from being shocked if he touches the device.

Signal ground, or common ground, is circuit AB on pin 7. This circuit establishes a common ground reference potential for all interchange circuits, except protective ground. Hence, it is also referred to as *common* or *return ground*. Circuit AB is normally connected to the protective ground circuit via an internal wire strap, In certain situations, frame noise can be introduced by this strapping and the removal of the strap will normally alleviate this situation.

Data circuits

The data circuits are the paths through which data is transmitted and received. The two data circuits are transmit data (circuit BA on pin 2) and receive data (circuit BB on pin 3).

The direction of data flow on circuit BA is from the data terminal to the associated communications equipment: DTE to DCE. The signals for this circuit are thus generated by the data terminal. When the terminal is not sending data or is between transmitting characters, a negative voltage is applied to this interface lead to denote a marking or binary 1 signal condition. The interchange voltages that represent the mark and space signaling conditions are denoted in Figure 9.1.1.

Although data is transmitted on the BA circuit, certain control signals must be activated or set to the ON condition before data transfer can occur. The hand shaking routine illustrated by the control signals in Figure 9.1.2 must first occur. As illustrated, the data terminal equipment initiates a request-to-send signal. If the data circuit-terminating equipment has control of the telephone circuit, it will send a clear-to-send signal back, as well as a data-set-ready signal. In response, the terminal will transmit a data terminal ready signal to the DCE. These control circuits are covered in more detail later.

The received data lead (circuit BB, pin 3) passes signals from the DCE to the DTE. For data to flow over this circuit, the DCE obtains a signal from the transmitting device via the connecting line. The DCE converts or demodulates that signal for transmission

TABLE 9.1.1 EIA 232-C/D Circuit Assignment by Operation

Circuit operation	Circuit	Pin number	Description	GND	Data From DCE	Data To DCE	Control From DCE	Control To DCE	Timing From DCE	Timing To DCE	Testing From DCE	Testing To DCE
Ground	AA	1	Protective GND (shield)	X								
	AB	7	Signal GND	X								
Data	BA	2	Transmitted data			X						
	BB	3	Received data		X							
Control	CA	4	Request to send					X				
	CB	5	Clear to send				X					
	CC	6	Data set ready (DCE ready)				X					
	CD	20	Data terminal ready (DTE ready)					X				
	CE	22	Ring indicator				X					
	CF	8	Received line signal detector				X					
	CG/RL	21	Signal quality detector (Remote loopback)				X					
	CH	23	Data signal rate selector (DTE)					X				
	CI	23	Data signal rate selector (DCE)				X					
Timing	DA	24	Transmitter signal element timing (DTE)							X		
	DB	15	Transmitter signal element timing (DCE)						X			
	DD	17	Receiver signal element timing (DCE)						X			
Secondary Data	SBA	14	Secondary transmitted data			X						
	SBB	16	Secondary received data		X							
Secondary Control	SCA	19	Secondary request to send					X				
	SCB	13	Secondary clear to send				X					
	SCF	12	Secondary rec'd line signal detector				X					
		9	Reserved data set testing									
		10	(Power for Reserved data set testing test sets)									
		11	Unassigned									
Testing	LL	18	Unassigned (local loopback)									X
	TM	25	Unassigned (test mode)								X	

Description in parenthesis unique to RS-232-D

	INTERCHANGE VOLTAGE	
	-3 TO -25V	+3 TO +25V
BINARY STATE	1	0
SIGNAL CONDITION	MARKING	SPACING
FUNCTION	OFF	ON

Figure 9.1.1 Interchange voltage and conditions on interface leads.

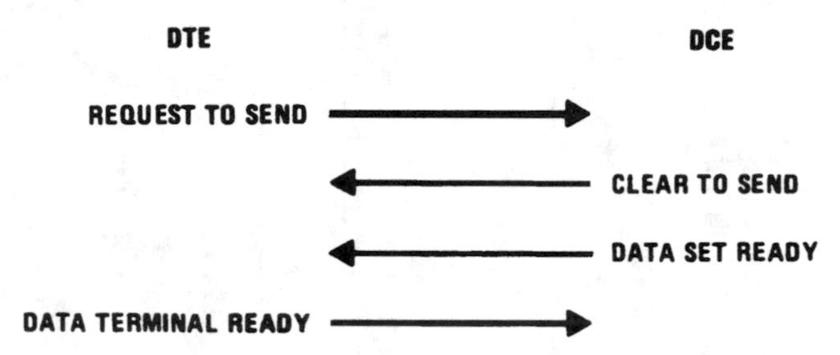

Figure 9.1.2 Handshaking prior to transmitting data.

via circuit BB to the receiving business machine. This received-data circuit can have its state varied by the received line signal detector (circuit CF, pin 8) control circuit. When circuit CF is OFF, this condition forces circuit BB to a marking condition.

If transmission is half-duplex, the received-data circuit is in an inactive or marking state whenever the attached DCE has its request-to-send lead in ON. This is because, in half-duplex transmission, a device can either transmit or receive data at one time, but not perform both operations at the same time. To permit a transmission to be completed prior to turning the line around—in a half-duplex mode—for data reception, the received data lead is held in a marking condition for a short time after the request-to-send signal is turned OFF.

Control circuits

Of the nine EIA-232-C/D control circuits, request to send, clear to send, data set ready, received line signal detector, data terminal ready, and ring indicator are considered the basic control circuits.

Request to send (circuit CA, pin 4) is transmitted from the DTE and functions as a signal to prepare the local DCE, such as a modem, for transmission when half-duplex transmission is used. This signal controls the transmission direction of the local DCE. The effect of the request-to-send signal on half- and full-duplex transmission is illustrated in Figure 9.1.3. Here, changing the condition of the circuit from OFF to ON places the DCE in the transmit mode.

This signal can be turned ON at any time as long as the clear-to-send circuit is in the OFF condition. Once the request-to-send signal is activated, if the associated DCE is ready to transmit, it will turn on the clear-to-send circuit. Conversely, when the request-to-send signal changes states from ON to OFF, this change informs the DCE to complete its transmission, then places the equipment in a nontransmit or receive mode, depending on the device's operation.

Clear to send (circuit CB, pin 5) is transmitted from the DCE to the DTE. When this circuit becomes active, it informs the terminal equipment that the DCE is ready to transmit data. The clear-to-send signal occurs in response to data-set-ready and request-to-send ON conditions and is combined with those two signals and a data-terminal-ready signal to indicate that data is ready to be transmitted. When the clear-to-send signal is OFF, its absence indicates to the terminal that a data transfer via the transmit data circuit should not occur because the communications equipment is not ready.

The data-set-ready signal, called *DCE ready* under EIA-232-D, (circuit CC, pin 6) is passed from the DCE to the terminal device and indicates to the terminal the status of the local communications equipment. When in the ON condition, this circuit indicates that the local DCE is connected to a communications channel, that any timing functions required by the transmission medium have been completed, and that the local DCE is not in a talk, test, or dial mode of operation. When the data-set-ready circuit is in an OFF condition, this status informs the local DTE to ignore all circuit signals with the exception of a ring indicator "signal." If the data-set-ready signal should switch to an OFF condition during a call prior to the data-terminal-ready signal turning OFF, the data terminal equipment will interpret this situation as a lost or broken connection and terminate the call. For those modems that have voice adapters, interrupting a data call by transferring to voice coordination causes the data-set-ready signal to go OFF and remain OFF while voice is present.

The received-line-signal-detector signal (circuit CF, pin 8) indicates whether or not the signal being received falls within the frequency range associated with the DCE-to-DTE transmission. An ON condition signifies that the data being received is a signal suitable for acceptance by the DCE.

SIGNAL STATE	TRANSMISSION MODE	
	HALF-DUPLEX	FULL-DUPLEX
ON	INHIBITS RECEIVING OF DATA	MAINTAINS DCE IN RECEIVE MODE
OFF	INHIBITS TRANSMISSION OF DATA	MAINTAINS DCE IN TRANSMIT MODE

Figure 9.1.3 Effect of request-to-send signal on transmission.

As an example, the ON condition on circuit CF could indicate that a 4800-bit/s modem connected to a transmission line is receiving a signal that can be demodulated by that modem. Conversely, an OFF condition indicates that either an unsuitable signal or no signal is being received by the modem from the transmission line. In addition, an OFF condition on the CF circuit will put the received data circuit (BB) in a marking or OFF condition.

For operation in the half-duplex mode, EIA-232-C/D places the CF circuit in the OFF condition whenever request-to-send is ON, and for a brief time after that signal goes OFF. Thus, appropriate delay times are built into DCE to prevent the loss of a line signal by ensuring that a signal is turned ON in response to one going OFF. Also, sufficient time elapses between the two to prevent misinterpretation of the signals.

The data-terminal-ready signal, called *DTE ready* under EIA-232-D, (circuit CD, pin 20) is transmitted from the DTE to the DCE. This signal can be ON at any time and it controls the switching of the DCE to the telephone line. Here, an ON condition prepares the DCE for connection to the telephone line and then helps to maintain that connection. When the signal returns to an OFF state, it removes the DCE from the line. When transmission is over the switched network, the data-terminal-ready signal cannot be turned ON again until the data-set-ready signal is turned OFF.

The ring indicator signal (circuit CE, pin 22) is transmitted from the DCE to the DTE and denotes that a ringing signal is being received from a remote station. This signal is normally used only when transmission is over the switched telephone network. All modems designed for operation over the switched network have a ring indicator circuit. For manual answering, this circuit is disabled.

To see the interaction of the previously these signals, examine the operation of these circuits when transmission is accomplished over one particular medium using a specific type of modem.

9.2 Circuit Operation, Secondary, and Timing Channels

The operation of the circuits in an EIA-232-C/D modem-to-terminal interface depends on many factors, including the transmission mode, the type of modems used, and the transmission medium.

For an example, investigate the circuit operations between a 113D-type modem and a data terminal when transmission is over the switched network. The 113D modem is “answer only,” single mode. Its transmitter and receiver are tuned to specific frequencies so that, by convention, an “originate only” modem’s transmitter operates at the same frequency as the 113D’s receiver, and the “originate only” receiver operates at the same frequency as the 113D’s transmitter.

Assuming that the 113D modem has the automatic-answer feature set, when a terminal operator at a remote site dials the 113D telephone number at the computer, a ringing voltage is placed on the connected line. In response to this, indicator circuit CE is turned ON while the ringing voltage is present. If the data-terminal-ready circuit CD is then turned ON by the computer, the modem will be connected to the line after the end of the ringing interval. When connected, the data-set-ready circuit CC is turned ON by the modem.

In response to the data-set-ready signal, the computer, if ready, will raise its request-to-send CA circuit, and the modem will transmit a mark frequency at 2225 Hz. After the receipt of this tone, the originating station will transmit a 1270-Hz tone.

If this signal is not received after 14±4 seconds from the beginning of the connection attempt, an abort timer will automatically disconnect the 113D modem from the line. If the 113D receives a continuous 1270-Hz mark tone, however, the modem turns on the received line signal detector CF and clear-to-send CB circuits, which in turn enable the transmit data BA circuit. At this time, the computer can send and receive data.

This sequence of circuit operations is illustrated in Figure 9.2.1. In this illustration, time starts at the top of the Time axis, and the elapsed time increases as one goes down this axis.

Secondary channel circuits

Also known as *auxiliary circuits*, secondary circuits are equivalent in operation to their primary circuits. The transmission direction of secondary channels is independent of the primary channel and is controlled by the set of secondary control interchange circuits.

In comparison to a primary channel, transmission speed on a secondary channel is considerably slower because that channel has a much narrower band than the primary channel.

Several types of asynchronous and a few synchronous modems incorporate secondary and reverse channels. In comparison to a secondary channel, whose data flow can be either direction, transmission on a reverse channel is always opposite the

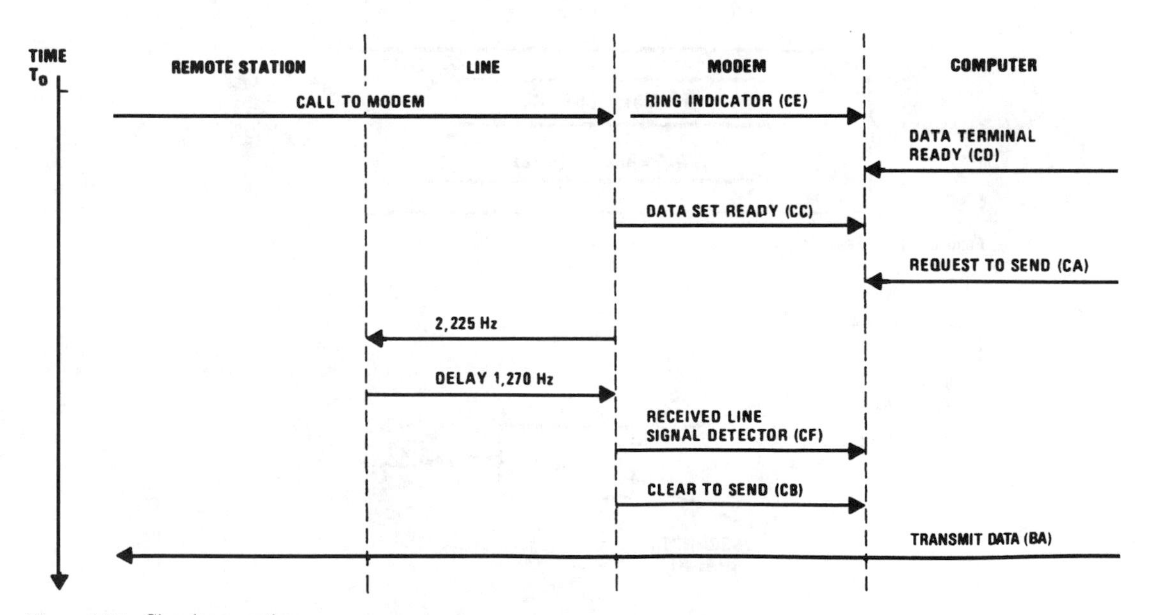

Figure 9.2.1 Circuit operation.

transmission direction of the primary channel. For synchronous transmission, the reverse channel can be used to acknowledge data blocks received without having to turn around the direction of primary transmission. Secondary and reverse channels are illustrated in Figure 9.2.2.

Secondary channels normally operate at speeds up to 300 bit/s. Their incorporation into modems permits low-speed terminals, such as a teleprinters, to share circuits with higher-operating-speed terminals. Figure 9.2.3 shows a modem with a secondary channel serving a remote batch terminal and a teleprinter.

If the secondary channel is used as a reverse channel, it is usually for circuit assurance. For example, as data blocks are received correctly, the carrier secondary channel switches between ON and OFF. Because the direction of the secondary channel is opposite that of the primary channel, an ON condition on circuit CA (primary request-to-send) will inhibit that condition from occurring on circuit SCA at the same time. Data is normally not transmitted when the secondary channel is used as a reverse channel. Instead, the presence or absence of carrier is used to denote an interrupt condition. As an example, as data blocks are received correctly, the carrier signal remains ON (ACK). However, if a data block is received in error and a retransmission is thus required, the carrier is turned OFF (NAK) to provide the same

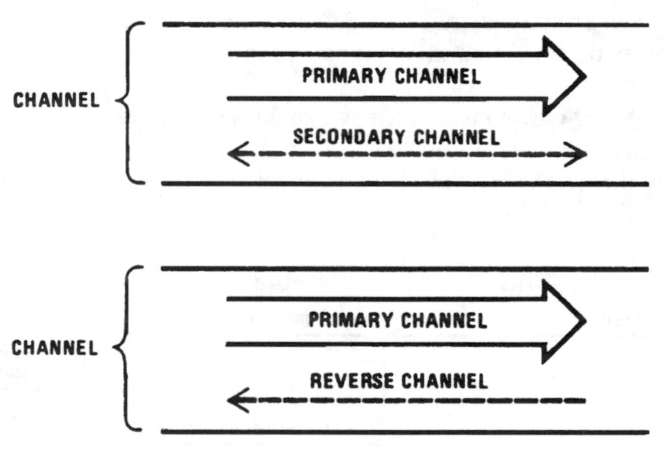

Figure 9.2.2 Secondary and reverse channels.

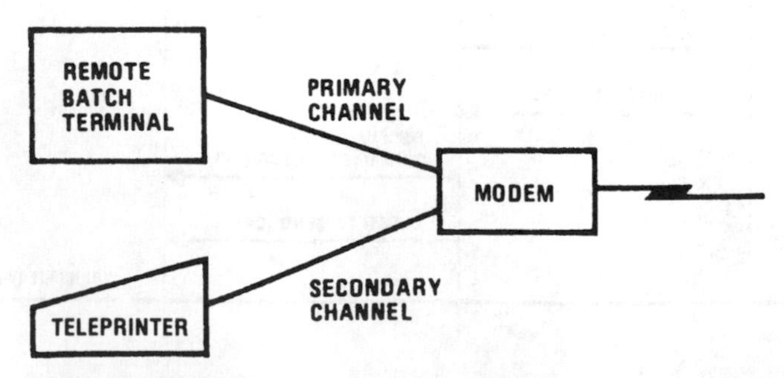

Figure 9.2.3 Secondary channel permits two terminals to share the line.

function as transmitting a negative acknowledgment. Because a reverse channel only requires the presence or absence of carrier, secondary clear to send, transmitted data, and received data do not perform any functions in a reverse channel role.

Timing circuits

In the EIA-232-C/D circuit assignments, three circuits are used for timing. Transmit signal element timing (circuit DA, pin 24) is transmitted from the DTE to the DCE; transmit signal element timing (circuit DB, pin 15) is transmitted from the DCE to the DTE; and receiver signal element timing (circuit DD, pin 17) is transmitted from the DCE to the DTE.

Timing signals are supplied either by the modem or by the business machine to synchronize the modem with the terminal device. A timing signal is only required for synchronous transmission because in asynchronous transmission, each start-stop bit sequence in a character synchronizes the transmission. Thus, pins 15, 17, and 24 are normally unassigned when asynchronous modems are used.

For synchronous modems, the timing source can be internal or external, so long as the frequency of the connected clock is an appropriate multiple of the baud rate needed to obtain the desired data transfer rate. Circuit DB is used for internal clocking, while circuit DA provides the modem with external clocking. The third timing circuit, DD, provides the local terminal with received timing information. Whenever the receiver line signal detector is ON, this circuit is ON and stays ON for a short time after the receiver line signal detector signal goes OFF. The DD circuit is active whenever its voltage is positive, and turns OFF when it is negative.

Test circuits

Under EIA-232-D, pins 18 and 25, which had been unassigned under EIA-232-C, were dedicated to testing functions. In addition, pin 21, which had been reserved for signal-quality detection under EIA-232-C, can be used for that function or for a remote-loopback function.

The local-loopback signal under EIA-232-D (circuit LL, pin 18) is transmitted from the DTE to the DCE. This signal places the attached DCE in a test mode of operation.

In response to either remote loopback (circuit RL, pin 21) or local loopback (circuit LL, pin 18) being active, the test-mode signal (circuit TM, pin 25) will become active. Similar to the local loopback signal, the test-mode signal is applicable only to EIA-232-D operations.

9.3 EIA-449 General-Purpose Interface Standard

This standard, together with EIA standards EIA-422 and EIA-423, was originally intended as a replacement for EIA-232-C for the interface between data terminal equipment (DTE) and data circuit-terminating equipment (DCE) using serial binary data interchange. This standard is now intended primarily for data applications using analog networks.

In the late 1970s, it was felt that EIA-232-C was in need of replacement in order to specify new electrical characteristics and to define several new interchange circuits.

New electrical characteristics were needed to accommodate advances in integrated-circuit design, to reduce crosstalk between interchange circuits, to permit greater distances between equipment, and to permit higher data rates.

Although EIA-449 was expected to replace EIA-232-C by the mid-1980s, by the late 1980s, less than five percent of communications equipment was manufactured with this interface. The actual reason for EIA-449 not becoming well established remains a matter of conjecture. However, its specification of two different connectors that differ from the near-universal 25-pin connector is deemed by many to be the primary cause of its lack of adoption. EIA-449 specifies 37-pin and 9-pin connectors, neither of which is compatible with the 25-pin connector. As a result, organizations installing EIA-449 equipment were almost always forced to obtain connector adapters to permit 25-pin computer ports and other devices to operate with EIA-449 devices.

In spite of its lack of universal adoption, the EIA-449 standard still has a respectable base of equipment, and information is presented in this chapter to make readers aware of its use.

In March 1987, the Electronic Industries Association issued EIA-530, which includes a specification for the use of the D-shaped 25-pin interface connector. The foreword of the EIA-530 standard states that it is intended to gradually replace EIA-449. However, the associated EIA-422 and EIA-423 standards will survive: Those standards specify the electrical characteristics of the interface and are referenced by EIA-530, as well as by other EIA standards.

With the expected increase in use of standard electrical interface features between many different kinds of equipment, it became appropriate to publish the electrical interface characteristics in separate standards. Two electrical interface standards were published for voltage digital interface circuits:

EIA Standard -422,

Electrical Characteristics of Balanced Voltage Digital Interface Circuits.

EIA Standard -423,

Electrical Characteristics of Unbalanced Voltage Digital Interface Circuits.

With the adoption of EIA-422 and EIA-423, it became necessary to create a new standard that specified the remaining characteristics (i.e., the functional and mechanical characteristics) of the interface between data terminal equipment and data circuit-terminating equipment. That is the purpose of the EIA-449 standard.

The basic interchange circuit function definitions of EIA-232-C were retained in this standard. However, there are a number of significant differences:

1. Application of this standard was expanded to include signaling rates in excess of 2,000,000 bits per second.
2. Ten circuit functions were defined in this standard that were not part of EIA-232-C. These include three circuits for control and status of testing functions in the DCE (circuit LL, local loopback; circuit RL, remote loopback; and circuit TM, test

mode), two circuits for control and status of the-transfer of the DCE to a standby channel (circuit SS, select standby; and circuit SB, standby indicator), a circuit to provide an out-of-service function under control of the DTE (circuit IS, terminal in service), a circuit to provide a new signal function (circuit NS, new signal), and a circuit for DCE frequency selection (circuit SF, select frequency).

In addition, two circuits were defined to provide a common reference for each direction of transmission across the interface (circuit SC, send common; and circuit RC, receive common).

3. Three interchange circuits defined in EIA-232-C were not included in this standard. Protective ground (EIA-232-C, circuit AA) was not included as part of the interface, to permit bonding of equipment frames, when necessary, to be done in a manner in compliance with national and local electrical codes. However, a contact on the interface connector is assigned to the shield of interconnecting cable.

   The two circuits reserved for data set testing (EIA-232-C contacts 9 and 10) were not included in order to minimize the size of the interface connector.

4. Some changes were made to the circuit function definitions. For example, operation of the data-set-ready circuit was changed and a new name, data mode, was established because of the inclusion of a separate interchange circuit (test mode) to indicate a DCE test condition.

5. A new set of standard interfaces for selected configurations was established. In order to achieve a greater degree of standardization, the option in EIA-232-C that permitted the omission of the request-to-send interchange circuit for certain transmit-only or duplex-primary-channel applications was eliminated.

6. A new set of circuit names and mnemonics was established. To avoid confusion with EIA-232-C, all mnemonics in this standard are different from those used in EIA-232-C. The new mnemonics were chosen to be easily related to circuit functions and circuit names.

7. A different interface connector size and interface connector latching arrangement was specified. A larger size connector (37 pin) is specified to accommodate the additional interface leads required for the 10 newly defined circuit functions, and to accommodate balanced operation for 10 interchange circuits.

In addition, a separate nine-pin connector is specified to accommodate the secondary channel interchange circuits. The 37-pin and nine-pin connectors are from the same connector family as the 25-pin connector in general use in equipment conforming to EIA-232-C. A connector-latching block is specified to permit latching and unlatching of the connectors without the use of a tool. This latching block also permits the use of screws to fasten the connectors together. The different connectors also serve as precautions with regard to interface voltage levels, signal rise times, failsafe circuitry, grounding, etc. because an adapter is needed before equipment conforming to EIA-232-C can be connected to equipment conforming to the new EIA-449. The pin assignments have been chosen to facilitate connection of equipment conforming to this standard to equipment conforming to EIA-232-C.

EIA-449, like EIA-232-C/D, applies to both synchronous and asynchronous data communications networks. The standard also applies to all classes of data communications service, including nonswitched-, dedicated-, leased- or private-line service of either 2- or 4-wire circuits, and switched-network service.

With the key mechanical difference between EIA-449 and EIA-232-C/D being the interface connectors, users wishing to interconnect EIA-449 equipment to equipment with EIA-232-C/D connectors require interface adapters. Such adapters cross-connect the appropriate circuits from, for example, a 37-pin plug of EIA-449, assuming no secondary channel is used, to the appropriate circuits on the 25-pin RS-232-C/D plug. Figure 9.3.1 illustrates the DTE connector face and 37-pin plug of EIA-449. An adapter is shown that connects terminal equipment built to that standard with communications equipment (such as a modem) built with a 25-pin plug under the EIA-232-C/D standard.

Interchange circuits

The interchange circuits of the RS-449 standard fall into four general classifications: ground or common return circuits, data circuits, control circuits, and timing circuits. A list of the interchange circuits showing the circuit mnemonic, circuit name, circuit direction, and the circuit type is contained in Table 9.3.1.

Equivalency

A list of the EIA-449 interchange circuits showing the nearest equivalent EIA-232-C/D and CCITT identification in accordance with recommendation V.24 is listed in

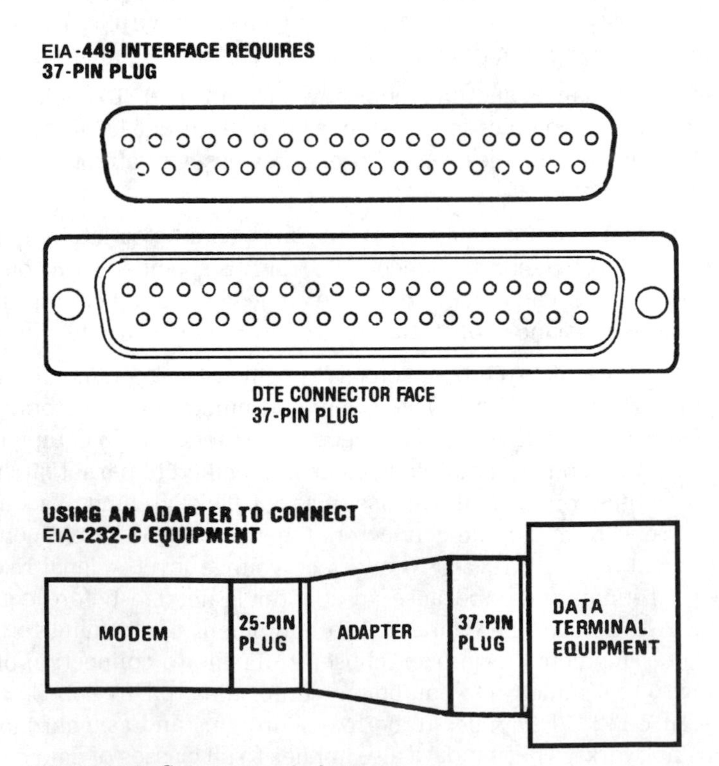

Figure 9.3.1 Connector and adapter.

TABLE 9.3.1 EIA-449 Interchange Circuits

Circuit mnemonic	Circuit name	Circuit direction	Circuit type	
SG	Signal ground	–	Common	
SC	Send common	To DCE		
RC	Receive common	From DCE		
IS	Terminal in service	To DCE	Control	
IC	Incoming call	From DCE		
TR	Terminal ready	To DCE		
DM	Data mode	From DCE		
SD	Send data	To DCE	Data	
RD	Receive data	From DCE		
TT	Terminal timing	To DCE	Timing	Primary channel
ST	Send timing	From DCE		
RT	Receive timing	From DCE		
RS	Request to send	To DCE	Control	
CS	Clear to send	From DCE		
RR	Receiver ready	From DCE		
SQ	Signal quality	From DCE		
NS	New signal	To DCE		
SF	Select frequency	To DCE		
SR	Signaling rate selector	To DCE		
SI	Signaling rate indicator	From DCE		
SSD	Secondary send data	To DCE	Data	Secondary channel
SRD	Secondary receive data	From DCE		
SRS	Secondary request to send	To DCE	Control	
SCS	Secondary clear to send	From DCE		
SRR	Secondary receiver ready	From DCE		
LL	Local loopback	To DCE	Control	
RL	Remote loopback	To DCE		
TM	Test mode	From DCE		
SS	Select standby	To DCE	Control	
SB	Standby indicator	From DCE		

Table 9.3.2. It should be noted that the EIA-449 circuit definitions can vary to a degree with the equivalent EIA-232-C/D or CCITT recommendations. Consult the EIA-449 standard for additional information.

EIA-530

As previously mentioned in this section, EIA-530 is intended to gradually replace the EIA-449 standard. Like EIA-449, EIA-530 addresses higher data rates than EIA-232. This is accomplished by specifying the use of balanced signals at the expense of several secondary signals and of the ring indicator signal provided in EIA 232. Here, the term *balanced signals* means that each signal, such as transmit data or receive data, uses two wires with opposite polarities to minimize distortion.

Table 9.3.3 summarizes the 25 EIA-530 signals. Notice the absence of a ring indicator signal, which suggests that this standard is not intended for use in dial-up applications.

TABLE 9.3.2 EIA-449 Equivalency

EIA-449		EIA-232C/D		C.C.I.T.T. recommendation V.24	
SG	Signal ground	AB	Signal ground	102	Signal ground
SC	Signal ground			102a	DTE common
RC	Receive common			102b	DCE common
IS	Terminal in service				
IC	Incoming call	CE	Ring indicator	125	Calling indicator
TR	Terminal ready	CD	Data terminal ready	108/2	Data terminal ready
DM	Data mode	CC	Data set ready	107	Data set ready
SD	Send data	BA	Transmitted data	103	Transmitted data
RD	Receive data	BB	Received data	104	Received data
TT	Terminal timing	DA	Transmitter signal element Timing (DTE source)	113	Transmitter signal element Timing (DTE source)
ST	Send timing	DB	Transmitter signal element Timing (DCE source)	114	Transmitter signal element Timing (DCE source)
RT	Receive timing	DD	Receiver signal element Timing	115	Receiver signal element Timing (DCE source)
RS	Request to send	CA	Request to send	105	Request to send
CS	Clear to send	CB	Clear to send	108	Ready for sending
RR	Receiver ready	CF	Received line signal detector	109	Data channel received line signal detector
SQ	Signal quality	CG	Signal quality detector	110	Data signal quality detector
NS	New signal				
SF	Select frequency			128	Select transmit frequency
SR	Signaling rate selector	CH	Data signal rate selector (DTE source)	111	Data signal rate selector (DTE source)
SI	Signaling rate indicator	CI	Data signal rate selector (DTE source)	112	Data signal rate selector (DCE source)
SSD	Secondary send data	SBA	Secondary transmitted data	118	Transmitted backward channel data
SRD	Secondary receive data	SBB	Secondary received data	119	Received backward channel data
SRS	Secondary request to send	SGA	Secondary request to send	120	Transmit backward channel line signal
SCS	Secondary clear to send	SCB	Secondary to send	121	Backward channel ready
SRR	Secondard receiver ready	SCF	Secondary received line signal detector	122	Backward channel received line signal detector
LL	Local loopback	(LL)	(Local loopback)	141	Local loopback
RL	Remote loopback	(RL)	(Remote loopback)	140	Remote loopback
TM	Test mode	(TM)	(Test mode)	142	Test indicator
SS	Select standby			118	Select standby
SB	Standby indicator			117	Standby indicator

NOTE: Unique EIA-232-D signals in parentheses

9.4 Resolving Interface Problems

Ideally, an interface standard, such as the Electronic Industries Association's EIA-232-C/D, should guarantee the interconnection compatibility of all data communications hardware designed to its specifications, regardless of the manufacturer. In practice, though, complete electrical compatibility does not always exist—even if the connectors physically mate. The modems and terminals covered by EIA-232-C/D can run into interface problems, most often because of improper signal timing or the lack of an expected command or response. Sometimes, too, difficulties can arise with products built to international standards because the international standards do not entirely match U.S. counterparts.

TABLE 9.3.3 EIA-530 Summary

Pin no.	Circuit	Signal description
1	—	Shield
2	BA	Transmitted data
3	BB	Received data
4	CA	Request to send
5	CB	Clear to send
6	CC	DCE ready
7	AB	Signal ground
8	CF	Received line signal detector
9	DD	Receiver signal element timing (DCE source)
10	CF	Received line signal detector
11	DA	Transmit signal element timing (DTE source)
12	DB	Transmit signal element timing (DCE source)
13	CB	Clear to send
14	BA	Transmitted data
15	DB	Transmitter signal element timing (DCE source)
16	BB	Received data
17	DD	Receiver signal element timing (DCE source)
18	LL	Local loopback
19	CA	Request to send
20	CD	DTE ready
21	RL	Remote loopback
22	CC	DCE ready
23	CD	DTE ready
24	DA	Transmit signal element timing (DTE source)
25	TM	Test mode

Once the reasons for the problems are understood, however, the solutions often turn out to be very simple—perhaps just cutting a line or adding a jumper. (Interface problems involving software or operating modes are another matter because no standards exist in these areas.)

The EIA-232-C/D standard defines signal functions and characteristics, cable length (limited to 50 feet), and pin assignments on standard 25-pin connectors. Compatibility of electrical characteristics, such as voltage, current, or impedance usually is not a problem, because most manufacturers use standard drivers and receivers designed to meet interface standards. But there are many differences among signals in different types of modems because of speed and transmission differences.

A leading cause of incompatibilities is mistiming of signals or misinterpretation of commands and responses between two units, which can occur because the EIA-232-C/D standard is open to different interpretations by designers with different perspectives on equipment design.

One problem arises from attempts to interconnect simple devices that fail either to produce or respond to some of these signals. A second problem occurs when unassigned lines in the interface cables are used for special functions and connected to equipment that lacks corresponding special-function pin assignments. A third group of incompatibilities is caused by a lack of understanding of international specifications set forth by the CCITT.

According to the EIA-232-C/D specification, the circuits for transmitted and received data (pins 2 and 3) should be complementary because the standard was originally intended for the interface between modems and terminals. Thus, the driver circuit that transmits data on pin 2 of a terminal should be connected to a receiver circuit feeding off pin 2 of a modem. But some of today's devices cannot clearly be categorized as either terminals or modems, giving rise to the possibility that connecting these pins might not link two circuits in a complementary manner. In such a case, the driver output of one might be connected to the driver output of the other and receiver input might be connected to receiver input, as shown in part A of Figure 9.4.1. This problem can be overcome simply by cross-connecting the wiring between pins 2 and 3, as shown in part B.

For asynchronous operation, few if any of the control signals defined by EIA-232-C/D are required. If there are no control signals, an asynchronous device operating full-duplex can use only three of the 25 pins: pin 2 (transmitted data), 3 (received data), and 7 (signal ground). This presents a problem if the asynchronous device is connected to a synchronous unit.

One handshaking mode specified by EIA-232-C/D requires that data be transferred when the receiving device produces a clear-to-send signal on pin 5. (The side of the interface without the clear-to-send capability has a modem.) The problem occurs when the modem-interfaced device is not ready to receive, a state it signals by not producing a clear-to-send signal. If the terminal device cannot sense the absence of this signal and respond accordingly, then it might transmit data that will be lost at the modem. Obviously, the only way two such devices can be successfully interfaced is to add clear-to-send circuits to the terminal device.

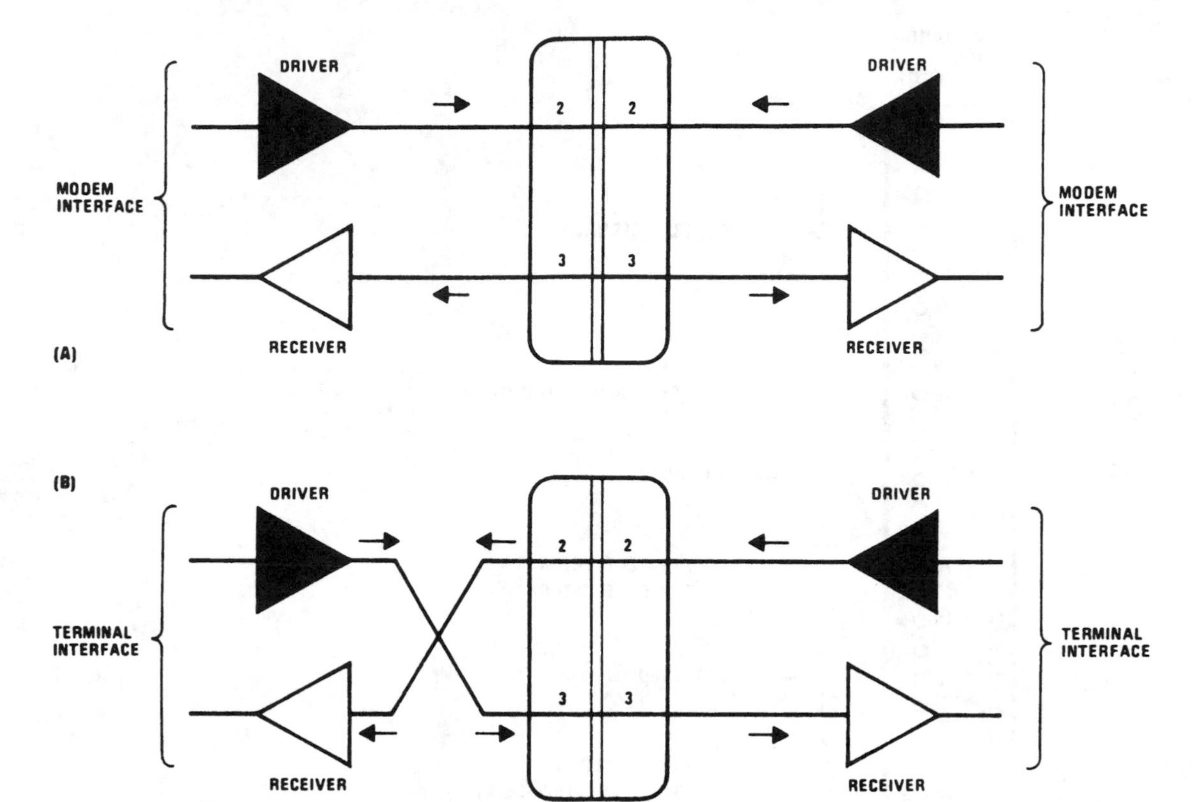

Figure 9.4.1 Cross connecting.

Unassigned pins

A clever designer's trick that sometimes plagues users is to utilize unassigned pins—11, 18, 25, and any others not required for specific applications—for testing and special operating modes. For instance, if a modem does not provide a reverse channel capability, then all of the secondary lines (12, 13, 14, 16, and 19) are available for other assignments. If, however, a device connected to such a modem has different lead assignments, then some aspects of its operation will be disrupted.

Such a disruption could occur when an EIA-232-C/D connector is fitted to a low-speed, asynchronous terminal with a current-loop interface, like a teletypewriter, which requires signals only on pins 1 through 8 and on 20.

Figure 9.4.2 shows how a popular acoustic coupler is wired to such a connector. Here, the substitutions made for EIA-232-C/D signals on pins 9, 10, 16, 17, 23, and 24 convert the connector into an asynchronous current-loop interface. But if the terminal at the other end happens to be a synchronous/asynchronous device operating in the asynchronous mode, then it can be transmitting unneeded clock signals on line 24, making the acoustic coupler appear faulty. The solution is simply to cut the wire on the terminal's pin 24.

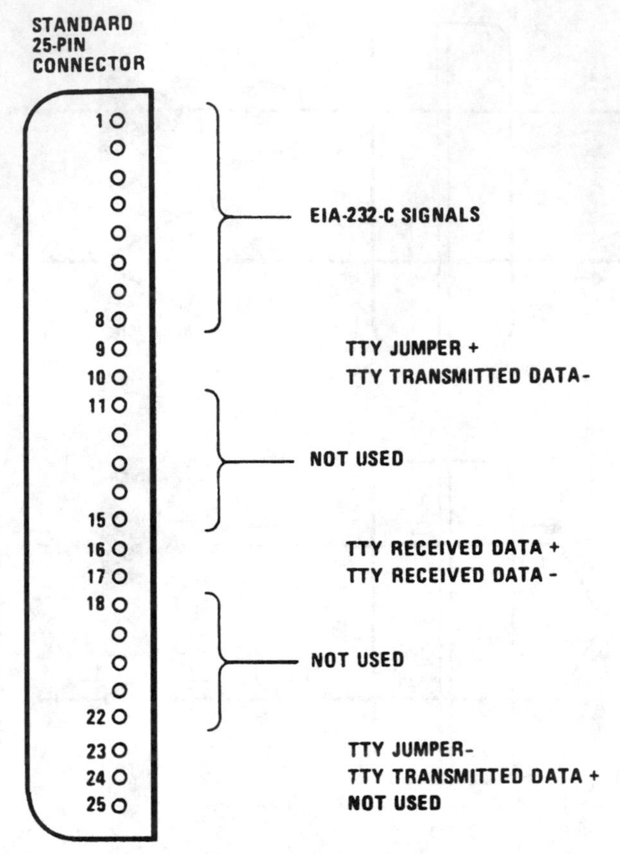

Figure 9.4.2 Current-loop modification.

Not quite the same

A common belief is that the foreign counterpart of EIA-232-C/D, the CCITT's V.24, is identical to the U.S. standard. This can be true in theory, but is seldom true in practice because many countries require modems that conform to national standards that are slightly different.

One difference between the two standards is the definition of the data-terminal-ready signal (pin 20). In EIA-232-C/D, this signal indicates to the modem that the terminal is connected and ready to interact with the telephone line, so that when the telephone rings, the modem will answer if it has an automatic answer feature. When the signal disappears, the modem will "hang up the telephone." The problem is that many nonprogrammable terminals obtain the data-terminal-ready signal by simply wiring pin 20 to a positive voltage level so that the terminal is always ready if it is connected to the modem and the power is ON. This approach is practical for EIA-232-C/D.

However, CCITT V.24 has two modes of operation for pin 20, specified in paragraphs 108.1 and 108.2. The 108.2 approach is virtually the same EIA-232-C/D, but 108.1 is intended only for private lines. If a terminal in which pin 20 is always positive is used on a private line, then the terminal will always be connected to the line.

This could be undesirable because the terminal cannot be taken off line. It is possible, however, to use the 108.1 mode on dial-up lines, provided the terminal can control the signal on pin 20.

To minimize compatibility problems and to facilitate connector modifications, the user should take the following precautions:

- When connecting two EIA-232-C/D devices, ensure that the EIA-232-C/D connection from one device is a DCE interface and the connection from the other device is a DTE interface. EIA-232-C/D was defined to connect DCE to terminal equipment and cannot be used, as an example, to connect two terminals directly to each other.
- Check whether an EIA-232-C/D compatible device has pin functions that truly meet those of the specifications.
- Look for a method to disable unused signals either by switches, jumper options, or software.
- For EIA-232-C/D terminals that will be used outside the U.S. and Canada with CCITT modems, look for a way to cross-wire signals within the EIA-232-C/D connector housing.

9.5 Equipment Options

No discussion of interfacing data communications devices would be complete without talking about equipment options. Like the interface between devices, where signals from one component will affect another, options available on both DTE and DCE must be considered in tandem when interfacing such equipment.

Modem options

Most modems shipped to customers are configured with standard factory settings designed for optimum normal operation. Because different applications require different functions, users might have to change factory settings to obtain peak performance. In some situations, where a modem might have a dual mode of operation, such as originate or answer on Bell System series 100 data sets, the mode option enables the customer to specify the operating frequencies. Hence, for a point-to-point application one modem will be in the originate mode while the other modem will have its operating frequencies assigned for the answer mode. Table 9.5.1 lists some of the common options available for user consideration on many sets.

- *Automatic answering.* For the automatic answering option, if NO is selected at installation, incoming calls must be answered manually. This can be accomplished by depressing the associated locking line key on a telephone set connected to the modem and then lifting the telephone handset. The data set can then be connected to the line by momentarily depressing the data key. Some manufacturers denote this option as key-controlled answering.
- *Loss of carrier disconnect.* YES option will cause the modem to terminate a data call when a prolonged loss of received carrier energy is detected. For Bell System

TABLE 9.5.1 Common Modem Options

Operating mode	Originate, Answer
Automatic answering	Permanently wired, No
Loss of carrier disconnect	Yes, No
Receive space disconnect	Yes, No
Send space disconnect	Yes, No
Data-set-ready (CC) indication	Early, Delayed
Data-set-ready (CC) indication for analog loopback	On, Off
Common grounds	Yes, No
Received data squelch	156, 9 millisecond, or none
Clear-to-send delay	180, 60, 30, 8 millisecond
Soft carrier turn off	In, Out
Transmitter timing	Internal, External, Slave

100 Series data sets, the data set ready (CC) circuit will turn OFF 200 to 350 msec after the carrier falls below the carrier detector threshold, disconnecting the telephone line. If the carrier interruption is less than 100 msec, a disconnect will not occur. A carrier interrupt in excess of 250 msec will always cause a disconnect.

- *Receive space disconnect.* With YES option installed, the modem will disconnect a call after receiving approximately two seconds of continuous spacing.
- *Send space disconnect* With the data terminal ready (CD) circuit being turned OFF, the YES option will cause the modem to transmit approximately three seconds of spacing signals to the modem at the opposite end of the connection before disconnecting itself. This spacing signal causes the far-end modem to go "on hook" if the receive space disconnect feature is installed at that location.
- *Data set ready (CC) indication for analog loop.* If the data terminal equipment connected to a modem requires an ON condition of the CC circuit to transmit and receive test data during an analog loopback test, the CC indication for analog loop ON option must be set.
- *Data set ready (CC) indication.* For some Bell System 100 series modems, circuit CC is turned ON when the data set enters the data mode if "CC-indication early" is installed. If the "CC-indication delayed" is installed, CC is turned ON when carrier from the called data set is detected.
- *Common grounds.* Through the use of the ground wire of a modem's power cord, a protective (frame) ground circuit is established. This cord also provides grounding of the modem housing and chassis. The signal ground circuit on the 25-pin connector is the common reference potential for all circuits on the interface. On most modems, signal ground and protective ground circuits are tied together by an option in the modem housing, installed as the common grounds—YES op-

tion. Tying these two circuits together is intended to provide additional protection against power line noise.

- *Received data squelch.* This option is available on a few Bell System Series 200 modems, such as the 202S and 202T. When a modem that is transmitting in the half-duplex mode on a two-wire circuit turns its request-to-send circuit OFF, the telephone line may echo or reflect signals back to the transmitting device for a period up to the round-trip delay of the circuit. This round-trip delay or signal propagation time within the continental United States is less than 100 msec. The received data squelch option prevents the receiver of the modem that has been transmitting data from delivering the reflections as data to the received-data circuit.

These reflections, or echoes, of the transmitted signal are created when a change of impedance occurs on the line. The circuit impedance acts as a mirror and reflects the signal back to the transmitting station at a reduced amplitude. Echo conditions normally occur on the public switched telephone network. However, echo suppressors are used to stop the signal reflection by attenuating the echo and inserting loss into the echo's return path to the transmitter. Although echo suppressors work well with voice transmission, data transmitted in the half-duplex mode could encounter significant problems if the echo suppressors are not disabled. Thus, when a call is placed on the switched network and a connection established, echo suppressors on the circuit must be disabled before data transmission can start. Modems can disable the suppressors by transmitting a frequency tone at 2000 to 2250 Hz for approximately 400 msec, which appears to the operator as a high-pitched whistle and informs him that the line is ready for data transmission.

During half-duplex operations, each change in direction will require between 50 to 250 msec for the echo suppressors to reverse direction, and this delay contributes to the total transmission turnaround time. Because no turnarounds are required in full-duplex transmission, the echo suppressors do not have to reverse direction and the transmission time is reduced.

For AT&T 202S and 202T modems, the 156-msec option is recommended for public switched telephone network, two-wire private-line, and four-wire private-line facilities. When transmission distances on two-wire facilities are less than 50 miles, the 9-msec option can be used. For distances over 50 miles, the 9-msec option is used only if the associated data terminal can ignore echoes. If the data terminal is able to ignore echoes, the no-squelch option can be used on four-wire facilities and in certain subscriber-engineered applications on two-wire facilities.

For transmission over the switched telephone network, it is impractical to attempt to optimize turnaround time if the data terminal cannot ignore echoes by using the 9-msec or no-squelch option in conjunction with 60-, 30-, or 8-msec clear-to-send options that will be covered later in this section. This is because the propagation time and echo delay vary widely on the switched network as a result of the alternate routing capability that makes the transmission distance for two calls between the same points differ. In addition, even if the data terminal can ignore echoes, it is difficult to optimize the turnaround time on the switched network because echo suppressors can be used in the connection and their turnaround time can be as long as 100 msec. If a terminal can ignore echoes and keep echo suppressors disabled by using the reverse channel of the modem, then the turnaround time can be optimized by using the 9-msec or no-squelch option in conjunction with the 60-, 30-, or 8-msec clear-to-send delay.

- *Clear-to-send delay.* The availability of this option is similar to that of the received-data-squelch option. This option concerns the time delay between the ON condition of a request-to-send circuit and the ON condition of the clear-to-send circuit. Bell System 202S and 202T modems permit delay settings of 180, 60, 30, or 8 msec. The delay option selected must be chosen to be compatible with the remote modem's squelch and receive-line signal-detector acquisition timing and for soft carrier turn-off operations on two-wire media.

  The 180-msec option is recommended for use on the switched telephone network, four-wire leased line facilities with talkback, and two-wire leased-line facilities. This option is required when a 202S or 202T modem has the 156-msec squelch option installed. When used on the switched network, the 180-msec delay ensures that the echo suppressors on the circuit are turned around prior to data being transmitted.

  For four-wire point-to-point and multipoint facilities requiring fast startup, the 30- and 60-msec options should be used. The 8-msec option can be used in certain modems for full-duplex multipoint applications that require a fast modem startup.

- *Soft-carrier turn-off.* At the conclusion of a data transfer, when a terminal turns request-to-send OFF, transients can cause spurious spacing signals to be received at the other end of the circuit. When the soft-carrier turn-off-IN option is used, the data set will transmit a soft-carrier frequency at 900 Hz for either 8 or 24 msec for 202-type modems after the request-to-send circuit is turned OFF. The 8-msec option is used when the distant modem has a fast mode carrier detection option that permits a fast response time to the soft-carrier turn-off. When the OUT option is selected, the carrier is turned OFF in less than 1 msec after the request-to-send circuit is turned OFF.

- *Transmitter timing.* This option is available on synchronous modems and it provides a method of bit timing to the modem's transmitter. When the internal timing option is selected, the transmitter timing will be provided by the modem. An external timing option will use the DTE as the clocking source. If the modem has a slave timing option, the transmitter will be driven by the receive clock from circuit DD.

- *Terminal options.* Terminal options vary considerably, depending on the manufacturer and type of terminal used. For a simple teleprinter, options include an answerback station identification that is automatically transmitted in response to receiving a special control character to the type of stop bit (one, one and a half, or two elements) that it can accept. New terminals have up to 30 user-selected options, including interpreting received control characters and performing predefined functions for maintaining high, low, or no voltage on certain interface circuits that can be used to modify the EIA-232-C/D transmission sequence.

9.6 Importance of Turnaround Time

For applications where a large volume of data is to be transmitted between a terminal and a computer, or where many terminals share a common circuit, turnaround time becomes extremely important and can seriously reduce transmission efficiency. Turnaround time is the time it takes a modem to switch from receiving data to transmitting data and vice versa, and consists of a number of components. These components include the modem's internal delay time, which is the time it takes between a digital signal entering a device until the first modulated tone is put on the line; the propagation time for the signal to travel to its source; and the request-to-send/clear-to-send delay, which is the time between a terminal raising its request-to-send signal and the modem informing the terminal that data can be transmitted by returning a clear-to-send signal.

For transmissions over the public switched telephone network, request-to-send/clear-to-send delays between 100 and 200 msec are typically available. These timings are greater than for modems designed to operate on private lines, because that time on the switched network must be allowed for echoes and transient signals to decay.

Modems designed for operation on private lines have short request-to-send/clear-to-send times, ranging from about 5 to 60 msec. When designed specifically for multipoint private line operation, modems with short request-to-send/clear-to-send times are also commonly referred to as *fast-poll modems*.

Effect on data transfer

To get a feel for the turnaround time effect on data transfer, let us assume our data transfer rate is 2400 bit/s. Suppose you are transmitting data synchronously and your data blocks consist of 50 8-bit characters, or a total of 400 bits per transmitted block. When this block is transmitted, it takes

$$\frac{400 \text{ bits}}{2400 \text{ bit/s}} = 166.6 \text{ milliseconds}$$

for the last bit in the block to leave the transmitter. If you assume that the modem's request-to-send/clear-to-send delay time is 100 msec, then the transmission overhead is

$$\left(\frac{100}{100 + 166.6}\right) \times 100 = 37.5\% \text{ overhead}$$

Based on this equation, it is interesting to note that the 37.5% overhead is actually a minimum overhead time because the transmission control-character overhead had not been considered. This overhead time can be reduced by enlarging the size of the data blocks to be transmitted. Thus, if the data block size is doubled to 800 bits, the time to transmit the block will double to 333.2 msec if the transmission rate remains at 2400 bit/s. Because only one turnaround time is required, overhead decreases to

$$\left(\frac{100}{100 + 333.2}\right) \times 100 = 23.1\% \text{ overhead}$$

Although this blocking approach can be used to increase data transmission efficiency, the reader should be aware that the larger the data block, the longer it takes to retransmit that block when it is received in error. Thus, the user must consider additional factors, including the expected error rate of the medium, in order to determine an optimum block length for transmission.

Questions

9.1 Into what basic groupings can the EIA-232-C/D interface circuits be arranged?

9.2 What is the difference between protective ground and signal ground circuits?

9.3 What effect does the request-to-send signal have on transmission in the half-duplex mode and full-duplex mode?

9.4 State the transmission directions of the following circuits:

A. Request to send
B. Clear to send
C. Data set ready
D. Received line signal detector
E. Transmit data
F. Receive data
G. Data terminal ready
H. Ring indicator

9.5 How does a reverse channel differ from a secondary channel?

9.6 Why transmission speed on a secondary channel at a much lower speed than that possible on a primary channel?

9.7 What are the key differences between EIA-232-C and EIA-232-D?

9.8 What is the key mechanical difference between the EIA-232-C/D and EIA-449 standards?

9.9 What is the purpose of a mode option on some AT&T Series 100 modems?

9.10 If an automatic answering NO option is selected for modems at a computer site, what effect does this selection have on terminal-to-computer transmission?

9.11 If data is to be transmitted at 4800 bit/s with blocks of 200 8-bit characters and the modem's request-to-send/clear-to-send delay time is 80 msec, what is the transmission overhead in percent?

9.12 If a fast-poll modem with a request-to-send/clear-to-send delay time of 10 msec is used, what happens to the overhead of question 9.11

9.13 What effect on transmission overhead (question 9.11) does an increase in the size of the data block to 300 8-bit characters have?

9.14 If the transmission speed in question 9.11 is raised to 9600 bit/s, what happens to transmission overhead?

Chapter

10

Introduction to Local Area Networks and Internetworking

Few areas in the data communications world have seen as much technological innovation and as many new commercial offerings as has that of local area networks. This development has been in response to users' demands for greater transmission speed and capacity. Common carrier communications is adequate for most remote-job-entry (RJE) terminals, interactive terminals, and clustered terminal controllers. However, in many cases, it is highly desirable to provide independently programmable computers and conventional terminals located within a relatively small geographical area with methods of exchanging data and of sharing access to one or more mainframe computers.

This chapter first examines the different topologies associated with local communications, beginning with the channel interfaces provided by computer manufacturers to link mainframes, minicomputers, and microcomputers. Next, it focuses on two popular access methods that are used to control the placement of data on a LAN. Because most local area networks are confined to installation within a building, the various options that should be considered for installing wiring within a building are also covered. Concluding this chapter are the IEEE's Ethernet and 802.5 token ring access standards and IBM's token ring, one of several popularly implemented and standardized methods for transmitting data on a local network.

10.1.1 Topology

In elementary forms, local networks have been around for some time. Since the early 1960s, computer manufacturers have provided "channel couplers" between large-scale processors, "channel interfaces" between mainframes and minicomputers, and "parallel interfaces" between minicomputers, as illustrated in Figure 10.1.1. Exam-

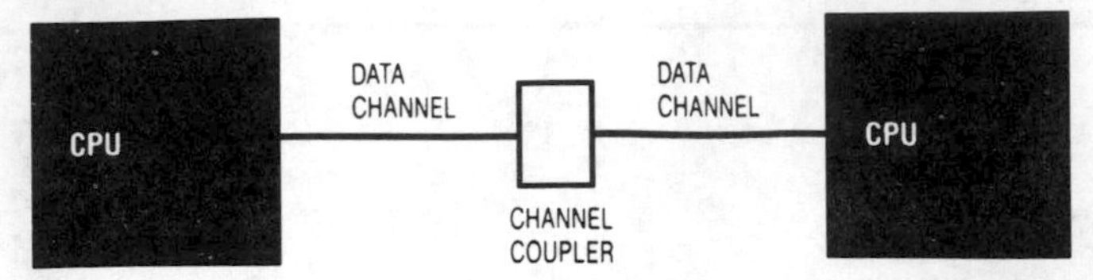

Figure 10.1.1 Local communications.

ples of these are the IBM Channel-to-Channel Adapter that couples the input/output channels of two IBM central processing units (CPUs), Digital Equipment Corporation's (DEC's) DX-11 interface that allows a PDP-11 to appear to be a peripheral on an IBM selector or byte multiplexer channel, and the DEC DA-11B interface that allows two PDP-11s to exchange blocks of data at channel speed of roughly 1.5 Mbit/s.

These types of connections offer two main advantages. The processor's data transfer rate is about as fast as can be expected, and the incremental cost of adding the "network" is small relative to the cost of the processors themselves. There are, however, several inherent disadvantages. First, the distance covered by these local links is small—generally no more than 200 feet, and often less than 100 feet. Also, such a link is designed to have only a certain type of processor at each end. Because most manufacturers are reluctant to support digital interfaces that attach to a competitor's processor, there is generally a limited inventory of available channel interfaces.

Separate entities

A significant step in the development of local networking was the network's becoming a facility separate from the discrete computers attached to it, rather than just a series of links connecting individual computers. The initial design objectives (which remain critical even today) were threefold:

1. The local network connections should allow data to move directly from one node to another without intervention by intermediate host processors. This essentially establishes local network communications on a peer-to-peer basis, where no single CPU is responsible for all communications.
2. Data enters the network in a standard format. This means that adding a new type of processor to the local network requires a hardware and software interface between the new host and the local network. With this interface, the new processor should be able to communicate with any network node.
3. Because both efficient throughput and high responsiveness are goals of the local network, it must be able to allocate data transmission capacity on a demand—rather than a partitioned—basis. Traffic between any two nodes on a local network tends to be "bursty," in the same sense that data traffic volume between a CPU and a disk drive varies greatly over a period of time. At the same time, the network must ensure that a single node with high traffic requirements does not lock out other network nodes.

Enter the ring

Incorporating these separate design objectives gave birth to the "ring networks." Probably the most publicized was built at the University of Cambridge in England, beginning in 1974. The ring itself, which is illustrated in Figure 10.1.2, consisted of a set of "repeaters" connected in a loop by a pair of twisted-pair wires. The loop continuously circulated a pair of "packets" at a raw bit rate of 10 Mbit/s. Each packet contained two bytes of information plus routing and control information, which was constantly examined by each repeater.

If a node on the ring wanted to transmit, it waited until it detected a passing packet that was empty. It then forwarded the packet while inserting its own routing information and data. The packet then made a complete circuit on the ring and returned to the originator, where it would have been marked as copied, or rejected, by the receiving node. The original sender then marked the packet as empty and forwarded it on the ring so that another node could use it.

The repeaters were not powered by the processor nodes, but by a dc voltage that was injected at several points onto the network. This minimized the possibility of a repeater failure. A station unit informed the repeater when data was to be sent and received, and provided the data, in parallel digital form, to the "access box," which interfaced to the host computer.

The ring architecture realizes most goals of the local network. Data can proceed more or less directly to the intended destination at about 3 or 4 Mbit/s, which is adequate for most applications. Users found that they achieved higher network performance by improving the communications link between the repeaters—typically by

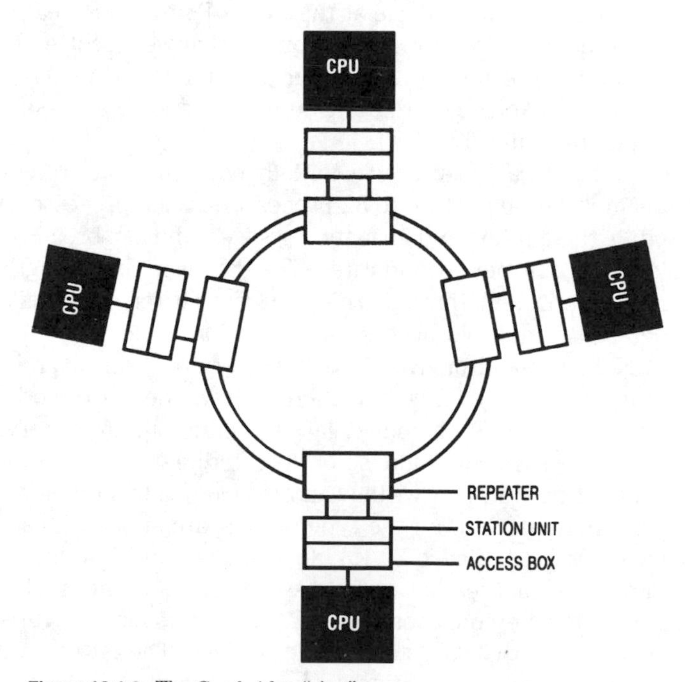

Figure 10.1.2 The Cambridge "ring".

replacing the twisted pairs with coaxial cable. Depending on the adaptability of the access box, adding a new processor to the network interrupts network service only briefly.

The major disadvantage of the ring network is that a failure of any single repeater or communications link will cause the entire network to fail. Furthermore, these repeaters and links are often geographically distributed, so it is not always possible to immediately repair or circumvent their failures. Various refinements have since been added to the Cambridge ring to detect the failure point; other network designs have attempted to centralize the ring management so that all repairs can be performed from a single location.

The bus connection

Another type of local network topology that has achieved commercial acceptance is the bus. Network nodes, connected by a linear "bus," can broadcast data directly to the destination. All other nodes simply ignore the broadcast data and await their turns to broadcast. A principal advantage of the bus architecture is that all nodes on the bus are "passive"; that is, if one node fails, it will simply fail to send or receive its signals—all other activity remains uninterrupted. Each device connected to a bus network has an interface that drives an external data bus. The computer node can request the use of the bus and place a word of data on it, along with its destination address. Interfaces generally allow each attached computer to send or receive a block of data to or from any other computer on the bus. One of the earliest developed bus-structured local area networks is Xerox's Ethernet.

Ethernet was developed for internal use at the Xerox Palo Alto Research Laboratories in 1976 to link a set of single-user minicomputers that were scattered throughout the research center. Ethernet provided a means for these minis (and their associated researchers) to exchange programs and data, and to access specialized peripherals rapidly and dynamically.

The Ethernet serial-bus local network was initially constructed to run on a single coaxial cable that is multidropped at up to 32 processors. Each processor has a network interface module that buffers and formats messages, and can broadcast data on the cable in bursts at megabit-per-second rates. The data is in fixed-length packets, and contention is accomplished through the use of a "listen-while-talk" feature (CSMA/CD—carrier sense multiple access with collision detection). Whenever a node is broadcasting, it also monitors the signal for interference from another node. If two nodes pick the same time to transmit, then both will detect the other's interference and cease transmission. Each node delays for a random time interval and retransmits if the other node's transmission is not detected first.

Either coaxial cable or twisted-pair wiring can be used in a bus topology, with the actual wiring permitted, dependent on the type of local area network used and the type of network adapters used. Figure 10.1.3 illustrates a typical method of attachment to a bus-structured local area network. In some LANs, the network adapter is a card that is inserted into a personal computer. LAN vendors manufacture adapters on cards as well as adapters that are stand-alone devices. The latter permit dumb terminals, stand-alone printers, and storage devices to be added to the network.

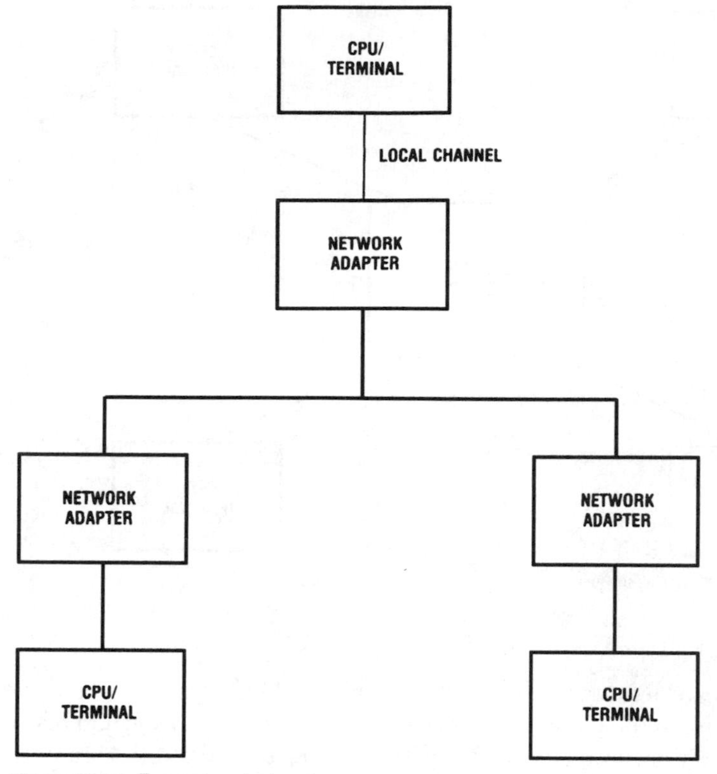

Figure 10.1.3 Bus network structure.

Star network

Perhaps the oldest local area network topology is the star because it provided a LAN capability prior to the actual development of local networks. This capability was originally furnished by the switchboard, which permits any device connected to the switchboard to communicate with any other device connected to it.

Figure 10.1.4 illustrates the star local area network topology. Today, the primary use of this topology in local area networking is accomplished by the use of the data Private Branch Exchange (PBX) and port selector that serve as a centralized switching device. Their use permits data terminals and personal computers to easily access computer ports and other devices connected to one or more lines on those devices.

10.2 Access Methods

The access method governs the placement of data onto a network, analogous to a traffic light regulating the vehicle flow from an entry ramp onto an expressway. Two of the most common LAN access methods are carrier sense multiple access with collision detection (CSMA/CD) and token passing. This section examines each access method, including its strengths and weaknesses and its utilization in commercially available LANs.

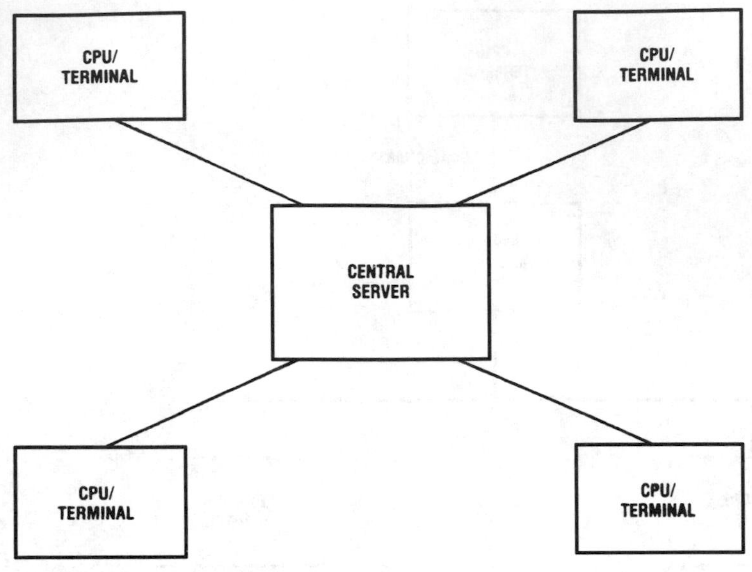

Figure 10.1.4 Star topology.

CSMA/CD

CSMA/CD can be categorized as a "listen-then-talk" access protocol. Its development can be traced to the pioneering work of Dr. Norman Abramson and his colleagues at the University of Hawaii during the 1970s. It was then that Dr. Abramson developed a packet radio transmission scheme to connect geographically distributed island sites to a common communications channel. The resulting technique, known as *Aloha*, represented the first practical utilization of shared-channel and packetized-transmission technology.

During that period, Dr. Robert Metcalfe was working at the Xerox Palo Alto Research Center (PARC) in California. Tasked with interconnecting a large number of computers at that location, he developed the technology that has come to be called *Ethernet*. One of Ethernet's key elements was its use of the CSMA/CD access protocol.

In CSMA/CD, each station "listens" for the presence or absence of a carrier signal prior to transmitting data. If no carrier signal is sensed, the station is free to transmit. Because two (or more) workstations wanting to transmit could sense the absence of a carrier signal, they could simultaneously (or nearly so) begin to transmit data. This would result in a "collision." Figure 10.2.1 illustrates this signal-collision process.

In Figure 10.2.1A, both stations A and B desire to transmit, listen, and sense no carrier signal. In Figure 10.2.1B, station A begins transmission and its data starts to flow onto the bus-structured network. In Figure 10.2.1C, station B, not yet sensing a carrier signal, begins its own transmission. The two transmissions collide, as shown in Figure 10.2.1D.

The first station to detect a collision transmits a jam signal, whose duration equals the time required for a frame to be placed on the medium. This jam signal alerts other network stations to the collision occurrence and not to attempt to transmit normally. Instead, each station having data to transmit applies a random time delay prior to again listening to the network for a carrier signal and attempting to retransmit.

Because the key to CSMA/CD is carrier sensing, you might ask just how a station accomplishes this. The original Ethernet network, which was standardized by the IEEE 802.3 committee as 10BASE-5, uses digital baseband signaling. Data is encoded using Manchester coding, which ensures that each zero bit and one bit results in a signal transition (Figure 10.2.2).

In Manchester coding, each binary 1 is represented by a low-to-high signal transition, while each binary 0 is represented by a high-to-low transition. Each signal transition results in the use of one half of the bit interval, and the previous state fills the prior half of the bit interval.

This coding method ensures continuous transitions with network data flow. The transition function is equivalent to a carrier signal associated with analog transmission. That is, by noting Manchester code transitions, a station knows that another station is transmitting.

Token passing

In token passing access, one or more bits in a field within the "token" frame is used to control a station's decision to transmit data. If the token bit(s) indicate a free (available) token, a station having data to transmit can acquire the token. The sta-

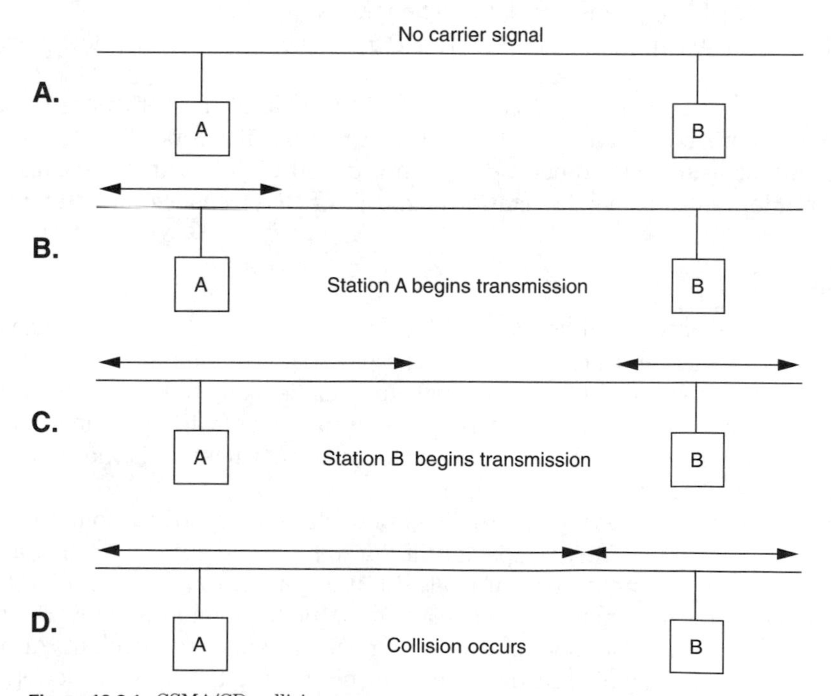

Figure 10.2.1 CSMA/CD collision occurence.

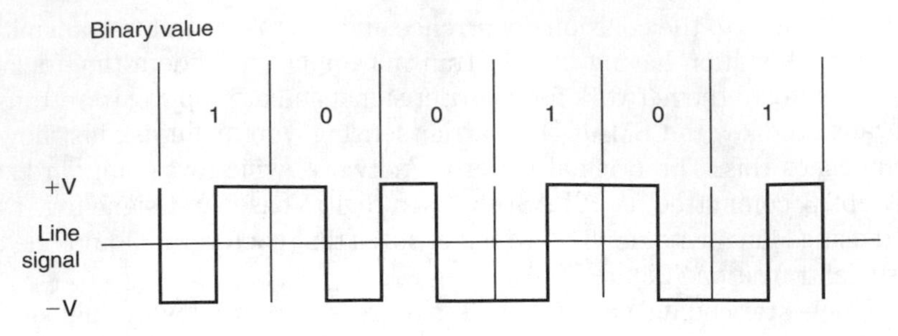

Figure 10.2.2 Manchester coding.

tion first converts the token's value to indicate that it is in use. Then, the station adds information to the frame: the recipient's destination, the originator's address, and the data to be conveyed to the recipient.

In comparing LAN access methods, you can call CSMA/CD *probabilistic* and token passing *deterministic*. In a CSMA/CD-based network, it is very difficult to forecast performance: an increase in network utilization increases the probability of a collision occurrence. And the random backoff interval used by stations after a collision occurrence worsens the difficulty of forecasting the effect of an increase in network utilization on the network's capacity to transport data.

By comparison, a token passing network is much more predictable. Here, models can be developed to calculate the effect of an increase in network utilization on the network's ability to transport data. In general, at low network utilization levels (less than a 30 to 40 percent range), CSMA/CD and token passing provide equivalent performances.

As network utilization increases, a CSMA/CD-based network's performance will worsen more rapidly than that of a token passing network. The reasons include growing collision occurrences, delays introduced by jam signal transmissions, and the random time delays that stations incur prior to retransmission attempts.

10.3 Building Wiring

Data is transmitted over so many different types of cable—coaxial, twinaxial, even triaxial, shielded and unshielded twisted pairs, and optical fiber—that the coexistence of multiple cable types in a single location is likely to be the rule, rather than the exception. And the age-old debate over what type of wiring to run in a building—to accommodate current and future voice and data needs—is more complex and controversial today than ever before.

Installing and operating a multiplicity of cable types clearly is something to avoid whenever possible: effective space utilization, record keeping, and cost accounting all become increasingly more difficult. But it is getting tougher to inhibit the proliferation of transmission media, especially with the boom in local network offerings.

Installation and planning, however, are necessarily interdependent. Although the cost of any initial cabling installation must be taken into consideration, the impact and eventual cost of future changes must also be factored into the analytical equations used to select a local wiring scheme. It is not unusual today for 50 percent of the telephones and data terminals already in place to be moved each year, as offices and equipment are relocated. Moreover, the obstacles to connectivity—increased wiring complexity and loosely organized wiring plans—inherent in modern communications device designs, invariably cause degradation in a network's effectiveness.

To make the task more manageable and predictable, careful early planning is an important step in selecting and installing cost-effective wiring plans that anticipate the present and future needs of voice and data users. The successful communications managers are learning this.

Key ingredients

Proper installation planning starts with the understanding that, to be used to its best advantage, a building's wiring space must accommodate the tenant's short- and long-term communications requirements. Many other issues and factors, of course, will determine how well the job gets done. These factors include the type and scope of planning resources available, time, the qualifications of the installer, and the frequently overlooked contractual relationship between the building's owner and the tenant.

Because the cost of wiring an existing building goes up with the height of the ceiling, and rises even higher after the tenants have moved in, making the right decisions as early as possible can significantly alleviate future costs. Aside from the serious loss of productivity that results from interfering with normal working patterns, dangling cables and drifting dust create safety hazards and generally demoralize employees.

The increasingly popular method for ensuring current and future economies is prewiring—installing cabling or raceways (cable conduit) during building construction. Not only does this reduce installation costs, but it also permits careful labeling and recording of cable runs and circuits, thus making future network changes much easier to implement. The final appearance or utility of wiring installed after building construction rarely equals the quality that is attainable through prewiring.

Building types

A building's design reflects and influences the type of activities that it houses. Communications station density, building codes, and communications cable distribution are all interrelated. They affect, and are affected by, the design, construction, and use of each building.

Planners should provide the highest possible degree of flexibility in cabling installations for general-purpose office buildings. Tenant activities generally exhibit high station density and medium-to-heavy communications requirements that are greatly affected by frequent changes of tenants and rearrangement of offices.

Retail buildings, by comparison, have fairly low station density—even though stations are often clustered in strategic locations within the available floor space. Cable distribution can be regarded as somewhat more stable, though it still must be flexible enough to accommodate occasional floor-space rearrangements that are intended to improve the sales environment.

Manufacturing buildings, in contrast to other types of buildings, do not usually require that the cable be hidden. Communications cabling in manufacturing environments might be comparatively stable, but this will depend on the equipment, the product line, and product-cycle factors. Because cabling is frequently run overhead or along walls, communications wiring in many manufacturing structures tends to be quite flexible.

Residential communications cabling can be characterized as accommodating telephone, television, and perhaps security equipment. For the most part, this equipment is contained in living areas of fixed size that rarely change. Cabling installed in walls and ceilings can generally be considered permanently in place.

Health-care facilities typically house heavy users of communications equipment. Communications equipment and networks in these environments are subject to frequent inspection and rigid enforcement of building and safety codes that regulate the installation and operation of primary, as well as backup, communications facilities. Network and wiring changes are common and extensive in this environment, and highly flexible cabling plans are essential.

Building codes

A careful examination of local and national building construction codes is a vital prerequisite to the planning process. Local codes can override or preempt nationally accepted codes, and regulations that governed the construction of existing buildings might have changed significantly by the time new construction starts.

Although prewiring might be the preferred approach, it isn't always the most economically attractive. The first issue that a network planner must resolve is: should a new facility be constructed and prewired, or should an existing one be rewired? In many cases, this decision will not be made unilaterally by the communications managers, though their input would likely play a major role in company determinations. Such decisions will likely also determine many subsequent issues that are within the purview of communications managers. They include:

- Choice of networks or technologies.
- Types of terminals and equipment.
- Relative and real costs of installations.

Whether the network is installed in an existing building or in new construction, the need for its modification will eventually arise. Before choosing a network technology or topology, planners should determine the degree of flexibility needed for future changes. Specifically, they should answer these questions:

- How will existing equipment, users, and peripherals be moved and new equipment be added?
- What capabilities will be needed to expand the number of user stations?
- Will prewired new construction use raceways, risers, or point-to-point cabling?

Plan for growth

Network growth is as inevitable as it is difficult to predict. There are, however, three basic components of such growth:

1. Expanded network utilization by the original users.
2. Extension of the network to new users.
3. Physical plant or campus modifications.

The natural growth of a network should be a primary planning consideration. This includes estimates of expansion caused by business activity. A second aspect of natural growth is the increase in network utilization that comes with user experience and familiarity. For example, within a three-year period after initial activation, a network's inquiry and response activity per person can be expected to double.

Perhaps the biggest cause of expanded network use comes from the addition of new users. Predicting growth rates for this factor is generally more difficult than for other causes of expansion, but it is not impossible. Classic market research techniques, combined with hard data based on specific analyses, as well as the experience of other companies with similar products and services, are all good planning tools.

What is called a *substitution effect* usually accompanies the introduction of a newly installed communications network. If, for example, telephone inquiries from a sales office to headquarters are replaced by terminal queries, a one-for-one substitution of the traffic will occur initially, and will then probably grow only gradually. The general growth pattern for terminal-to-host activity tends to follow conventional market penetration curves (as illustrated in Figure 10.3.1). A slow initial acceptance is followed by rapidly accelerated usage, which then levels off to more steady, predictable growth.

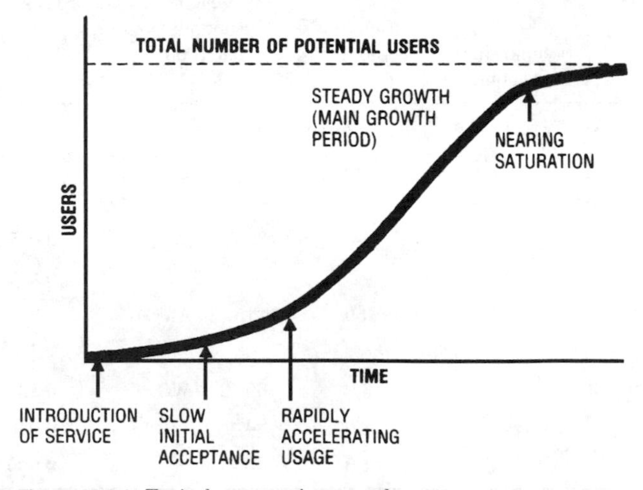

Figure 10.3.1 Typical new service growth pattern.

There are three media alternatives today for use within buildings for general-purpose communications distribution:

- Twisted pairs of copper wire, including shielded twisted pair.
- Coaxial and twinaxial cable.
- Optical fiber.

Other frequently used transmission media, which essentially are composites or variations of these, include flat, or under-carpet, cable and mixed-conductor cable.

Each cable type serves some application or supports certain transmission techniques better than do the others. Each, moreover, has its own cost and performance benefits.

Microwave transmission, though widely used, is unsuitable for intrabuilding data or voice communications. Infrared lightwave transmission is also making an appearance, but its use in the intrabuilding environment is still largely developmental.

In evaluating the appropriate cable, the physical characteristics of each type should be studied with respect to its pull strength, bending radius, and weight. Table 10.3.1 lists the physical characteristics of telephone cable and three types of common data cables.

Twisted pair

Twisted copper wire, the twisted pair, is better known as common telephone cabling. Although used ubiquitously for telephony, twisted pair has several disadvantages for data transmission. It is extremely susceptible to electrical interference, which might come from sources including nearby electric typewriters, air-conditioners, X-ray equipment, or a host of other external sources. Although such interference, or noise,

TABLE 10.3.1 Physical Characteristics of Communications Cable Types

Cable type	Pull strength (pounds)	Bending radius* (inches)	Weight (pounds per 1000 feet)	Cross-section area (square inches)
	Telephone cable			
3-pair	18	.22	—	0.0375
4-pair	24	—	—	—
6-pair	—	—	35 (24 AWG)	0.075
25-pair	150	—	—	0.220
50-pair	300	.69	215 (24 AWG)	0.375
	Data cable			
Shielded twisted pair (2-pair)	55	3.0	60 (22 AWG)	0.125
Coaxial	22	1.0	34	0.046
Twinaxial	26	1.5	65	0.096

*Normal bend radius is typically from eight to ten times the outside diameter

does little to impede conventional analog voice signals, it causes two problems for data transmission that are interrelated: limitations to both speed and distance.

The sensitivity of twisted pair to noise limits the speeds at which data can be sent. This is because a burst of noise that would garble only a few bits of low-speed data will destroy many bits of high-speed data.

Twisted pair also exhibits severe distance limitations for data. An electrical signal (voice or data) grows weaker, or attenuates, as it travels. Although electrical signals will attenuate on any conducting medium, twisted pair's vulnerability to noise is complicated by its capacity to act as an antenna. The longer the length of twisted pair, the more noise it gathers. After a while, the accumulated noise obliterates the attenuated signal.

Shielding the wire and periodically repeating the signal can reduce twisted pair's vulnerability to noise. Shielding—enclosing one or more twisted pairs within a metallic shell—makes twisted pair much less affected by noise, but it also adds significantly to its cost.

Repeaters, which retransmit signals from one length of wire onto another, increase the distance that a signal can travel. They are, however, also expensive and can greatly increase the cost of running twisted pair.

Twisted pair, shown in Figure 10.3.2A, is best for low-cost, short-distance local networks, especially for small networks linking, for example, microcomputers. This type of cable can carry data at rates in excess of 10 Mbit/s over distances up to several hundred feet without repeaters. The photo shows three basic twisted-pair cable types: the fat, 25-pair (50-conductor) version is the mainstay of the telephone industry. The one in the middle is a shielded length of three twisted pairs that are surrounded by a metallic foil. And the last is a basic unshielded cable that contains three twisted pairs—a total of six conductors.

Figure 10.3.2B shows the type of termination connectors that are most commonly used with the twisted pair. The 25-pair connector (left) is a telephony standard. The middle connector, similarly, is a standard telephone wall jack: in this case, terminating three pairs—six conductors. Most data applications use the ubiquitous EIA-232 connector, which is used to terminate unshielded or shielded twisted pair. In the photo, the EIA-232 connector uses only six leads and terminates a three-pair twisted-pair cable. Most EIA-232 devices can be accommodated with as few as six leads, and many require only four.

Coaxial cable

Coaxial cable, or coax, until 1990, was the most widely used medium for data transmission in local networking. The cable is available in several forms, each suited to a different kind of application. All forms of coaxial cable have the same general structure (Figure 10.3.2C). A central conductor—the part of the cable that carries the signal—is surrounded by a dielectric, or nonconducting, insulator, then by a solid or woven metal shielding layer. The outside of the cable is encased in a protective coating. All these layers are concentric; hence, the term *coaxial*.

Of the three basic coaxial cable types shown in Figure 10.3.2C, the first is encased with a shielded twisted-pair cable. The second type is essentially two coaxial cables,

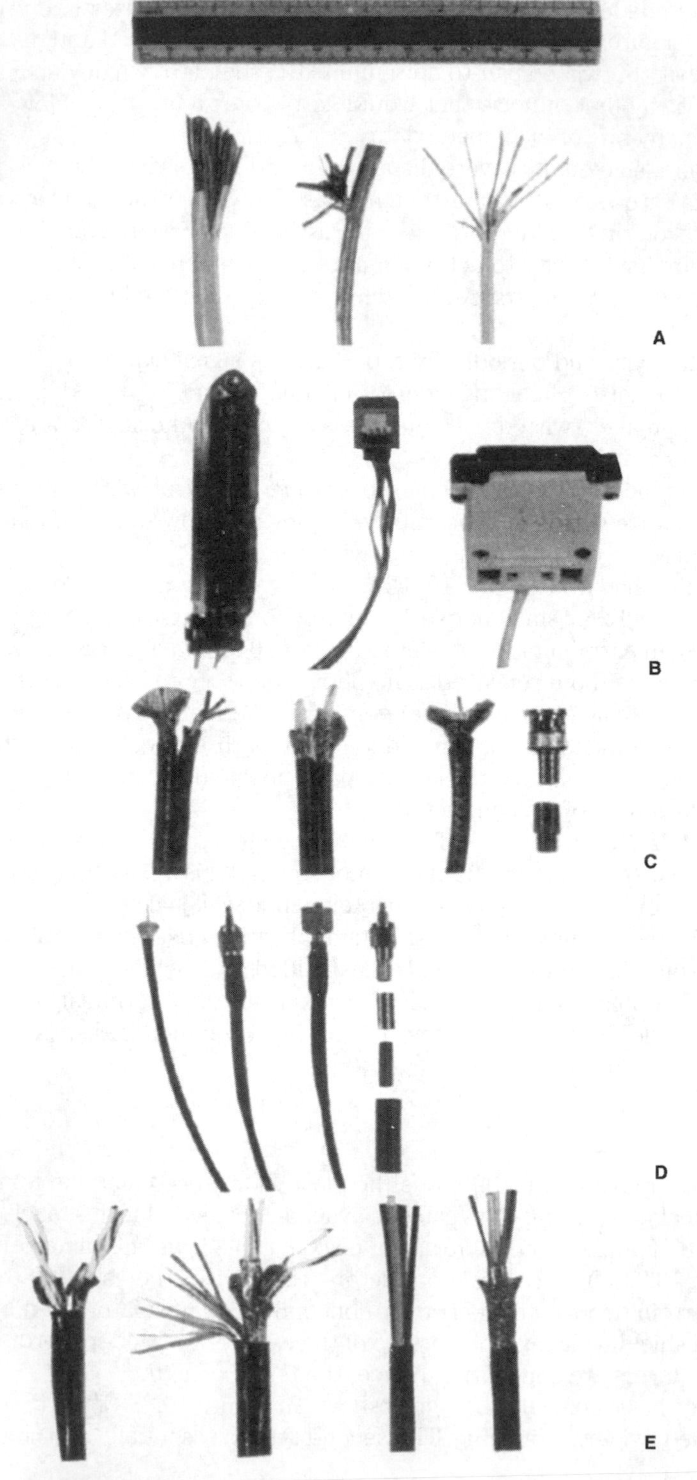

Figure 10.3.2 Wiring variations.

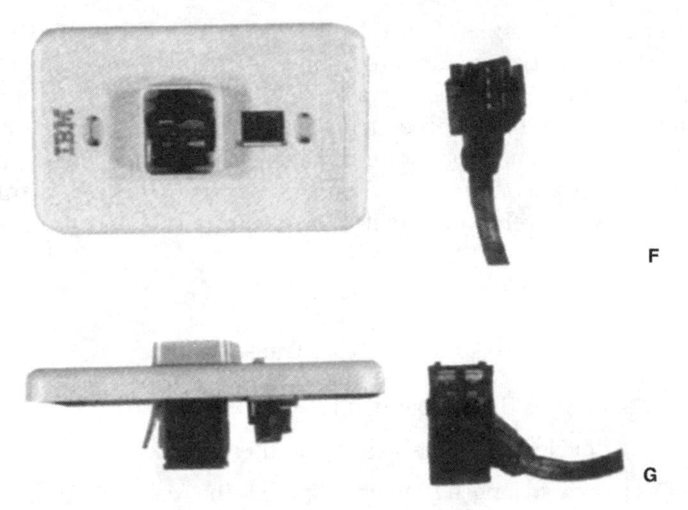

Figure 10.3.2 Continued.

which constitute a twinaxial cable. The third type, a single coaxial cable, has next to it a BNC connector, which is commonly used to terminate a length of coaxial cable, especially for data transmission. The cable types in the photo all have a fine wire mesh shielding, which is less expensive—but also provides less shielding from noise—than coaxial cable that has solid, extruded-metal sheathing.

Because it is shielded, though to varying degrees depending on the specific type, coaxial cable is largely immune to electrical noise. This cable can carry data at higher rates over longer distances than twisted pair. Also, because it has been around for several decades, the techniques used to install coaxial cabling are well-established.

There are two general classifications of coaxial cable, named for the transmission techniques they support; *baseband* and *broadband*. For years, controversy has flared over the relative merits of these two transmission techniques. The truth of the matter is that each serves quite different applications.

Baseband coaxial cable carries only one digital signal at a time—though typically at rates ranging from 1 to 10 Mbit/s. This bandwidth enables the time-division multiplexing of many signals. With baseband transmission, a signal is broadcast in both directions along the cable.

Broadband coaxial cable can carry many radio frequency (RF) or analog signals at a time, with each signal occupying a different frequency band on the cable. Currently, practical data rates on any single channel of a broadband coaxial-cable-based network are somewhat lower than those supported by baseband transmission—generally between 1 and 10 Mbit/s. But the availability of 20 to 30 channels on a single cable—combined with time-division multiplexing—greatly increases the composite amount of data that the medium can carry.

As a result of the nature of the coupling and amplification hardware that carries the RF signal, broadband signals are effectively unidirectional. Broadband transmissions, moreover, require RF modems to convert the analog signal carried over the cable to and from its original digital form.

A leading example of a broadband-based network is the IEEE 802.3 10BROAD36 LAN. This Ethernet-type network standard defines a 10-Mbit/s operating rate on 75-ohm coaxial cable, with cable segments up to 3600 meters long. The topology of a 10BROAD36 network resembles a tree structure, with a "reconverter" at the top (head-end) of the tree. Each network station uses a radio frequency (RF) modem to modulate data for transmission. The head-end reconverter receives a transmitted signal at one frequency and rebroadcasts it to other stations at another frequency.

Baseband limitations

Studies indicate several drawbacks to baseband transmission:

- Lower aggregate data rates than broadband.
- Serious problems from electromagnetic and RF interference.
- Higher signal loss per cable length than with optical-fiber cable at comparable data rates.

Nonetheless, baseband transmission is well-suited to small- to medium-sized data processing or office automation environments. Though more distance-limited than broadband, baseband networks can adequately cover a single building or a small campus. Broadband networks work most effectively in larger geographic configurations, where economies of scale can justify the generally higher broadband maintenance costs. A broadband network can connect the buildings on a large campus or even those in a good-sized city.

Another key advantage of a broadband network is being able to use common cable for voice, data, and video applications. The use of RF modems operating at different frequencies enables the addition of voice and video. The data is modulated and placed on the cable at different frequencies.

For dispersed network applications, hybrid broadband/baseband configurations are useful. In such configurations, broadband trunks carry data among several local baseband subnetworks, along with such specialized transmissions as local video-conferencing. Hybrid networks take advantage of the best features of both technologies: They use baseband transmission to handle most of the data traffic, and they use the more accommodating broadband technology as a backbone for the multiservice network.

Optical fiber

Optical-fiber (Figure 10.3.2D) uses light signals, pulsed in fine glass strands (about 0.005 inches in diameter) to carry a stream of data at extremely high modulation rates. A lightweight optical-fiber cable of 144 strands can carry as many conversations as a large copper-conductor cable that would need to contain 20 times as many copper wires.

Figure 10.3.2D shows an unterminated single-fiber cable, along with male and female ends of a terminated optical-fiber cable. An exploded view of one type of optical-fiber termination connector is also shown.

There is a relative immunity of the light signal to external noise. Also, there is less loss of the lightwave signal, compared with electrical-wire transmission over equivalent distances and with comparable modulation rates.

Light can be modulated, which is necessary for encoding data, in shorter-duration and lower-powered pulses than can electrical signals. This is inherently because of the shorter wavelength of light, as compared with the wavelengths used for electrical analog-signal or pulse-coded transmission. The light-sensitive receivers used with optical-fiber transmission can, moreover, accurately decode light pulses consisting of much less residual energy than that required for, for example, an analog telephone-line modem. All of these factors contribute to a much greater bandwidth—or data-carrying potential—over optical-fiber cable than with electrical-wire transmission. Optical transmission capacities today are limited only by the speed of the conventional electronics required at either end of the fiber to generate and receive the rapid on/off light pulses.

In typical configurations today, fiber can serve as a viable replacement to wire for even relatively low-speed and short-distance data transmission of the EIA-232 type. The signal integrity of the light pulses traveling through optical-fiber cable enables communications at data rates ranging from 19.2 kbit/s to megabits/second over a 1000-meter length of optical-fiber cable. This is in contrast to the 1.2-kbit/s maximum data rate supported over a 914-meter EIA-232 cable.

In a LAN environment, optical fiber is used primarily to extend cabling distances between LAN segments via electro-optical (E/O) repeaters. An E/O repeater receives an electrical pulse and generates a light pulse by means of a laser or LED. LAN cabling distances vary, depending on vendor products and the type of network segments connected. Typically, E/O repeaters permit network segment connections over a few thousand meters of cable length.

Another use of optical fiber cable is as the transmission medium for the Fiber Distributed Data Interface (FDDI) network, a token-passing type operating at 100 Mbit/s. The cost of FDDI network adapter cards and cable are currently between three and 10 times the cost of the equivalent Ethernet or token ring network components. However, FDDI networks have grown in popularity as a mechanism to interconnect lower-speed networks. In this application, the FDDI network forms a backbone similar to an interstate highway: the lower-speed networks serve as "ramps" onto the FDDI network for their stations to access data from a network station connected to another "ramp."

The principle advantages of fiber optics are the sturdiness and inherent security. Optical fiber is immune to both physical and electrical influences from the environment. Whereas copper corrodes, glass does not. And optical-fiber cable is almost impossible to tap surreptitiously. Current technology within the military enables operators to isolate a break, or even a significant movement, in an optical-fiber cable to within a single inch over a mile or more of cable.

New technologies are simplifying the problem of connecting to an optical-fiber trunk. Future technologies are likely to simplify operations even more. Although it is the newest medium in the commercial local-distribution communications-cabling market, optical fiber in the long run has the most potential.

Wiring costs

Costs for communications distribution cabling are subject to a broad range of variables, and comparisons are somewhat difficult because of their essential differences. In addition, as the market grows and changes, costs for certain new transmission technologies will continue to decline rapidly. A few general rules apply to pricing for local communications distribution:

- Established communications technologies are usually less expensive than state-of-the-art and experimental technologies.
- Off-the-shelf equipment costs less than a customized network configuration.
- A direct correlation exists between level of service and overall network price.
- Price, speed, and performance move upward together.

Average price per workstation connection is sometimes used as a comparison tool. But the price per connection for, for example, local networks can vary greatly, depending on what components and services are included (software, for example) and the type of transmission medium used. As a method of comparison, therefore, price per connection is likely to be useful only when one vendor's price is matched against another's for an identical network product. Even then, factors such as installation can make all the difference.

The relative cost of the transmission media is another way of measuring and comparing local network cost. This approach is especially valid because media has a major influence on both long-term operating costs as well as initial network price. Table 10.3.2 compares historic and current prices of the transmission media that are available for communications distribution wiring.

The most readily available of all transmission media—because it is in constant, high-volume production for voice telephony—is twisted pair wiring. This cable is the easiest to install and can be run along a baseboard in a matter of minutes. Prices today range from $0.05 to $0.25 per foot, depending on wire gauge, number of wire pairs, insulation, and so on.

For the foreseeable future, copper-wire prices will remain reasonably stable. Although ironic, the widespread deployment of optical fiber could open a new source of raw copper supply—the vast amounts of copper cable that are likely to be replaced by fiber. This might actually result in a small drop in copper prices, but this would not significantly alter the role of either transmission medium in the future.

TABLE 10.3.2 Price per Foot of Cable (Cents) 1981–1994

Type of cable	Year									
	1981	1982	1983	1984	1985	1986	1987	1988	1989	1994
Coaxial	10	10.1	10.2	10.3	10.5	10.7	10.9	11.0	11.1	12.0
Fiber optic	17.2	17.0	14.0	12.0	11.0	8.7	7.7	6.5	6.0	6.0
Flat cable	6.25	6.5	6.8	7.1	7.25	8.0	8.1	8.5	8.9	9.0
Twisted pair	—	—	—	—	—	—	—	—	—	—

Coaxial cable is currently the medium of choice for many local network applications, but its use for local general-purpose voice and data communications distribution is virtually nonexistent. Still, promoted for years as the IBM mainstay, coaxial cable for data is now in wide use. Available varieties offered today by a variety of vendors support services from point-to-point to shared-channel wideband.

Still, baseband coaxial cable, which costs from $0.50 cents to $3 per foot, is much more expensive than broadband cable, costing typically between $0.35 and $1 per foot. Baseband coax, moreover, might have to be run in hard conduit (such as a metal pipe) in order to comply with fire regulations in some areas. Broadband cable, on the other hand, comes already clad in a rigid aluminum sheathing. Also, baseband cable cannot extend more than a few thousand feet without expensive digital repeaters, whereas broadband cable can span many kilometers using only inexpensive amplifiers.

Although baseband cable is easy to install and require almost no maintenance, broadband data networks require staffs of trained RF technicians for design, day-to-day maintenance, and careful tuning of components to handle the specific range of frequencies. Both the cable itself and the broadband connecting hardware are extremely sensitive to temperature changes and variations in humidity.

The small diameter of optical-fiber cabling is bound to make it the leading contender where severe space limitations exist. By comparison, the ducting demands of coaxial cable might, in the end, cost more in architectural demands than its potential flexibility warrants.

Wiring management

As terminals and peripherals begin to appear in nearly every workspace, their interconnection becomes increasingly complex. Overcrowded raceways, time lost as a result of downtime, and increasing installation costs are forcing alternative approaches. The solution might be called *wire management*.

Wiring, or cabling, management plans currently in use fall into three general categories: direct cabling, wired office furniture, and distribution centers or networks.

Direct, or point-to-point, cabling is the oldest and most common type of wire management. Sometimes called the *spaghetti-model*, point-to-point cabling is an evolutionary solution, dating back to a time before the introduction of such concepts as distributed processing or networking. The logic of direct cabling is impeccable: given a computer with a certain number of ports located in one place, and the same number of peripherals scattered about in various locations, simply run a cable from Port A to Peripheral A—regardless of cost, obstacles, or distance.

More to accommodate cosmetic and safety concerns than from any functional imperative, this method usually then entails the installation of false floors, walls and ceilings and many ducts as well as the generalized construction of an internal life-support system to contain the corporate data processing organism.

Practical though it might be, direct cabling almost always necessitates running a new cable to the latest location of the workstation that needs access to a computer port. Given the labor cost of retrieving, measuring, resplicing, and reinstalling existing cable—often significantly more than the value of the cable—long cable runs are left buried in what is affectionately called "left-over spaghetti."

Considering how frequently workstations are moved, with the cost of relocating just one terminal as much as $1500—not counting associated downtime while the work gets done—point-to-point wiring is something less than the optimal wiring management solution.

Wired office furniture

A variant of the direct-cabling solution has been developed by office furniture vendors. Conventional desks, tables, computer workstations, and printer stands now are commercially available with factory-installed or designed-in wire channels or wiring.

Wired furniture helps clean up the immediate workstation area itself, but it does little else that is different from the spaghetti approach in solving the intrabuilding cabling problem.

Wiring closets

Sometime shortly after World War II, telephone companies, taking a lesson from electricians, began to run wiring into central distribution points within an office or a building. Individual telephones were then connected to distribution panels within a wiring closet, thereby saving a long wire run to a singular external connection point. The wiring-closet concept offers many advantages:

- It allows communications network installation, alteration, and maintenance without affecting normal business operations or the appearances of offices or buildings.
- Noise generated by communications equipment or by those working on it can be contained and controlled.
- The internal environment of the wiring-closet area can be more easily managed and controlled.
- Network security is easier to ensure—the closets can be locked.
- During initial building construction, wiring closets can be located in areas that will provide optimum, or shortest-length, cable distribution to user equipment.

The floor area served by any given wiring closet is limited by the space available for cables to enter and leave the closet. This usually limits the area that can be served to about 10,000 square feet. In any case, no more than about 50,000 square feet of floor area should be served by any one closet.

Having multiple closets on each floor, while each then serves a smaller area, helps reduce cable-drop lengths, facilitates the addition of more drops, and generally helps make the rearrangement of cabling installations easier.

The wiring closet should be located at the point offering the best access to the cabling distribution. Within the closet, carpeting or uncovered concrete should be avoided. Carpets create static electricity, and concrete is a source of dust. The light source should be overhead and arranged to distribute a shadowless light. All doors should open outward and be equipped with locks.

Data communications wiring closets generally mount patch panels on EIA-standard 19-inch racks with universal hole spacings. These are often bolted to the floor, but can alternately be mounted on the walls. A minimum-sized closet provides one rack for each 100 drops, along with space for one additional rack for future expansion.

Raceways and ducts

Cabling between closets, workstations, and other devices is linked according to a cable-distribution plan. Cable distribution is generally achieved via one or more of the following techniques: underfloor ducts with headers, cellular floors with trench headers, raised or unlimited-access floors, open or closed ceilings (depending on codes), conduit, surface raceways, grid systems, and cable trays.

Several different cable-distribution methods have been developed and introduced—most notably by such industry heavyweights as IBM and AT&T. Some, such as AT&T's, take advantage of existing telephone-line wiring and ducting. Others are designed to be installed when the electrical, PBX and telephone, or other wiring is run.

IBM cabling system

Although the type of cabling medium prescribed is often directly related to a specific local-network technology and/or product, most cable-distribution schemes being installed today are based on twisted-copper pairs. A few installations feature hybrid combinations of the three most popular transmission media and address combinations through adapters. This, in large part, is the approach taken with IBM's Cabling System.

The IBM Cabling System is designed to let users permanently wire buildings with two types of balanced twisted-pair cables [Figure 10.3.2(E)] in a star topology. It then eliminates the need to run cables every time new terminals or small computers are installed or moved. The Cabling System provides an outlet into which peripherals can be plugged, using connectors of IBM's proprietary design [Figure 10.3.2 (F and G), front and side views]. The office outlets connect to a distribution panel in a wire closet via shielded twisted-pair wire. The panels, in turn, connect to the central computer room.

Connection of devices is accomplished by plugging the device into the wall connector, thus eliminating a special run of cable for each station. Changing the device requires that the unit be unplugged, moved to its new location, and plugged in again. Patch cords at the distribution/patch panel enable easy changes from one port to another.

The IBM cabling types [Figure 10.3.2(E)] are configured as two or six twisted-copper pairs—two pairs, within the metallic foil shielding, are reserved for data; the others are for voice communications. Versions with one or two optical fibers, with and without additional twisted pairs, are also available. Coaxial and twinaxial adapters are available from IBM to attach the Cabling System to currently installed equipment.

Able to run at multimegabit rates when it is part of the IBM token ring local network, IBM says, the cable is suitable for transmitting not only voice and data, but, at some point in the future, video-image data as well.

Prices for the transmission cable used in the IBM Cabling System start at $0.50 per foot. The fiber and shielded twisted-pair combination cables can cost as much as $4 per foot, and availability may be a concern. Faceplates with connectors for the outlets [Figure 10.3.2(E and F)] currently sell for less than $5 each, while 64-wire distribution panels cost $273. Devices to convert from coaxial cable to the Cabling System's shielded twisted pair range in price from $45 to $155, depending on the type of device to be connected.

IBM's Cabling System calls for integrated voice and data communications via the combined shielded and unshielded twisted-pair cable combinations. At the wiring closet (illustrated in Figure 10.3.3), these would be separated, with the two shielded pairs routed to the distribution/patch panel. The voice pairs would be routed to a telephone-type termination block.

Other firms have predicated their cable-distribution plans on well-established and inexpensive unshielded twisted-pair cabling.

The twisted-pair distribution scheme uses ubiquitous RJ11-type and EIA-232 telephone connectors (Figure 10.3.4) and twisted-pair cabling consisting of three or four pairs, plus common 25-pair cabling for distribution. It can readily be installed in an existing building, or during the construction of new facilities.

One wiring plan uses a multiplug octopus wiring harness to link a multiport computer with a distribution panel, usually located in or near the computer room. Each harness connects 16 computer ports to standard telephone modular connectors, coaxial or other connectors at the wall-mounted patch panel.

Wires terminating in modular connectors run from the central location to distribution blocks installed in selected areas within the facility. Discrete wiring and modular outlets can be installed in any suitable location—floor, ceiling, wall, post, or furniture. Plug-compatible wires are connected to the outlets at each workstation. Interconnection is made at the floor and central stations by unplugging cables or devices and replugging them into the appropriate plugs.

By using EIA-232, BNC, FJ11, and optical-fiber modular connectors, such a cabling approach can allow a high degree of flexibility between existing and future telephone, computer, and terminal equipment.

Whatever alternatives the user evaluates, it is worth remembering that transmission media and installation will play an important role in network cost, expansion, and flexibility.

10.4 Ethernet Networks

The original efforts of Dr. Robert Metcalfe and his colleagues at Xerox PARC in developing a CSMA/CD-based network resulted in the development of a standard jointly supported by Digital Equipment, Intel, and Xerox (DIX). That DIX network standard defined an operating rate of 10 Mbit/s over 50-ohm coaxial cable. The standard was based on the use of five hardware components: coaxial cable, a cable tap, a transceiver, transceiver cable, and an interface board.

Figure 10.4.1 illustrates the interrelationships of an Ethernet LAN's hardware components. Coaxial cable was used because of its shielding properties, which minimized the possibility of electrical interference. The interface board—also known as

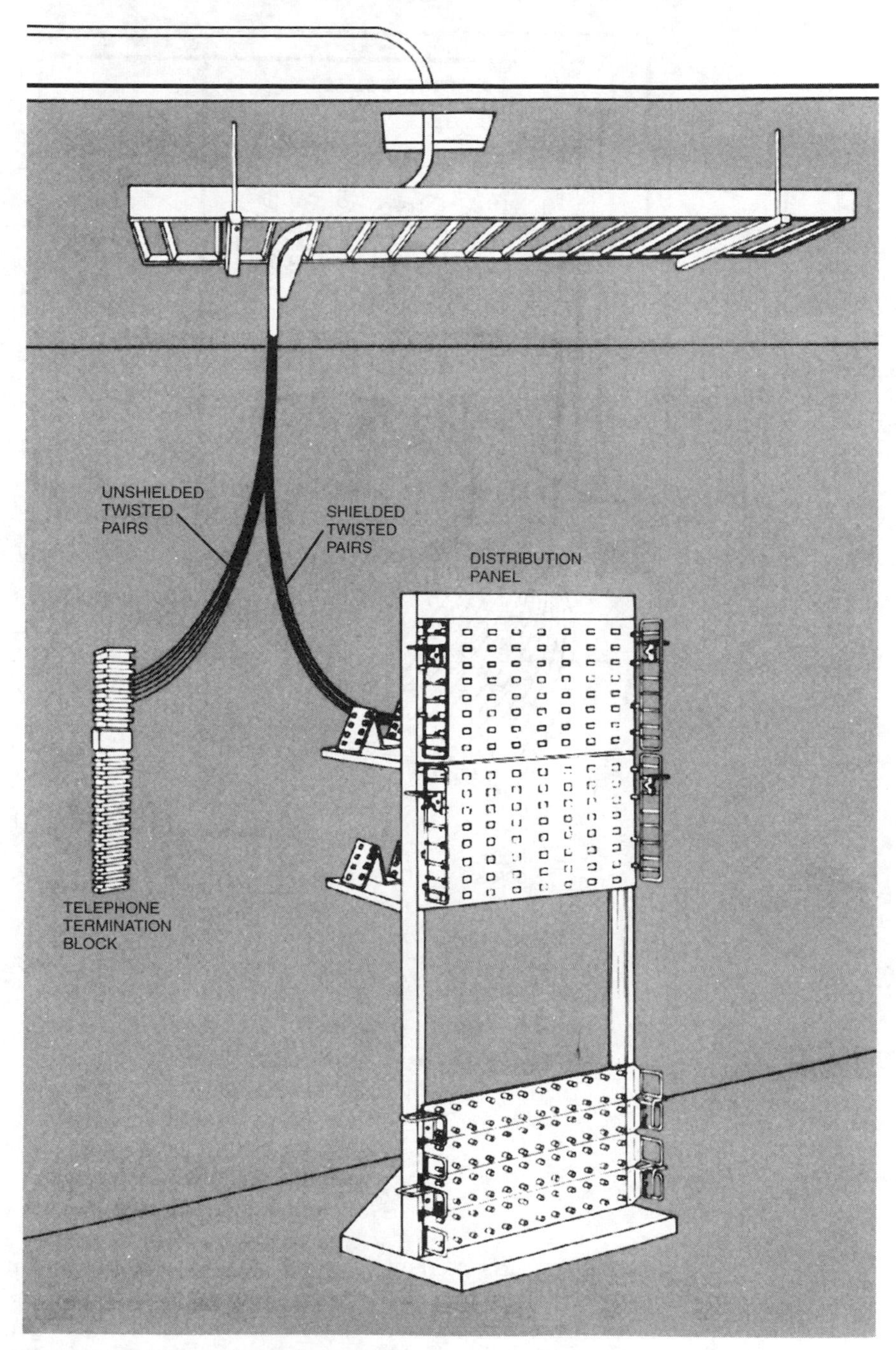

Figure 10.3.3 Voice and data. Source: IBM

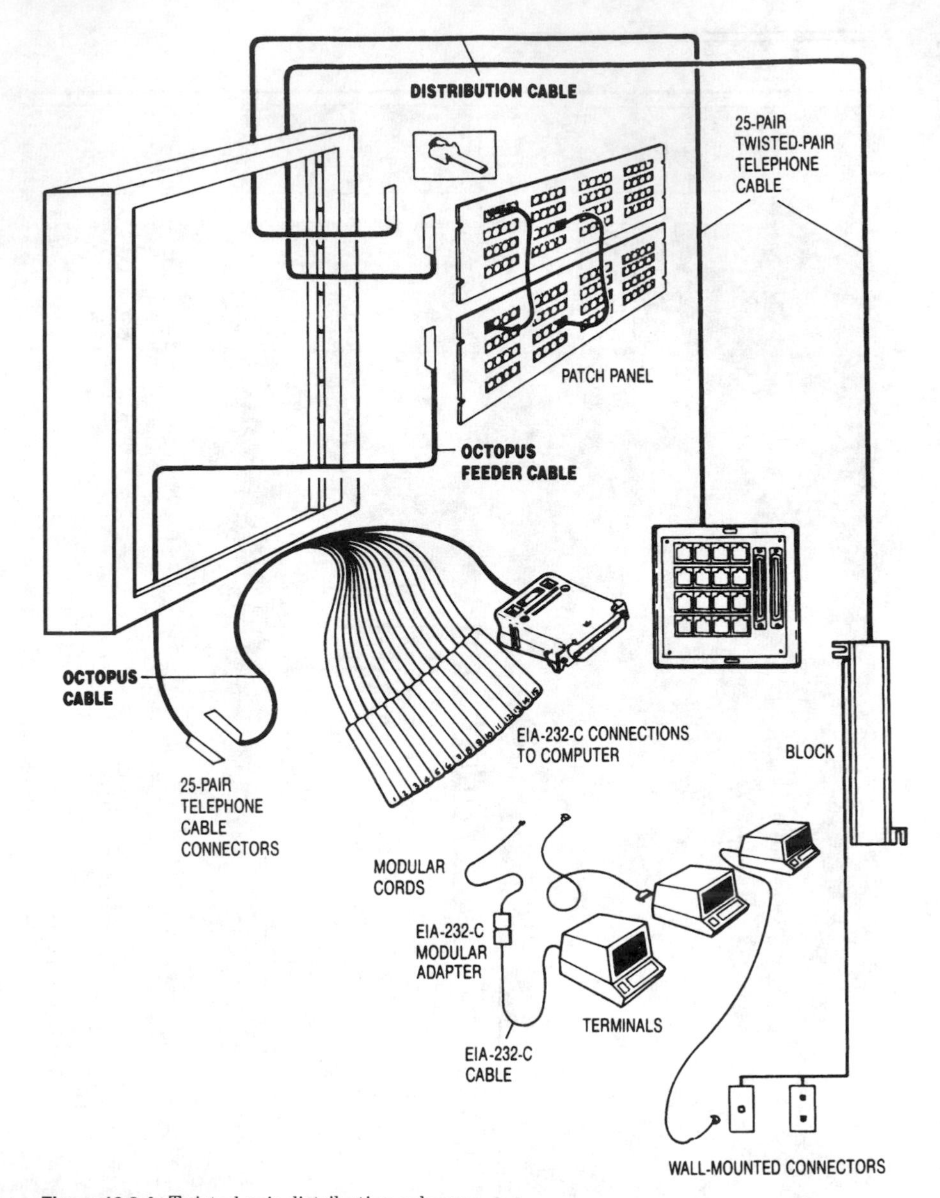

Figure 10.3.4 Twisted pair distribution scheme.

an *Ethernet controller*—was inserted into a computer's system unit to convert computer data to Ethernet's frame format. The transceiver cable connected the interface board to a transceiver. When the transceiver transmitted data, it encoded information according to the Manchester code (previously described). The part of the transceiver that connected to the coaxial cable was the cable tap, which physically penetrated the cable until it made contact with the cable core.

The cost of "thick" 50-ohm coaxial cable and its relative wiring inflexibility resulted in the adoption of a second type of Ethernet cable. That type is a "thin"

50-ohm coaxial cable, commonly called *cheapnet* or *thinnet*. This cable enabled the use of standard BNC connectors to connect interface adapters through the use of T-connectors to a common bus.

Although thinnet cable facilitated Ethernet LAN construction, it also limited the number of connections per cable segment. With thick coaxial cable, up to 100 stations could be attached to a 500-meter segment. In comparison, thinnet limited the number of stations to no more than 30 connected to a maximum-length cable segment of 185 meters.

When the IEEE 802 committee standardized the CSMA/CD access protocol, the Ethernet nomenclature was replaced by "S Type L," where S is the operating rate in Mbit/s, Type is replaced by BASE for baseband or BROAD for broadband transmission, and L indicates the maximum segment length in meters.

10BASE-5

The IEEE 10BASE-5 standard was based on the original Ethernet's use of thick 50-ohm coaxial cable. Most of the differences of 10BASE-5 with Ethernet are in ter-

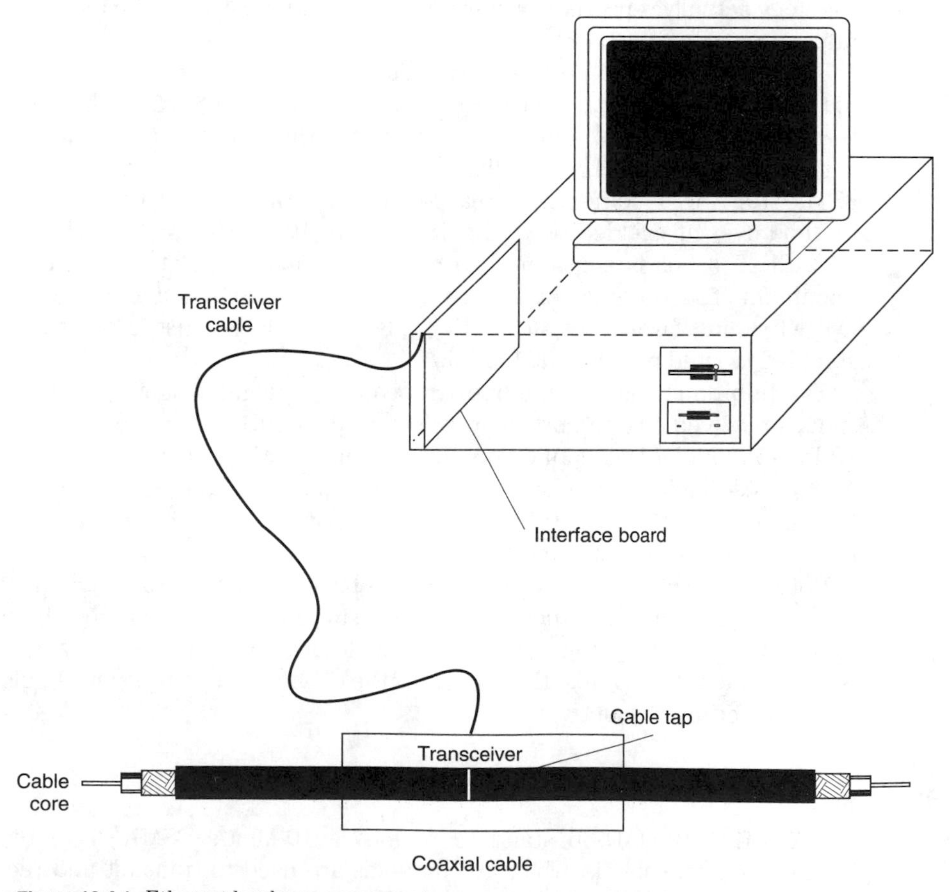

Figure 10.4.1 Ethernet hardware components.

minology: the adapter card is called a *Network Interface Card (NIC)*; the transceiver is called a *Media Attachment Unit*; and the transceiver cable is called an *Attachment Unit Interface (AUI)*. The only substantive differences between Ethernet and 10BASE-5 are the frame fields (covered later in this chapter).

10BASE-2

The IEEE 10BASE-2 network is equivalent to the Ethernet network that uses thin coaxial cable. That is, 10BASE-2 uses CSMA/CD for transmission on a 50-ohm thin coaxial cable. A maximum of 30 stations can be connected on a segment up to 185 meters long.

1BASE-5 and 10BASE-T

The 1BASE-5 standard is based on AT&T's original Ethernet modification—operation on unshielded twisted pair wiring at 1 Mbit/s. Unlike 10BASE-5 and 10BASE-2, which use a bus topology, 1BASE-5 introduced hubs to which a group of stations are cabled. In turn, hubs are cabled to each other to expand the network. Thus, the topology actually represents a star-bus configuration because the interconnection of hubs forms a bus.

The use of 1BASE-5 networks was essentially made obsolete by the introduction of the 10BASE-T standard (covered next). However, 1BASE-5 can be considered extremely valuable for its contribution to proving the ability of unshielded twisted pair wiring to be used as a LAN medium.

The 10BASE-T standard increased the operating rate of the CSMA/CD access method over unshielded twisted pair media to 10 Mbit/s. Under both 1BASE-5 and 10BASE-T, a hub is considered to be a cable segment. Both standards support a maximum of 12 stations per hub. By 1994, the 10BASE-T standard had become the overwhelming favorite of all IEEE 802.3 standards because of its support of unshielded twisted pair wire at 10 Mbit/s.

The hub functions as a multiport repeater, receiving, retiming, and regenerating signals received on one port to other ports. Most 10BASE-T hubs have eight, 10, or 12 RJ-45 modular jack connectors for attaching stations via unshielded twisted pair wiring. Also included are BNC and DB-15 connectors: The BNC connector permits the hub to be cabled to a 10BASE-2 coaxial network, while the DB-15 connector permits the hub to be connected to a 10BASE-5 network.

Both Ethernet and IEEE 802.3 networks have a variety of cabling constraints. For example, in a 10BASE-T network, no two stations can be separated by more than four hubs connected together by five cable segments. Fortunately, most vendor products include a manual that indicates the cabling constraints you should consider in constructing your network.

10BROAD-36

The IEEE 10BROAD-36 standard defines a 10-Mbit/s CSMA/CD network using 75-ohm coaxial cable in which RF modems are used to transmit and receive data.

This broadband network supports a maximum cable segment of 3600 meters and has a topology represented by a tree structure.

Frame composition

Although Ethernet and IEEE 802.3 networks support similar frame structures, the differences are significant enough to preclude compatibility between the two. With IEEE 802.3 standards in place for quite some time, frame compatibility becomes an issue only when early Ethernet networks require expansion. Fortunately, the cost of IEEE 802.3 network interface cards has declined to the point where 10BASE-T NICs are $100 or less. This makes a migration from Ethernet to 10BASE-T very economical.

Figure 10.4.2 illustrates the composition of Ethernet and IEEE 802.3 frames. The preamble announces the frame and consists of the repeating sequence 1010... The Ethernet frame repeats that sequence for eight bytes; the IEEE 802.3, for seven bytes and six bits into the "Start of Frame Delimiter" field. The last two bits of the six are set to 11. Thus, there is a difference of one bit in the first eight bytes of the two frame formats.

A second difference between the Ethernet and IEEE frame formats involves the destination and source address fields. Ethernet uses six bytes for each field; IEEE 802.3 supports both two- and six-byte addressing.

In the original Ethernet frame format, a two-byte Type field was used to specify the protocol conveyed in the frame. Under the IEEE 802.3 frame format, the Type field was replaced by a Length field that specifies the number of bytes that follows as data. Notice that both frame formats require a minimum data field of 46 bytes and specify a maximum of 1500 bytes. If there are less than 46 data bytes, the remainder (fill) is set to PAD characters.

The last field in each frame format is the Frame Check Sequence (FCS). The FCS field contains a 32-bit CRC for data integrity.

Both source and destination address fields consist of two or more subfields (Figure 10.4.3). The one-bit I/G subfield—applicable only to IEEE 802.3 networks—is set to a binary zero to indicate an individual address; a binary one indicates a group address. The U/L subfield, applicable to both frame formats, is set to a binary zero to indicate universally administrated addressing; a binary one indicates locally administrated addressing.

Ethernet

Preamble	Destination address	Source address	Type	Data	Frame check sequence
8 bytes	6 bytes	6 bytes	2 bytes	46–1500 bytes	4 bytes

802.3

Preamble	Start of frame delimiter	Destination address	Source address	Length	Data	Frame check sequence
7 bytes	1 byte	2/6 bytes	2/6 bytes	2 bytes	46–1500 bytes	4 bytes

Figure 10.4.2 Ethernet and IEEE 802.3 frame composition.

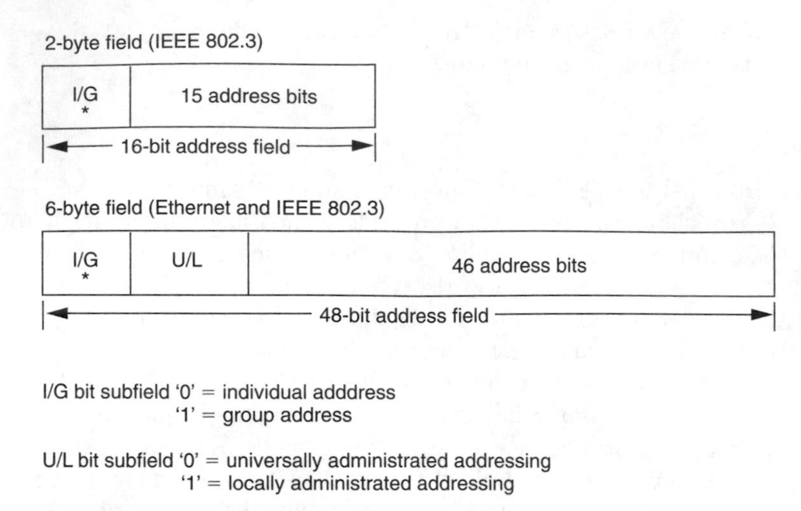

Figure 10.4.3 Ethernet and IEEE 802.3 source/destination address fields.

Under universally administrated addressing, a unique address that is embedded in ROM on each NIC is used for addressing. The first three bytes of the 46-bit address (Fig. 10.4.3) identifies the manufacturer of the adapter; the remaining bits identify the adapter card number. Under locally administrated addressing, the user is responsible for configuring the source address of each workstation.

Although this configuring adds an administrative burden, it provides a potentially important degree of flexibility. For example, consider a gateway that links LAN workstations to a mainframe computer. With universally administrated addressing on the LAN, the addition of a workstation and NIC requires a mainframe-software modification to recognize the new address. This modification normally requires the reprogramming, recompilation, and reloading of the mainframe's communications controller or front-end processor software—which mandates the deactivation of the network supported by the communications controller. Because this deactivation calls for the communications controller modifications to be reserved for third-shift time periods, support for new LAN workstations is delayed, preventing them from gaining immediate access to the mainframe.

In comparison, locally administrated addressing enables programmers to reserve a pool of predefined LAN addresses in the communications controller. Then, the support of a new workstation requiring mainframe access via a gateway can be accomplished almost immediately: The workstation is assigned a locally administered address equal to one of those in the communications controller's address pool.

10.5 IBM's Token Ring

IBM representatives have been very active on the IEEE 802 subcommittee that drafted the token ring specification. In addition, the company (in early 1988) announced a cabling system to wire buildings so that IBM customers can interconnect many of their IBM products.

The cabling scheme also accommodates voice communications, and IBM describes it as an alternative to coaxial, twinaxial, twisted-pair, and other special-purpose cabling. The devices that are able to attach to the cabling system include terminals and processors/controllers such as:

- 3270 Display System
- 4300 Systems
- 5250 Display System
- System/38
- System/36
- 8100 Information System
- Series/1 minicomputer
- 5520 Administrative System
- 3600/4700 Finance Communication System
- 3680 Programmable Store System

Adapters are used to attach IBM Personal Computers (PCs) and the Displaywriter to the cable. The large IBM mainframes—IBM Enterprise Systems, 309X, 308X, 303X, System 370, and 4300 Systems—interface indirectly to the cable through an IBM 3725/3745 communications controller or a 3174 controller. The basic components of the cabling system include:

- Transmission cables
- Data and telephone connectors
- Distribution panel and rack grounding kit
- Patch panel and patch cables
- Receptacle faceplates and surface-mounted receptacle devices
- Surge suppressor
- Cable identification labels and cable location charts
- Cable assemblies for connecting coaxial, twinaxial, financial loop, and other networking products
- A new "loop-station connector" and "loop-wiring concentrator" to connect multiuse loop products

Why a token ring?

The token ring access method lets multiple nodes/stations on a local network gain equal use of the network through a "deterministic" access method, where the next "logical" station is also the next "physical" station on the ring. In this context, "deterministic" means that the user can predict and control the maximum length of time a node will wait before gaining access to the network. Nodes gain access to the ring for the transmission of information by capturing the token, which is a unique fixed-bit pattern. When a node is done transmitting, it attaches a token to the end of its last frame. Thus, there is only one token on the ring at a time. All active nodes on the network are continually "looking" for the token bit pattern. The token is always passed to the next logical node, or station, which is also the next physical node. If the station receiving the token has nothing to transmit, it simply passes the token on to the next node.

The nodes on a token ring network are serially connected to the network medium. Each station can be in either an active or a bypass mode. Data is transmitted serially from one station to another. In bypass mode, the station is essentially disconnected from the network. In the active mode, each station receives and regenerates each frame as it passes, bit by bit. Figure 10.5.1 shows an eight-node token-ring configuration with all of the stations active, except station B, which is in bypass mode.

The addressed destination station, or stations, "copy" the information as it passes. The source station—the one that originally transmitted the information—then removes the frame from the ring when it finally receives back the same information that it sent. Because each node/station regenerates the information, the length of the ring does not affect transmission characteristics. A token-holding timer controls the maximum amount of time that a station can use the network before transmitting, or "passing," a token. Each transmission can consist of one or more frames.

Nonequal access

The token ring access method provides for up to eight priorities of service, which can be oriented toward specific types of traffic or exception conditions. Different service priorities may be designated, for example, as synchronous/real-time, asyn-

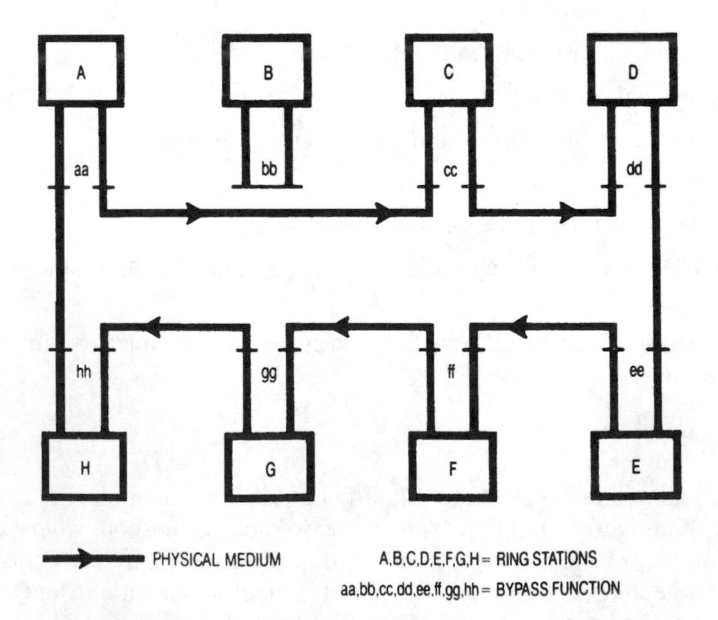

Figure 10.5.1 Typical token ring.

chronous/interactive, or immediate (such as for network recovery). When a frame is passing a node that has a message or data to transmit, the node can "reserve" a token for a specific priority-level transmission.

Token ring access method provides for three basic patterns, or formats, to transmit information (as illustrated in Figure 10.5.2). These are:

- Token Format, which is used only to pass the token from one node to the next;
- Abort Sequence, which consists only of starting and ending delimiter fields. It can be sent at any time to abort a transmission;
- Frame Format, which is the basic transmission mechanism used for token ring MAC, LLC, and network management messages. An information field might be present.

The information field can contain any number of bytes, as long as the time to transmit the frame is less than the time the station is allowed to hold the token. This period is defined in, and controlled by, the token-holding timer.

A technique called *Differential Manchester Encoding* is prescribed in the IEEE specification as the method used to encode information on the network. With Differential Manchester, the sequence of line signal element polarities is completely dependent on the polarity of the previous signal element. With this approach, the encoding of nondata symbols, as well as the conventional "0" and "1," is allowed. Two additional encodings, which are referred to as nondata "J" and nondata "K" symbols, are possible.

As in the case with the NRZ (non-return to zero) encoding convention, the polarity of the leading signal element (or "boundary") is opposite that of the trailing signal element for binary zero ("0"). For binary one ("1") in each case the algorithm is reversed, and the polarity of both elements is the same. In either case, polarity transition occurs at the midpoint between boundaries.

The algorithms for "J" and "K" symbols are similar to those for "0" and "1," except that there is no polarity transition at the midpoint between boundaries. The polarity of the leading edge is the same as that of the trailing edge of the preceding symbol for "J" and the opposite for "K."

Frame formats

The "J" and "K" symbols are used only in the starting and ending field "delimiters," so their presence always signals a delimiter field. The "T" bit in the access control field (Figure 10.5.2A) identifies a token frame: It is "0" for the token format and "1" for a normal information-transfer frame. The "PPP" network priority bits in the token frame establish the network priority for the next transmission. A station that passes the token sets the three "P" bits of the token to match the three "R" (reservation) bits of the last transmission frame it removed from the network. The reservation bits are set (values are plugged in) by nodes with messages waiting to be transmitted as they regenerate the frame around the network.

As a frame is transmitted around the ring, each node that has data to send looks at the "RRR" bits in the frame's access control field. If its priority is higher than the

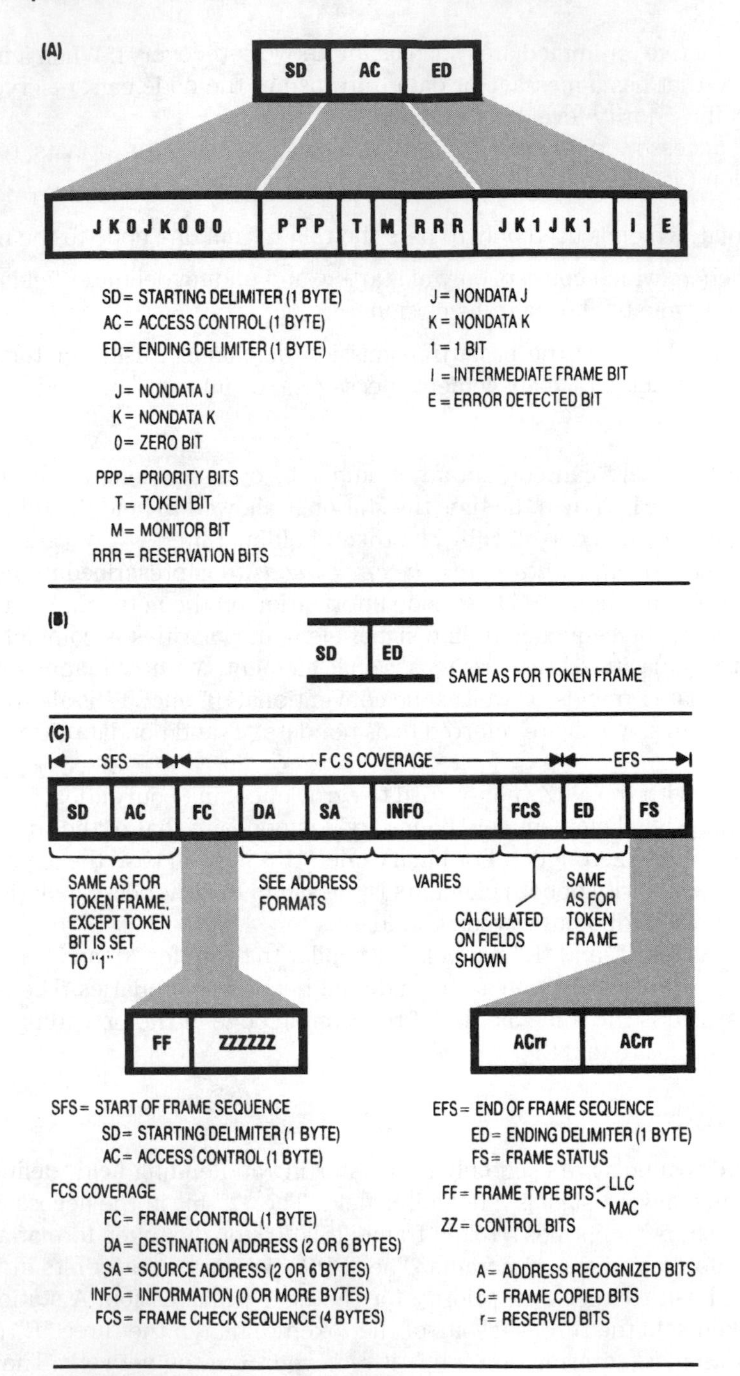
(A)
SD
AC
ED
J K 0 J K 0 0 0
P P P
T
M
R R R
J K 1 J K 1
I
E
SD = STARTING DELIMITER (1 BYTE)
AC = ACCESS CONTROL (1 BYTE)
ED = ENDING DELIMITER (1 BYTE)
J = NONDATA J
K = NONDATA K
0 = ZERO BIT
PPP = PRIORITY BITS
T = TOKEN BIT
M = MONITOR BIT
RRR = RESERVATION BITS
J = NONDATA J
K = NONDATA K
1 = 1 BIT
I = INTERMEDIATE FRAME BIT
E = ERROR DETECTED BIT
(B)
SD
ED
SAME AS FOR TOKEN FRAME
(C)
SFS
F C S COVERAGE
EFS
SD
AC
FC
DA
SA
INFO
FCS
ED
FS
SAME AS FOR TOKEN FRAME, EXCEPT TOKEN BIT IS SET TO "1"
SEE ADDRESS FORMATS
VARIES
CALCULATED ON FIELDS SHOWN
SAME AS FOR TOKEN FRAME
FF
ZZZZZZ
ACrr
ACrr
SFS = START OF FRAME SEQUENCE
SD = STARTING DELIMITER (1 BYTE)
AC = ACCESS CONTROL (1 BYTE)
FCS COVERAGE
FC = FRAME CONTROL (1 BYTE)
DA = DESTINATION ADDRESS (2 OR 6 BYTES)
SA = SOURCE ADDRESS (2 OR 6 BYTES)
INFO = INFORMATION (0 OR MORE BYTES)
FCS = FRAME CHECK SEQUENCE (4 BYTES)
EFS = END OF FRAME SEQUENCE
ED = ENDING DELIMITER (1 BYTE)
FS = FRAME STATUS
FF = FRAME TYPE BITS
LLC
MAC
ZZ = CONTROL BITS
A = ADDRESS RECOGNIZED BITS
C = FRAME COPIED BITS
r = RESERVED BITS

Figure 10.5.2 Three IEEE 802.5 formats.

one already encoded in the "RRR" bits, it raises the value to its level. This essentially reserves the next token for its use, assuming its message is the highest-priority one waiting to be transmitted. If another node has a message of even higher priority to be transmitted, the value it sets in the reservation bits will be higher. Thus, the priority that is set in the "PPP" field of a token (taken from the "RRR" field of the previous frame) is that of the highest-priority message waiting to be transmitted.

When a station raises the priority of the reservation bits, it must store the old value in a "stack" (an area of memory) so that it can return the network priority to its original value after it sends its higher-priority message. A stack is required because a station might have to set reservation bits more than once before its message is transmitted. Because there are eight priority levels, the priority set in the reservation bits could conceivably change several times before the station captures a token with a priority equal to or lower than the priority of its message.

When the token is passed to a station with information to transmit, that station must compare its priority with the value of the "PPP" bits in the token's access field. If its priority is equal to or greater than the priority set in the "PPP" bits, it can capture the token and transmit its message. If its priority is lower, it must pass the token to the next node.

Network monitoring

From the network's point of view, or actually from the network interface, all nodes on a token ring local network are equal and can perform the same functions. This is not the case, however, from the user's point of view. Connected to the actual "network nodes," which the user accesses, are computers, terminals, and "servers" of various kinds, such as file, print, communications, and network-management stations, or consoles. For the purposes of this section, these user-level devices are distinguished from the network nodes, which are the interfaces that connect user devices to the network.

The token ring network has no controller: network control is distributed among all the nodes, and all can perform the same control functions. For network malfunctions and network recovery, however, one node is designated the "active network monitor." All the other network nodes operate as backup monitors, and any one of them can become the active monitor if the current one fails.

Any node can access the network and transmit information to any other node on the network. Messages transmitted on the network can be broadcasted to all the other nodes, addressed to an individual node, or addressed to a group of nodes. Thus, each node must be able to recognize a broadcast address, group addresses, and its own individual address.

Addressing

All information-transfer frames (Figure 10.5.2) include both source and destination addresses, which can be either 16 or 48 bits in length. These conventions are shown in Figure 10.5.3.

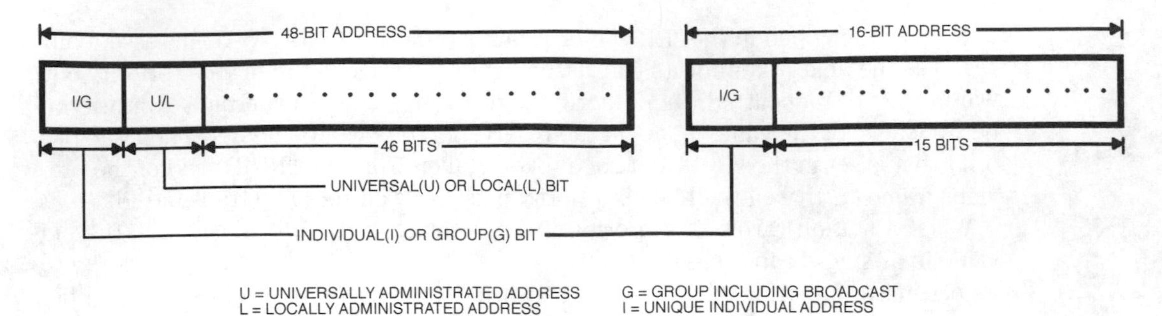

Figure 10.5.3 Addressing.

The first bit of any address designates whether the address is that of an individual node or a group of nodes ("I" or "G"). An address consisting only of "I" nodes signals that the message is being broadcast to all active nodes. The second bit in the address designates whether the address is universal or local ("U" or "L"). Here, universally and locally administrated addressing function in the same manner as previously described for Ethernet and IEEE 802.3 networks.

When a node is transferring a frame of information to another node, which is the most common type of transmission, a two-bit field at the beginning of the "frame-control" byte (Figure 10.5.2C) denotes whether the frame is used for logical link control (LLC) or media access control (MAC). This "frame-type" indicator determines how the nodes at each end of the transmission treat the first few bytes of the information field to control the flow of messages.

For MAC frames, the first two bytes of the information field are used to encode higher-level addressing information, referred to as the *destination service access point (DSAP)* and the *source service access point (SSAP)*. As shown in Figure 10.5.4, the DSAP and SSAP addresses refer to the local node only. The SSAP defines where the information originated in the source node; the DSAP defines where it is to be sent within the node. The first bit of the SSAP byte, called the *C/R bit*, is used to designate whether the transmission is a command or a response message. Response messages are used for acknowledgment, as well as for other response messages.

LLC frames provide for two types of services. One is a simple "datagram" service. With this service, messages do not require acknowledgments. Datagrams are adequate for some applications that do not require the more complex "standard" network service. The standard service, a connection-oriented service comparable to that provided by HDLC in the OSI model, guarantees delivery of every data-link data unit. LLC provides these services for all the MAC methods that are defined by the IEEE 802 committee, and they are described completely in the IEEE 802.2 LLC standard.

Recovery operations

The token ring MAC protocol provides for adding nodes to the network, removing nodes from the network, isolating faults, and recovering from errors. Again, the network has one node that is designated as the active monitor at all times, and all others are to operate as standby monitors. This active monitor makes use of various timers to ensure that the network is functioning properly and that it provides the synchronization timing for the entire ring network. Other nodes set, or reset, their timers so that they are in sync with the active monitor.

Periodically, the monitor sends out an active monitor present (AMP) frame to notify all the other stations that the network has an active monitor. When this occurs, each station creates and readies a standby monitor present (SMP) frame, which is transmitted the next time it captures the token. The SMP frames are used to identify the address of each station's upstream neighbor. Each station stores this address, which is used in identifying problems on the network.

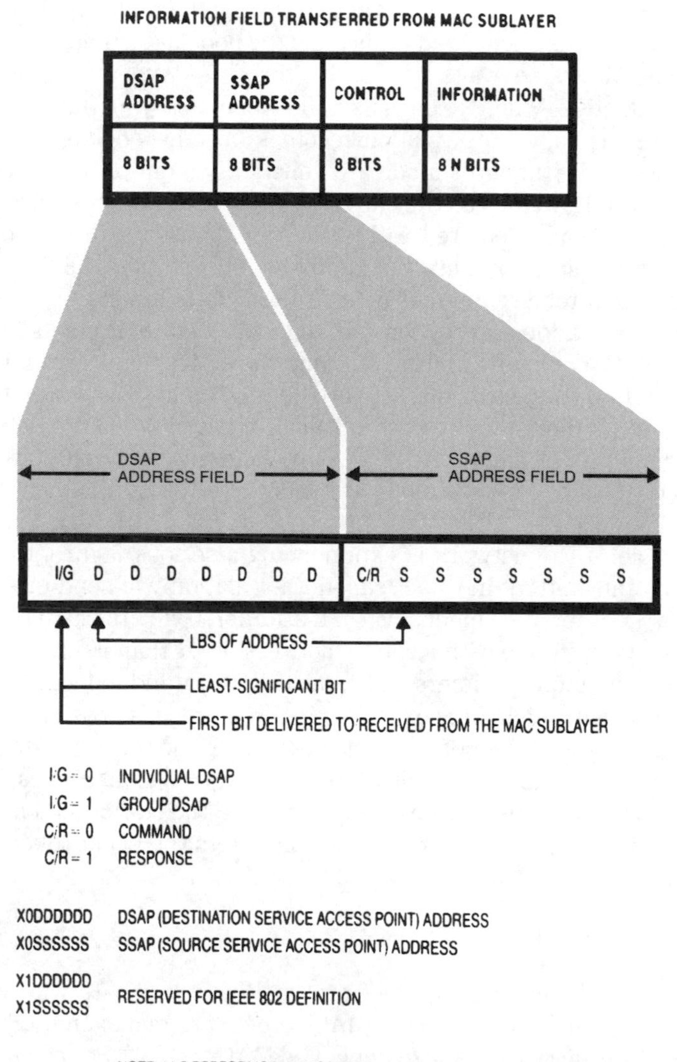

Figure 10.5.4 Subaddressing.

When a station comes on line, it synchronizes its receive clock with receive signals, then repeats the received signals on the line until either an AMP frame or a "purge" frame (covered later) arrives. If either is received, the station readies what is called a *duplicate address test (DAT) frame* and waits to capture the token. The station also then resets a timer-standby monitor (TSM) counter and enters the initialize state.

The DAT frame (Figure 10.5.2C) is sent out with the destination address equal to the source address and the "A" (address recognized) and "C" (frame copied) bits in the Frame Status Field set to "O." If the frame comes back with the "A" bit set to " 1," another node has the same address as the one being initialized, in which case the station notifies network management and it returns the node to a bypass state.

If the DAT frame returns unchanged (indicating the node has a unique address), the station readies an SMP frame, waits for a usable token, and resets its timer-no token (TNT) and TSM counters. When it sends its SMP frame, notifying the next node that it has a new upstream neighbor, the station becomes an active node on the network.

Stations in the standby monitor mode are constantly monitoring the ring by observing tokens and AMP frames as they are repeated on the ring. If token and AMP frames are not periodically detected within the time limits set in the station's no-token and standby-monitor counters, a standby-monitor station will begin to transmit a string of "claim-token" frames. When the claim-token frame returns from around the ring, that station moves to an active monitor state. It resets its no-token timer and "purges" the network. During purging, the active monitor sends out a series of purge MAC frames to clear the network before sending out a new token.

It is possible for another standby station to also transmit a claim-token frame. If the station receives such a frame, with the source address larger than its own, that station returns to standby monitor mode and lets the other station become the active monitor.

It is also possible, if the network is experiencing a serious failure, that no AMP frames are being transmitted. In this case, another station will send out a series of "beacon" frames (another MAC mechanism). If a station trying to claim the token receives instead a beacon frame with a source address other than its own, it resets its no-token and standby-monitor timers and returns to the standby monitor mode.

If a station's upstream neighbor is experiencing a failure, it is possible that the station's no-token timer will expire before its claim-token frame returns. When this happens, the first station begins transmitting supervisory beacon frames. When the beacon frame returns with the source address equal to the station's address, signaling that the ring is again in operation, the station enters the claim-token mode and resumes transmitting claim-token frames.

Boundary services

The various layers of token ring network protocols communicate with each other through boundary services: MAC to LLC; MAC-to-network management; physical to MAC; and physical-to-network management. The MAC-to-LLC service allows a local LLC "entity" in one node to transfer data to a single remote "peer" entity or a group of remote peer entities in another. Conversely, the MAC sublayer can also receive data from an LLC entity in a remote node and pass it on to a single LLC entity, or to a group of entities, in its node. The MAC service also permits status information concerning the success or failure of a request to be passed between the MAC and the LLC sublayers.

The MAC-to-network management service provides for initializing the network, requesting and reporting network status, and defining and confirming data transfers between the two levels. The setting of parameters allows this service to reset and reconfigure the network. It establishes the individual MAC addresses, group MAC addresses, the addresses of all stations that are on the ring, time-out values for all timers maintained by each station, priority of the AMP frame, and other conditions associated with data transmission between the two levels.

The physical-to-MAC service, provided by the physical layer, allows the local MAC sublayer entity to exchange data units with remote peer (MAC) sublayer entities. The physical layer communicates with the MAC sublayer through the exchange of information symbols one at a time.

The physical-to-network management service allows the network-management logic to control the operation of the physical layer. It allows network management to request that the physical layer insert or remove a station from the ring, and the physical layer to inform network management of its status.

Questions

10.1 What were the three initial design objectives of the LAN?

10.2 Name at least two distinguishing features of the early Cambridge Ring.

10.3 Of what topology is Ethernet an example? Name at least two of its distinguishing features.

10.4 What is a third—and probably oldest—LAN topology?

10.5 How does one Ethernet station determine that another station is transmitting? Contrast this with how a token-passing station makes a similar determination.

10.6 Why is it difficult to forecast the performance of a CSMA/CD-based network? Why does its performance decrease more rapidly than a token passing network's as network utilization increases?

10.7 What are today's three media alternatives for network wiring of a building? Which is most widely used today?

10.8 Name two beneficial features of optical fiber.

10.9 What is the chief reason for using optical fiber cable in a LAN environment? What device reinforces its use?

10.10 Which LAN was designed to use optical fiber as its transmission medium? What is its access method?

10.11 Name two popular methods of wiring management (cabling). Which is preferred for orderly growth?

10.12 What wiring topology does the IBM Cabling System use?

10.13 The use of what device distinguishes the operation of a broadband-based network from a baseband one?

10.14 In the IEEE 802.3 network, explain the format and feature differences of a universally administrated address and a locally administrated address.

10.15 Name three distinguishing features of the token-ring LAN?

10.16 How is network control achieved on the token ring?

Chapter

11

Functional Networking Relationships

Regardless of the application, a data communications network's fundamental task is to economically and reliably connect remote terminals to a host computer, which might perform data processing activities, or to a message switch that routes messages between terminals. Once management has defined a network's operational requirements, the network analyst will find available numerous options in equipment, software, and operating procedures. Until the 1980s, a network analyst's efforts were primarily focused on designing an infrastructure for the movement of data from conventional terminal devices to mainframe computers. With the growing acceptance of LANs, the network analyst's job has become significantly more difficult in satisfying the requirements of many private organizations, government agencies, and academia. Some organizations still perform data transport via a wide area network (WAN) to one or more mainframe computers. In comparison, other organizations use LANs exclusively, or a combination of LANs and WANs.

With a LAN, its expansion or connection to a WAN requires the use of a new series of communications devices. Thus, the network analyst has had to become familiar with these new devices, as well as the facilitating software.

When constructing the communications infrastructure for moving data over a WAN, the major choices generally involve decisions about:

- Terminals
- Transmission devices used (couplers, modems, and service units)
- Transmission medium
- Multiplexers and remote data concentrators
- Software and error-control procedures
- Front-end processors
- Fault isolation and backup features

When the use of a LAN is considered, the network analyst is faced with a much different series of implementation choices. Those choices generally involve decisions about:

- Network Interface Cards
- Repeaters
- Hubs
- File servers
- Print servers

If a LAN must be segmented because of workstation or geographic constraints, a new series of implementation choices become relevant. Some major choices include decisions about:

- Local bridges
- Remote bridges
- Routers
- Gateways

In this list, the gateway decision is applicable when connecting a LAN to a mainframe or minicomputer. Collectively, all of the implementation equipment in the list are commonly referred to as *internetworking devices*. Their utilization provides the implementer with the means to interconnect LANs, as well as to connect LANs to conventional mainframe- and minicomputer-based networks.

Before any design decisions are made, some type of analysis must clearly establish the economic and performance benefits of the planned network. This feasibility study is conducted to determine if the economic and performance benefits would sufficiently outweigh the cost and effort involved, or if alternative approaches might be preferred.

Next, given a positive result of the feasibility study, a communications plan is formulated. Covered in the plan are: the kinds of transactions to be processed and their urgency and volume, geographic disposition of company sites, expected growth rates, and the accuracy needed.

To develop a practical communications plan, the network analyst must know the functional networking relationships of terminals, transmission devices, transmission media, and central and remote branch office computer-site hardware. In addition, hardware devices and techniques to optimize networks must be known to permit the design to be fine tuned. The focus of this chapter is on functional networking relations, and following chapters are designed to expand the reader's knowledge of communications devices and optimization techniques.

The relationships between networking devices and transmission facilities can be placed into three general categories: LANs, WANs, and internetworking. First, the relationships between equipment and transmission facilities used to construct mainframe- and minicomputer-based WANs are examined. This is followed by an examination of the functional networking relationships of LAN equipment and internetworking devices.

11.1 Wide Area Networks

Because WANs were established chiefly to provide access to mainframe computers, the focus here is on mainframe-based WANs. Thus, equipment that is primarily so associated is examined. However, certain information, such as the operation of a PC as a data terminal device, are applicable to its use in both WAN and LAN environments.

Terminals

Because the remote data terminal is the only device in a wide area data communications network that interfaces directly with business activity, it must be subject to several levels of critical evaluation. Management must be certain that terminals satisfy functional requirements within a prescribed cost. Operators must find them acceptable in the work environment. And planners must be sure the terminals are technically compatible with the rest of the network.

Fortunately, today, the user can select from many different kinds of terminals. They range from simple, low-cost keyboard/printers—basically Teletype machines and their emulators—to rather sophisticated, high-speed terminals, which can also serve as stand-alone data processors. Typical of the latter is the proliferation of "personal" business microcomputers. This variety is because of rapid advances in terminal design, the most significant of which is the addition of programmable microprocessors.

Programmability makes a general-purpose terminal suitable for tailoring to many different business applications. It enables the terminal to take over certain data manipulations, and so reduces the amount of data sent over the communications lines and relieves the central computer of some jobs unrelated to its basic work of data processing. That is, when needed in a given situation, programmability enhances terminal performance in the business environment and can reduce communications and computer expenses.

Business requirements dictate the nature and form of the information to be sent to the computer, the urgency with which this information must be processed, and the nature and form of the processed information delivered back to the sending site. Therefore, the first thing to consider when selecting a remote data terminal is the input/output data requirements of the application. This takes any of four forms:

- Enter transaction-oriented data.
- Enter batch data.
- Output batch data.
- Retrieve—or ask for—information from a database.

Input/output classifications. In transaction-oriented data entry, data is keyed in at the place where the transaction occurs, and goes at once to the host computer to update the database. For example, an operator at a remote branch warehouse might key in data describing a shipment and so update the central inventory-control database immediately. (Sometimes, though, data from several transactions is briefly stored in the computer mass memory and then processed in small batches.) Terminals suited for transaction-oriented inputs resemble those used in "interactive" or "conversational" applications, in which the operator engages in a dialog with the computer or with an operator at another terminal.

Being dependent on the operator's peak keying rate, idle time, and other performance factors, data enters the transaction terminal relatively slowly. Consequently, transaction-oriented functions use terminals that can be classified as having slow input and slow-to-medium output.

In batch-data entry, a large quantity of data is gradually accumulated on magnetic disks or some other machine-readable medium, and later sent to the host computer as one long message. The accumulated data, usually in fixed record lengths, could represent eight hours of local business operations, but the total batch could take perhaps just 10 minutes to transmit. This type of terminal is therefore classified as *slow input/fast output.*

In batch-data output, the host computer quickly transmits a quantity of data to the local site, where it is recorded as fast as possible on punched cards, magnetic tape, disk packs, or—most commonly—paper. A high-speed line-printer terminal, for example, could print out invoices generated by the computer based on batch data that had been entered earlier. For this application, the terminal would be classified as *fast input/fast output.*

Database inquiry is the retrieval of a fairly small amount of information from a centrally situated database, for display at the local site. The classic example is an airlines reservations system. Here, the terminal has a keyboard for requesting specific flight data, and a CRT screen for displaying the answers. Often database inquiry requires a hard copy—for example, a completely filled-in airline ticket, which is produced by a teleprinter. The inquiry function has much in common with the transaction function.

Although most airlines have replaced their conventional CRT-based terminals with personal computers, they are programmed to function primarily as enhanced terminal devices. The chief difference between the old terminals and their replacements is the use of the data storage and programming capability of the PC. This capability enables an airline to develop software that operates on the PCs in conjunction with a mainframe's databases.

For example, consider a person calling an airline reservation clerk/operator to inquire about flights from Atlanta to New York. And consider further that the airline has a PC-based terminal system that accesses a mainframe-based reservation system database. The operator's keyed query might result in several days' worth of flight information being downloaded to the PC's memory (although only one screen of information is displayed). Then, if the caller wants other information, it is immediately available to the operator via a program on the PC that scrolls through its memory.

To satisfy one or more of these basic data input/output requirements, manufacturers have developed several classes of terminals. Besides the keyboard/printers (or teletypewriters), alphanumeric CRT displays, and remote batch terminals already mentioned, they include such special units as point-of-sale and bank-teller terminals, as well as optical scanning units and other once-considered-exotic devices.

More distinctions. Terminals with slow manual inputs generally fall into the asynchronous class. They have keyboards much like those of typewriters and generate coded characters as responses to operator keystrokes. Therefore, their outputs are normally asynchronous in that they depend not on a fixed time base, but on the irregular performances of operators.

Machine-input terminals, and manual-input batch terminals that accumulate slow input data, generally deliver output data at a fast rate to the line, and use synchronous transmission. In synchronous transmission, long data blocks made up of a string of coded characters are transmitted, with each block—rather than each character—framed by special synchronizing codes.

Finally, remote data terminals can also be categorized as senders, receivers, and sender/receivers. Each has a controller and a buffer memory. The sender has a data-input mechanism, most commonly a keyboard, while the receiver has a data-output mechanism, most commonly either a character or line printer, or a CRT display. The sender/receiver consolidates the send/receive functions into one physical device as, for example, in a keyboard/printer terminal.

The buffer in an electronic terminal is a memory that stores enough bits to represent at least one character. Larger buffers store a word, a line, or even a whole message. With the development of "floppy" disks (diskettes) and integrated-circuit memories with their high-volume production, buffering cost has come down, so terminal makers can afford to install more internal buffering in their terminals. External buffers, too, like a disk pack, can be used to enhance a terminal's versatility.

Basic tasks. The controller is a logic device—it directs all the tasks that the terminal must perform to convert, for example, a keystroke into a sequence of bits for transmission on the communications link. Depending on the particular terminal, the controller might be hard wired and thus have been fixed in functions when the terminal was designed, or it might be programmable so that its functions and tasks can be suited to one or several applications.

Whether hard wired or programmable, the terminal's controller must perform several basic operations. One of its simpler tasks is to handle the few signals that govern the input or output device. Here, the controller sends or receives signals to start up or shut down such devices as keyboards, card readers, CRT screens, character and line printers, magnetic tape drives, magnetic cassettes, and punched-paper-tape drives. A control signal also initiates translation of data from human-readable form to machine-readable form, and vice versa.

As for buffers, the controller in a sender fills this memory with bits from the input device. Then, it empties it bit by bit in proper sequence onto a communications line. The reverse occurs, of course, when the terminal is a receiver.

Buffer control is quite complex when the controller simultaneously manages separate buffers for send and receive functions in the same terminal, or when the terminal uses double buffering to improve terminal speed. (In double buffering, one buffer is being filled while the other is being emptied; then the buffers reverse roles, and so on.) Control of simultaneous and double buffers, although complex, is a relatively fixed function and well within the capability of hard-wired logic.

A third task is code translation. The controller converts a character from the code form in which it was sensed by the input device into another code form, one suited to transmission and computation, or vice versa. In a sender, for instance, it might translate Hollerith (punched card) code into ASCII code. Most modern teleprinters and CRT displays contain internal logic circuitry that directly accepts ASCII-coded information sent by the computer and translates (or generates) the coded information into characters.

In the generation of error-detection codes, the controller adds a 0 or 1 parity bit, if necessary, to a coded character before it is sent to the receiving terminal. The controller there scans the received character, including the parity bit, and its logic tells whether an error has occurred during transmission.

The character (or block, or message, depending on the particular requirement) must be stored in the sender's buffer until it is told by the receiving terminal that the character has been received correctly. Otherwise, the transmission is repeated until it has been correctly received. Because the buffer cannot be filled with a new bit sequence until it is emptied, and time elapses while the receiving terminal evaluates the correctness of the received character, terminals often increase their data rates by using the double buffering mentioned earlier.

Automatic answering and transmission enable a sending terminal to be used with no operator present. Data can be "telephoned" in by an unattended remote batch terminal in response to a request made during the evening, when lower-cost night-rate telephone tariffs apply. The automatic answering equipment is not a part of every terminal's controller, but it can usually be obtained as an option.

The host computer's overhead. However much is done by the terminal's controller, the host computer is still left with the job of handling terminal signals. This is overhead as far as the computer is concerned because any time that the computer spends other than in processing data reduces its efficiency. Remote-data-concentration and front-end-processing equipment intervening between the terminal and the host computer lower the host computer's data communications overhead. However, regardless of whether the terminal interfaces with a remote data concentrator, a programmable front-end processor, or a host computer, one such device will have to provide the data manipulation services required to interpret the hard-wired terminal's output. In the following paragraphs, it will be assumed the terminal feeds directly into the host computer.

For character-by-character (asynchronous) transmission, the computer must continually observe the lines coming in from the terminals. It can do this by frequent polling—that is, by asking each terminal if it has any messages to send—or by watching for an interrupt signal, which forces the computer to read the interrupting terminal's message.

After the connection has been established, the computer must assemble each sequence of incoming bits into a full character, strip off the start and stop bits, translate the data from transmission code into computer-processable code, test the parity bit to see if an error has occurred, and place the character in the correct location in its memory for assembly into words and messages. Furthermore, the computer must determine whether the received character is a message character or a control character. A control character sets up a different level of activity within the computer. For example, it could indicate the end of a message, which allows the computer to disconnect the terminal and go on to other communications or data processing tasks.

In character-by-character transmission, the computer must perform each of these activities from six to 10 times each second for each terminal in the network in turn. Comparable activity occurs for outgoing characters, which have to be handled at up to 30 times each second for each terminal. All this is a considerable burden on the computer.

Batch terminals with hard-wired controllers also place a data manipulation load on the computer. Because such terminals produce blocks of bits and operate in a synchronous mode, they do relieve the computer of some of the burden of individual character detection, validation, and removal of sync bits. Even so, the computer still has much to do to support their operation and communications requirements.

Such batch terminals as punched-card readers and magnetic-tape readers transmit information on a record-by-record basis. For example, one punched card might constitute a record. The computer must read the record, test it for accuracy, and either accept it or ask for a retransmission. For a good record, the computer must then perform necessary code translation, interpret card data for special commands, and position the data correctly in its memory.

Note that the entire record of a punched card, usually 80 characters, must be transmitted by the terminal and processed by the computer, even though the card itself might only contain perhaps 20 or 30 meaningful characters. The balance of the 80 fields on the card might be blank, contain card "deck" identification, long sequences of zeros, and other extraneous characters. Most hard-wired batch-terminal controllers cannot distinguish meaningful from extraneous information. Thus, when the computer sends a record to a card-punch terminal, it must transmit all 80 fields—even though some are meaningless.

Most large host computers are designed for data processing, and not for such data communications tasks as terminal control, character and block analysis, and message assembly. Programmability has been added to terminals for one main reason: to take much of the specialized work away from the general-purpose computer and do it instead at the terminal site. As a side benefit, the programmable terminal—sometimes called a "smart" or "intelligent" terminal—also reduces the data load on the communications links.

Benefits of programmability. What has made programmability economical and therefore feasible are the advances in semiconductor memory technology. It is now easy to build special memories, called *read-only memories (ROMs)*, program them for a fixed application, and insert them into the terminal. The ROM is basically an expanded equivalent of hard-wired logic, which adds significantly to the functions that the controller can perform.

Some terminals even contain stored-program computers—minicomputers or microprocessors with functions programmed by software, just as in the large host computers. Such a programmable terminal, with its own mass memory or other peripherals, then becomes very flexible and powerful. It can serve as a standalone data processor in a local environment, besides communicating with the host computer. It also permits the development of distributed data processing networks.

The concept of the stored-programmable terminal has been applied to more-modern laptop and notebook computers. Recent electronics advances have resulted in the development of PCMCIA (Personal Computer Memory Card International Association) cards and card slots built into laptop and notebook computers. The PCMCIA card is about the size of a pack of cigarettes, but it is only one-tenth as thick. These cards can contain built-in functions, such as terminal emulation capability, LAN interface, modem, or other communications devices.

The stored-program terminal, therefore, depending on the amount of electronic equipment it contains or on the creativity of its programmer, can perform a selection of the following functions:

- *Formatting input and output.* Many normal business applications require information to be displayed or entered in a highly structured form. Cases in point are invoices or insurance applications that could best be prepared on preprinted forms, and customer records that could do with a standardized form for display on a CRT screen.

  From a careful description of the form, a receiver-terminal controller can be programmed to generate the necessary spaces, tab stops, carriage returns, line feeds, and other device-control characters to properly position the carriage on a character printer or a cursor on a CRT display.

  Similarly, with suitably described field formats for a sender-terminal, an operator simply enters the necessary information in each field. If the information does not fill a fixed-format field, the operator then keys in a delimiter symbol, and the controller fills in the field automatically with characters that are meaningless in the context—dollar signs, zeros, or spaces. The computer is programmed to ignore field-filling characters, but as a check will count all characters to make sure that the field is filled to its prescribed length—no more and no less.

- *Data compression.* Instead of transmitting entire field-filling and other extraneous character sequences, the controller in a programmable terminal simply sends a two- or three-character number to indicate how many filler characters were removed from the data stream. The receiver then skips that many characters. Or the terminal can send a short code giving the position (or address) in the record at which the next meaningful character or sequence starts. Removing sequences of extraneous zeros or spaces is a simple application of data compression.

  More significant compression can occur if the terminal stores several fully expanded (English-language) text messages to be displayed or printed on receipt of a short code corresponding to each piece of text. Such texts might be the specific payment terms to be printed on an invoice, an account credit status to be displayed on a CRT terminal, or a full product description to be printed on an inventory status report. Because some of these texts are quite long—perhaps 50 characters—the ability to store them at the terminal, rather than at the host computer, and call them up with a short address code, reduces line traffic significantly.

- *Content-oriented error detection.* Though all receiver-terminal controllers check parity bits to determine whether errors have occurred during transmissions, parity checking cannot recognize whether the correct data was sent. However, programmable logic at the terminal, particularly if it is based on a minicomputer, can undertake a more thorough check for errors. It relies on the content and context of the data being transmitted. The programmable terminal can make sure that received data conforms to prescribed format rules. It can, for example, prevent a field that should contain alphabetic characters from accepting numeric characters. It can count characters to make sure a field is filled and, if the field is not filled, will ask for a retransmission. It can perform quantitative checks, making sure a received number does not fall outside a prescribed minimum-maximum range.

- *Buffering multiple batches.* An effective use of the terminal's buffer storage is the accumulation of character-by-character input data, as from a manually operated keyboard. This accumulation might last for as little as a line or as much as a page of information. The terminal assembles characters into messages, inserts redundancy-check characters where appropriate, interprets control characters, and responds as necessary to the display or printer to let the operator know that his input has been correctly received at the computer. Here, each time a line or a page is completed, it is immediately sent out as one long continuous message at high speed to the computer. In short, the programmable terminal performs those functions that a central computer might otherwise do for character-by-character transmission, but much less time is taken on the communications link and the computer's overhead is substantially reduced.

  At a more complex level, one programmable terminal can handle batches accumulated from several input devices (keyboards). In this situation, it operates as a gathering place for what amounts to multiple keypunch operations being performed away from the central computer site. The input from each keyboard can be assigned its own area on a disk pack, with each keyboard operator perhaps performing a different processing task. When all work is finished at the remote site, or on demand from the computer, information stored on the disk is transmitted at high speed to the computer.

- *Local editing of data.* When the programmable terminal contains a large local buffer, a complete line or page can be reviewed by the operator, who then edits the message to correct keying and other errors. When she spots an error in a displayed message, she simply backspaces the cursor to the error position and keys in the correct data. Once satisfied with the message's accuracy, she causes the buffered data to be transmitted. Local editing thus reduces the communications link and computer overhead that would otherwise have been used to handle wrong data.

- *Handling simple computations.* With programmability in the controller, the terminal can perform computations on accumulated data locally, rather than having to send all raw data to the host computer. Typical examples of such computations are price extensions from unit prices and quantities, quantity discounts, and tax amounts. With such local computations, the terminal can then prepare an invoice locally.

- *Appending local constant data.* During a business day, constant data is added to each document or business transaction serviced by the terminal. The controller can be programmed to include terminal location and identification, operator identification, date and time, and security code to allow only authorized access to the terminal.

Such data is automatically added to the variable data being inputted by the operator. In a sense, appending local data is a form of data compression.

- *Control sequence numbering.* To avoid confusion within a business operation, it often is important to consecutively number messages, invoices, and other documents. If the terminal is programmed to assign and print numbers in sequence and to use a number only when it produces a valid document, no skips will occur in numbering sequences.
- *Automatic restart and recovery.* From the standpoint of operating integrity, one attractive use of a programmable terminal is to accumulate a local log, with checkpoints along the way, of all transactions passing through it. Then, if the data communications/data processing network malfunctions, operation can be backed up to the last checkpoint and restarted, and the logged data automatically transmitted up to the point at which the malfunction occurred. This eliminates the need for the operator to find source documents and rekey information because the terminal's "audit tape" simulates the operator's actions. When recovery transmission has been completed, the operator continues the transmission with new manual inputs.
- *Local operator guidance.* Some programmable terminals are designed to handle a single type of operation, but others, particularly ones based on a minicomputer, can be changed by the operator to handle several different assignments during the day. In either case, the input has to be entered in a structured form. Often, the operator will make mistakes either in keying in good data or in entering the wrong kind of data for a particular field in the prescribed format. To prevent these situations and thus to improve overall efficiency, terminals can be cleverly programmed to display a specific message on a CRT screen that will tell the operator what to do next. And if the operator does it wrong, the program can even lock the keyboard, preventing any further input, and flash a message telling the operator what he did wrong and what to do instead.

Remote data concentrators

A data communications network can grow so complex that it needs a remote data concentrator to interface clusters of terminals with the long line to the host computer. This concentrator is designed to combine many low-speed lines into one of more high-speed lines for transmission to the host computer. The functions performed by remote data concentrators vary considerably, and include simple data concentration code conversion, code compression, speed conversion, traffic smoothing, and error control. Today, a number of communications components on the market perform one or more of the previously mentioned functions. These devices are known as *multiplexers* (frequency division and time division), statistical or intelligent multiplexers, wideband multiplexers, and concentrators.

In hardware form, the differences between multiplexers and concentrators have narrowed considerably. Today, the functions performed and capabilities of such devices can be the same as a result of advances in hardware technology and the incorporation of microprocessors into multiplexers.

Initial developments. In the late 1960s and early 1970s, concentrators not only combined many low-speed lines into a high-speed line or lines, as conventional time-division multiplexers did, but also handled more terminals than a TDM could. With its computer architecture, the concentrator could perform such functions as line servicing, code and speed conversion, traffic smoothing, and error control—in short, all the data communications overhead functions necessary for improving line and host-computer utilization.

Concentrators. In hardware form, concentrators are implemented with a general-purpose stored-program minicomputer or with a communications-control computer designed specifically for the purpose. The difference between them, essentially, is that a minicomputer (or microcomputer) includes certain hardware, like an arithmetic multiplier or floating-point hardware, not needed in data communications, and it usually uses software to service communications lines. Conversely, the communications-control computer contains no superfluous equipment, and uses hardware for multiplexing, logical programming, and data manipulation at the line interface. Either, however, can be programmed to implement specific functions required in a particular network and can be readily modified to suit changes and growth. And each contains extensive memory in which input data can be temporarily stored.

Code and speed conversion. Any fairly large data communications network contains a variety of terminals operating at different speeds and in different code formats. Transmitting this intermix over a high-speed line to different ports in the host computer again places a severe overhead burden on both line and computer. Instead, the programmable concentrator can convert both speeds and codes at a relatively local level. Concentrators contain a hardware interface that accommodates such standard terminal speeds as 110, 134.5, 150, 300, and 1200 bits/s.

This hardware, in combination with suitable software, detects the incoming speeds prior to reading the data, and executes code and speed formatting. For example, it will strip out start-stop bits from asynchronous codes—a procedure that by itself improves line utilization by about 20%—and will arrange mixed codes into a synchronous bit stream, with a fixed format suited to the host computer. In this manner, one computer port can service a variety of terminal speeds, instead of a separate expensive port being needed for each speed.

The concentrator's code- and speed-conversion capability has several advantages. It allows users in remote cities to dial up the computer, through the concentrator, and insert data from various terminals operating at any common speed. It permits the concentrator to undertake message switching between terminals operating with different speeds and codes. Furthermore, it allows for network expansion without disturbing the host computer's software or hardware, and permits a new, more efficient type of terminal to be substituted for another without any modification at the host computer installation.

When implementing line servicing and code and speed conversion, the concentrator uses a special form of memory called a *buffer*. Each input channel to the concentrator has its own buffer. This memory accumulates one or more characters that are then read into the concentrator's fast internal-core memory, which performs immediate processing of the data stream.

The concentrator also has mass memory—usually disk packs or drums—in which it holds data for long time intervals, as might be needed for traffic smoothing and for saving data in case of temporary outages of the line. The concentrator also has output buffers to interface the accumulated high-speed data stream onto the line between the concentrator and the computer. Often, these output buffers serve instead of software to implement error-control procedures. Consequently, how much and what kind of memory is needed depends on two major functions of remote data concentration—traffic smoothing and error control—both of which have to be carefully matched to user requirements.

Traffic smoothing. Conventionally, network planning includes an analysis of traffic to be carried from the terminals to the computer and to other terminals. Peak load occurs at certain times of the day, and it will determine the number of high-speed lines needed to carry the peak traffic. But if this peak-load determination assumes all terminals will operate simultaneously at their full rated speeds, the network might be overspecified.

Many terminals, particularly the manually operated low-speed keyboard types, can be connected to the line, but there will be scattered intervals during which operators will not be hitting the keys. Thus, the network's actual data throughput will be less than the rated capacity. Therefore, with the traffic-smoothing function made available by adding mass memory to the concentrator, the fully utilized average speed on the line to the computer can be less than the sum of the rated speeds of the terminals going into the concentrator.

For traffic smoothing, the mass memory acts as a temporary reservoir for bits entering the data concentrator when the total input rate is higher than the output rate to the line. When traffic slows, stored messages leave the memory for their destinations. That is, properly designed traffic smoothing takes care of the effect of random variation in terminal traffic, and assembles the message completely in the memory before transmission to ensure full utilization of the high-speed line.

Even though the memory size might be economically selected for some assumed peak traffic load, this load might be exceeded from time to time. In this case, the concentrator, under software control, will have to be able to raise a busy signal to incoming terminals to prevent them from transmitting.

Besides providing maximum utilization of the line to the computer during peak load periods, the mass memory can aid integrity. It can store one of more consecutive outgoing data blocks as needed for a request-for-retransmission error-control procedure, and it can store incoming terminal messages during an outage on the line between the concentrator and the computer.

Outages occur for many reasons. For instance, the common carrier might have to switch the primary leased line to a backup line when trouble arises. A Bell System document defines two types of outages; a hit, lasting less than 300 msec (as might happen from a lightning strike), and a dropout, lasting 300 msec or more. A dropout of 300 msec on 4800-bit/s lines means a loss of 1440 bits. The loss of so much data, which could have originated at many different terminals, could upset company operation—unless the bits are retained within the concentrator's memory for retransmission in case of an outage.

The mass memory can also be made large enough to implement a store-and-forward feature, particularly desirable in networks handling considerable traffic between terminals. With this feature, the sending terminal does not have to wait for a busy receiving terminal to finish a call and hang up. The store-and-forward memory acts as a temporary receiving terminal. Then, for example, the line protocol can allow each "free" terminal to request messages addressed to it, if any, that are stored in the concentrator's mass memory. The feature is also desirable for data communications networks requiring especially high operational certainty. Here, the need might be not only to forward messages, but also to store all traffic for up to 24 hours, to provide repeats of messages that might have gotten lost, or to retain data until an audit of the day's work has been completed.

Error control. One of the main benefits of using a remote data concentrator is that it can check data coming in from all terminals for errors and add a checking code to traffic between the concentrator and the host computer. That is, if the concentrator detects an error, it can request retransmission of a message over a short link, without involving the host computer. Moreover, the corrected messages can be sent in long, economical blocks over the high-speed line to the computer.

Error detection and control can be implemented using several techniques, such as automatic request for repeat (ARQ) and forward error correction (FEC).

One thing all these procedures have in common is that extra coding bits are added to the data block. These redundant bits, which can range from a simple parity bit to perhaps one-third redundant bits in a forward error-correction method, require memory to store messages temporarily while they are being analyzed for proper coding and decoding. An extensive memory is available in remote data concentrators at a relatively low cost per bit.

Error control can be implemented by software or hardware. For the commonly used cyclic redundancy check, for instance, special dedicated hardware is assigned to each line. Although such hardware raises the concentrator's installed cost, its advantage is that checking is done in real time with no software overhead required at the data concentrator. Software implementation reduces the concentrator's cost, but adds a burden to the computer within the concentrator. For example, a single 4800-bit/s link controlled by a concentrator with a 1.8-µsec memory cycle can use up to 10% of available real time for performing the checking function by software. The decision whether to use hardware or software error control will fall out as part of the overall technical and cost analysis of the network requirements.

Multiplexers. As previously mentioned, there are four types of multiplexers: frequency division, time division, wideband, and statistical or intelligent. Although there are significant differences between the first three types of multiplexers and concentrators, statistical multiplexers are based on microprocessors; today, they perform the same functions of most concentrators at a fraction of the cost. Also, the use of multiplexers maintains network transparency—not requiring any programming changes. This is usually not true for concentrators. Although the specific capabilities and differences between devices are covered later, what is most important is to recognize the types of data sources that can be concentrated at remote locations for transmission to the host computer, as illustrated in Figure 11.1.1

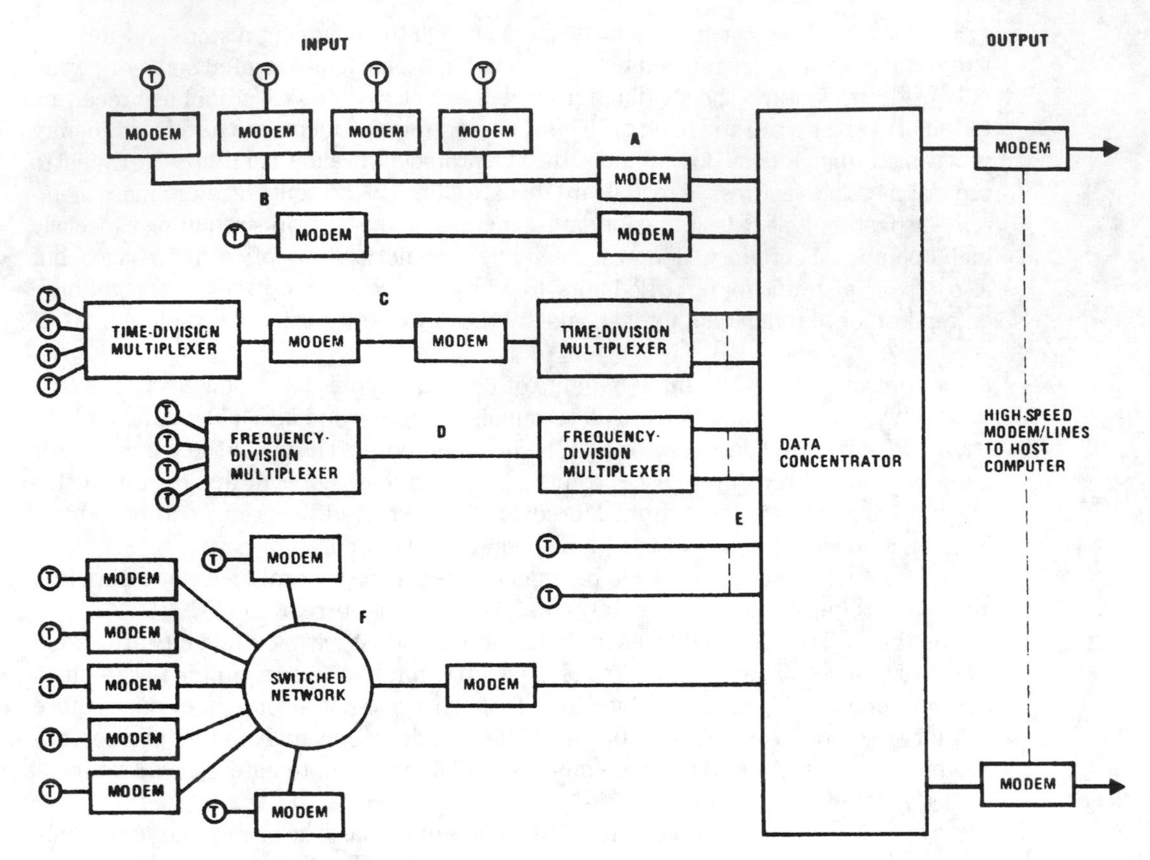

Figure 11.1.1 Concentrating data.

Concentration sources. Figure 11.1.1 shows six examples of data sources that can be concentrated. In part A, a multiport line connecting four remote terminals can be serviced by the data concentrator. In part B, a terminal on a point-to-point line is directly interfaced to the data concentrator. Parts C and D illustrate the concept of "remultiplexing" or "recombining." In parts C and D, data is first concentrated by time-or frequency-division multiplexers for transmission to a hub area where the data concentrator is located. At the hub area, additional data sources are combined with the previously multiplexed data into one or more high-speed composite channels for transmission to the host computers. Local terminals can be directly connected to the concentrator, as illustrated in part E. Part F shows how many terminals can connect to the concentrator by using the public switched telephone network as the initial transmission medium.

Front-end processors

When computer-based data communications networks began coming into use in the 1960s, the preferred interface between the host computer and the communications lines was the hard-wired transmission controller—notably IBM models 2701, 2702, and 2703 (or 270X). Occasionally even then, small computers, called *data communications preprocessors*, were installed at the front of the host computer by such computer makers as Burroughs and Digital Equipment Corp.

In 1969, several microcomputer makers began to advocate actually replacing the hard-wired units with programmable front-end processors. The idea took hold so well that IBM developed the model 3705—and more recently, the 3725 and 3745—series of programmable communications controllers.

A programmable front-end processor (PFEP) today might cost less than a hard-wired controller. Its real advantage, however, is its ability to free the host computer's internal memory, software, and execution time of much of the burden of data communications. First, compared with the hard-wired unit, the PFEP makes much less demand on the host computer for line control. Second, being far more versatile than the hard-wired unit, the PFEP takes over such data communications tasks as polling, code conversion, formatting, and error control.

In a well-designed data processing configuration, with the computer operating in a batch data processing mode, average computer utilization is about 75% to 80%. The balance of 20% to 25% has to be left available as a reserve for servicing peak loads. But adding a data communications mode increases the average load by perhaps another 20%, eliminating the reserve capacity.

In this situation, if the extra load is handled by upgrading the host computer to a larger mainframe or more memory or both, thousands of dollars will be added to the cost of the setup. But when retaining a peak-load reserve is accomplished by adding a PFEP, costs are kept relatively low.

Today, only the simplest data communications networks use hard-wired controllers. But the line-control discipline implemented by the hard-wired controller is so thoroughly embedded in the technology of computer-based data communications that the front-end processor was initially programmed to emulate the hard-wired controller.

Later, the evolutionary development of front-end processors resulted in the transfer from the host computer to the FEP of most, if not all, of the software related to specific data communications functions—as illustrated in Figure 11.1.2. The result of this transfer was to make the network managed by the processor appear to the host computer almost like a single input/output device that is ideally compatible with the design and operation of the host computer.

Serial bits into parallel bits. A typical data communications network includes several types of terminals, as well as multiplexers and perhaps a remote data concentrator, operating asynchronously and synchronously and with several codes and speeds. Thus, when all bits in all characters in all messages converge at the central computer site, the host computer is confronted with a random, interleaved, and intermittent data stream from all online terminals and other devices.

The host computer is not designed to process these diverse inputs directly. The bits composing a character are entered one after the other at the terminal and continue as a series along the transmission link to the computer. But the host computer takes in all the character bits in parallel, at one time.

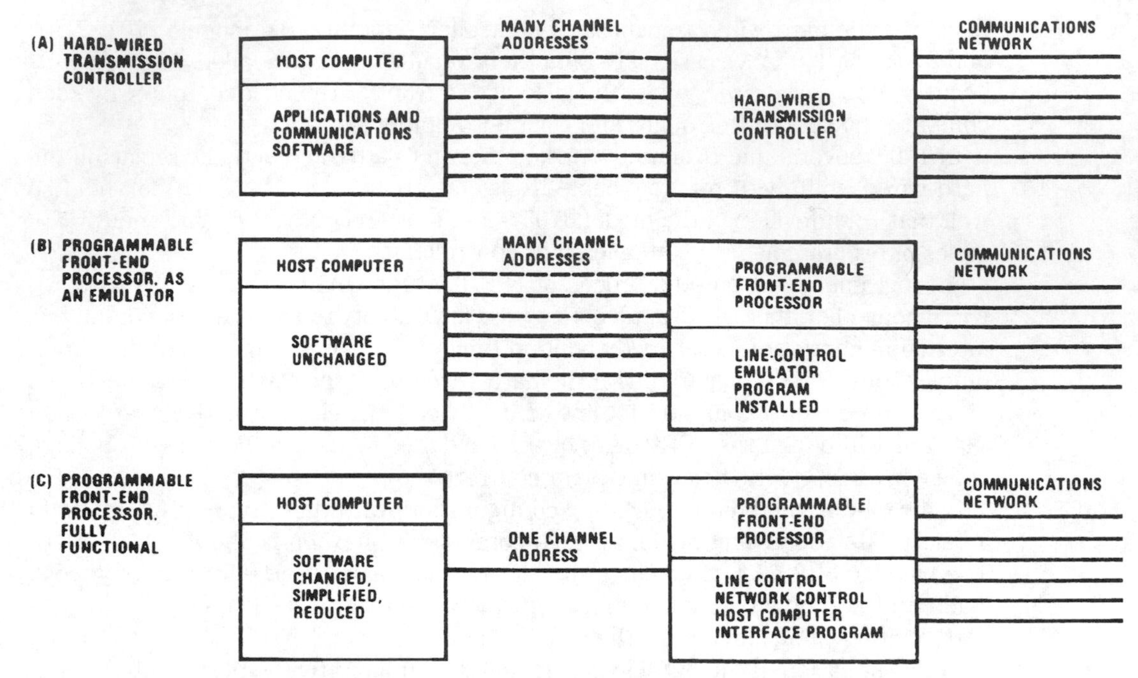

Figure 11.1.2 Relieving the host.

In computer terms, this bit-parallel character is called a *byte*. Moreover, the host computer can accept bytes at a rate of at least 1000 times faster than characters come off a high-speed voice-grade line. It would be wasteful for such a machine to spend its time, worth perhaps $5 to $20 a minute, on slowly accumulating bits and converting them into bytes.

Instead, this job is done by hardware registers in the interface. There, each serial-bit character coming from a given terminal is directed into that terminal's interface channel and converted into a parallel byte for the computer. The computer, in turn, temporarily stores that byte in a dedicated address space in its internal memory, from which the byte, or character, moves either to another area in internal memory or to a secondary memory that accumulates words and messages. One address space is needed for each line.

All this the hard-wired controller does under control of communications programs stored in the host computer, as illustrated in Figure 11.1.2A. When the processor is used only as an emulator (Figure 11.1.2B), it operates under control of the same communications software and has as many channel addresses between the host computer and the processor as there are lines into the processor.

For the true front-end processor, note that just one channel (Figure 11.1.2C) is used by the processor to communicate with the host computer. This means that only one address space is needed in the host computer, leaving more space for use by other peripherals. Also, the access-method program, which "connects" input lines to the application program, is simplified because the host computer interfaces with only one line to the processor.

Acting solely as an emulator, the processor offers no real economic advantage because the data communications overhead still burdens the host computer. This overhead consists of extensive data manipulation, or preprocessing, that must be accomplished by the host computer before it can actually operate on the messages. These preprocessing programs require host-computer execution time and internal memory, both of which might be in short supply in a given application.

When most of the overhead is transferred out of the host computer and into the processor, the front-end processor will, among other things:

- Poll the input and output devices to determine whether an information transfer should take place.
- Restructure incoming data to more compact forms to increase host-computer input efficiency.
- Convert data to the code most suited to the host computer.
- Check incoming data for errors. Reject data blocks if errors are present.
- Route messages from one terminal to another. That is, perform message switching without the data having to enter the host computer at all.

No channel limitation. Conserving the host computer's internal memory and execution time is not the only reason for preferring a front-end processor to a hard-wired transmission controller. Another is that the processor's programmability makes it easier to change the network—when, for example, substituting one type of remote terminal for another—without having to make actual wiring changes at the host computer. A third reason is that a minicomputer, the kernel of the front-end processor, is designed to handle more, and more-varied, types of input and output devices than is the large computer, which is primarily designed to process large batches of data rapidly.

Thus, in its data communications mode using a hard-wired controller, the host computer can be restricted in the number of lines it can handle either by the limit of the computer's channel addresses (address space) or by the fixed number of channels available on a particular hard-wired controller. In either case, the installation is channel limited, not throughput limited.

The minicomputer, on the other hand, is not channel limited, but is throughput limited. By way of example, if the minicomputer has an instruction execution time of 1 μsec, it can perform 1 million instructions each second. Suppose that sampling a line to tell if a pulse is a 1 or a 0 takes 25 instruction times and that to define the start and stop edges of a bit's pulse, each bit is sampled eight times. Thus, each bit uses 200 instruction times. Therefore, the minicomputer can service this throughput as eight (input or output) lines at 600 bit/s, 30 lines at 150 bit/s, 45 lines at 110 bit/s, or some combination of speeds that does not exceed 5000 (1 million ÷ 200) bit/s. More than likely, assuming a random traffic pattern, the processor could handle even more lines. Moreover, if network analysis indicates that the instantaneous data rate exceeds 5000 bit/s, then the peak data can be stored at the processor, in a mini disk pack, for inputting (or outputting) when traffic slows down. This small mass memory can also augment the host computer's larger memory.

However, most minicomputers have inexpensive hardware interfaces to perform serial-to-parallel conversion of bits to bytes. Depending on the particular minicomputer, throughput can thus be increased by 5000 to 20,000 bytes, or characters, per second.

The whole business of a programmable front-end processor unburdening a host computer is easy to grasp. The practical problems lie in performing a technical and economic analysis to ensure that a processor of adequate size, including all necessary peripherals, is specified. Then, you must manage software modifications to keep abreast of changing network requirements, such as new terminals that need different protocol support. The hardware and software relationship between a front-end processor and a host computer is illustrated in Figure 11.1.3.

WAN integration

For the network analyst to plan efficiently for the addition of terminals to an existing network, or to design a new network, the provisions for integration of the components and their limiting factors must be examined. To do so, begin by exploring the methods used to connect remote terminals to a host computer.

Individual service. As the first example, let us assume that every terminal is given individual access to the host computer, as illustrated in Figure 11.1.4. If each terminal transmits and receives large volumes of data and must be connected on line to the computer most of the working day, then this configuration could be an optimum one. However, very few, if any, terminals are connected on line during the entire working day, and alternative approaches can be considered to service such terminals, as illustrated in Figure 11.1.5.

Access contention. In comparison to the first configuration, access contention in Figure 11.1.5 uses a multipoint circuit and a rotary switch to reduce the number of ports required on the front-end processor to service a larger number of terminals than in the first example. These ports or channels are the physical interfaces through which communications devices are connected to the front-end processor. Not only are such ports expensive, but there is a physical limit to the number of such ports per front-end processor. Once that number is reached, the network planner

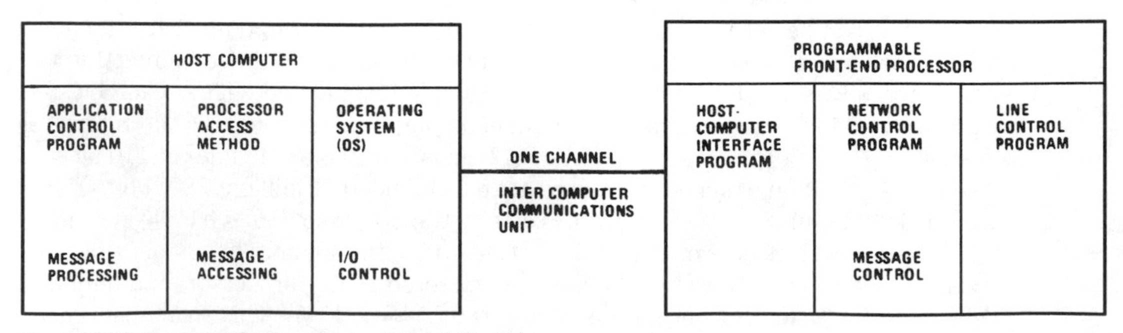

Figure 11.1.3 Processor-host computer interrelationship.

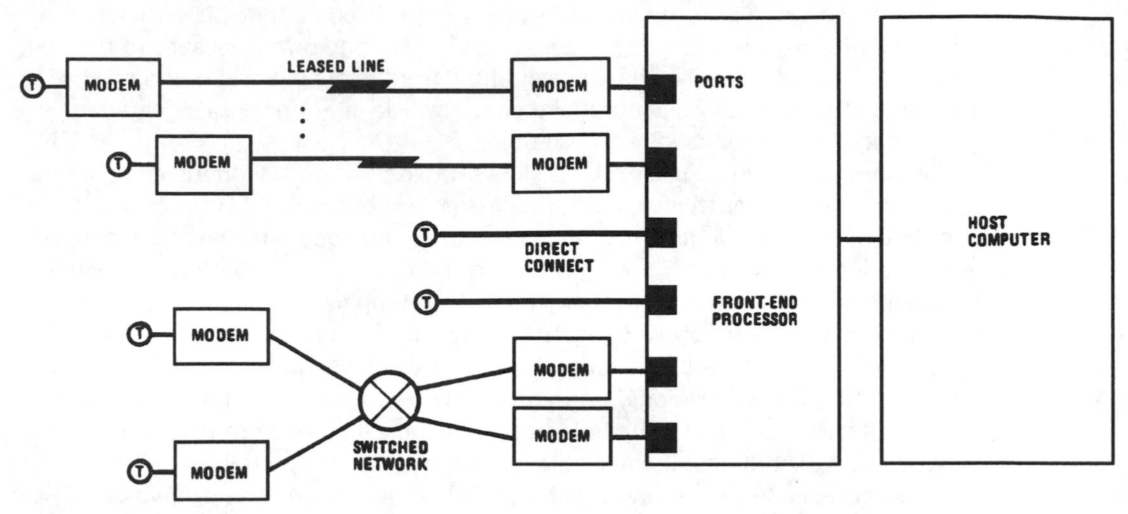

Figure 11.1.4 Individual access.

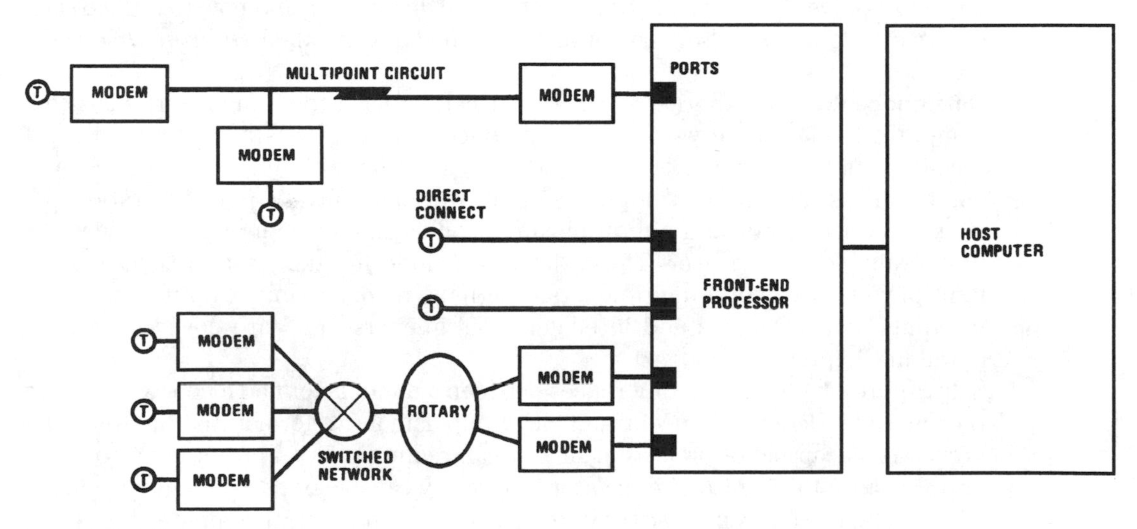

Figure 11.1.5 Contending for access.

must either plan for an additional processor or implement ways to reduce port requirements. All this time, the planner must continue a high level of service to each terminal that requires access to the network.

By the use of appropriate poll and select software, the leased lines of Figure 11.1.4 can be converted into a multipoint circuit, as shown in Figure 11.1.5. If the locations of the terminals are right, not only will port requirements be reduced from two to one, but the total length of the leased line required can be reduced. Because a poll and select protocol is now used, each terminal on the multipoint line contends for access to the computer, which is the price that you must pay for this type of connection. If the terminals used are CRTs operating at 2400 or 4800 bit/s, then the terminal operator will experience "direct access" to the computer because of the high data transfer rate and fast-polling power of the front end. Only when perhaps five to 10 or more terminals share a multipoint circuit will an operator notice degraded performance, such as increased response time.

The lower portion of Figure 11.1.5 shows a two-position telephone company rotary switch installed at the front-end processor local business lines to connect those lines to two dial-in modems. This rotary facilitates a number of terminal operators dialing the local numbers for a connection to the computer via the public switched telephone network. Basically, the rotary acts as a stepping device. If the number that is dialed is busy, the rotary will step the calling party to the next number on the rotary, alleviating the calling party from hanging up and dialing the next number. (Refer to Section 5.4, Operations.) Although only two telephone numbers would be associated with the two dial-in modems shown, imagine the operator inconvenience if there were 10 or 20 dial-in lines to try for each connection attempt.

Assuming each terminal that uses the switched network only transmits data for a portion of the business day, then some number over two (for example, four or five) terminals can use the two dial-in lines on a contention basis before waiting times become excessive. The methodology to determine delay times and how to reduce such delays by adding more "servers" or dial-in connections is called *queuing theory*.

Data concentration. Another method that can be used to connect terminals to a computer through a network is by first concentrating the data from a number of terminals for retransmission over one or more common high-speed lines to the computer, as illustrated in Figure 11.1.6. Although only two concentration devices, a concentrator and a multiplexer, are illustrated, a number of devices are now available for the user. These devices include frequency- and time-division multiplexers, wideband multiplexers, multiport modems, line- and modem-sharing units, concentrators and intelligent multiplexers—and are covered in more detail in Chapter 13.

In Figure 11.1.6, the key differences between a multiplexer and a concentrator are recognizable. Here, the concentrator can group data into a format that the front-end processor is capable of processing, and it can require only one front-end-processor port. Because the front-end processor is normally capable of processing the concentrated data, no complementary device is required at the central computer site.

When a multiplexer is used, normally the routine for data concentration is not compatible with the software in the front-end processor, so another multiplexer, functioning as a demultiplexer, must be used at the computer site. This demultiplexer reverses the data concentration process and regenerates the individual terminal data streams from the composite high-speed line, feeding each data stream into a separate port on the front-end processor.

Although the use of a concentrator can reduce the physical number of ports required by the front-end processor, the reduction will be accompanied by additional software processing requirements for the processor. These and additional tradeoffs between the use of different data concentration devices are covered in Chapter 13.

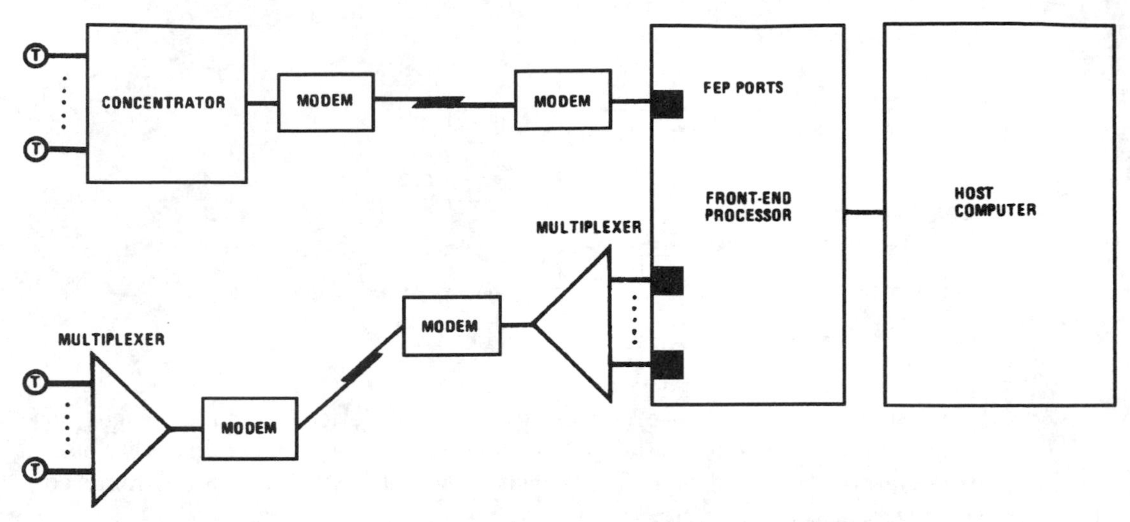

Figure 11.1.6 Data concentration.

11.2 Local Area Networks

The construction and expansion of a local area network is more restrictive than that of a wide area network. The restrictions are caused by the variety of cabling restraints that apply to the installation of each type of LAN, such as an Ethernet or token ring network. The constraints include the number of stations that can be connected to a network, the length of cable to connect a station to the network, and the spacing between cable connections. These restrictions are usually specified in the installation manuals accompanying LAN hardware products and should be carefully heeded when constructing or expanding a LAN.

Although you might be tempted to increase a cable length beyond its restrictive maximum, such a procedure should be avoided; it could result in the entire network becoming inoperative. To better understand this situation, focus on the operation of a repeater that is used to extend a LAN's cabling distance.

Repeater

The repeater is a device that operates at the physical layer of the OSI reference model and is transparent to data content. As digital data travels on a LAN, the cable's resistance, inductance, and capacitance distort the shape of the digital pulses (Figure 11.2.1A). After a certain cable distance, the pulse becomes unrecognizable. This explains why the extension of a LAN cable beyond the stated maximum limit can result in an inoperative network.

The repeater functions as a data regenerator. That is, when it senses the occurrence of a pulse (disregarding the distortion) on the cable, it simply regenerates that pulse. Figure 11.2.1B illustrates the repeater's operation.

Although a LAN's cable length can be extended by using a repeater, there are limitations associated with this device. Because the repeater takes a finite time to sample

A. Pulse distortion

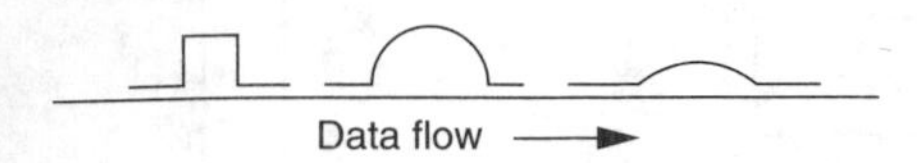

B. Repeater pulse regeneration

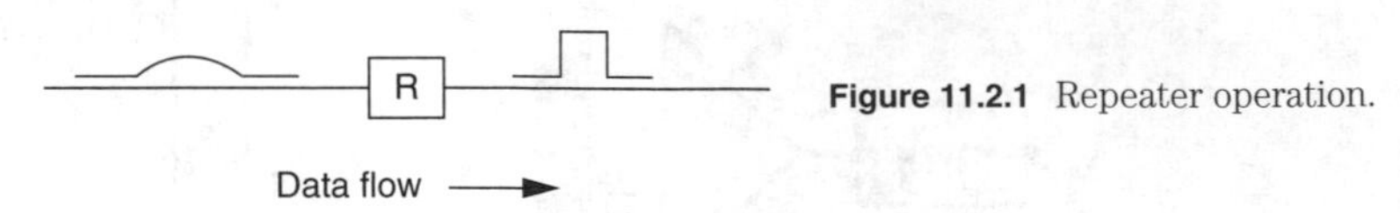

Figure 11.2.1 Repeater operation.

a pulse rise and to regenerate the received pulse, the device introduces a slight pulse delay—known as *jitter*. As jitter accumulates, it adversely affects the ability of network stations to receive data. This explains why each LAN type has constraints concerning the use of repeaters.

Network interface card

The NIC is responsible for the transmission and reception of data to and from the LAN. The NIC operates at the physical layer of the OSI reference model, providing a physical connection to the LAN cable.

The NIC contains a series of integrated circuits that format data into the LAN transport protocol supported by the card. Besides the LAN transport protocol it supports, the type of NIC used is based on its LAN and computer workstation interfaces. Examples of LAN transport protocols include Ethernet and token ring; examples of LAN cable interfaces include DB-15, BNC, and RJ-45 connectors. The computer workstation interface references the NIC fabricated to operate in the system unit of a specific type of computer. Examples include the system expansion unit slots of the Apple Macintosh, IBM Microchannel, IBM PC 16-bit ISA (Industry Standard Architecture), and IBM 32-bit EISA (Extended ISA).

Hub

As noted in Chapter 10's coverage of Ethernet and token ring, 10BASE-T and token ring networks are hub-based LANs in which workstations are first cabled to a hub. The expansion of each network type is accomplished by the use of additional hubs. This expansion process continues until either the maximum number of stations supported by the LAN is reached, or network traffic reaches a level that warrants network segmentation into two or more separate entities. If stations on one network segment have to communicate with stations (or file or print servers) on another network segment, a method for interconnecting LAN segments is required. Two commonly used devices that provide such a capability are bridges and routers. Prior to examining the operation of those devices, first focus on the functions performed by file and print servers.

File server

A file server can be considered a data repository, required when a client-server network operating system (NOS) is used on a LAN. In comparison, a peer-to-peer operating system results in each workstation communicating directly with other workstations, eliminating the necessity for having a file server. Examples of file server-based NOSs are Novell's NetWare and IBM's LAN Manager; an example of a peer-to-peer NOS is Lantastic.

In a file server-based operating system, each workstation operates a "shell" program. This program is responsible for examining commands entered by the workstation operator. The shell program is also responsible for communicating with the NOS executing on the file server when it recognizes a network command.

The workstation shell program is normally loaded via a command entry in a PC's AUTOEXEC.BAT file and serves as a filter. It examines command entries and passes local PC commands to the PC's operating system. For example, a DIR command to obtain a directory listing of a local PC drive would be passed to DOS (disk operating system). The resulting directory listing would be displayed on the computer's screen, as if no shell program were loaded in the computer.

If the operator enters a network-related command—such as NDIR (to receive a directory listing of a network drive)—the shell program first recognizes that a network command was entered. Instead of passing the command to the PC's operating system, the shell program recognizes that the command must be communicated to the NOS executing on the file server. Thus, the shell program formats a request to the NOS and passes the request to another module operating on the PC as either part of the shell program or as another program loaded through an entry in the AUTOEXEC.BAT file. This file transmits the request through the NIC onto the network. The NOS on the file server receives the request, processes it, and transmits a response back to the workstation. In this example, the response would be the directory listing for a network drive.

In addition to executing the NOS, the file server also functions as a storage repository for application programs and data. When a workstation executes a program on the file server, this method of data processing is normally called a *client-server operation*. Actually, client-server processing is a term liberally used to refer to several types of workstation and file server operations. For example, the execution of a spreadsheet program on a workstation and the storage of the spreadsheet file on the file server can be considered client-server processing. Similarly, the execution of a program on a workstation that operates in tandem with a program executing on the file server can also be considered a client-server operation. However, the term *cooperative processing* is probably more accurate in describing this workstation-file server operational relationship.

The ability of a file server to concurrently execute the NOS and one or more application programs is based on the NOS's support of a multitasking capability. In fact, most NOSs represent tens to hundreds of man-years of program development and hundreds of thousands to millions of lines of coding. All this effort provides the basis for enabling workstation users to simultaneously perform different network-related operations.

Print server

If you consider the file server to represent a data repository, you can consider the print server to be the focal point for outputting data from a LAN. Most LAN operating systems support the use of print servers as a software module operating on either a file server or on a dedicated computer. For either case, the printing software accepts print jobs routed over the network and controls the printing of data on printers connected to the print server.

Figure 11.2.2 illustrates the relationships among workstations, a file server, and a print server on a bus-structured network. A print job directed from workstation A to network printer B will first flow to the file server for processing by the NOS. The NOS temporarily stores the print job in a print queue on the file server. When the job's turn arrives, the NOS sends it in packets to the print server. The print server directs the print job to the appropriate network printer. In addition, the print server controls the print job's formatting to work in tandem with the connected printer.

The primary rationale for using one or more individual print servers, instead of operating print server software on a file server, is based on performance and security. From a performance perspective, the buffering and print-control operations can adversely affect the file server's ability to handle network-related requests. With regard to security, most organizations prefer to place their file servers at locations that are not readily accessible, because of critical information that might be stored on that device. Because it is desirable to have printing results readily accessible to network users, the operation of separate print servers facilitates accessibility without compromising network file server security.

Bridge

As previously mentioned, a bridge is used to join segmented LANs. Operating at the data link layer of the OSI reference model, a bridge examines the source addresses in frames flowing on the LAN connected to one of its ports. It then constructs a table of addresses associated with the network's workstations. As it builds its table of addresses, the bridge also examines the destination addresses of the frames and compares them to those in its table of source addresses. If it does not find the destination

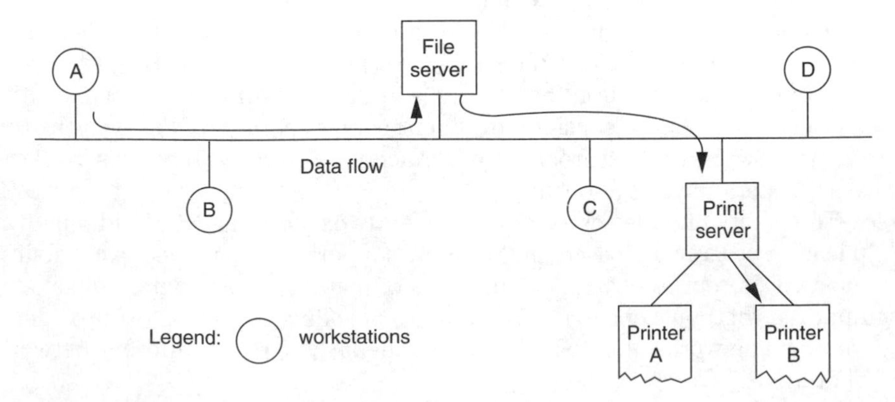

Figure 11.2.2 File and print server operational relationship.

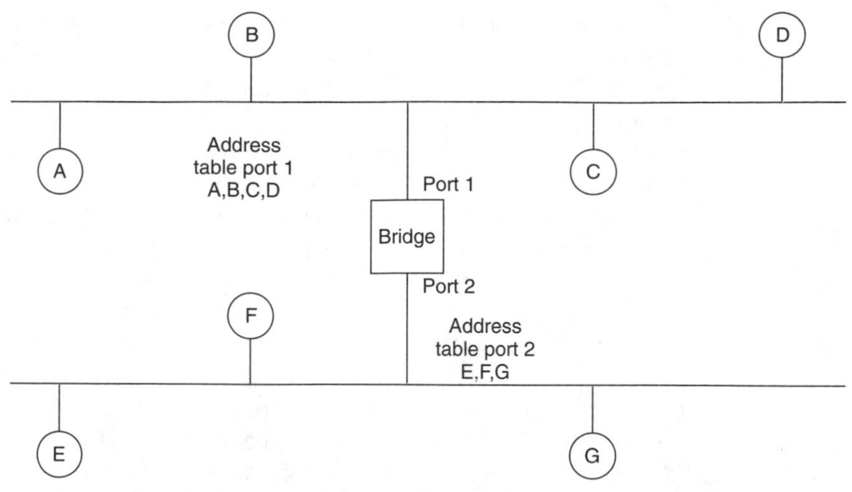

Figure 11.2.3 Basic bridge operations.

address in its source address table, the bridge considers the frame to be destined for the network connected to its other port. The bridge then rebroadcasts or forwards the frame onto that network.

Figure 11.2.3 illustrates the operation of a local bridge connecting two bus-based LANs. Assume that the source addresses of stations on one network are A, B, C, and D; on the other network, E, F, and G. As each station transmits data, the bridge examines the station's frame that is flowing on the network and constructs a table of source addresses associated with each port. This table-building process is called *filtering*. The rate at which a bridge examines frames flowing on a connected network is called the *bridge's filtering rate*.

When a bridge encounters a frame with a destination address that does not match any of the source addresses in its address table, it considers the frame destined for another LAN. So, the bridge then forwards that frame onto the other LAN to which it is connected. The maximum rate at which a bridge can forward frames is called its *forwarding rate*. Thus, the filtering and forwarding rates define a bridge's performance level.

When a bridge connects two similar networks, such as the bus-based ones in Figure 11.2.3, it simply copies a frame transparently from one network onto the other. When the two networks are dissimilar, such as an Ethernet network and a token ring network, the bridge must be able to translate from either format to the other. This bridge type is called a *translating bridge*.

Both transparent and translating bridges have certain operational restrictions. The primary restriction is that the resulting topology cannot form an operating closed loop. In a token ring environment, bridges can form a closed loop because the routing of information between bridged LANs can occur based on source routing information carried within each frame.

In a source routing environment, a workstation transmits a discovery packet that flows between bridged networks to its destination. It elicits a response in which the route taken between numbered LANs and numbered bridges is included in the responding frame. By accepting the first responding frame (multiple frames can be received, as frames are broadcast over different paths), an optimum path is assumed ("discovered"). That route is used when a workstation on one LAN wants to communicate with a station on another network.

To enable Ethernet and token ring networks to be linked together without sacrificing the source routing capability of token ring networks, vendors developed source-routing transparent bridges. Such a bridge operates as a transparent translating bridge when sending data from a token ring network to an Ethernet network; as a source-routing translating bridge when sending in the reverse direction.

Until now, this chapter has focused on local bridges, which are directly attached to two LANs. Another bridge type the network analyst must consider is the remote bridge. This type replaces the second LAN interface with a WAN interface. In addition, the remote bridge must use software that converts the LAN frame to a WAN transport protocol, such as HDLC. This conversion enables the effective transport of the LAN frame over the WAN. Figure 11.2.4 illustrates the use of a pair of remote bridges to interconnect two geographically separated LANs.

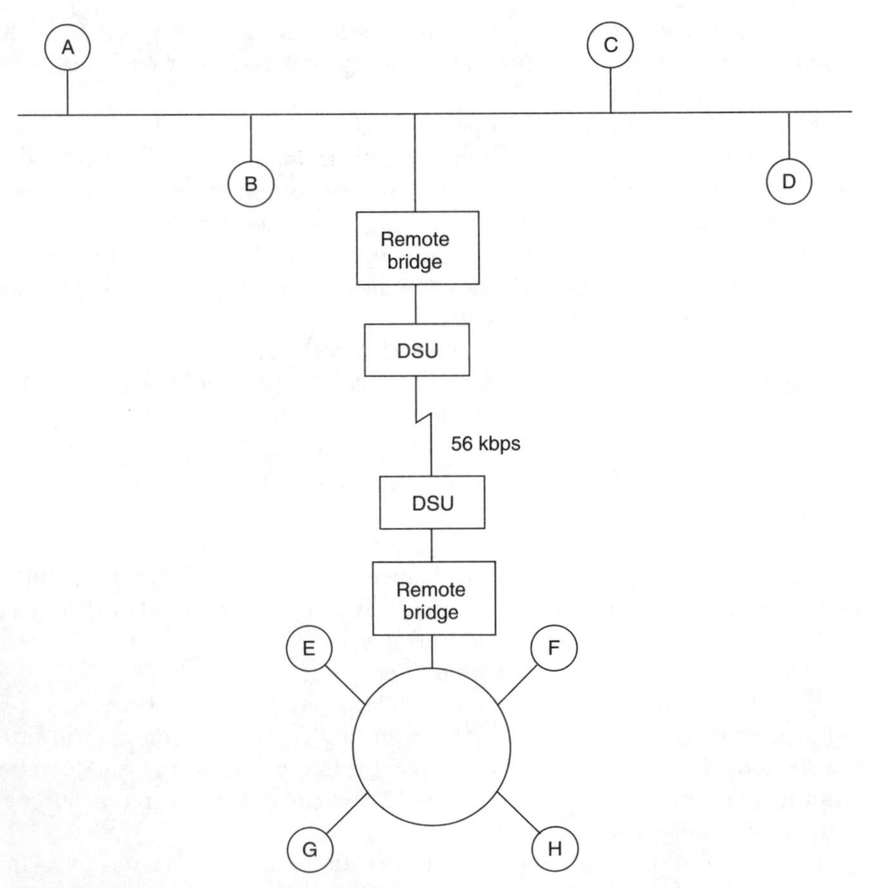

Figure 11.2.4 Using remote bridges to interconnect geographically separated LANs.

Notice that the WAN connection shown in Figure 11.2.4 is a 56-kbit/s digital circuit, which is normally the minimum data rate required to provide a LAN interconnection. In determining the WAN facility to use for interconnecting geographically separated LANs, the network analyst must carefully estimate the potential data flow between networks. If data flow is underestimated, transmission delays can adversely affect user productivity; if overestimated, the result could be a partially used expensive facility, such as a fractional T1 line.

Router

A router operates at the network layer of the OSI reference model, compared with the bridge operating at the data link layer. Although this difference might initially appear trivial, in effect it results in a considerable difference between the operational capabilities of the two.

Operating at the network layer permits a router to use network addresses to make routing decisions. The result is that a router can direct data over different paths (enabling load balancing), make decisions about the use of different paths, and perform other routing decisions for the transport of data from one LAN to another.

Figure 11.2.5 illustrates the use of three routers to interconnect five LANs located at three separate geographic areas. One difference between the use of bridges and routers concerns the ability of routers to segment traffic onto different paths. In this example, data is to be transmitted from a workstation on LAN A to one on LAN B. Packets could be transported either on the route A-B-C or on A-C. This ability of a router to select different paths means that it could also reroute data around an inoperative or failed path. Thus, a network analyst is provided with the ability to introduce a degree of network backup into the WAN connection.

Another difference between bridges and routers is the ability of routers to segment frames for transmission between LANs. For example, a 4-Mbit/s token ring LAN can support a data field of up to 4500 bytes; Ethernet, up to 1500 bytes. When a bridge is used to connect the two networks, software must be set at each token ring station to limit all frames to a maximum of 1500 bytes, to accommodate their flow onto a bridged Ethernet network. (Note that the software can be set to limit only those frames destined for an Ethernet network, while still allowing the full token ring capability.) In comparison, a router can divide a token ring frame into two or more Ethernet frames, eliminating the necessity of resetting software on each token ring workstation.

The additional capabilities of routers affect both cost and performance. A router typically costs at least twice as much as the least costly bridge and has a filtering and forwarding rate about one-half to two-thirds that of a bridge. However, when interconnecting a number of geographically distributed LANs, the use of routers considerably expands the ability to provide greater network availability. Therefore, the use of routers deserves serious consideration by the network analyst.

Gateway

A *gateway* is a communications device that provides a translation service for the data on one network and the format required by a mainframe or minicomputer ap-

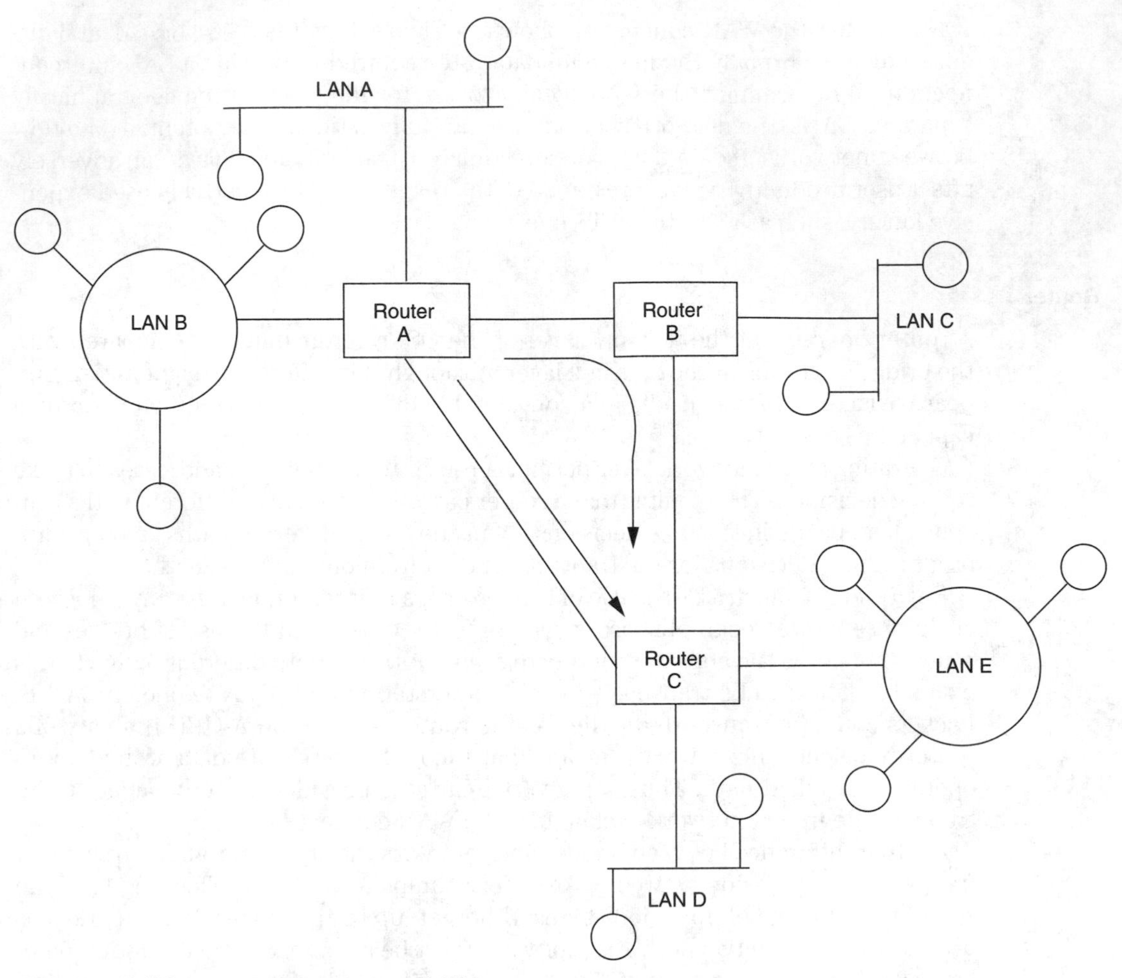

Figure 11.2.5 Using routers to interconnect LANs.

plication on another network. To provide this translation capability, a gateway normally operates through all seven layers of the OSI reference model.

To best understand the gateway's operation, study one typical application. Figure 11.2.6 illustrates the use of a gateway to provide a connection between token-ring-network workstations and an IBM host computer. Here, the gateway translates token ring frames into SNA SDLC-transported information for processing by the IBM mainframe. Physically, a gateway consists of a PC as a base. The PC can be configured with a token ring adapter card and an SDLC adapter card, to provide physical connectivity between LAN and WAN communications facilities.

From a software perspective, frames flowing on the token ring network must be translated into SNA-formatted packets transported by SDLC frames. In addition, the contents of the token ring frames must be examined by the gateway. Certain information, such as screen attribute codes and colors of specific portions of data on a

PC's display, must be translated into the codes supported by the mainframe. Here, the latter function represents the gateway's operation at the application layer, as information is translated between the token ring and SNA networks.

In examining Figure 11.2.6, note that the gateway PC is shown connected via a WAN transmission facility to a port on an IBM 3i74 control unit. This type of gateway connection is but one of several types possible when connecting a LAN to an IBM mainframe. Other popular methods include the use of a token ring or Ethernet adapter card on the 3174, or the use of a token ring interface coupler (TIC) on a 3745 communications controller.

The use of a token ring or Ethernet adapter card in a control unit or communications controller unit enables information from that unit to flow directly onto a token ring or Ethernet network. The unit (3745 in Figure 11.2.6) then becomes a participant on the LAN. The software in the unit replaces software in the gateway PC in performing required translation services.

Questions

11.1 What is the difference between a feasibility study and a communications plan?

11.2 List some of the functions an "intelligent" terminal might perform that normally a "dumb" terminal cannot.

11.3 What has made the network analyst's job more difficult since the 1980s?

11.4 What are the three major categories that depict the relationship of networking devices to transmission facilities?

11.5 How might an "intelligent" terminal be used to reduce the host computer's overhead?

11.6 What general functions does a concentrator perform? What additional functions might a concentrator be programmed to do?

11.7 When we talk about front-end processors, we normally speak about the functions they perform in reference to off-loading the software burden of the host computer. Discuss some of these functions that can be off-loaded.

11.8 Discuss the key differences between a multiplexer and a data concentrator.

11.9 What are some of the common functions of multiplexers and concentrators?

11.10 Why is the construction and expansion of a LAN more restrictive than that of a WAN?

11.11 What is the purpose of a repeater on a LAN? Why is there a finite limit to its use?

11.12 For what is the NIC responsible on a LAN? What determines what type of NIC is used?

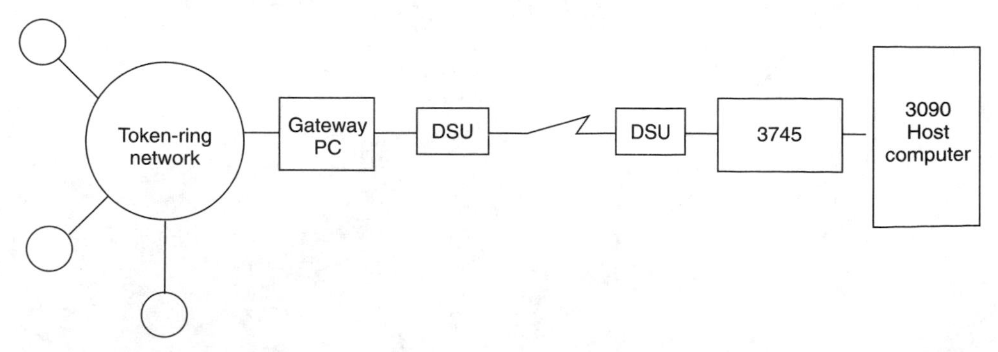

Figure 11.2.6 Using a gateway PC.

11.13 What are the two major functions of a file server? How does it relate to the term *client-server*?

11.14 Why would a network designer use one or more individual print servers rather than have the print server software operate on a file server?

11.15 How do the operational capabilities of bridges and routers differ?

11.16 What is the function of a gateway? Describe a typical use of this device.

Chapter

12

Personal Computer/ Workstation Communications Software

12.1 Introduction

The growth in personal computing has resulted in a revolution in the use of communications by small businesses and persons at home, as well as a dramatic increase by large corporations. Today, it is quite common for persons to have a computer on their desk that is linked to a local area network for communications within a building and which has a modem for accessing information utilities, commercial electronic mail firms, bulletin board systems (BBS), stock brokerage order entry networks, and similar facilities.

Although the capabilities of the personal computer and the modem used are important communications considerations, of equal importance is the software that operates on the computer. This chapter examines in detail those basic functions that are normally essential to establishing a high level of communications capability.

Without question, people want their personal computers to communicate with other computers. The best way to accomplish this goal, however, is not always readily apparent. Although communications software packages offer an increasingly popular approach to the problem, they must be chosen with care. To help readers make intelligent choices, this chapter presents the key features of communications packages, in particular those designed to work with the EIA-232 physical interface commonly used between personal computers and modems.

Evaluating EIA-232 communications packages can be a complicated process. Buyers must weigh carefully the jobs they require the software to perform against the product's cost. Reviews and surveys of communications packages appear frequently in the popular computer publications, and they can help. But surveys consisting mainly of charts in which, for example, 100 popular software packages are compared on the basis of, for example, 20 arbitrarily chosen features are unavoidably superficial. Moreover, many articles concentrate on frills and conveniences while skipping over such key issues as connection establishment and maintenance, terminal emulation, and file transfer.

12.2 The EIA-232 Package

A software program that runs on such PCs as the IBM PC, IBM clones, Apple II, or Macintosh, an EIA-232 communications package can communicate asynchronously with remote computers through EIA-234 serial ports—either directly or through modems. Unlike specialized products that emulate synchronous terminals in the IBM mainframe, Wang, or similar networks, the EIA-232 communications package does not require the use of special adapters and connections.

Communications packages for PCs must have two fundamental capabilities: terminal emulation and data transfer. The former connects the PC to a minicomputer or mainframe as though the PC were a terminal, thus making it possible for users to conduct timesharing sessions or to access an application on a central, shared system. Data transfer lets the user exchange information between a PC and a minicomputer or mainframe (or another PC).

To accomplish terminal emulation, the user needs only the PC-resident software; for error-free data transfer, however, the receiving computer must have a companion program. The cost for commercial minicomputer- or mainframe-based communications programs—typically $1000 to $100,000—is much higher than it is for PC versions.

Making the connection

Besides terminal emulation and data transfer capabilities, a PC communications program must support the operating system or operating environment used by the computer with which you want to communicate. For an Apple Macintosh, the PC communications software must work with System 7 or an earlier version of the Macintosh's operating system. On an IBM PC or compatible computer, the use of DOS-compatible communications programs—which, in 1990, accounted for over 90 percent of all communications programs—has rapidly decreased in market share because of the growing popularity of Microsoft Corporation's Windows operating environment.

Today, over 60 percent of all communications programs designed for use on IBM PC or compatibles are designed to provide a graphical user interface (GUI). They also require the use of Windows to operate.

Initially, what impresses the user most about any program is its interface. Before communications software can connect computers to each other, it must communicate with a human—the user. This aspect of the program, known as the *user interface*, includes the prompts, commands, menus, function keys, and other features that let the user talk to the program; it also includes the displays through which the program communicates with the user.

There are many styles of user interface: command line, interactive prompt and command, menus and arrow keys, mice, and windows. The fundamental tradeoff;

ease of learning versus ease of use. Ease of learning is important if many people will be using the package infrequently or if there is rapid personnel turnover so that relatively little time need be "wasted" in learning and training. A user interface designed for ease of learning presents all the choices in menus. The penalty is that menus for everything are always shown, which slows down the expert.

At the other extreme are programs that favor the expert, providing only terse and cryptic commands, sometimes with no way for a novice to get help—short of reading the manual. A compromise, "menu on demand," lets the expert issue rapid, terse commands, while still allowing the novice to see a menu at any point by entering a special help key.

There is an oft-neglected aspect of the user interface that falls into the ease-of-use category: Can the package be used by people with disabilities like motor impairment, blindness, or deafness? If you can depress only a single key at a time, how can you enter complicated Ctrl-Alt-Shift key combinations? If a PC is connected, for example, to an ASCII-oriented speaking device or a Braille terminal, how can multicolor animated graphics screens be deciphered? How will a person who cannot hear know when the package is beeping or whistling to signal an important event? Often, the fancier the user interface, the less it lends itself to use by the disabled.

Another item, of minor importance, but the absence of which can be a nuisance, is the ability to access system functions without actually leaving the program. To change directories, list files, display a file, or delete a file, the user should not have to exit the communications program and then restart it afterward. This can be time consuming, especially on floppy-disk-based systems, and even more so when settings—or the connection itself—must be reestablished.

Commercial communications packages tend to place great emphasis on the appearance and style of the user interface, primarily for marketing reasons. But for most people, the user interface should not be a key factor in evaluating a product. It lets the user specify only the real work to be done, and it should take up a relatively small proportion of the total time spent with the package. Ultimately, it is much more important to know whether, and how well, the product can perform the required tasks.

Although the user interface is important to many persons, perhaps the most important aspect of any communications package is its set of mechanisms for establishing a connection—that is, matching communications parameters to the communications medium on the other end, monitoring the connection once established, and breaking the connection.

Any communications program should allow control over such communications parameters as bits per second, parity, duplex, flow control, and the number of data, start, and stop bits per character. Each of these parameters is important to a satisfactory connection between computers. In selecting a communications package, be sure that it supports all the parameters and settings required by all the computers with which it must communicate.

For example, most minicomputers and mainframes from Digital Equipment Corp. (DEC) use X-on/X-off full-duplex flow control to prevent data overruns; if the communications package for a particular PC does not support X-on/X-off, then data transferred between the PC and the DEC system could be lost.

IBM mainframe ASCII TTY (teletypewriter) line-mode connections, however, are half-duplex, and they exercise a line-turnaround "handshake" discipline. Thus, a transmission sent to the IBM mainframe before it has sent a special handshake character, such as Control-Q, will not be accepted.

Certain popular mainframes and minicomputers, as well as public data networks like SprintNet and MCI/Tymnet, use even, odd, or mark parity; they will not recognize characters unless the right parity is applied. And if a communications package cannot distinguish parity bits from data bits, the wrong characters will be displayed on the screen. Table 12.2.1 shows typical EIA-232 communications parameters for various equipment.

12.3 Modem Support

Some communications packages support only a limited range of transmission speeds. They may be designed to work only for dial-up connections at speeds up to 9.6 or 19.2 kbit/s. In this case, if there is ever a need to connect two computers directly, the transmission would be limited to those speeds.

Even if dial-up transmissions are the only connections foreseen, some communications programs do not assess the effect of data compression technology built into most modern modems. For example, a V.32bis modem has an operating rate of 14.4 kbit/s. However, when the modem's V.42 data compression feature is enabled, the modem can accept data for short periods at 57.6 kbit/s when the data is compressible at a 4:1 ratio.

To successfully use the V.32bis modem at its optimum transfer rate (when its compression feature is enabled), the communications program must be capable of performing three interrelated functions. First, it must send the appropriate command

TABLE 12.2.1 Typical Communications Parameters

Computer	Front end	Duplex	Flow control	Parity	Terminal
Data General MV	None	Full	X-ON/X-OFF	None	Dasher
DEC PDP-11	None	Full	X-ON/X-OFF	None	VT52, VT100
DEC VAX	None	Full	X-ON/X-OFF	None	VT52, VT100
DECSYSTEM-20	PDP-11	Full	X-ON/X-OFF	Even	VT52, VT100
Honeywell DPS8	DN335	Half	X-ON Handshake	None	VIP7300 7800
HP-1000,3000	None	Full	ENQ/ACK	None	HP262x
IBM 370 Series	3705 TTY	Half	X-ON Handshake	Mark	TTY
IBM 370 Series	7171 P.E.	Full	X-ON/X-OFF	Even	Various*

P.E. = 3270 Protocol Emulator, TTY = ASCII Linemode Connection.

*Delivered with support for 13 popular terminals, configurable for more.

codes to enable compression in its V.32bis operating mode. Next, the modem must support a mutually acceptable method of flow control between the computer and the modem. Once these two functions are accomplished, the software must transfer data at 57.6 kbit/s to the modem.

The support of a mutually agreeable method of flow control is very important to prevent the overflow and subsequent loss of data. When the modem performs data compression at a 4:1 ratio and operates at 14.4 kbit/s, the data transferred from the computer is modulated by the modem and placed on the switched network without delay. If the compression ratio decreases, the modem cannot keep up with the data flow from the computer. Characters are placed in a small buffer in the modem. As the buffer fills, the modem sends a flow control signal to the computer, which prevents buffer overflow and loss of data. That signal must be recognized by the communications program, which then suspends data transfer from the computer until the modem disables its flow control signal.

Microcomputers, such as the IBM PC/AT and the Macintosh, can drive their EIA-232 ports to speeds of 38.4 kbit/s or faster, and two such PCs connected back to back can actually transfer data at these speeds. But the higher the speed, the more important it is to have an effective flow-control mechanism supported by the machines on each end of the connection.

The key to the transfer rate supported by a PC is the type of Universal Asynchronous Receiver/Transmitter (UART) built into its serial port. UARTs used in the original IBM PC were limited to storing one character and would lose characters when data was transferred at speeds above 19.2 kbit/s.

More modern PCs, such as the IBM PS/2 series, use a UART with a multiple character storage buffer, enabling transfer rates up to 57.6 kbit/s without data loss. Hayes Microcomputer products markets an Enhanced Serial Port (ESP) that is fabricated on an adapter card. This card uses a modern UART with appropriate circuitry to support transfer rates beyond 100 kbit/s.

Unless computers are hardwired together with dedicated transmission lines, the communications connection to other computer gear is probably made with asynchronous dial-up modems. These devices communicate special control information to the PC via EIA-232 modem signals, such as data-terminal-ready, data-set-ready, and carrier-detect. Most communications packages can control and monitor these signals, thereby detecting when the connection is broken or initiating the break and hanging up the phone. When a package is used interactively, modem control is largely superfluous because a broken connection is obvious. For unattended operation, however, modem control is important to avoid the excessive telephone bills that could result when the communications package fails to notice a broken connection and leaves the phone "off-hook."

Modems can be either external or internal. The external devices are controlled in a consistent way to accommodate EIA-232; they rarely pose a problem to communications software. Although generally more expensive than internal modems, the external units are interchangeable among different computers.

Internal modems are built specifically for certain computers and sometimes they require special software. A particular software package will not necessarily operate correctly with a specific internal modem, so buyers should understand a package's compatibility limitations. It is also important to check whether the networked modems will support the same modulation techniques and speeds.

Some communications packages are designed to be used only with modems. To make these devices communicate when two computers are connected directly by a cable, certain modem signals must be "faked" by cross connections or jumper wires within the cable connectors. Such a fake-out cable is called a *null modem* or *modem eliminator*—readily available either from a computer supply house or in the form of an adapter that can be connected to the modem cable. But different devices might require different signals connected in different ways, so users should be prepared to experiment using a breakout box. Such tinkering can be avoided if the communications software package can be configured to ignore modem signals when two computers are directly connected.

Dialer control

Many PCs can be connected to other computers only by telephone. So-called smart modems, like those manufactured by Hayes, are able to dial the telephone if they receive commands in the right format from the PC. This means that the communications package must understand the dialing language of the modem. Although the Hayes "AT" language has become a de facto industry standard, not all autodial modems conform to it. Be sure that the communications package supports the modem's dialing language.

Dialing software simplifies connection establishment, thus concealing the details of the dialing language. The user merely tells the program what number to call. Some packages go a step further and include a telephone directory so that the user needs to remember only names, not phone numbers.

12.4 Debugging Communications Parameters

Often, a user can only guess the right combination of speed, stop bits, parity bits, data bits, duplex, and flow control for a particular connection. What if the guess is wrong? What tools does the communications package offer to determine the offending parameters.

If the package allows the parameters to be set independently, they can be varied until the connection works; the number of combinations, however, might seem endless. To reduce the guesswork, the communications package should include debugging tools, such as special display or logging of received characters (preferably including their eight-bit numeric values). If examination of the log reveals a byte with a numeric value of 193 (= 11000001 binary) where you would expect an ASCII "A" (= 01000001 binary), then a good guess might be seven data bits with odd or mark parity, rather than eight data bits with no parity. The package should also include a troubleshooting guide in the manual. Table 12.4.1 illustrates sample entries from a troubleshooting guide.

Saving parameters

After discovering the proper settings for communicating with a particular machine, the next task is saving them for later use. To alleviate the tedium of setting five or 10 com-

TABLE 12.4.1 Sample Entries from a Troubleshooting Guide

Symptom	Possible cause	Cure
Bland, dark screen	PC turned off	Turn on PC
Total garbage on screen	Wrong speed	Try another speed
Spurts of garbage on screen	Noise	Hang up and redial
Uniform mixture of good and bad characters on screen	Parity	Select a different parity
Typed characters appear twice	Duplex	Select full duplex
Typed characters don't appear	Duplex	Select half duplex
Random gaps in screen text	No flow control	Use flow control or a slower speed

munications parameters for each communications session, the package should allow settings to be collected into configurations that may be saved under mnemonic names.

Some packages come with a set of configurations for popular dial-up services, such as Dow-Jones, Compuserve, and MCI Mail. These built-in configurations shield the user from having to know anything about data communications parameters. But when connecting to a service that the package doesn't recognize—for example, from a home office to a mainframe—it should be possible to manipulate the communications settings and save them. The appropriate settings, for example, for a company's DEC VAX, IBM 3090, Harris 800, plus a local SprintNet PAD could be accomplished by typing just the associated configuration name.

Script language, unattended operation

Just as a communications package stores communications settings or telephone numbers, it can also allow repetitive interactive tasks, like log-in sequences, to be automated via scripts. Such short "programs" look for specific outputs from the remote computer and provide appropriate responses. When the package, or the underlying operating system, allows a script to be executed at a predetermined time, then it is possible to carry on a canned dialog with no human operator present. For instance, a PC might be programmed to "wake up" at midnight, set the proper communications parameters, dial the office minicomputer, log in, deposit the day's transactions, fetch and print the day's mail, log out, and hang up. Script languages vary from the primitive and cryptic to full-blown programming languages, complete with variables and conditional branching. Figure 12.4.1 shows a simple script for dialing a Hayes modem to establish a connection to a Unix and then log in. It illustrates how a script can be used in place of built-in dialer control.

The "output" commands send the indicated text strings to the Unix machine ("\13" is a code for carriage return), and the "input" commands search the incoming data for the indicated strings. If any of the input commands fail, the script is automatically terminated. However, through the use of conditional statements that are supported by many script languages, you can construct a program that will retry one or more commands a predefined number of times, or even redo the entire dialing sequence at different times. The ability to conditionally test the results of commands

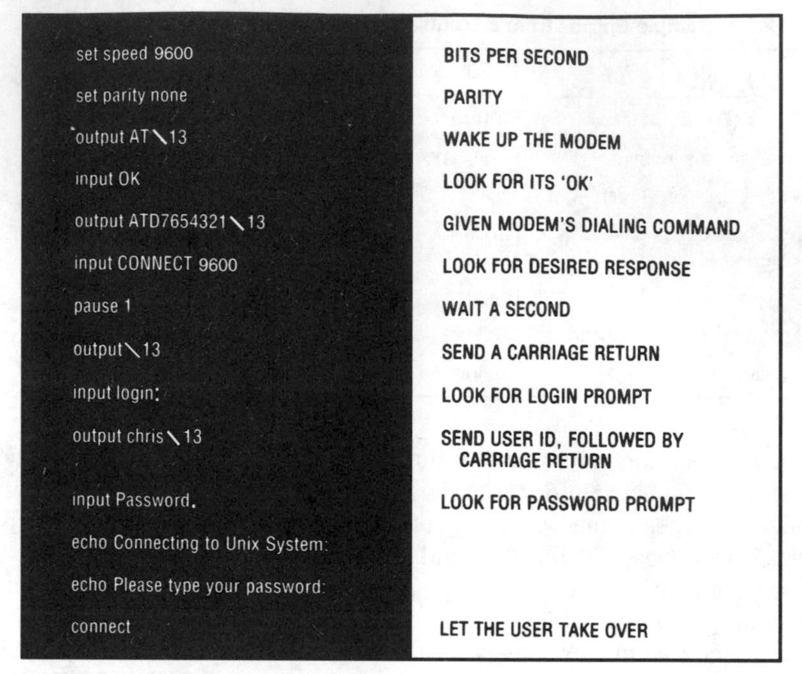

set speed 9600
BITS PER SECOND
set parity none
PARITY
output AT\13
WAKE UP THE MODEM
input OK
LOOK FOR ITS 'OK'
output ATD7654321\13
GIVEN MODEM'S DIALING COMMAND
input CONNECT 9600
LOOK FOR DESIRED RESPONSE
pause 1
WAIT A SECOND
output\13
SEND A CARRIAGE RETURN
input login:
LOOK FOR LOGIN PROMPT
output chris\13
SEND USER ID, FOLLOWED BY CARRIAGE RETURN
input Password.
LOOK FOR PASSWORD PROMPT
echo Connecting to Unix System:
echo Please type your password:
connect
LET THE USER TAKE OVER

Figure 12.4.1 Script language example.

can be an important feature of a script language. Suppose, for instance, that a nightly script sends the day's work to a mainframe and then deletes it from the PC's hard disk. If the data cannot be successfully transmitted, then the script should not forge ahead stubbornly and destroy all of the day's work.

Scripts are essential for unattended operation; they are also useful in setting up procedures for relatively unskilled operators, such as data-entry clerks. For the typical interactive user, however, scripts are a minor convenience, rather than a necessity.

12.5 Terminal Emulation

The PC has increasingly replaced the terminal in many organizations. In addition to its other capabilities, a properly programmed PC can act like, or emulate, a terminal. Thus, it can be used to conduct a dialog with a remote computer. Keystrokes are transmitted via the communications port, and characters that arrive at the port are displayed on the screen. On half-duplex connections, the sender's keystrokes are also echoed locally by the PC on the screen. On full-duplex connections, terminal emulation can be a tricky business because characters can arrive at the port at the same time as the user is typing; communications programs vary in their ability to handle both events at once, especially at higher speeds.

Terminal emulation does not provide any automatic error control—any more than a real terminal would. Bare characters are sent back and forth with absolutely no error-recovery mechanism. If a package claims to supply error-checking data transfer, the buyer should understand that this claim applies to the package's file transfer functions and not to its terminal emulator. A noisy telephone line would probably leave garbage on the screen during terminal emulation, even though files could be transferred successfully.

In addition to sending and displaying characters, a terminal emulator attempts to imitate the repertoire of special effects of particular ASCII video display terminals, such as the DEC VT 100 series, the IBM 3101, the Televideo 920, or the ADM3A. This means that the program responds to screen control sequences sent by the host just as the real terminal would. For example, the ASCII sequence "ESC [5; 7 H" sent to a DEC VT100 positions the cursor at a row 5, column 7; "ESC [0 J" clears the screen; so a PC programmed to emulate a VT100 would understand the same sequences and perform the same actions. When emulating a terminal, the package should also provide mapping between the function keys of the terminal and those of the PC so that they transmit the same sequences. If the VT100 PF1 key sends "ESC 0 P," the IBM PC's F1 key might be programmed to send that sequence.

Today's video display terminals possess a formidable array of features for tabbing, highlighting, partitioning the screen, erasing and inserting text, positioning the cursor, drawing figures, changing colors, switching character sets, activating printers, and so forth—all controlled by host-transmitted escape sequences. A package can emulate such a terminal completely, or it can emulate a subset of its functions. Some terminals have features that cannot be emulated by certain PCs. For example, the DEC VT100 allows switching between 80- and 132-column modes, but an IBM PC can only display 80 columns. To get 132 columns on the IBM machine, a special board might be needed. Another example is the VT100's "smooth scrolling" feature, which allows a file to glide slowly up the screen. Some PCs equipped with appropriate video adapter cards or chipsets do have this capability. Other PCs with less-sophisticated video adapter cards or chipsets do not provide this capability.

The emulation provided by a communications package should be complete enough to allow access to any host-resident software that is designed to control the appearance of the terminal's screen; full-screen text editors like EDT on VAX/VMS or GNU EMACS on a Unix machine are good tests. Another is IBM 3270 protocol emulation as performed by the IBM 7171 or other protocol converter. If emulation is incomplete, the screens might appear to be fragmented and jumbled, characters or lines might overwrite each other, or gaps and transpositions might occur.

Terminal emulation is a key function for users who engage in a lot of interactive dialog with a remote computer, especially when screen control is involved. In this case, it is essential that the communications package be capable of emulating a terminal that the remote computer supports, such as a DEC VT100 or -200 series with DEC VAX/VMS or a Data General Dasher with DG minicomputers. Terminal emulation is less important for brief or noninteractive encounters, such as occasional sessions primarily for file transfer.

Redefining and translating

Because a PC keyboard might have a different layout than that of the emulated terminal, it might be helpful to "move" the misplaced keys to their familiar locations (no, not with pliers). For instance, the Escape (ESC) key (important to much host-resident software) is notoriously mobile, appearing in many different locations even on PCs from the same maker (the IBM and DEC keyboards spring to mind). If you are accustomed to finding ESC immediately to the left of the "1," but the PC has "`" (accent grave) in that position, the "`" could be redefined to transmit ESC (and vice versa). Similarly for function keys; the VT100 PF keys are on the right, whereas the IBM PC's F keys are on the left or at the top of the keyboard; some VT100 users might find it more convenient to assign the PF keys to the PC's numeric pad.

A package might also make it possible to assign any arbitrary character string to a key so that the user could transmit commonly typed items (name or log-in sequence, for example) with a single keystroke. Such many-to-one assignments are called *keyboard macros*, and there are limits to the number of characters that can be represented by a single key.

Key redefinition is important if a user's application requires frequent switching among terminals and PCs having different keyboard layouts. It is also helpful when switching the same PC between different hosts. If the user is accustomed to typing the Backspace key to erase a character, but one host uses ASCII Rubout for this function while another uses Control-H, the suitable character can be assigned to the Backspace key.

12.6 Other Features to Consider

Character sets

The ability to handle European and non-Roman character sets (keyboard input as well as screen output) is important for those who deal in languages other than English. It is common practice in Germany and Scandinavia, for instance, to assign umlaut, slashed, or circled vowels to the ASCII bracket positions. PCs and host computers must agree on these conventions in order for characters to be displayed as intended, rather than in Anglo-American ASCII. Translation of outbound and arriving characters is, therefore, an important function of the communications package. To satisfy the needs of multinational companies, or users with international business dealings, the package should not be restricted to seven-bit ASCII, but it should allow for eight-bit international character sets in line with such International Organization for Standardization (ISO) Recommendations as 2022 ("ISO 7-Bit and 8-Bit Coded Characters Sets—Code Extension Techniques") and 6937 ("Coded Character Sets for Text Communication").

Text screen memory

A special advantage of emulating a terminal on a PC is that the PC can surpass the capabilities of the terminal. The PC's memory can be used to hold hundreds of lines that have scrolled off the top for later recall. Current or previous screens can be dumped to a disk file or printer at the touch of a button.

Graphics

For PCs that have color monitors, the communications program should be able to set the fore- and background text screen colors. A well-chosen color scheme can reduce operator fatigue or even wake up the user during the less exciting hours of the day.

To access graphics-oriented applications on the mainframe or minicomputer, such as SAS Graph, SPSS Graphics, Plot 10, Tella-Graf, or various computer-aided design packages (not to mention certain dial-up shopping services), the communications package must emulate a graphics terminal or a standard known to the application, such as Tektronix 4010, 4014, or other model; DEC ReGIS, HPGL, GKS, GDDM, or NAPLPS. Graphics-terminal emulation is found in only a few communications packages (usually at an additional cost); and certain PCs might require a special monitor and graphics board.

In recent years, graphics have tended to be done directly on the PC by such packages as Lotus, MacPaint, and others. It is not usually possible to connect one PC to another in order to access the remote PC's graphics applications, though certain specialized packages do allow this type of connection. Merely running Crosstalk from PC A to PC B will not result in Lotus on PC B putting a color pie chart on PC A's screen. More commonly, the graphics package exists on both PCs, and their data files are moved from one computer to another using a file transfer protocol built into the communications package. Those packages are commonly called *remote control* communications programs. By loading a remote control program on a distant computer, you can use a similar program on your PC to control the distant computer via a switched-telephone-network call. Once you enter an appropriate password, you can remotely operate the distant computer and have its screen display mapped to your PC's screen.

This type of program is extremely useful for providing after-hours assistance: LAN users still at work can dial the network server and "kill" print jobs, establish new-user accounts, and perform similar network operations without having to travel to the remote computer's site.

12.7 File Transfer

". . . transfers your data over phone lines at speed of light!" was a claim that once appeared in an advertisement for a communications package. Although it's true that electricity travels through wires at near light speed, it is not (yet) true that one electron is equivalent to one bit of data. In fact, at the most common speed used for dial-up data communications, 2.4 kbit/s, a single bit is pretty big—about 75 miles long! A character (generally represented in transmission by 10 bits) is 750 miles long; four characters, like OK, would span the North American continent.

Spurious advertising claims notwithstanding, transmission speed is a technological issue; data transfer is a software issue that includes such questions as how to make effective use of the transmission medium and how to smooth over the differences between computers. Specific areas to watch out for include the following:

- Can binary files be transferred?
- Can text file formats be converted to useful form between dissimilar elements in a network?
- Can a group of files be sent in a single operation?
- Can file-name collisions be avoided?
- Can a file transfer be cleanly interrupted?

ASCII vs. error-checked protocols

Communications packages offer two basic types of data transfer: raw and error-checked. Transmitting raw data, the most common form of transmission, is usually referred to as using the ASCII protocol. This means that the data is sent as is—as ASCII characters—from one computer's communications port to the other. The advantage to this method is its simplicity. No special software need be resident on the remote computer, beyond its text editor or a Type or Copy command. The disadvantages, however, explain why error-checked protocols have evolved. They are as follows:

- The data sent using the ASCII protocol will be corrupted if there is noise on the communications line.
- Data will be lost if the receiving computer cannot keep up with the sender.
- Binary (nontextual) files generally cannot be transferred this way, because many computers will either ignore the parity bit or act on control characters, rather than accept them as data: Control-C, Control-S, and Control-Z are frequent culprits.
- This method works for only one file at a time.

A refinement of ASCII protocol incorporates X-on/X-off or some other method of flow control to reduce the chances of data loss. In this case, both computers must support the same method of flow control, but corruption of the data (including the flow control signals themselves) remains a problem—as does file limitation and the restriction on binary files.

To achieve reliable, correct, and complete transmission of files between computers, neither the ASCII nor the X-on/X-off protocol are sufficiently trustworthy. The communications package must include a true error-correcting file transfer protocol. Error-checked data transfer requires cooperating programs on each end of the connection to exchange messages, which are called *packets*, according to agreed-upon formats and rules (similar to those used on the telephone: one person dials; another person, hearing the phone ring, picks up and says "hello;" the caller is identified. The two parties take turns talking. If one doesn't understand what the other says, a repetition is requested. Good-byes are exchanged, and the phones are hung up). Key to the success of this interaction, of course, is that the conversation is conducted in the same language. A file transfer protocol operates similarly: The two processes "connect," identify the files that are being transferred, request retransmission of lost or damaged information, identify the end of the file, and then disengage.

A special caveat: the fact that most newer modems provide error correction does not eliminate the need for file transfer software. An error-free data stream from modem to modem does not guarantee correct data from computer to computer. Issues of end-to-end flow control and error correction, file limitation, and format conversion must still be addressed with the computers themselves.

Error-checking protocols

Two well-known error-checking file transfer protocols are XModem and Kermit. Many commercial packages include one or both of these protocols (sometimes alongside proprietary protocols), but there are also hundreds of public-domain or freely sharable Kermit or XModem programs. The major advantage of XModem and Kermit is that they are ubiquitous. The protocol specifications are open and public, and large bodies of Kermit and XModem software are available. The cost to a large organization for these programs is minimal, compared with the per-CPU licensing fees required for commercial packages. Furthermore, chances are greater that Kermit or XModem protocol support will be included in almost every PC communications program.

In the case of Kermit programs, source code is included, which encourages their adaptation to a wide range of computers. Noncommercial versions of Kermit can be had for more than 250 different machines and operating systems, ranging in size from the smallest microcomputer to the largest supercomputer, and Kermit is included (at no extra charge) in about 100 different commercial software packages.

According to Telebit Corp. (Cupertino, California.), Kermit protocol has found its way into silicon. XModem is also available for a wide variety of computers, but it was designed primarily for microcomputer-to-microcomputer links. It is most widely know by its commercial implementations, and it is a fixture in programs, such as Crosstalk.

Kermit, XModem, and other error-checking protocols do not all offer the same features.

- XModem uses eight-bit binary bytes in its packet fields and, therefore, requires an eight-bit transparent communications link. It cannot function, even for text files, when parity is in use. Similarly, when any device in the communications path is sensitive to control characters, such as Control-Z or Control-S (which occur in the XModem packet control fields), XModem packets are subject to interference. For this reason, XModem cannot operate in conjunction with X-on/X-off or other in-band flow control. Kermit, on the other hand, encodes its packets as though they were lines of text and, therefore, does not have these restrictions.
- XModem packets are sent in only one direction. The responses are bare, unchecked control characters, such as Control-F for acknowledgment, Control-U for negative acknowledgment, or Control-X for cancel. Corruption of XModem responses into other valid responses is possible, and it can cause a file transfer to terminate prematurely or, worse, corrupt the file. Kermit uses fully error-checked packets in both directions and is more robust in the face of transmission errors.
- XModem uses fixed-length packets. There is no length field. If a file's length is not an exact multiple of 128 bytes, then extra bytes will be transmitted. Furthermore, if a computer, multiplexer, or other device cannot handle bursts of 132 characters, XModem packets will not get through. Kermit packets include a length field. Packets can be adjusted to accommodate small buffers, and a short packet can be sent at the end, so there is no confusion about the exact end of file.
- XModem includes no mechanism for transmitting the file name, so it has no way of sending multiple files in a single session. Kermit does this routinely.
- XModem makes no distinction between text and binary files. But because the conventions for representing text files on different computers can vary, the results of

an XModem text-file transfer between dissimilar file systems can be surprising. Kermit specifies a common intermediate representation for text files during transmission so that incoming text files can always be stored in a useful form. However, this places the burden on the user to select text or binary transfer mode.

- Both the XModem and Kermit protocols have seen a number of extensions over the years. XModem has no formal or consistent way to negotiate the presence or absence of given features, whereas negotiation is built into the basic Kermit protocol. A pair of variant XModem programs will not necessarily be able to communicate, whereas any pair of Kermit programs will automatically fall back to the greatest common set of options.

Among the more popular variants of XModem are the XModem-CRC, XModem-1K, XModem-G, YModem, YModem-G, and ZModem protocols.

The XModem-CRC protocol replaces the 8-bit checksum of XModem with a 16-bit cyclic redundancy check (CRC) character. The use of the CRC significantly reduces the probability of an undetected error occurrence.

The XModem-1K expands XModem's 128-byte data block to a maximum of 1024 bytes. This increases the protocol's efficiency in transporting files over a good-quality line connection, compared to the use of the XModem protocol: XModem-1K reduces the number of required acknowledgments. However, XModem-1K, like other XModem protocols (except for XModem-G, described next) operates half-duplex, requiring the suspension of transmission after each block until an acknowledgment is received.

XModem-G and YModem-G are "streaming" protocols. This means that data is sent continuously, block after block, without an acknowledgment returned for each block. You should use a streaming protocol only with a modem that is communicating with its error-detection-and-correction feature enabled. A streaming protocol, in effect, does not perform its own error detection and correction. Streaming protocols were designed to reduce the delays associated with both modems and computers performing error-detection computations during a file transfer operation.

YModem is a "batch" file-transfer protocol that supports the transfer of multiple files. Most YModem implementations support the use of wild-card characters. Thus, specifying *.BAS might allow you to download all files with the extension .BAS when using the YModem protocol that supports wild-card use.

Because the transfer of multiple files is supported, YModem and its variants include a header that enables the transfer of key file attributes, such as the file name and extension, and date and time of creation. Unfortunately, YModem, like XModem, is a half-duplex protocol that requires the transmitting computer to wait for an acknowledgment after sending a block prior to sending the next block.

Recognizing the limitations of XModem, YModem, and their derivatives, Omen Technology developed the ZModem protocol, which is now widely incorporated into communications programs. ZModem not only supports full-duplex transmission, but also automatically varies the block size in tandem with the number of negative acknowledgments, thus maximizing throughput.

Other ZModem features include a 32-bit CRC for extended error detection and the ability to resume transmission at the point where a transmission interruption occurred. The latter can be very valuable. For example, assume you were downloading a 5-Mbyte file and lost the line connection after transmitting for 20 minutes. With XModem or YModem, you would not be able to restart the download. However, with ZModem, the protocol would provide you with the ability to restart the download at the point of interruption.

XModem and Kermit protocol extensions include the following features:

- *Multiple files.* Modem7, YModem, and ZModem can transfer multiple files in a single batch, XModem cannot. The ability to perform multiple file transmission is built into the basic Kermit protocol.
- *Eight-bit data through seven-bit channel.* XModem, YModem, and ZModem do not have the ability to pass eight-bit data through a seven-bit channel. Kermit supplies this as a negotiated feature (commonly available).
- *Alternate checksums.* XModem-CRC uses a 16-bit cyclic redundancy check to achieve greater rcliability and tries to adapt itself to eight-bit checksum only. XModem-G and YModem-G do not use any error-detection method, and ZModem incorporates a 32-bit CRC. Kermit supplies an optional 12-bit checksum, as well as a 16-bit CRC, which is negotiated with automatic fallback to the single-character checksum.
- *File transfer interruption.* Both XModem and Kermit allow file transfer to be interrupted with no ill effects. Kermit also includes the ability to cancel the current file in a group and proceed to the next one. ZModem provides the ability to resume transmission at the point of a file-transfer interruption.
- *Compression.* Kermit programs can negotiate compression of repeated bytes. XModem and YModem lack a compression option. ZModem has compression built in.
- *Long packets.* XModem-1K allows 1-kbyte fixed-length packets for greater efficiency. Kermit extensions permit variable-length packets up to about 9 kbytes; they are negotiated with automatic fallback to regular-length packets.
- *Sliding windows.* Kermit programs can negotiate simultaneous and continuous transmission of packets and their acknowledgments on full-duplex links, with a window of up to 31 unacknowledged packets and selective retransmission of lost or damaged packets. (This option is not widespread among Kermit implementations.) Sliding windows are not possible in XModem because its responses carry no sequence number (an XModem variant called *WModem* simulates sliding windows, but this version will only work if there are no errors).

ZModem not only permits full-duplex transmission but will dynamically alter the length of packets based on the ratio of positive to negative acknowledgments. That is, an increasing number of negative acknowledgments signifies a deteriorating line condition. This causes ZModem to reduce packet length, which reduces the time required to retransmit a packet containing one or more bit errors. Conversely, as the number of negative acknowledgments decreases, ZModem expands packet length. This dynamic altering of packet length boosts throughput with improving line conditions, while minimizing packet retransmission delays associated with deteriorating line conditions.

- *File attributes.* YModem and ZModem transmit a file's name, size, and creation date. XModem does not. Kermit always transmits the name, and the ability to communicate a wide range of other file attributes can be negotiated (but, like sliding windows, this is not a widely implemented Kermit feature).
- *Checkpoint/restart.* Neither XModem nor Kermit has the ability to restart a file transfer after the connection is broken. ZModem, however, does.

Kermit also differs from XModem by including a fileserver mode of operation, in which the remote Kermit program receives all of its instructions from the PC Kermit in packet form. Kermit servers can transfer files and perform a variety of file management functions, such as deletion, directory listing, and changing directories.

Implementations of Kermit can be had for most PCs, minicomputers, and mainframes. XModem, YModem, and ZModem implementations are found mostly on PCs, rarely on minicomputers or mainframes. Basic XModem is somewhat more efficient than basic Kermit because the packets are slightly longer and there is less encoding overhead. The situation is reversed when the Kermit package is given the ability to perform compression, long packets, or sliding windows.

Kermit won't talk to XModem and vice versa. Each must be evaluated according to several criteria: is there a version of the protocol available for all the computers that must communicate? Can the protocol accommodate all the necessary communications parameters? Is the performance acceptable? Is the software affordable.

XModem is more properly called the *Christensen protocol* after its designer, Ward Christensen, who originally intended it only for communications between CP/M microcomputers. Christensen put his original 1977 Modem program into the public domain, and it was modified by others over the years; some protocol features were added, resulting in protocol variants with names like Modem2, Modem7, XModem, YModem, and ZModem.

The Kermit file transfer protocol was originally developed in 1981 at the Columbia University Center for Computing Activities for CP/M, MS-DOS, the DECSystem-20, and IBM mainframes with VM/CMS; that is, for use in microcomputer-to-mainframe applications. It was shared freely with other institutions, with sources and documentation included. Everyone was, and is, permitted and encouraged to copy and share, to make improvements, and to contribute new versions.

Kermit and XModem both transfer files between computers in blocks of data, or packets (as illustrated in Figure 12.7.1). They require a program running on each computer to compose, send, read, decipher, and act on the packets. Each packet is error-checked via calculated checksums, and retransmission is requested when packets have incorrect checksums. Deadlocks are broken by timeouts and retransmission. Missing or duplicate packets are caught using packet sequence numbers. Both protocols are half-duplex stop-and-wait, which means that the next packet is not sent until the current packet is acknowledged.

Most commercial EIA-232 communications packages claim to include XModem, Kermit, or both. In general, the commercial XModem implementations include none

(A) XMODEM

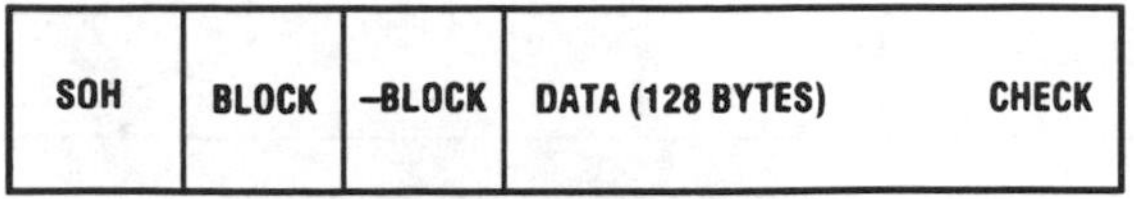

(ALL FIELDS ARE 8-BIT BINARY)

SOH = ASCII CONTROL-A (SOH, START OF HEADER)
BLOCK = 8-BIT BINARY 'BLOCK' (PACKET) NUMBER, 1-127 (RECYCLES)
–BLOCK = 255 MINUS THE BLOCK NUMBER (1's COMPLEMENT OF BLOCK NUMBER)
DATA = EXACTLY 128 BYTES OF UNENCODED 8-BIT DATA (A CP/M DISK BLOCK)
CHECK = AN 8-BIT BINARY CHECKSUM

(B) KERMIT

START	LEN	SEQ	TYPE	DATA. . .	CHECK	<CR>

EACH KERMIT PACKET FIELD EXCEPT DATA IS A SINGLE CHARACTER
EACH FIELD EXCEPT START IS COMPOSED ONLY OF PRINTABLE ASCII CHARACTERS. THE PACKET IS NORMALLY TERMINATED BY A CARRIAGE RETURN.

START = CONTROL-A (SOH), BUT IT CAN BE REDEFINED
LEN = PACKET LENGTH, 0-94, ENCODED AS A PRINTABLE ASCII CHARACTER
SEQ = PACKET SEQUENCE NUMBER, 0-63 (RECYCLES), PRINTABLE
TYPE = PACKET TYPE, S F D Z B Y N, ETC.
DATA = FILE NAME, FILE DATA, ETC., DEPENDING ON TYPE, PRINTABLE ASCII
CHECK = 8-BIT CHECKSUM, FOLDED INTO 6 BITS AS A PRINTABLE CHARACTER

Figure 12.7.1 Comparing Kermit and XModem packets.

of the Modem7, YModem, or ZModem options, but often do include support for cyclic redundancy checks. Thus, they can transfer only a single file at a time and only through transparent eight-bit communications channels. Commercial Kermit implementations vary from the bare-bones to the very advanced, but all versions can transfer text files through seven-bit links, and they can handle multiple files in a single operation. It is not always apparent from vendor literature exactly which options are supported. Buyers who are interested in these features should call the company and confirm whether the package includes them. After all, one of the advantages of commercial offerings is telephone support from the vendor.

12.8 More Asynchronous Protocols

XModem and Kermit are not the only asynchronous communications protocols available. Others include UUCP, Blast, MNP, X.PC, Poly-Xfr, DX, Compuserve, FAST, and DART. Most of these protocols are proprietary, and they are found primarily in commercial packages. They often include advanced capabilities, such as checkpoint/restart, bidirectional file transfer, and sliding windows.

But all proprietary protocols have these drawbacks: they must be purchased in conjunction with commercial packages, and if no package is available for a certain computer, then another protocol and set of packages must be found. Of the commercial packages, Blast probably comes closest to Kermit in covering a wide variety

TABLE 12.7.1 Sample worksheet

CONFIGURATION

Make a d model of your computer: ______
Operating system and version: ______
Operating environment: ______
Memory: ______ (K) Floppy drives: ______ Hard–Disk Capacity: ______ (M)
Communications interfaces: ______
Modem make and model: ______ [ ] Internal [ ] External
Name of communications package: ______
Communications package vendor: ______ Phone: ______
Package memory size: ______ (K) Package disk occupancy: ______
Before proceeding, be sure that the communications package is compatible with your computer's configuration!

COST

(a) What is the unit cost of the package? $ ______
(b) Is source code included, so that you can make changes and fix bugs? Is there an additional charge for source code? Cost of source code, if you want it: $ ______
(c) Is copying allowed? If so, go directly to (f).
(d) How many PCs will you need it for? ____ Is there a volume discount? If so, enter discounted cost: $ ______
(e) If a site license is available, what does it cost? $ ______
(f) Enter best total price for PC versions $ ______
(g) Do you also need minicomputer or mainframe versions? If so, enter total cost for mini or mainframe versions $ ______
(h) Total cost to your organization $ ______

DOCUMENTATION, TRAINING, AND SUPPORT

Is the manual . . .
[ ] thick and unmanageable?
[ ] thin and cryptic?
[ ] just right?

How important is the manual?
[ ] Must be consulted frequently
[ ] Occasional lookups required
[ ] Does the manual have a good index and table of contents?
[ ] Is training available?
[ ] Is training necessary?
[ ] Is telephone support available and included in the package price?

WHAT IS YOUR PRIMARY USE FOR THE PACKAGE?

[ ] Long interactive remote sessions. Communications parameter settings, terminal emulation, and key definition are the most important features.
[ ] Infrequent remote sessions mainly for the purpose of data transfer. Concentrate on the user interface, script language, and file transfer protocol.

USER INTERFACE

[ ] Is help available at all times?
[ ] Does the user interface favor the novice user? (Menus at all times)
[ ] Does it favor the expert user? (No menus)
[ ] Is the package equally convenient for both novice and expert? (Menu on demand)
[ ] Can canned procedures be set up for unskilled users? (Scripts, command files)
[ ] Can local operating system functions be accessed without leaving the package?
[ ] Can the package be used by the disabled?

COMMUNICATIONS PARAMETER SETTINGS (always important)

Bits/Second: 0, 110, 300, 1200, 2400, 4800, 9600, 19200, etc.
Maximum: ______

DUPLEX

[ ] Full (e.g. for DEC minis)
[ ] Half (e.g. for IBM mainframes)

ECHO

[ ] Remote (e.g. for DEC minis)
[ ] Local (e.g. for IBM mainframe linemode connections)

DATA BITS

[ ] 5 (Baudot)
[ ] 7 (ASCII)
[ ] 8 (national characters)

STOP BITS

[ ] 1 (for most connections)
[ ] 1.5 (rarely used)
[ ] 2 (used only for 110 bits per second or less)

PARITY SELECTIONS

[ ] None (all bits used for data)
[ ] Even (required by some mainframes, front ends, public networks, etc.
[ ] Odd (ditto)
[ ] Mark (ditto)
[ ] Space (rarely used, but sometimes handy)

CHARACTER SET SELECTION

[ ] 5-bit Baudot (used in Telecommunication Devices for the Deaf)
[ ] 7-bit US ASCII (most common in English-speaking countries)
[ ] 7-bit "national ASCII" (Norwegian, German, etc.)
[ ] 8-bit "extended ASCII" (e.g. use of IBM PC 8-bit character set)
[ ] Support for international standard non-Roman character sets
[ ] User-definable or downloadable character sets

FLOW CONTROL SELECTION

[ ] X-on/X-off (e.g. with DEC computers)
[ ] ENQ/ACK (e.g. with Hewlett-Packard computers)
[ ] RTS/CTS (for half-duplex modems)
[ ] Half-duplex line turnaround handshake (e.g. with IBM mainframes)
[ ] Other: ______________________
[ ] None (can flow control be turned off?)

DEBUGGING

[ ] Special display of all received and transmitted characters
[ ] Logging of all received and transmitted characters
[ ] Can you collect communications settings into recallable configurations

---

CONNECTION ESTABLISHMENT

Support for EIA-232 asynchronous modem signals (RTS, CTS, DSR, CD, DTR, RI):

[ ] Does the package monitor carrier detect (CD) and data-set-ready (DSR) from the modem?
[[Does the package assert data-terminal-ready (DTR)?
[ ] Can the package drop DTR to hang up the phone?
[ ] Does the package respond to ring indicator (RI) so that it can be called from outside?
[ ] If you have a half-duplex modem, does the package support RTS/CTS?
[ ] Does your PC have an internal modem?
[ ] Does the package support this internal modem?

DIALER CONTROL:

[ ] Does your modem provide automatic dialing?
[ ] What dialing language is used by your modem?

____

[ ] Does the package support automatic dialing?
[ ] Does the package support your modem's dialing language?
[ ] Does the package provide a phone directory?
[ ] Can the package operate over direct connections, without modems? That is, can it be told to ignore CD and DSR? (If not, you will need the "fakeout" (minimal) null-modem cable from Figure 1).

Script language for automatic login, unattended operation:

[ ] Access to all necessary package commands from script language.
[ ] Conditional execution/termination of script commands.
[ ] Fancy script programming features (variables, labels, goto's etc.)
[ ] Unattended operation (e.g. late at night, when phone rates are low).
[ ] Can the program be suspended and resumed without dropping the connection?

TERMINAL EMULATION:

What terminal(s) does the package emulate? __________

[ ] Is the maximum speed for full-duplex terminal emulation sufficient for your needs?
[ ] Does the package emulate a terminal that is supported by the computers you wish to communicate with?
[ ] Is the terminal emulated fully enough for use with all desired software applications on these computers?
[ ] Is any special hardware (like a 132-column board) required in the PC?
[ ] Does the package support fore- and background colors? (Do you need them?)
[ ] If a graphics terminal is emulated, does your application support it?
[ ] Screen rollback (view screens that have scrolled away)
[ ] Screen dump (save current or previous screens in PC files)
[ ] Printer control (copy displayed characters to printer, print whole screen)
[ ] Print or save text screens in alternate character sets
[ ] Print or save graphics screens
[ ] Function keys
[ ] Key redefinition
[ ] Keystroke macros
[ ] Translation of displayed characters, alternate character sets

---

FILE TRANSFER PROTOCOLS

[ ] ASCII (this is not an error-correcting protocol)
[ ] X-on/X-off (this is not an error-correcting protocol)
[ ] XModem [ ] YModem [ ] ZModem
[ ] Kermit
[ ] Proprietary (Blast, MNP, etc): ______________
[ ] Other: ______________________
[ ] Do the systems you're communicating with support the same protocol(s)?
[ ] Does the package transfer both text and binary files?
[ ] Do text files arrive on the target computer in useful form?

XMODEM OPTIONAL FEATURES

[ ] Modem 7-style transfer of multiple files
[ ] XModem-CRC for more reliable error checking
[ ] XModem 1-K packets for increased efficiency (half duplex)
[ ] YModem filename transmission
[ ] Checkpoint/restart (Zmodem)
[ ] WModem continuous transmission (full duplex)
[ ] Do the computers you wish to communicate with support the same XModem options?

KERMIT OPTIONAL FEATURES

[ ] 8-bit data through 7-bit links (e.g. links with parity)
[ ] Repeated character compression for improved efficiency
[ ] 12-bit checksum, 16-bit CRC, for more reliable error checking
[ ] File transfer interruption
[ ] Long packets (up to 9K) for improved efficiency (half duplex)
[ ] Sliding windows for improved efficiency (full duplex)
[ ] Transmission of file attributes
[ ] Server operation
[ ] Remote host commands and file management

of hardware, and it exceeds Kermit in many design and performance areas. The drawback is the cost: $250 for the PC version; $450 for a PDP-11 version; and more for larger minicomputers or mainframes.

Shopping for communications software can be a full-time job. The checklist in Table 12.7.1 is designed to help during the evaluation process. The first step is deciding what communications features are needed; then go out and find them by checking vendor literature or calling the company.

Questions

12.1 What is the difference between terminal emulation and data transfer?

12.2 What are some of the more common types of user interfaces by which a human interacts with a computer program?

12.3 What are two common methods used to implement flow control?

12.4 Discuss some of the key differences between internal and external modems that govern their selection for a particular personal computer?

12.5 Discuss how you could use a script language to automate a manual operation.

12.6 Why is the Kermit communications protocol in general more efficient than the XModem?

12.7 What are the advantages of a protocol dynamically altering its packet length? Discuss the criteria used by the ZModem protocol to alter its packet length.

12.8 What are five questions you should answer prior to selecting communications software for file transfer operations?

Chapter

13

Methods of Data Concentration

13.1 Introduction

Today, the network analyst can select from over 10 types of devices to concentrate many low-to-medium-speed data streams into one or more high-speed aggregate data streams for retransmission to a host computer. By combining a number of data streams from a geographical area into one high-speed link, you might be able to reduce either the number of long leased circuits or the expense of long-distance calls on the public switched network. If the cost of the special equipment required for data concentration is less than the savings that can be realized through a reduction in circuit transmission costs, then, from a cost standpoint, the use of such equipment is justified.

In addition to economics, performance parameters must be examined. In many uses, data concentration equipment can reduce the number of line errors. This is because the concentration equipment might use error-checking features. Here, data from a number of terminals is grouped into blocks of data for transmission, and a block check character is computed and appended to the block. At the receiving site, another device checks received data blocks by recomputing its block check character and comparing it with the transmitted block check character. If the check characters do not match, the receiving concentration equipment will request a retransmission (NAK) of the data block. This sequence is invisible to the terminal operator, who will not be aware that the long-distance circuit was experiencing line hits, and the same blocks of data were being transmitted more than once. If, instead, the terminal operator had a direct (i.e., no error control) long-distance transmission patch to the computer, these error conditions would become visible. This is especially damaging when using asynchronous terminals whose printed data would appear garbled because such devices have no provision for error checking and automatic retransmission before printing.

13.2 Multiplexers

One of the most common types of concentration devices is the multiplexer. Data communications users usually multiplex their high-capacity traffic on leased voice-grade lines between major data-accumulation centers so that they can transmit large amounts of data at reasonable costs. By multiplexing, one high-speed link can carry the same amount of traffic that several low-speed ones could handle without it.

By taking advantage of the rate structures for the various types of leased lines available from common carriers, the user can lease one high-speed link for much less than a number of low-speed lines would cost. More than likely, the savings realized from multiplexing traffic into one high-speed line would pay for the multiplexing equipment.

If a preliminary analysis indicates that multiplexing can save money, consideration must then be given to selection of either frequency-division multiplexing (FDM) or time-division multiplexing (TDM) equipment.

FDM represents the earliest type of equipment developed for permitting the shared utilization of a common line facility. Because of the higher efficiency obtained by the use of TDMs, by the mid-1980s they had replaced FDMs in almost all applications, except for certain multipoint operations for which FDM technology is still well suited.

The development of time division multiplexing forms the basis for several types of multiplexers. These devices include statistical multiplexers, T1 multiplexers, and fiber optic multiplexers, each of which will be covered in this chapter.

As the name implies, FDM achieves its concentration by dividing the telephone-line frequency band into smaller frequency segments. TDM, on the other hand, allocates time segments to the various traffic channels. Generally, FDM is used when one line must service many terminals in a local area through a multidrop arrangement. In contrast, TDM is usually used in long-distance point-to-point configurations, typically between major cities.

FDM

In FDM, a voice-grade line with a certain bandwidth, typically 3000 Hz, is split into narrower channel segments, often called *derived channels* or *data bands*. Each channel, tuned to a specific frequency, has a transmitter at one end and a receiver at the other end. The width of each frequency band determines the data-rate capacity of the subchannel. For example, the band can be made narrow enough to pass only 75-bit/s data streams or wide enough for 150 or 300 bit/s, if required by the terminal speeds. "Guard" bands separate the data bands to prevent data on one channel from interfering with data in an adjacent channel.

FDM normally provides full-duplex operation on a four-wire circuit. Each channel set (of equipment) has a transmit and a receive station. All transmit tones go out on one pair of wires, and all receive tones come back on a second pair. However, FDM can also operate full duplex on a two-wire circuit. For example, with 24 channels available on the line, one channel set is tuned for line channel 1 to transmit, and channel 13 to receive; another for channel 2 to transmit, and channel 14 to receive; and so on. Here, the number of data channels is halved, but this technique saves the difference in costs between four-wire and two-wire lines in a network servicing a small number of full-duplex terminals.

TDM

Time-division multiplexing's time segments utilize the full bandwidth of the line. A typical TDM network configuration is shown in Figure 13.2.1. Note the optional additional remote TDM. Several operator terminals are connected to the ports of the remote TDM, and the local TDM is connected to the host computer's ports. Alternatively, if the front-end processor is programmed to act as the local TDM, then just the remote TDM is required.

The high-speed-trunk line's (Figure 13.2.1) transmission rate equals the maximum aggregate of data rates available to all the terminal (or host) lines. For example, if the trunk capacity is 9.6 kbit/s, then no more than four 2.4 kbit/s terminal lines can be connected.

Each of the TDM's time segments is assigned a character (or bit) from each low-speed port. At the trunk's other end, the process is reversed. The bit stream is demultiplexed and the characters (or bits) are assembled into the messages as they appeared at the low-speed ports of the sending (remote or local) TDM.

The character-interleaved bit stream contains the bits in each character in a continuous data stream, the multiplexer having assembled a full character into one frame for transmission. In the bit-interleaved bit stream, the multiplexer takes one bit from each source device, in turn, and combines them into a frame for transmission. Because character interleaving requires the multiplexer to assemble bits into a character before multiplexing, the buffer area is larger than for bit interleaving.

The conventional TDM is a hard-wired (nonprogrammable) device that allocates a fixed portion of the high-speed trunk's bandwidth to each of its low-speed channels. Because terminal devices are rarely in operation full time, the fixed-allocation scheme uses the trunk's capacity inefficiently. When a terminal has nothing to send, its allocated time segments—carrying only null or fill characters—are still part of the data stream.

Configuring the network

A more efficient device—the intelligent or statistical TDM (STDM)—has just about obsoleted the conventional TDM.

A local STDM communicates with a remote one by using a higher-level full-duplex protocol, such as high-level data link control (HDLC), synchronous data link con-

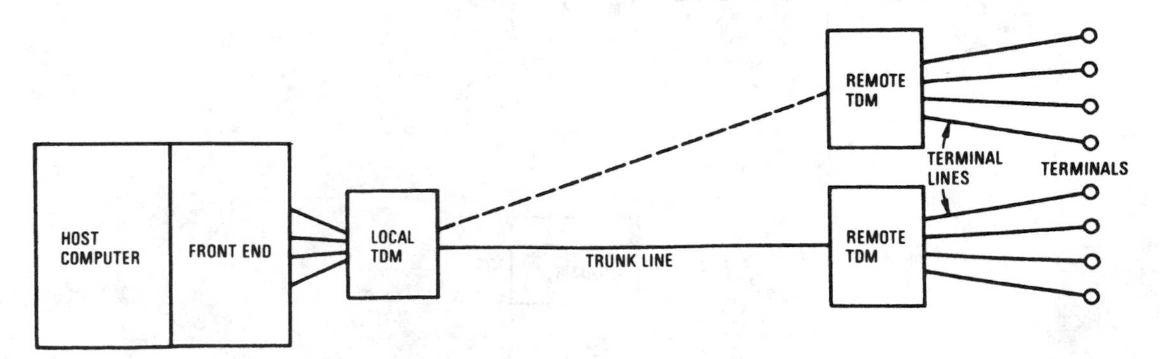

Figure 13.2.1 Typical multiplexer configuration.

trol (SDLC), or Digital Equipment's digital data communications message protocol (DDCMP)—or variations on one or another of them. The line is typically a 9.6 or 19.2 kbit/s trunk. The possible interface arrangements include:

- Two or more hosts to one local STDM
- One or more local STDMs to two or more remote STDMs (Figure 13.2.1)
- Use of the STDM as a network node, with connections to high-speed remote job entry (RJE) equipment, in addition to conventional user terminals and to other multiplexers

If a conventional remote TDM is connected to a 9600 bit/s trunk and services user terminals rated at 1200 bit/s, the configuration can support 9600 divided by 1200, or eight terminals. But it can be determined statistically that, at any one time, a certain number of terminals, for example 4, are going to be idle. As pointed out earlier, the conventional multiplexer continues to allot high-speed line segments to these idle terminals, sending nulls on the trunk to the local TDM, then on to the host for discard.

Statistical multiplexing eliminates the nulls by allocating trunk line segments only to terminals actually transmitting data. The TDM is driven by time; the STDM is driven by data. Because it was shown in the example that, statistically, four terminals are idle at any one time, at least four terminals can be added at the remote STDM without exceeding the capacity of the trunk.

STDM

Statistical multiplexers can support more data sources than traditional TDMs because they transmit data only when data sources are active. Figure 13.2.2 compares the servicing of terminals by a TDM and an STDM. In the top part of the exhibit, eight 1200-bit-per-second terminals serviced by a traditional TDM require a composite line speed of $8 \times 1200 = 9600$ bit/s to operate.

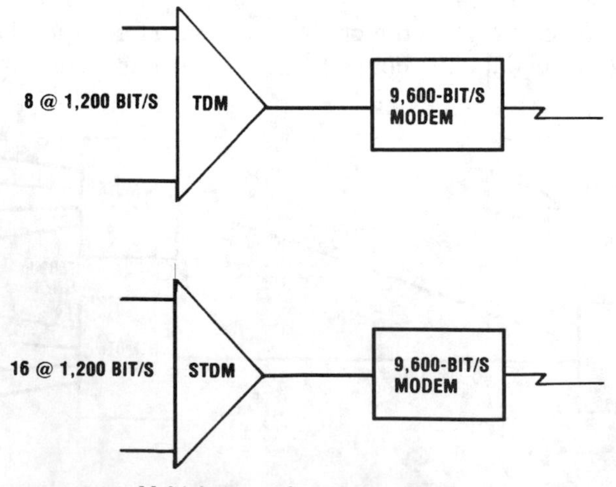

Figure 13.2.2 Multiplexer performance comparison.

In addition to taking advantage of the inactivity of data sources, a statistical multiplexer strips the start, stop, and parity bits from asynchronous data when it constructs the synchronous message frame.

The service ratio of a statistical multiplexer denotes its capability, that is, its overall performance level in comparison to a TDM. The lower part of Figure 13.2.2 shows an STDM with a service ratio of 2 to 1. Over a period of time, this multiplexer has twice the capability of a TDM because it can support twice the number of data sources on the average. If connected to a 9600 bit/s modem, the STDM typically can support sixteen 1200 bit/s terminals, because some will be inactive at any given time. However, if 10 terminal users simultaneously listed a long file, data would flow into the multiplexer at 12,000 bit/s (10×1200) and leave the multiplexer at a lower data rate (if the STDM is connected to a 9600 bit/s modem). This would cause data to fill the STDM's buffer, and if not checked, would cause data to be lost when the buffer eventually overflows.

Flow control

Multiplexer vendors have incorporated several flow control procedures into their products to prevent STDM buffers from overflowing. The most common procedures result in the STDM transmitting the XOFF character or dropping the clear-to-send (CTS) signal to inhibit data from flowing into the multiplexer. When the multiplexer's buffer is emptied to a predefined level, the STDM transmits an XON character or raises the CTS signal to enable transmission to the multiplexer to resume.

The transmission of the XOFF-XON sequence and the raising and lowering of the CTS signal applies only to asynchronous data sources. To control the flow of synchronous data, the STDM must lower and raise the clocking signal, a more difficult operation offered by only a few vendors.

To make transmission between STDMs even more efficient, data compression techniques allow information to be sent using fewer bits than when the data is not compressed, or "normal."

The simplest form of data compression calls for the remote STDM to scan the high-speed data stream, at some time prior to transmission, and pick out repeated characters, such as a string of periods or space characters. The STDM counts the number of times a character is repeated, then it sends the character once, followed by a short combination of bits that represents the number of times that the character is to be sent. The receiving STDM then restores the original data stream by performing data expansion.

A conventional multiplexer can accept and tag a received character as errored. But no provision is made to correct that error. The STDM, however, has the ability to correct errors by the automatic repeat request (ARQ) feature, using the cyclic redundancy check (CRC) code to detect errors.

In ARQ, the sending STDM stores each transmitted data frame while awaiting a positive or negative acknowledgment (ACK or NAK) from its correspondent STDM. On receipt of an ACK, the sending STDM discards the stored frame, and continues transmission. If the receiving STDM recognizes an error in the CRC code of a received frame, it sends back a NAK to the sending unit. The sending STDM then retransmits the errored frame and all following frames sent while awaiting the ACK/NAK. This process is repeated until the errored frame is accepted as correct, or until a frame transmission counter reaches a preset limit, and causes an alarm to go off.

Traffic buffering

As mentioned earlier, unacknowledged transmissions are stored until no longer needed for retransmission. The storage takes place in a buffer that allows dynamic (temporary) retention of a certain amount of transmitted traffic. The prospective user should check the buffer storage capacity of whichever STDM is being considered for his network. Typically, from 10 to 20 seconds of a user's data may be retained before buffer overflow is threatened. Overflow is likely to occur only during a trunk line outage, or during a noisy trunk line condition causing many repeated transmissions under ARQ, and warning of the overflow's imminence may or may not be provided to the terminals, depending on the sophistication of the circuitry.

One additional buffer consideration that you should be aware of is the relative "weight" assigned to each terminal port. Even when terminals of identical bit rates are connected to a remote STDM, it might be possible to program their ports so that one terminal can be treated at a higher priority than the others. Priority is accomplished by emptying the preferred terminal's port buffer at a faster rate than the others. In fact, a series of priorities may be assigned to establish different buffer-emptying rates for each port. Besides treating certain terminals preferentially, the weighting parameter can also be used to compensate for differing traffic volumes—in addition to what the statistical multiplexing feature accomplishes in this area.

Automatic rate detection

To make most efficient use of its bandwidth allocation characteristics, the STDM has a deluxe attribute of the conventional TDM: automatic data rate detection—also known as *autobaud* or *autospeed*. Under this feature, a terminal user seeking access to a host computer through a time-division multiplexer must transmit a short group of preliminary characters (usually up to three). The remote TDM analyzes them and determines the terminal's transmission rate. The bandwidth is then assigned to the terminal. Also, when the port on the host (or on its front-end processor, if one is used) is of the autospeed type, the local mux will supply the preliminary characters to the host, which will do its own rate recognition and will accept regular traffic.

Autospeed does not really apply to synchronous transmission because the applicable modems would require a range of data rates that would make the modem equipment economically impractical. And there does not appear to be demand for such an application. However, a minimal form of synchronous autospeed does exist. If a host port can accept, for example, either 2400 bit/s or 4800 bit/s, then the remote STDM has that much more latitude in assigning terminals to ports. For example, if all fixed-rate host ports of 2400 bit/s and 4800 bit/s are in use, and one multi-rate port for both speeds is still available, then a terminal of either speed can gain access to the host.

A corollary of autospeed is the automatic detection of code or character format. This feature enables the remote STDM to decipher the preliminary character or characters transmitted to detect the use of either the ASCII or the binary-coded decimal (BCD) code, as well as for data rate as described earlier.

Echoplex

An operator at a remote terminal usually creates a copy of what is being sent, either on the terminal screen or on an associated printer. This procedure provides no assurance to the operator of the data being received in that form.

Echoplex is a technique that attempts to provide such assurance. With the terminal's line connected directly to a host computer (or front-end processor, as in Figure 13.2.3A), the data as received from the terminal is returned to its printer or screen. This method enables the operator to verify that the transmission was received correctly.

Some STDMs offer the option of relieving the host of the processing time required by the echoplex function. If the local (host-end) STDM returns the received data (Figure 13.2.3B), then reception is verified for both the terminal and high-speed lines. If the remote STDM does the echoplexing (Figure 13.2.3C), verification is obtained for just the terminal line.

For low-speed terminals, such as those operating up to 300 bit/s, the delay between transmission and the echoplexed reception might be excessive—causing op-

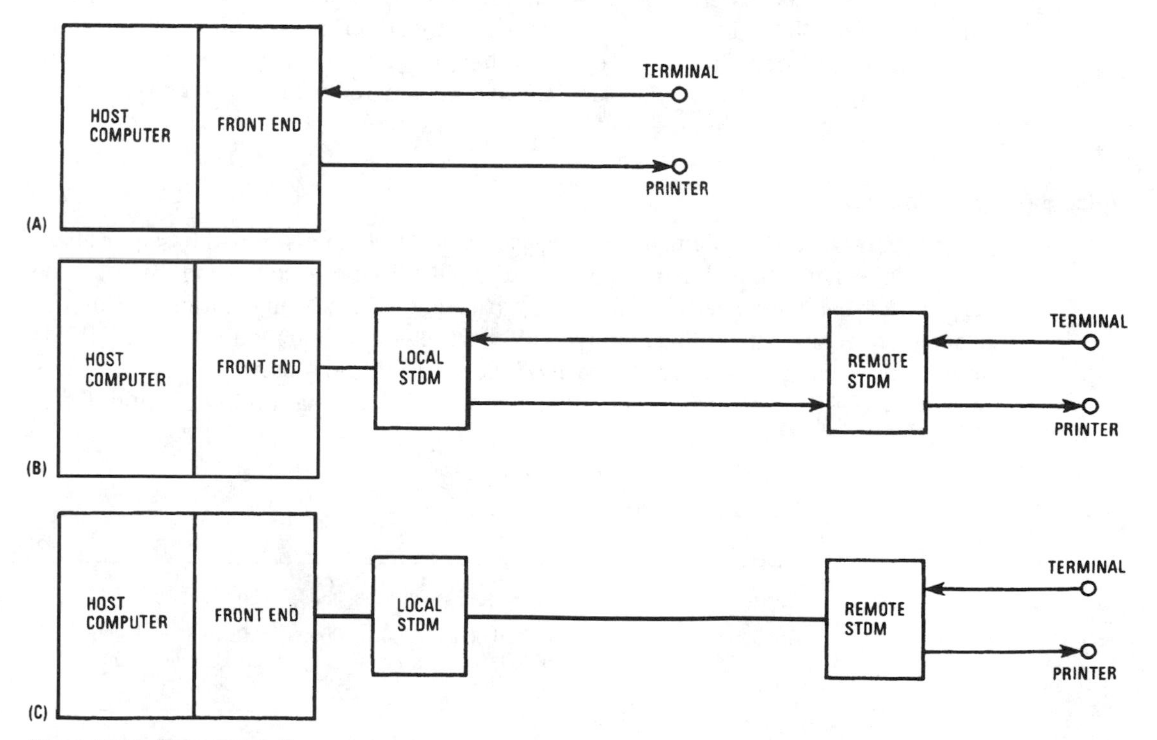

Figure 13.2.3 Echoplex choices.

erator disorientation. In such a situation, full verification can be sacrificed and echoplexing from the remote STDM accepted ("pseudo-echoplex").

When a network problem occurs, the STDM can also become a diagnostic tool. It assists in localizing and diagnosing the trouble with both its own indicators and the associated printouts. The STDM can initiate remote and local loopback testing of trunk and terminal lines under off-line or dynamic conditions. At the local site, the entire network can be monitored and abnormal conditions logged, with the times of such conditions noted.

Diagnostic routines are usually initiated by an operator at the local STDM. Some vendors offer a "hot" backup unit ready to take over for the primary unit when a failure is detected. The backup unit can be given the task of running noninterfering diagnostics automatically, until it is needed as a replacement.

Some STDM vendors also provide the modems for the terminal and trunk lines. These modems can occupy circuit board slots in the equipment, or could be external depending on the vendor's design and practice. The buyer/user should include modem costs when comparing STDM prices.

T1 multiplexers

Modern T1 multiplexers are microprocessor-based time-division multiplexers designed to combine data, voice, and video from various sources onto a single communications circuit that operates at 1.544 Mbit/s in North American and 2.048 Mbit/s in Europe. Figure 13.2.4 lists the typical input channel data rates accepted by most T1 multiplexers. Readers should note that although digitized voice is treated as synchronous input, its digitized data rate can vary considerably based on the type of optional voice digitization modules offered by the T1 multiplexer vendor.

Voice digitization modules

In a conventional PCM digitization process, 24 voice channels of 64 kbit/s per channel can be carried on a North American T1 facility. To increase the number of voice signals that can be carried on this facility, a variety of voice digitization techniques has been developed including adaptive differential pulse-code modulation (ADPCM) and continuous-variable-slope delta modulation (CVSDM), as well as several less widely used schemes, such as time-assigned speed interpolation (TASI) and differential PCM (DPCM).

Type	Data rates (bit/s)
Asynchronous	110; 300; 600; 1200; 1800; 2400; 3600; 4800; 7200; 9600; 19,200
Synchronous	2400; 4800; 7200; 9600; 14,400; 16,000; 19,200; 32,000; 38,400; 40,800; 48,000; 50,000; 56,000; 64,000; 112,000; 115,200; 128,000; 230,400; 256,000; 460,800; 700,000; 756,000
Voice	9600; 16,000; 32,000; 48,000; 64,000

Figure 13.2.4 Typical T1 multiplexer channel rates.

Adaptive differential PCM

The ADPCM technique uses a transcoder to reduce the eight-bit PCM samples into four-bit words while retaining the 8000-sample-per-second PCM sampling rate. This technique results in a voice digitization rate of 32 kbit/s, which is one-half of the PCM voice digitization data rate.

Under the ADPCM technique, the use of four-bit words permits only 15 quantizing levels; however, instead of representing the height of the analog signal, each word contains the information required to reconstruct the signal. This information is obtained by circuitry in the transcoder that adaptively predicts the value of the next signal, based on the signal level of the previous sample. This technique is known as *adaptive prediction*, and its accuracy is based on the fact that the human voice does not change significantly from one sampling interval to the next.

Continuous-variable-slope delta modulation

The CVSDM digitization technique compares the analog input voltage with a reference voltage. When the input voltage is greater than the reference voltage a binary "1" is encoded, and a binary "0" is encoded when the input voltage is less than the reference level. This permits a one-bit word to represent each sample.

The incoming bit stream at the receiver represents changes to the reference voltage and is used to reconstruct the original analog signal. Each 1 bit causes the receiver to add height to the reconstructed analog signal, and each 0 bit causes the receiver to decrease the analog signal in set increments. If the reconstructed signal were plotted, the incremental increases and decreases in the height of the signal would result in a series of changing slopes, hence the name of this technique: *continuous-variable-slope delta modulation*.

Because only changes in the slope or steepness of the analog signal are transmitted, a sampling rate higher than the PCM sampling rate is required to recognize rapidly changing signals. CVSDM usually samples the analog input at either 16,000 or 32,000 times per second and, using a one-bit word for each sample, transmits data at a rate of either 16 kbit/s or 32 kbit/s.

An interface to an organization's private branch exchange (PBX) in most T1 applications provides one or more tie lines through the use of two T1 multiplexers and a T1 carrier facility. Figure 13.2.5 illustrates a typical T1 multiplexer application in which voice, video, and data are combined in one T1 carrier facility. In this example, the PCM digitization channel modules used in the T1 multiplexer resulted in 10 voice channels on the PBX interface using 640 kbit/s of the available 1.544 Mbit/s T1 operating rate.

This example also required the organization's conference room to be connected to a distant location for video conferencing. The 700-kbit/s input to the T1 multiplexer in Figure 13.2.5 represents the required data rate for a digitized video conferencing signal. Furthermore, the organization has two data centers and a PBX at each end to permit computer-to-computer transmission to occur at 128 kbit/s. Finally, the organization's 12 data terminals, each operating at 4.8 kbit/s at one site, required access to the computer located at the other end of the T1 link. If the T1 multiplexer channels operate at 64 kbit/s, the video conferencing will require 11 channels, the 12 data terminals will occupy one 64 kbit/s channel, and the 10 voice conversations will require ten 64 kbit/s

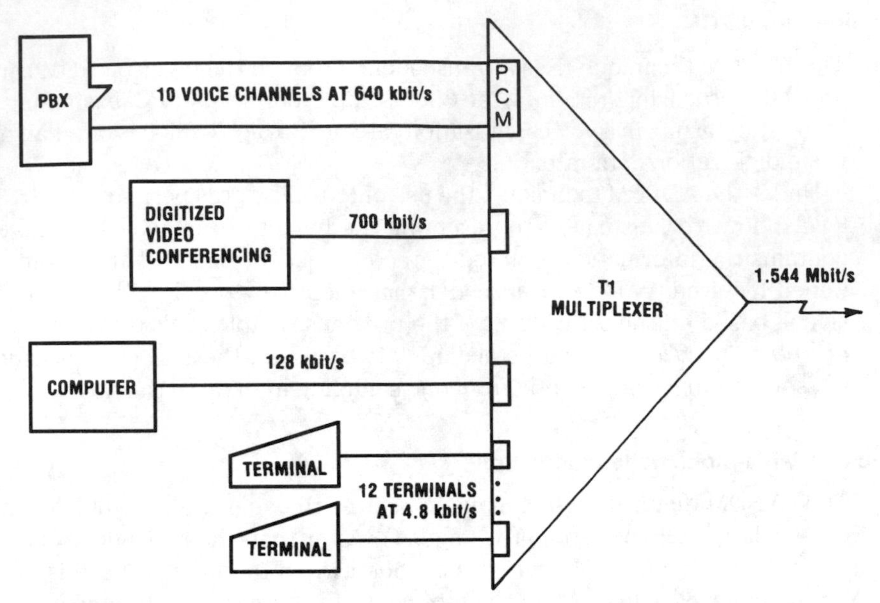

Figure 13.2.5 A typical T1 multiplexer application.

channels, leaving two 64 kbit/s channels for future requirements. Because the cost of 10 tie lines and a few leased lines to support the data terminal traffic usually equals the cost of a T1 carrier facility, the use of T1 multiplexers eliminates the bandwidth cost for video conferencing and wideband computer-to-computer transmission.

Fiber-optic multiplexer

The fiber-optic multiplexer is similar in functionality to a conventional time division multiplexer. It accepts the electrical input of several digital data sources and multiplexes them into one data stream for transmission over a single optical fiber. Included in most fiber-optic multiplexers is an optical modem that performs the electrical-to-optical and optical-to-electrical conversions necessary to transmit and receive data on a fiber cable.

The optical modem houses both an optical transmitter and an optical detector. The optical transmitter includes a laser or a light emitting diode, as well as appropriate control circuitry. This enables conversion of electrical pulses to light energy. The receiver in the optical modem consists of a photodetector and control circuitry. The photodetector, as its name implies, detects light energy, which the control circuits then convert into digital pulses.

The primary use of fiber-optic multiplexers is based on the large bandwidth and lack of shock hazard that optical cable provides.

Bandwidth

Optical fibers have wider bandwidths than metallic conductors. Because potential information capacity is directly proportional to transmission frequency, light transmitted over fiber cable affords very high data rates. Data rates of up to 10^{14} bits per second have been achieved on fiber-optic links, whereas telephone wire pairs are typically limited to 9.6 to 14.4 kbit/s. Fiber cable allows voice, video, and data transmission to be merged on one conductor. In addition, the wide bandwidth of optical fiber provides an opportunity for the multiplexing of many channels of lower speed, but still significantly higher data rates than are transmittable over telephone networks.

Electrical hazards

Use of light energy in fiber-optic networks minimizes the dangers of shocks and short-circuit conditions. The absence of sparks makes fiber-optic transmission particularly suitable in such potentially dangerous working environments as petrochemical operations, refineries, chemical plants, and grain elevators. A more practical benefit of optical fibers for most corporate networks is the complete electrical isolation they afford between the transmitter and receiver. This isolation eliminates the need for the common ground that is required for metallic conductors. In addition, because no electrical energy is transmitted over the fiber, most building codes permit optical fiber cable to be installed without being run through a conduit, which can result in a considerable savings.

Based on the bandwidth and lack of shock hazard afforded by the use of optical cable, fiber-optic multiplexers are commonly used by both communications carriers and end-user organizations. Communications carriers are routinely using fiber-optic multiplexers to place thousands of digitized voice conversations on a fiber-optic cable routed between carrier offices. End-user organizations normally use fiber-optic multiplexers to interconnect data centers located within a building. As a result of the high bandwidth of optical cable, it becomes possible to transmit millions of bits per second between data centers on a common cable, without requiring the use of a conduit. Because of the cost of copper cable, as well as the charges associated with installing a conduit, the use of an optical multiplexer can normally result in considerable economic savings.

13.3 Communications Processors

The communications processor (CP) is often not a single device, nor are its functions limited to one network location. Basically, a CP is used as 1. a front-end processor, 2. remote concentrator, and 3. message switch. The packetizing processor represents another CP use, but is considered a special form of the message switch communications processor.

Each of the three basic categories is shown in Figure 13.3.1, represented in somewhat simple network arrangements, and is explored later in more detail. As a front-end processor (FEP), the CP is located at its host computer. A prime function of the FEP is to relieve the host of network communications overhead.

When physically located at or near the distant terminal devices served, the CP is known as a *remote concentrator*. The connection back to the host can be direct (dashed line, asymmetrical) or through a similar CP acting as an FEP (solid line,

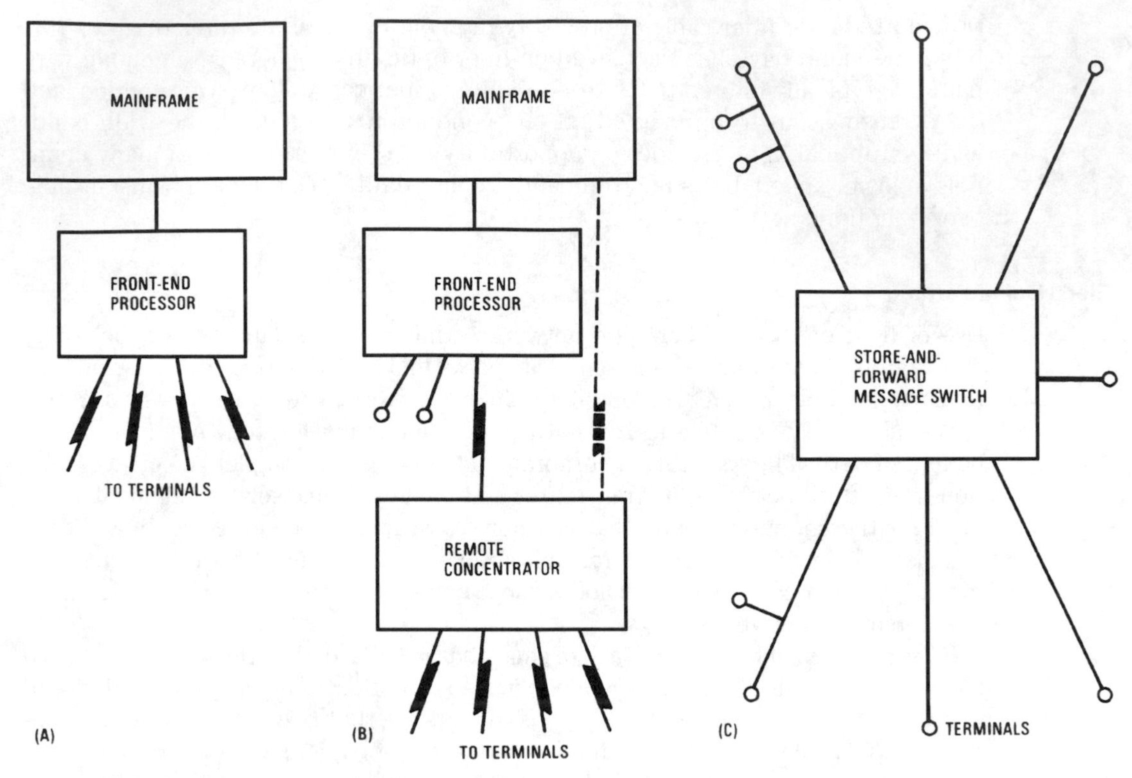

Figure 13.3.1 Processor variations.

symmetrical). (Many network applications, such as that of the remote concentrator, can be filled by statistical multiplexers.)

The prime purpose of locating the CP remotely is to reduce the cost of running multiple lines to distant terminals by running fewer to the remote CP. A combination of short runs between the terminals and the remote concentrator plus the CP connections back to the host is usually more cost effective than multiple long runs. In the symmetrical arrangement, when some terminals are located near the host, access to host applications can be made through the FEP, rather than through the remote concentrator, as shown in Figure 13.3.1B.

Another CP function is store and forward when used as a message switch. Elements of this function exist in the FEP and remote concentrator. But when switching is the primary function, the network can be represented more accurately by Figure 13.3.1C.

As supplied to the user, a communications processor will resemble a minicomputer or look like an entirely different unit. In the latter case, the equipment is supplied by an independent vendor who designs and builds his own processor. Many minicomputer vendors are also represented by their products in the CP field. Another group of companies increasingly represented are the specialized common carriers with their offerings of CPs originally developed for use on their own public packet networks.

Compatible

Some vendors aim almost exclusively at the IBM user market. Their front-end processors are plug-compatible replacements for IBM's 3705, 3725, and 3745 models) communications controllers, and are upgrades of IBM's hard-wired 270X transmission control units.

Where a user has a large network—one with several centralized host computers—using just one multihost front-end processor can be very convenient. Hardware costs are lower, and programming is less complex.

Even with one host, a front-end processor is effective in relieving a mainframe of dealing with each terminal desiring access to particular application programs. As indicated in Figure 13.3.2, the FEP responds to a terminal's application request and gains access to the proper host region. Shown are three typical IBM applications: timesharing option (TSO); information management system (IMS); and customer information control system (CICS).

Application programs and multihost support come together when each host is dedicated to a particular application. Then, a terminal is directed to its application

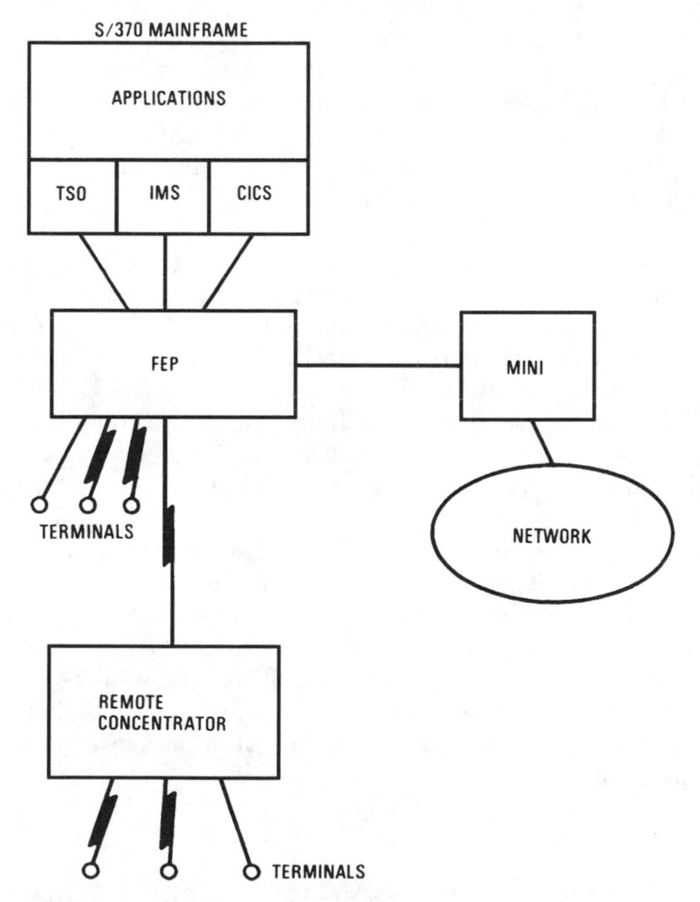

Figure 13.3.2 Accessing applications.

by the FEP switching the request to the proper host. If one host fails, then only that application is lost. Of course, to prevent loss of all applications if an FEP goes down, a back-up FEP should be included.

Networking

When a communications processor is used as a "pure" message switch (Figure 13.3.1C), no connection is needed to a host computer. Only when the network's terminals must also have access to a host's resources—such as particular applications—is switch-to-host connection required.

The CP's network functions—besides being a store-and-forwarding message switch—include buffering and queuing for load leveling. That is accomplished by pacing, otherwise known as *throttling*. The technique consists of the processor sensing an impending overflow situation at a receiving device, and either temporarily storing the data that would have caused an overflow, or stopping the sending device until the receiver is ready again.

A CP feature that exercises another form of traffic control is called *fast/slow poll*. Here, terminals are polled normally—fast—until a terminal does not respond or has no traffic to send. When such nonresponses occur, the CP automatically reduces the polling frequency of that terminal. When it again responds with traffic, the terminal is transferred back to a fast-poll mode. The algorithm for establishing polling frequencies and mode criteria is under user control. It is important to note that the operator of a remote terminal is not aware that he is dealing with an intermediate processor, and not directly with a host computer.

Another network function is logging and journaling. Such information is normally kept on magnetic tape or disk. Collecting and monitoring network statistics is yet another CP function. This data might include line and terminal usage, outages, numbers and types of messages, and repeat-request frequency. By keeping track of connect times, the CP can store—for access by the control console—accounting information, such as charge-back data.

Network maintenance is aided measurably by the CP. Software and hardware "bugs" can be detected, isolated, and analyzed, and their corrections assisted, by CP console action. Non-host-dependent diagnostic capabilities are featured in some communications processors.

One notable feature of the communications processor is the ability to deal with a variety of terminal types, speeds, and codes. Enabling each type to communicate with the others requires CP conversions, which are made possible by a combination of line interface devices and software. Standard speeds and codes are readily handled; nonstandard ones can be accepted, but usually require some reprogramming. Similarly, standard protocols—such as those for asynchronous and binary synchronous (BSC) communications—are accepted by the CP, as are, increasingly, the bit-oriented ones and X.25.

Gateway processor

As public networks proliferate, there is a growing need for a device on one network to communicate with devices on others. The need is becoming apparent with the various networks in different countries. But political considerations slow the implementation of some international data flow.

However, within the U.S., internetwork transmission is not difficult to implement. And with the proper software developed, and the guidance of standards, the CP can act as a gateway processor.

As a gateway processor, the CP provides access into the network of computational facilities it interconnects. Another term often used with gateway, but which has a different meaning, is *bridge*, which is a device used to provide a connection or link between two separate networks. Both gateways and bridges are most often associated with local area networks and are described more fully in the chapter devoted to that topic.

There are other situations when the CP, functioning as other than a switch, may operate without a host. For example, if the host computer shown in Figure 13.3.2 becomes disabled, the terminals—in the case of many processors—can still communicate with each other. Of course, there will be degraded service. But if the CP is so designed, it can queue application requests until the host is again operational and on line.

To relieve the host of as much of the network operational responsibilities as possible, the efficient CP manages such functions as message routing and network reconfiguration. Routing tables are maintained at the CP. When a table change or update is needed, some CPs require just a console entry for a simple change—such as a terminal identifier to correspond to a new location. But when the update is more complex, the new data is entered as part of an initial program load (IPL) or system generation.

Network reconfiguration data is entered similarly. This routing and configuration information is totally resident in the CP, with the host not involved.

One other network feature a communications processor might have is called *pass through*. As shown in Figure 13.3.2, if access is desired from a terminal to another network (mini-controlled in the figure), the FEP will recognize the request and pass the connection through to the other controlling hardware. This feature effectively widens the numbers of applications and terminals to which a CP-controlled device has access.

As a network node, the CP combines intelligence with the switching capabilities of hundreds of lines. Line speeds range typically up to 1.544 Mbit/s.

Interfacing with controllers

The typical hardware components included in a CP used in a concentration role are illustrated in Figure 13.3.3. The single-line controllers (SLCs) provide the necessary

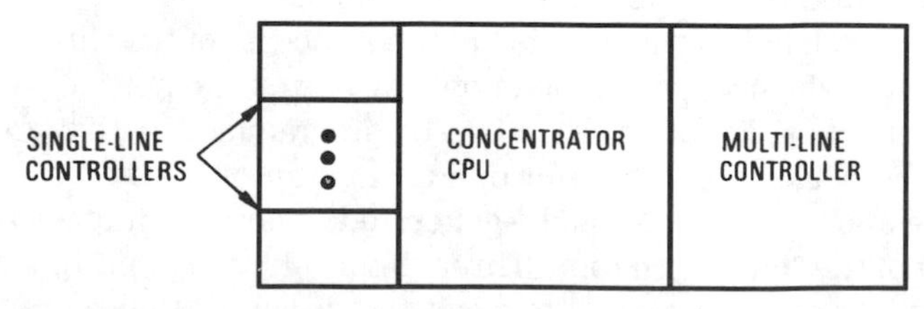

Figure 13.3.3 Concentrator hardware.

control and sensing signals that interface the concentrator to individual circuits. Although single-line controllers can be asynchronous or synchronous, the majority are the synchronous type. The preponderance of synchronous transmission is caused by the SLC's normally providing only one—or at most a few—high-speed transmission links from the concentrator to another concentrator or to a host computer (front-end processor).

Because the support of numerous lines would be expensive and would take up a lot of space if implemented with single-line controllers, most communications support for the concentrator-to-terminal links is implemented through the use of multiline controllers. Multiline controllers (MLCs) can be categorized by capacity (number and speed of lines supported), as well as by operation: hardware- or software-controlled.

Hardware-controlled multiline controllers place no burden on the concentrator's CPU, the hardware MLC requiring much less operating software than the programmed controller. On the other hand, programmed controllers have lower per line costs because hardware in the interfaces and the controller is held to a minimum, although a larger burden is placed on the processor. For a programmed controller, all sampling control, bit detection, and buffering is performed by the processor through software control. The amount of processing time required by the operational program is the main factor limiting the number of lines that can be connected to the concentrator via software-controlled MLCs.

To reduce the complexity of circuits in hardware multiline controllers, as well as to reduce software overhead of programmed controllers, incoming lines are often arranged in groups. These groupings are by bit rate, code level, and the number of stop bits for asynchronous terminal support. Figure 13.3.4 illustrates a typical grouping by channel for a multiline controller. This controller requires a minimum of four channels per group-all four channels of the same terminal class (same bit rate and code). Groups 1 and 3 are of the same class. The MLC might have any mixture of classes, until the number of groups multiplied by four equals the total number of channels supported by the controller (64 in the example).

Pure contention

In essence, a pure-contention concentrator is a port selector. In this function, the concentrator connects any of M input lines to any of N output lines as one of the N output lines becomes available. The M input lines are commonly called the *line side of the concentrator*, whereas the N output lines are referred to as the *port side-to-interface*, the ports of a front-end processor. The basic hardware components of a contention concentrator are illustrated in Figure 13.3.5.

Incoming data on each line of the line side of the device is routed through the concentrator's processor, which searches for a nonbusy line on the port side to transmit the data to. The determination of priorities can be programmed so that groups of incoming lines can be made to contend for one or a group of lines on the port side. When all ports are in use, messages can be generated to notify terminals attempting to access the host of the "busy" situation. Through the addition of peripheral storage devices, incoming jobs can be batched to await the disconnection of a user from a

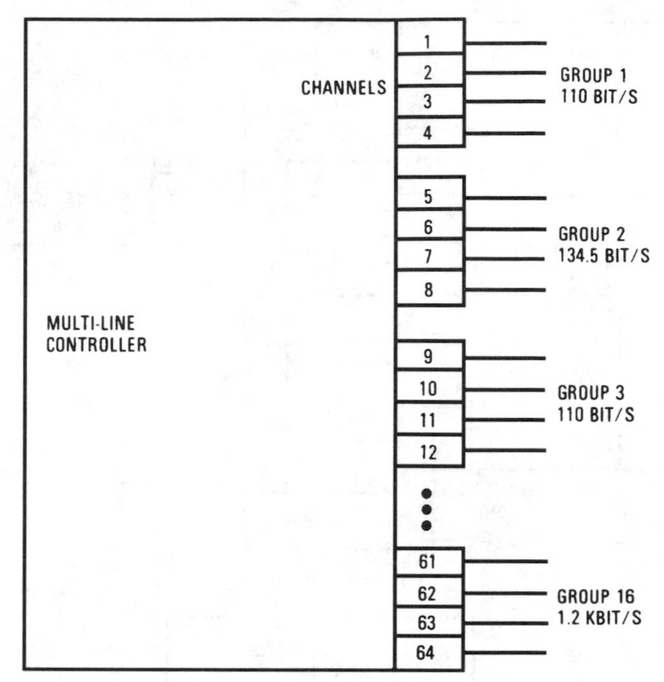

Figure 13.3.4 Groupings by channel.

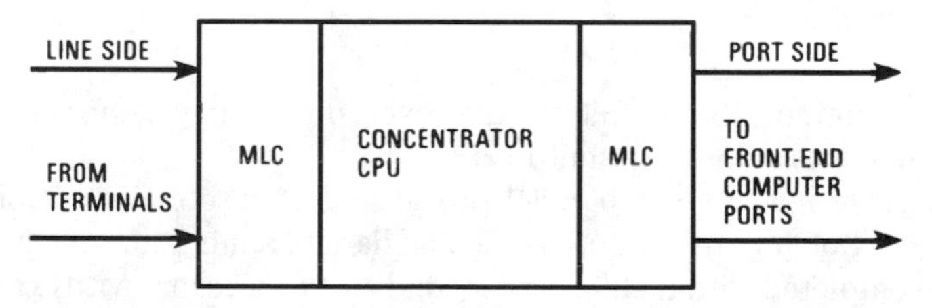

Figure 13.3.5 Contention concentrator.

host. Then connection to the newly available port side line is made to gain entry to the computational facility, and the stored job is transmitted.

If you need store-and-forward message concentration, emphasis should be placed on peripheral equipment, data transfer rates, and software appropriate for this particular application. In addition, if the application is critically time dependent, examining hardware reliability by itself might not suffice, and the user will most likely want to consider a redundancy configuration.

As shown in the redundant store-and-forward arrangement of Figure 13.3.6, both configurations are directly connected to each other by an intercomputer communications unit and share access to incoming and outgoing lines and peripherals via electronic switches. During operation, one configuration is considered the operational processor or master, and the other is the slave or standby, monitoring the master. Upon a hardware failure or power interrupt, the master signals the slave via the

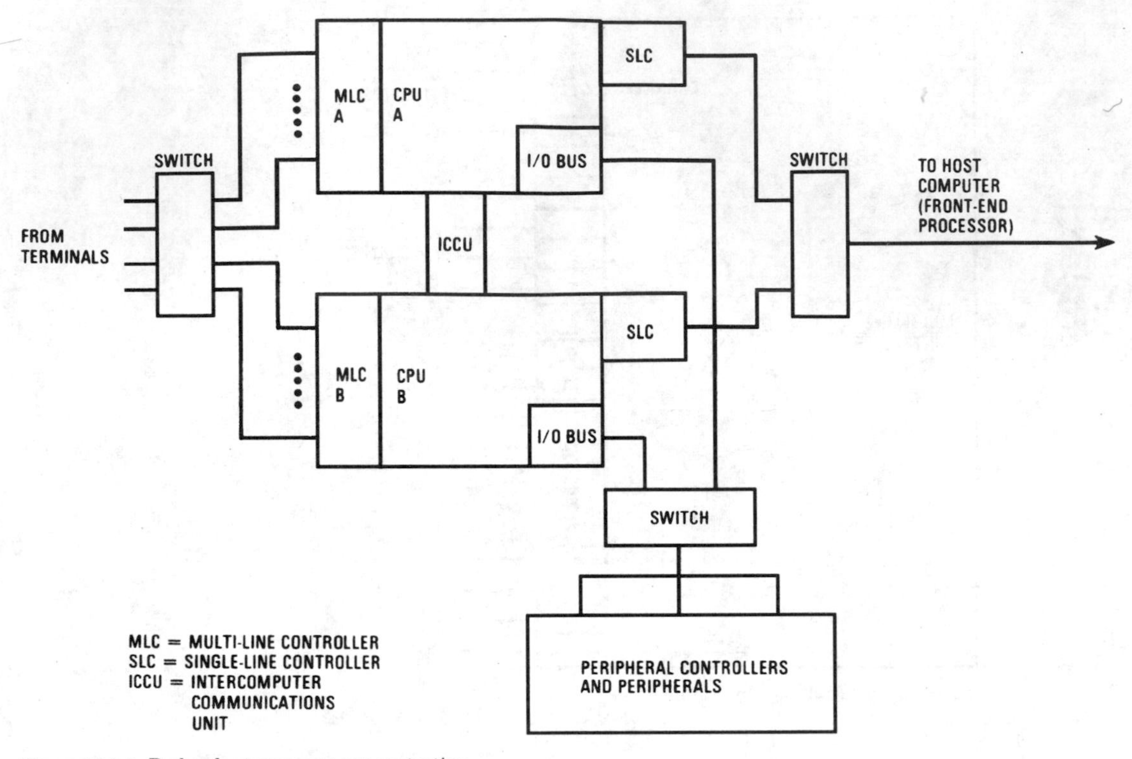

Figure 13.3.6 Redundant message concentration.

intercomputer communications unit to take over processing, generates an alarm message, and conducts an orderly shutdown.

Because the slave has been in parallel processing, it resets all controls and becomes the master, holding the potential of losing data to a minimum. This procedure can usually be completed within 500 microseconds for processors with cycle times of 750 nanoseconds or less. In this case, 666 cycles or more (500 microsec ÷ 750 nanosec) are available within that time slot to execute the required instructions to transfer control and effect the orderly shutdown. Actually, just two to four computer cycles are needed to execute such instructions.

Switching the message

To effect message switching, incoming data is routed to a central point where messages are concentrated for processing. Then, based on some criteria, messages are routed over one or more lines connected to the processor. Once a message has been processed, the destination data it contains is acted on. So, with message switching, any terminal on a network can communicate with any other terminal on the network.

The hardware required for a message switch is quite similar to that required for a store-and-forward message concentrator. The primary differences between switches and concentrators are in application software. Also, with a switch, incoming messages are not stored, but are processed and then routed over one or more lines; and accesses to peripherals, such as disks are via direct memory access (DMA) channels instead of through a processor's I/O bus.

The interface used to transfer data to communications controllers and peripherals is usually determined by the necessary I/O transfer rate. Interfacing can occur at the computer's I/O bus, or through direct memory control (DMC) or DMA devices. Data transferred on the I/O bus is bit-serial and under control of a program. In the DMC mode, data transfers are independent of program control, and data blocks are transferred on a word basis (bit-parallel) to and from any portion of main memory. The DMC mode is used for medium-speed data transfer, and requires a starting and ending address, as well as a count of characters to be transferred. Although similarly word-oriented and a direct-to-memory medium, the DMA mode requires only the starting and ending address. For high-speed data transfers, the DMA mode is used, but at a cost higher than with DMC. The speed at which DMA permits transfer of data is such that a computer using DMA on a high-speed channel can exchange data with several devices (peripherals) and controllers concurrently on a timeshared basis.

Figure 13.3.7 illustrates a typical data communications network consisting of several different types of concentrators and a front-end processor. This network combines examples of much of what was covered previously.

At location 1, a standard concentrator is used to concentrate the traffic from 32 terminals onto a high-speed line for transmission to the front-end processor. Because location 2 has requirements for remote batch processing and for connecting

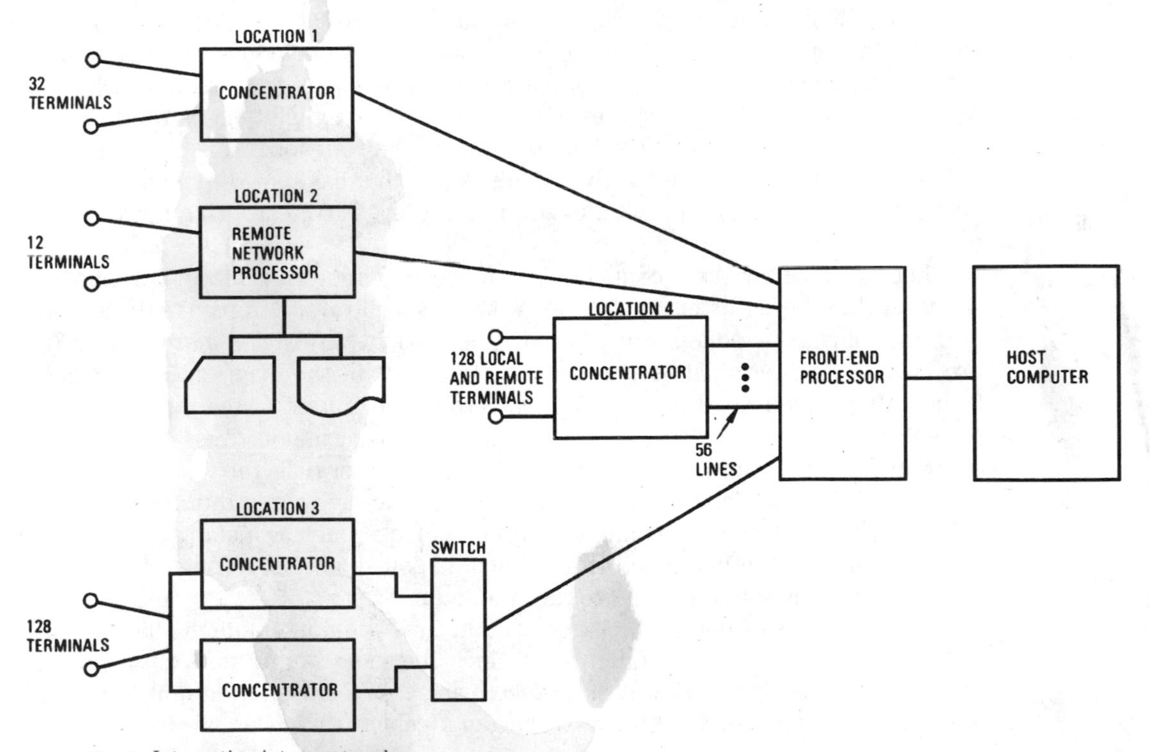

Figure 13.3.7 Integrating into a network.

12 terminals to the host computer, a remote network processor has been installed to perform these two functions. And because location 3 has a significant number of terminals doing important applications, a redundant store-and-forward message concentrator was installed.

Terminals remaining in the network (location 4) total 128. However, it was felt that at most only 56 would ever become active at the same time. Therefore, to economize on front-end processor ports, a contention concentrator was "front-ended" to the front-end processor, making the 128 channels connected to terminals contend for the 56 front-end processor ports.

13.4 Modem- and Line-Sharing Units

It might be evident that a single communications link is less costly than two or more. What is sometimes less obvious is the most economical and effective way to make use of even a single link.

Multiplexing is usually the first technique that comes to mind. But there are many situations where far less expensive, although somewhat slower, equipment is quite adequate. In those situations, terminals can be polled one by one through a "sharing device" acting under the instructions of the host computer.

Typically, the applications where this method would be most useful and practical would be those where messages are short and where most traffic between host computer and terminal moves in one direction at a time.

The technique, which can be called *line sharing* (distinct from multiplexing), can work in some interactive situations, but only if the over-all response time can be kept within tolerable limits. The technique is not as a rule useful for remote batching or remote job entry, unless messages can be carefully scheduled, so as not to get in each other's ways because of the long run time for any one job.

Line sharing, then, is comparatively inexpensive, but has some limits to its usefulness in situations where a multiplexer, most likely a TDM, can bring in more economic leverage.

A line-sharing network is connected to the host computer by a local link, through which the host polls the terminals, one by one. The central site transmits the address of the terminal to be polled throughout the network by way of the sharing unit (Figure 13.4.1). The terminal assigned this address (01 in the diagram) responds by transmitting a request-to-send (RTS) signal to the computer, which returns a clear-to-send (CTS), to prompt the terminal to begin transmitting its message (ABCD in diagram). When the message is completed, the computer polls the next terminal.

Throughout this sequence, the sharing device merely routes the signals to and from the polled terminal and handles supporting tasks, such as making sure the carrying signal is on the line when the terminal is polled, and inhibiting transmission from all terminals not connected to the computer.

There are two subspecies of device used in this technique-modem-sharing units and line-sharing units. They function in much the same way to perform much the same task—the only significant difference being that a line-sharing unit has an internal timing source, and a modem-sharing unit gets its timing signals from the modem it is servicing.

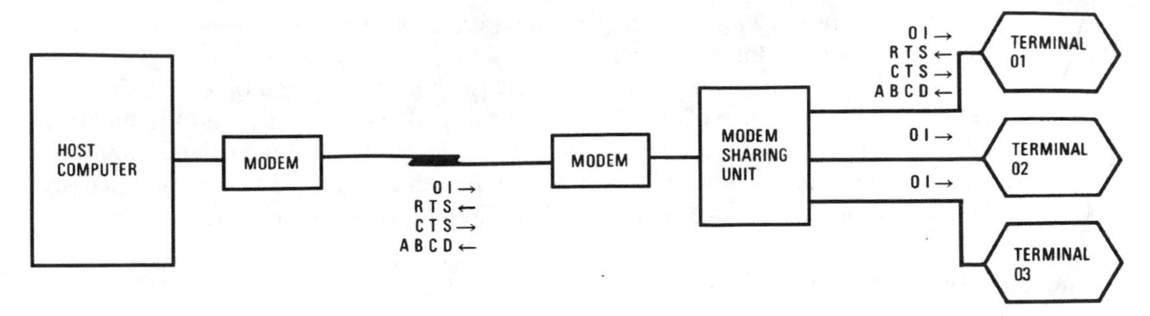

Figure 13.4.1 Line sharing.

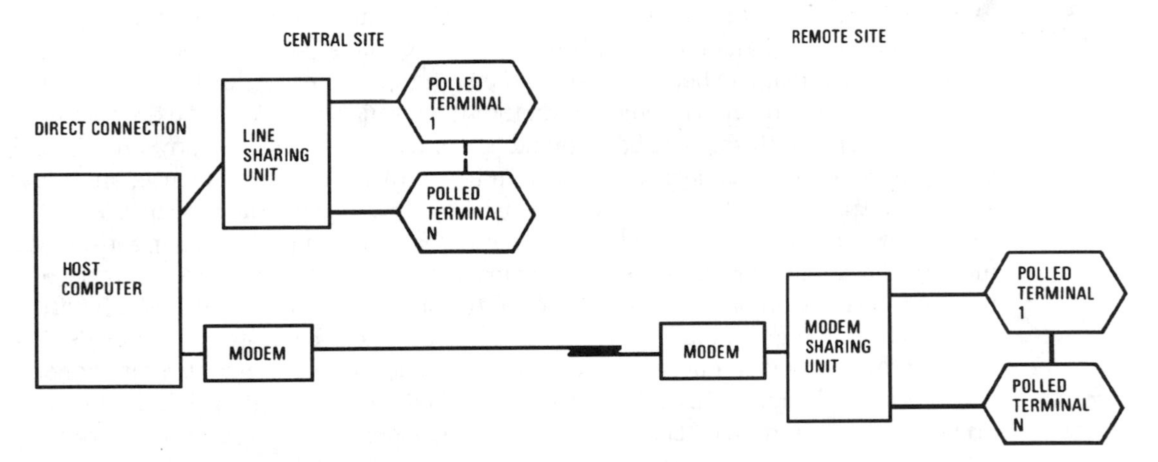

Figure 13.4.2 Line sharing and modem sharing.

Remote operation

A line-sharing unit is mainly used at the central site to connect a cluster of terminals to a single computer port (Figure 13.4.2). It does, however, play a part in remote operation. When a data stream from a remote terminal cluster forms one of the inputs to a line-sharing unit at the central site, it is possible to run the network with a less expensive single-port computer.

In a modem-sharing unit, one set of inputs is connected to multiple terminals or processors (Figure 13.4.2). These lines are routed through the modem-sharing unit to a single modem. Besides needing only one remote modem, a modem-sharing network needs only a single two-wire (for half-duplex) or four-wire (usually, for full-duplex) communications link. A single link between terminals and host computer allows all of them to connect with a single port on the host, a situation that results in still greater savings.

If multiplexing were used in this type of application, the outlay would likely be greater because of the cost of the hardware and the need for a dedicated host computer port for each remote device. A single modem-sharing unit, at the remote site, is all that is needed for a sharing configuration; but multiplexers usually come in pairs, one for each end of the link.

The polling process makes sharing units less efficient than multiplexers. Throughput is cut back because of the time needed to poll each terminal and the line turnaround time on half-duplex links. Another problem is that terminals must wait their turn. If one terminal sends a long message, others might have to wait an excessive amount, which can tie up operators if unbuffered terminals are used.

Sharer constraints

Sharing units are generally transparent within a communications network. There are, however, four factors that should be taken into account when making use of these devices: the distance separating the data terminals and the sharing unit (generally set at no more than 50 feet under EIA-232 interface specifications); the number of terminals that can be connected to the unit; the various types of modems with which the unit can be interfaced; and whether the terminals can accept external timing from a modem through a sharing unit. Then, too, the normal constraints of the polling process, such as delays arising from line turnaround and response, and the size of the transmitted blocks, must be considered in designing the network.

It is advisable to check carefully into what types of modems can be supported by modem-sharing units because some modems permit a great deal more flexibility of network design than others. For instance, if the sharing unit can work with a multiport modem, the extra modem ports can service remote-batch terminals or dedicated terminals that frequently handle long messages. Some terminals that cannot accept external timing can be fitted with special circuitry through which the timing originates at the terminal itself, instead of at the modem.

Questions

13.1 Name five data sources that could be candidates for data stream multiplexing.

13.2 What are the two basic techniques commonly used for multiplexing, and how do they differ?

13.3 Discuss the equipment differences between multiplexers that use bit interleaving and those that use character interleaving.

13.4 If a multiplexer frame contains two SYNC characters, what is the overhead when the frame's data characters are:

A. 2 **C.** 6
B. 4 **D.** 8

13.5 Why is the use of a multiplexer as a front end not common?

13.6 Using an intelligent multiplexer with a 4:1 asynchronous compression ratio will permit how many 1200-bit/s asynchronous circuits to be multiplexed if the composite high-speed line is to operate at 4800 bit/s?

13.7 What is the advantage of using a voice digitalization module using Adaptive Differential PCM instead of a module that uses conventional PCM digitalization with a T1 multiplexer?

13.8 Discuss two features of optical cable that encourage the multiplexing of data onto that cable.

13.9 Discuss the limitations with multiport modems in comparison with multiplexers.

13.10 What is the difference between a concentrator and a remote network processor?

13.11 Discuss the difference between a modem- or line-sharing unit and a multiplexer.

13.12 Where would a line-sharing unit normally be physically placed? Where would a modem-sharing unit be located?

Chapter

14

Network Topology and Optimization

Before you can apply the devices covered in Chapter 13 to the optimization of a network, you must first examine the different types of network structures that can be developed. Once the different structures, known as *network topologies*, are covered, you can explore how different network layouts can be developed, and some of the tools available for the network analyst to use in optimizing the design.

After the preceding has been accomplished, this chapter explores how you can develop a minimum-cost network. To do so, you will have to consider not only the cost of the communications medium, but also such factors as the types and costs of terminals, transmission equipment, and central-site hardware, including front-end-processor equipment.

This chapter devotes an entire section to a case study that applies the information previously covered to a requirement representing a typical user request for service. Through an analysis of the requirement, and the application of knowledge, several network possibilities will be developed. A minimum-cost network design is proposed as a solution to the requirement.

Notice again that, as in Chapter 8, all costs and means of establishing them in this and the next chapter, such as those detailed in the rate tables, should be considered as typical, not necessarily current, and are to be used primarily for the examples and exercise questions. If you have actual current applications, refer to the latest figures obtainable from the appropriate vendors or government organizations.

14.1 Network Structure

The arrangement of data links and nodal points (where data is concentrated) is called *topology*. There are many varieties of network structures. As will be seen, the topology used has a direct bearing on the operation, reliability, and operating cost of a network.

To develop a specific topology, you can use different types of data links, such as point-to-point and multipoint, combining them in some manner, with or without specialized data communications equipment.

Centralized (star)

A centralized, or star, network configuration is illustrated in Figure 14.1.1. When this topology is used, each communications link is directly connected to the host computer. Data movement is directly from each terminal to the computer. There are no delays caused by concentrator buffering times or the internal delay of a multiplexer because no such devices are used. Moreover, if any link should fail, only the terminal on that specific link will be affected by the line outage.

The star topology is suitable where each terminal has a large volume of data traffic and must operate at 9600 bit/s or another high data rate on a leased line, or each terminal has a low traffic volume and a correspondingly low connect time via the public switched telephone network. Otherwise, this configuration results in economic inefficiencies that can be corrected by the use of a hierarchical, or tree, structure.

Hierarchical (tree) structure

In a hierarchical, or tree, structure, data is transferred through intermediate points where one or more operations can be conducted prior to the data continuing on to the host computer. Figure 14.1.2 is an example of a hierarchical, or tree, topology. Here, the concentration equipment could be a multiport modem, multiplexer, modem-sharing unit, or concentrator. When this configuration is used, the traffic from a number of low-to-medium-speed terminals is combined for retransmission to the host computer via a high-speed link. Such an arrangement can result in a substantial reduction in circuit costs because one or a few lines from the concentration device replace numerous terminal lines to the computer that would be required without concentration.

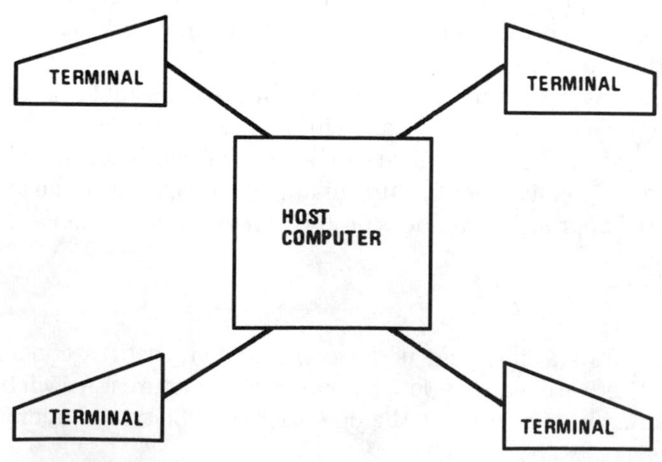

Figure 14.1.1 Centralized or star.

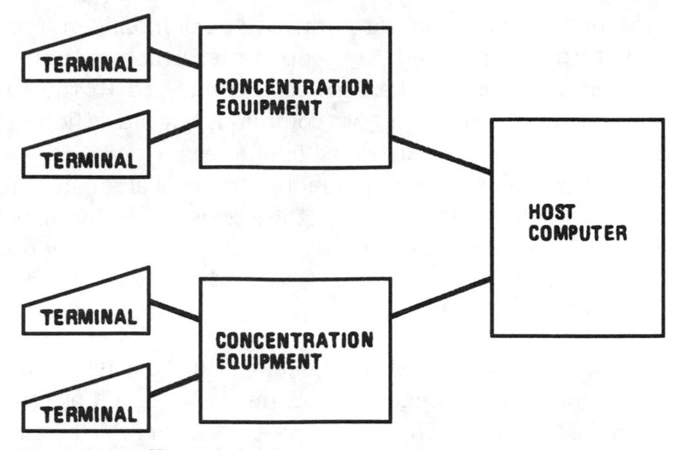

Figure 14.1.2 Hierarchal or tree structure.

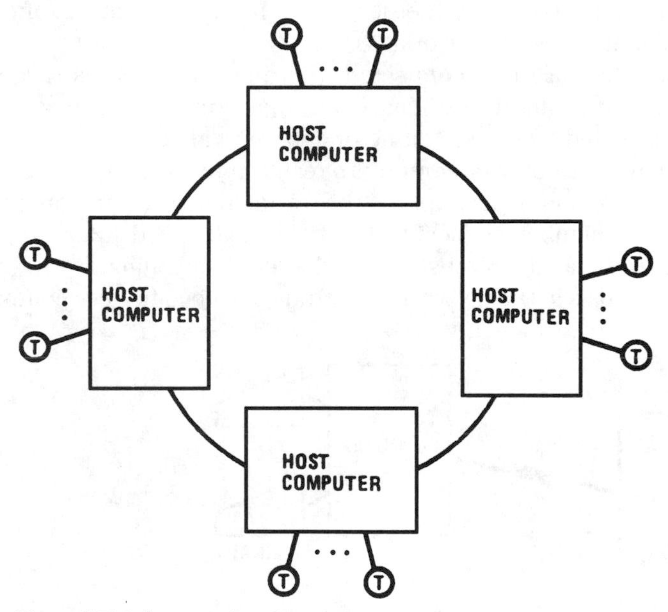

Figure 14.1.3 Loop or ring structure.

One of the primary disadvantages of such an arrangement is that if either the concentration equipment or high-speed line should fail, all terminals connected via that arrangement might become inoperative, unless alternative data links are available to reestablish a connection to the computer.

Loop, or ring, structure

One topology normally used to link computers rather than terminals is the loop, or ring, structure illustrated in Figure 14.1.3. Here, terminals might first be connected via a star (or tree) structure to a computer or other type of intelligent (programmable) device, which, in turn, is connected via a loop or ring structure to the host computer. In this arrangement, most of the terminals connected to the ring do not communicate directly with the intended host computer, but have their data looped around the ring until it reaches the proper host computer.

This structure is economical when many remote terminals and computers are located close to one another. If remote terminals are geographically dispersed over long distances, line costs could become very expensive for such a structure.

Distributed or multistar

As data processing requirements have evolved, many firms have had to decentralize portions of their computational equipment, relocating it to branch or regional sites. This movement of processing power to the field offers considerable advantages with respect to communications costs by reducing the distance and charges for terminal-to-computer communications, as illustrated in Figure 14.1.4. Although all three computers are shown connected to each other, this is not a requirement of distributed processing nor of a distributed network.

When the computers are interconnected, there are more paths through which data can flow. Hence, the failure or outage of any one circuit between computers can be temporarily alleviated by the rerouting of data onto an alternate path. Although the extra circuits to interconnect computers result in extra communications costs, the additional transmission paths improve the overall network performance.

To determine the number and capacity of the high-speed intercomputer data links, a network traffic analysis must be conducted. Depending on the number of computers and terminals in the network, this study can become very complex. Other

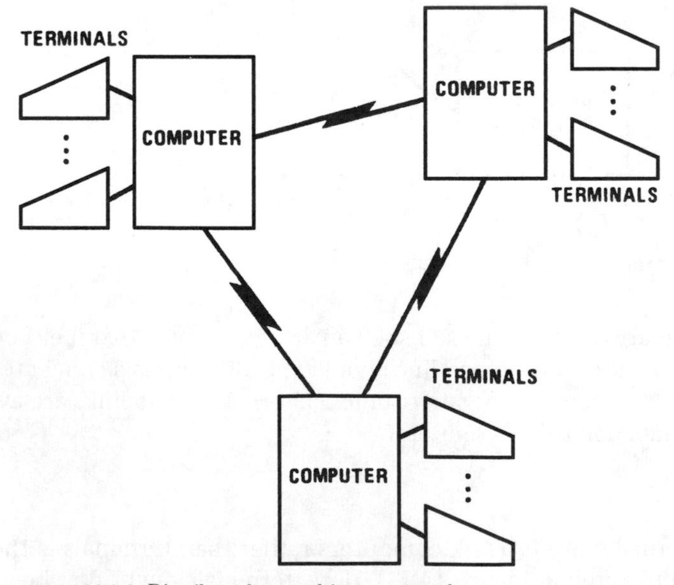

Figure 14.1.4 Distributed, or multistar, network.

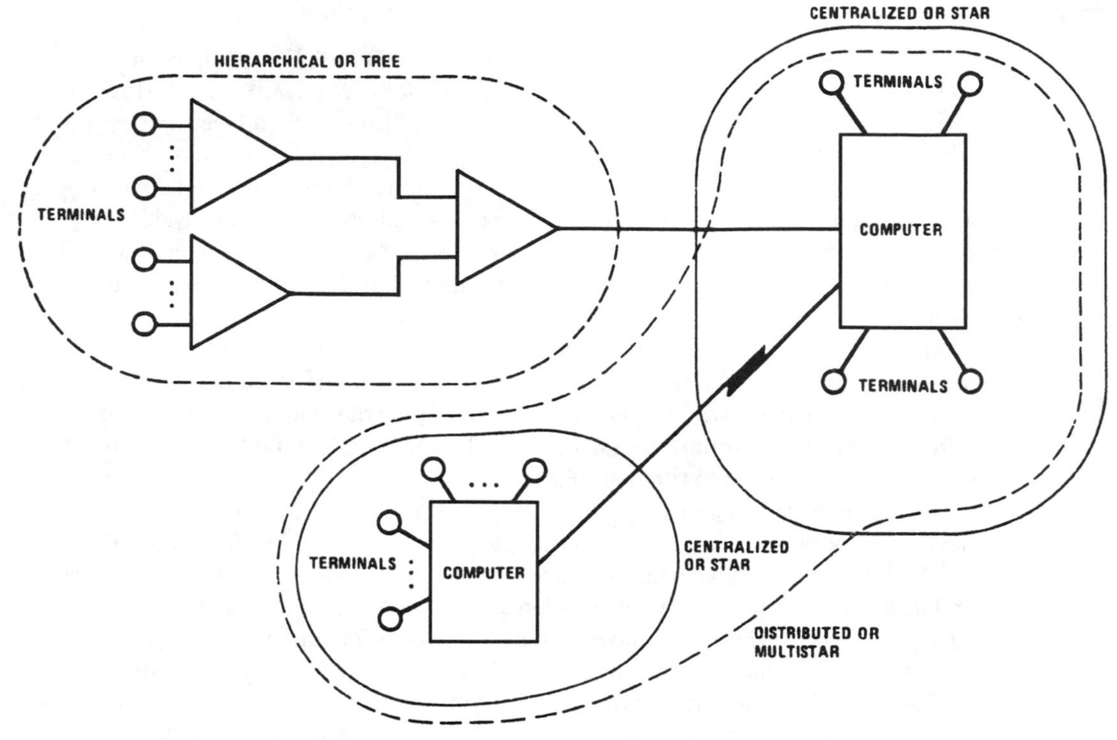

Figure 14.1.5 Mixed structure.

factors must also be considered, including computer software modifications to resolve the problems of linking distributed data bases. Very few truly distributed networks have been implemented because of the cost of modifying application software. This cost can be quite high in comparison to that of the communications mediums required. A number of computer manufacturers have introduced distributed computer networks that include the required software to interconnect computer databases, relieving users of this development expense.

Mixed structure

The topology of a network can vary considerably, as illustrated in Figure 14.1.5, where a mixed structure configuration is shown. In this example, a hierarchical or tree structure, using multiplexers or other types of data concentration equipment, can be used to concentrate many data sources within a geographical area for retransmission to a host computer. Because of particular operating characteristics or traffic volumes, a number of terminals can be directly connected to the computer in a centralized or star arrangement, as shown in the upper-right-hand portion of Figure 14.1.5.

And another star structure connects terminals directly to the computer located in the lower portion of Figure 14.1.5. The latter computer, when connected to the other computer, forms a distributed or multistar structure.

14.2 Sizing

The information in this section is excerpted and adapted, with permission, from Gil Held's *Practical Network Design Techniques* (John Wiley & Sons, 1991). The authors recommend that the reader desiring the additional detail that is beyond the scope of this book refer to that text.

Sizing is the process of ensuring that the configuration of a selected device—such as a port selector, multiplexer, or concentrator—will provide the desired level of service. This service level is the foundation on which the availability level of a network is built and is often directly related to the number of dial-in lines connected to the device.

Although most telephone traffic formulas were developed during the 1920s, many are still applicable to the common problem of determining the number of dial-in lines required to service terminal users. Another applicable common problem is determining the number of ports or channels that should be installed in communications equipment connected to the dial-in lines.

Two basic methods can be used to configure the size of communications network devices: experimental modeling and the scientific approach. The first method involves the selection of the device configuration based on a mixture of previous experience and intuition. Normally, the configuration selected is less than the base-plus-expansion capacity of the device. This allows the device size to be adjusted or upgraded without a major equipment modification if the initial sizing estimate is inaccurate.

The second or scientific method ignores experience and intuition. This approach is based on a knowledge of data traffic and the application of mathematical formulas to traffic data. The method is simplified considerably by a series of tables generated by the development of appropriate computer programs. Thus, many sizing problems could be reduced to one of a simple table lookup process.

Theoretically, there should be a much higher degree of confidence and accuracy in the selected configuration when the scientific method is used. However, it requires a firm knowledge or accurate estimate of the data traffic. At some organizations, unfortunately, this data could be difficult to obtain. Therefore, in many cases, a combination of two techniques will provide an optimum result. For such situations, sizing can be conducted using the scientific method, with the understanding that the configuration selected might require experimental adjustment.

Traffic measurement

Telephone line activity or traffic can be defined as the product of the calling rate per hour and the average holding time per call. This measurement can be expressed mathematically as

$$T = C \times D$$

where T = traffic
C = calling rate per hour
D = average duration per call.

Using this formula, traffic can be expressed in call-minutes (CM) or call-hours (CH), where a call-hour represents one or more calls having an aggregate duration of one hour. If the calling rate during the busy hour of a particular day is 500 and the average call duration is 10 minutes, the traffic flow or intensity would be 500×10 or 5000 CM, which would be equivalent to 5000/60 or about 83.3 CH.

The preferred measurement unit in telephone traffic analysis is the *erlang*, named after A.K. Erlang, a Danish mathematician. The erlang is a dimensionless unit and represents the occupancy of a circuit: one erlang of traffic intensity on one traffic circuit represents a continuous occupancy of that circuit.

A second term often used to represent traffic intensity is the *call-second (CS)*. One hundred call-seconds may be written 1 CCS. Assuming a one-hour unit interval, these terms can be related to the erlang as follows:

$$1 \text{ erlang} = 60 \text{ call-minutes} = 36 \text{ CCS} = 3600 \text{ CS}$$

Grade of service

If a subscriber attempts to originate a call from one city to another when all trunks are in use, that call is said to be *blocked*. Given the traffic intensity and number of available trunks, and based on mathematical formulas, the probability of a call being blocked can be computed. The concept of determining the probability of blockage can be adapted readily to the sizing of data communications equipment.

In a network design, you can specify the number of calls that you are willing to have blocked during the busy hour. This specification is known as the *grade of service*—an important concept in the sizing process—and represents the probability of having a call blocked. For example, if you specify a grade of service of 0.05 between two sites, you require a sufficient number of trunks so that only one call in every 20, or five in every 100, will be blocked during the busy hour.

14.3 Network Layout

Prior to initiating the design of a network, several items must be considered by the network planner. One of the first is the type and anticipated usage of terminals to be connected via a data communications network to the computer.

Numerous approaches can be used to obtain the required information, with these approaches being covered in considerable detail in Chapter 19. For the initial investigation into the concepts of network design, the problem will be limited by considering only a few alternative transmission media and assuming that the terminal operating statistics are given.

VH coordinate system

In the United States, the distance of the telephone connection is one of the cost variables that has to be considered for all land-based transmission methods, with the exception of packet switching. Although the communications carrier routes both leased lines and connections over the public switched telephone network over any desired path, charges are normally independent of actual routings and are based on the straight-line airline distance between locations connected.

First filed as a tariff by AT&T, the VH coordinate system of distance measuring is now used by almost every communications carrier in the United States. This system divides the continental United States by vertical and horizontal grid lines so that every location is capable of being expressed as a pair of vertical (V) and horizontal (H) coordinates that can then be used to calculate distances between locations.

Once the VH coordinates for two locations are known, by the Pythagorean theorem, the distance between those locations can be computed as:

$$Distance = \sqrt{\frac{(V_1 - V_2)^2 + (H_1 - H_2)^2}{10}}$$

In using the VH system and the theorem for distance computation because fractional miles are considered as full miles, the result of taking the square root will be rounded to the next higher number.

As an example of the use of the theorem, consider determining the mileage between Reading, Pennsylvania ($V = 5258$, $H = 1612$) and Atlanta, Georgia ($V = 7260$, $H = 2083$). Substituting in the previous equation, the distance between the two cities becomes

$$D = \sqrt{\frac{(5258 - 7260)^2 + (1612 - 2083)^2}{10}}$$

$$= 650.37 \text{ miles}$$

which becomes 651 miles when rounded to the next higher whole number.

Although a table of airline mileages might appear to be easier to use than obtaining the VH coordinate of every terminal location, then computing the distances from each terminal to possible data concentration equipment locations, VH coordinates are useful for computerization. Today, several consulting firms offer subscribers the use of their optimization programs. Here, the user enters the terminal type and location as a city and a state, and the program then assigns from its database the corresponding VH coordinates. The program computes an optimum path for transmission based on the terminal's activity factors and the location and activity factors of other terminals in the network.

IXC mileage

IXC, or intercity exchange channel mileage, is the airline distance between locations. Although tables of VH coordinates are available from a number of sources, unless the subscriber wishes to computerize her network design, VH coordinates can be cumbersome to work with. Instead, a subscriber can contact his AT&T account representative and provide that person with a list of cities for which she desires the IXC mileage. The AT&T representative will then access a special timesharing program, which will produce a tabulation of the distances between locations. If the cities of concern are, as listed in Table 14.3.1, the IXC program would produce a series of tables showing the mileage from each location to all other locations. Portions of the first and last two such tables are shown in Tables 14.3.2 through 14.3.4.

TABLE 14.3.1 Terminal City Locations

City No.	City	City No.	City
1	Denver, Colo.	34	Montrose, Colo.
2	Boise, Idaho	35	Durango, Colo.
3	St. Anthony, Idaho	36	Grand Junction, Colo.
4	Burley, Idaho	37	Pueblo Dam, Colo.
5	Salem, Oregon	38	Salida, Colo.
6	Forest Grove, Oregon	39	Loveland, Colo.
7	Spokane, Wash.	40	Albuquerque, N.M.
8	Ephrata, Wash.	41	Farmington, N.M.
9	Yakima, Wash.	42	Cuba, N.M.
10	Grand Coulee, Wash.	43	Oklahoma City, Okla.
11	Sacramento, Calif.	44	Austin, Texas
12	Willows, Calif.	45	El Paso, Texas
13	Fresno, Calif.	46	Amarillo, Texas
14	Auburn, Calif.	47	San Angelo, Texas
15	San Bernardino, Calif.	48	Harvey, N.D.
16	Shasta Dam, Calif.	49	Garrison, N.D.
17	Folsom Dam, Calif.	50	Bismarck, N.D.
18	Tracey, Calif.	51	Watford City, N.D.
19	Goleta, Calif.	52	McCook, Nebr.
20	Altadena, Calif.	53	Grand Esland, Nebr.
21	Boulder City, Nevada	54	Billings, Mont.
22	Carson City, Nevada	55	Great Falls, Mont.
23	Reno, Nevada	56	Fort Peck, Mont.
24	Phoenix, Ariz.	57	Kalispell, Mont.
25	Parker, Ariz.	58	Huron, S.D.
26	Tucson, Ariz.	59	Watertown, S.D.
27	Salome, Ariz.	60	Rapid City, S.D.
28	Yuma, Ariz.	61	Courtland, Kansas
29	Page, Ariz.	62	Riverton, Wyo.
30	Salt Lake City, Utah	63	Casper, Wyo.
31	Provo, Utah	64	Laramie, Wyo.
32	Dutch John, Utah	65	Washington, D.C.
33	Logan, Utah		

TABLE 14.3.2 IXC Mileage from Point 1 to All Other Points

From	To	Mileage
1	2	636
1	3	450
1	4	496
1	5	985
.	.	
.	.	
.	.	
1	63	224
1	64	112
1	65	1489

TABLE 14.3.3 IXC Mileage from Point 63 to All Other Points

From	To	Mileage
63	64	112
63	65	1547

TABLE 14.3.4 IXC Mileage from Point 64 to All Other Points

From	To	Mileage
64	65	1511

14.4 Minimal-Cost Network Design

Consider the situation in which the computer center is located in Denver, Colorado, and you want to connect a number of terminals in the western region of the United States to it. For this example, limit the choice of transmission media to the public switched telephone network and to leased lines, and eliminate any expansion requirements from consideration. (In actual situations, planned expansion must be considered.)

To develop a minimal-cost network, the first priority is to compute the most economical method of data transmission from each terminal location into Denver. To do this, direct-dial long-distance costs are computed and compared with the costs of leased lines on a monthly basis. Assume 22 working days in a month, and that the number, operating speed, and location of terminals requiring a connection to the computer center in Denver are as listed in Table 14.4.1.

To develop a cost table of public-switched-telephone network vs. leased-line cost, you must determine the average daily connect time per terminal and the distance between each terminal location and the computer center. This information is listed in Table 14.4.2.

Notice that the daily terminal connect time could be determined from a communications survey. Or, the data could be extrapolated from terminal usage of other locations that are using the computer to solve similar problems for which the six new locations in Table 14.4.2 can be installed. The mileage to Denver could come from numerous sources, such as the telephone company's IXC program, or a table of airline mileages between points.

With the information in Tables 14.4.1 and 14.4.2, a comparison of leased vs. public-switched-telephone network usage is made for each terminal location. This comparison is contained in Table 14.4.3. In columns 3 and 4, the monthly connect time is divided into the number of initial one-minute connections and the number of additional minute connections because the toll charge per long-distance call is based on this division. Because 22 working days per month are assumed, a further assumption of only one call per day enables the monthly connect time costs in columns 3 and 4 to be readily computed.

As an example, for Billings, Mont., where the daily connect time (column 2) is 15 minutes, assuming one call per day results in 22 (column 3) initial minute costs per month. The total monthly connect time is 15 minutes/day × 22 working days/month, or 330 minutes/month. Subtracting the 22 initial minutes produces a total of 308 additional minutes per month of connect time, as shown in column 4.

The mileage to Denver comes from Table 14.4.2. The cost-per-call data contained in columns 6 and 7 were entered for illustrative purposes only and are intended to represent the Intra-U.S. Rate Table for telephone calls over the public switched telephone network based on mileage, type of call, and call duration. For the data in columns 6 and 7, it is assumed that calls are made during the day on a station-to-station basis. Next, the monthly switched network cost per terminal was computed (column 8) by multiplying column 3 by column 6 and adding the result to that obtained by multiplying column 4 by column 7.

Column 9 shows the monthly cost of the line. Because each leased line is assumed to be a point-to-point structure, two station terminals will be required per line at a monthly cost of $36.05 per station terminal, for a total of $72.10 per month, as shown in column 10. The total monthly leased line cost is then obtained by adding the line cost (column 9) and the station terminal cost (column 10). The total is shown in column 11. Lastly, the more economical method was determined by comparing the monthly switched-network cost (column 8) with the monthly leased-line cost (column 11). The results were placed in columns 12 and 13. All costs are for illustrative purposes only.

Network layout

Once a tabulation of the switched- and leased-line costs per terminal has been completed, a network layout can be constructed. Use an appropriate map of the locations to be connected to the computer center. Then, via a progression of steps in conjunction with other information (detailed here), obtain a minimum-cost network design. In the design of the minimum-cost network, the data communications devices listed in Table 14.4.4 and their associated monthly costs are used.

TABLE 14.4.1 Remote-Terminal Data

Location	Operating speed	Quantity
Billings, Mont.	300 bit/s	1
Salt Lake City, Utah	1200 bit/s	4
Boise, Idaho	300 bit/s	1
Boulder City, Nev.	1200 bit/s	4
Phoenix, Ariz.	1200 bit/s	4
Casper, Wyo.	300 bit/s	1

TABLE 14.4.2 Terminal Connect Time and Distance to Computer-Center Terminal

Location	Daily connect time (hrs.)	Mileage to Denver
Billings, Mont.	0.25	451
Salt Lake City, Utah	2.0	370
Boise, Idaho	4.0	536
Boulder City, Nev.	4.0	595
Phoenix, Ariz.	3.5	583
Casper, Wyo.	0.5	224

TABLE 14.4.3 Comparing Leased Lines and Public Ones

1	2	3	4	5	6	7	8	9	10	11	12	13
		Monthly connect time			Cost/call day rate						Most economical	
Location	Daily connect time (minutes)	No. of first minute (1)	No. of additional minutes	Mileage to Denver	Initial minute	Each additional minute	Monthly switched network cost/ terminal	Monthly leased line cost	Monthly station terminal cost	Total monthly leased line cost	Type	Monthly cost
Billings, Mont.	15	22	308	451	.62	.43	146.08	676.77	72.10	748.87	SWITCHED	146.08
Salt Lake City, Utah	120	22	2,618	370	.59	.42	1,112.54	506.72	72.10	578.82	LEASED	578.82
Boise, Idaho	240	22	5,258	536	.62	.43	2,274.58	756.67	72.10	828.77	LEASED	828.77
Boulder City, Nev.	240	22	5,258	595	.62	.43	2,274.58	812.13	72.10	884.23	LEASED	884.23
Phoenix, Ariz.	210	22	4,585	583	.62	.43	1,990.78	706.94	72.10	779.04	LEASED	779.04
Casper, Wyo.	30	22	638	224	.58	.39	261.58	463.39	72.10	535.49	SWITCHED	261.58

NOTE: (1) Assumes one call/day

TABLE 14.4.4 Data Communications Equipment for Network

Device	Monthly cost
2400-bit/s modem	$ 65
4800-bit/s modem	135
7200-bit/s modem	210
9600-bit/s modem	235
4-channel multiplexer	100
4-channel expansion	50
8-channel statistical multiplexer	175

With the data in Table 14.4.1, we can first place the terminal quantity and operating data at each location, using as an example the symbol 1 X 300 to denote 1 terminal operating at 300 bit/s at a given location. This is illustrated in Figure 14.4.1. Next, the more economical method of transmission per terminal, as computed in Table 14.4.3, can be indicated on the network layout, by placing an S for switched and an L for a leased line, and denoting the monthly cost of the connection. This additional information is illustrated in Figure 14.4.2.

Once the cost per terminal is placed on the network layout, examine the utilization of data concentration equipment, such as multiplexers. One of the best places to initialize a multiplexer location is at a site with a large number of terminals, with each terminal having a large monthly communications cost from the location to the computer site. Assume that the cost of a 4-channel asynchronous multiplexer is $100 per month, and that a 4800-bit/s modem rents for $135 per month. Using two multiplexers (one at a remote site and one in Denver to demultiplex data), the cost of the two pieces of equipment and the two high-speed modems necessary to service four terminals becomes ($100 + $135) × 2 = $470 per month. Using this equipment cost, you can modify Figure 14.4.2, as illustrated in Figure 14.4.3. In this illustration the cost of installing four-channel multiplexers at Phoenix, Boulder City, and Salt Lake City are compared with the cost of the previously computed most economical method. As an example, in Phoenix the cost of the multiplexing equipment and modems plus the cost of one leased line, becomes $1249 per month. This cost is now compared with the $779/month cost of a leased line from Phoenix to Denver. If you assume that the terminals in Phoenix cannot be polled, then four such lines would be required, for a total monthly cost of $3116. Because the multiplexing cost of $1249 per month is less, you can initially position a 4-channel multiplexer in Phoenix.

You can conduct a similar analysis for Boulder City and Salt Lake City, with the results indicating the 4-channel multiplexers can be initially positioned at those sites as well. Modify the network layout to indicate the initial multiplexer placements, as illustrated in Figure 14.4.3.

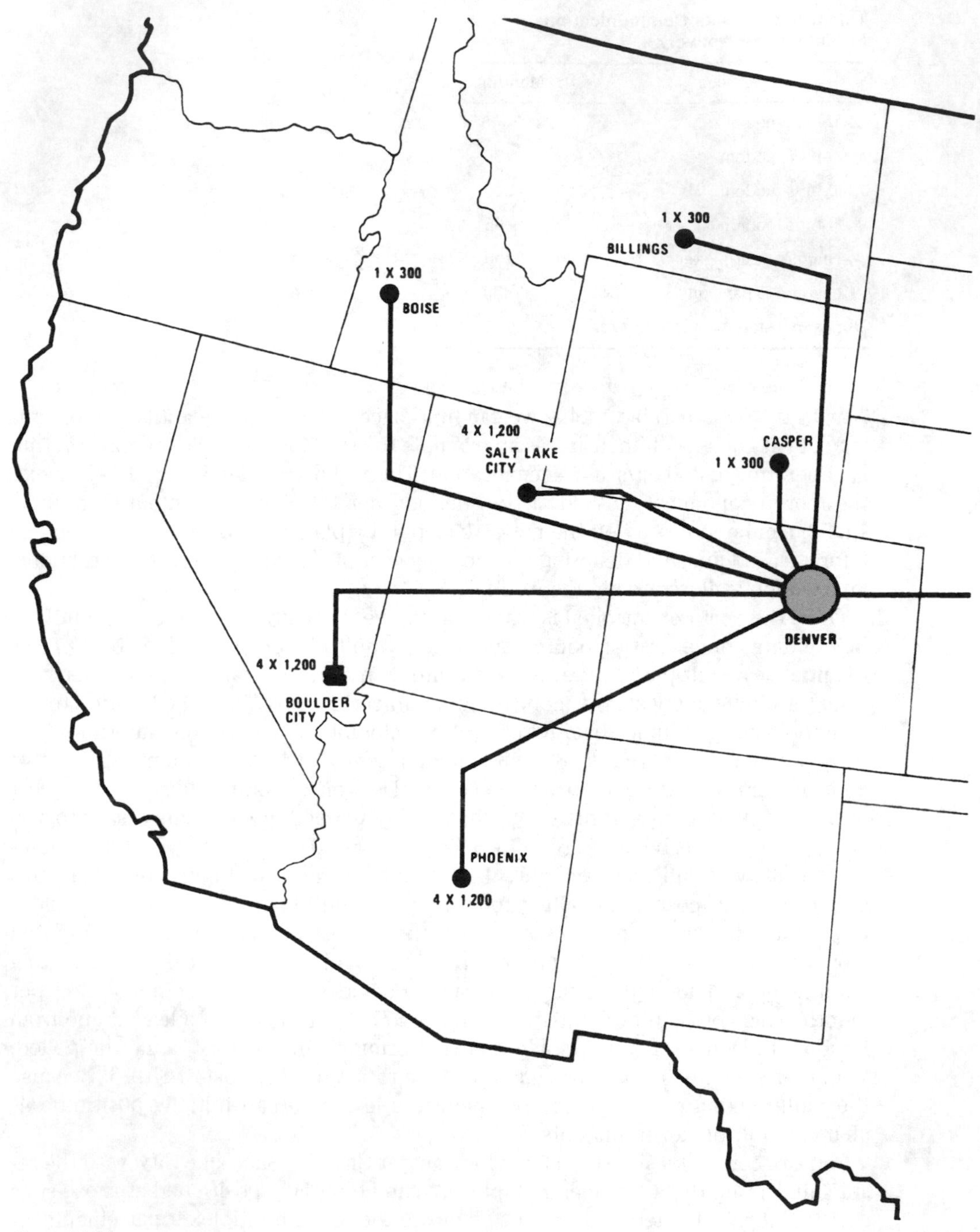
1 X 300
BILLINGS
1 X 300
BOISE
4 X 1,200
SALT LAKE
CITY
CASPER
1 X 300
DENVER
4 X 1,200
BOULDER
CITY
PHOENIX
4 X 1,200

Figure 14.4.1

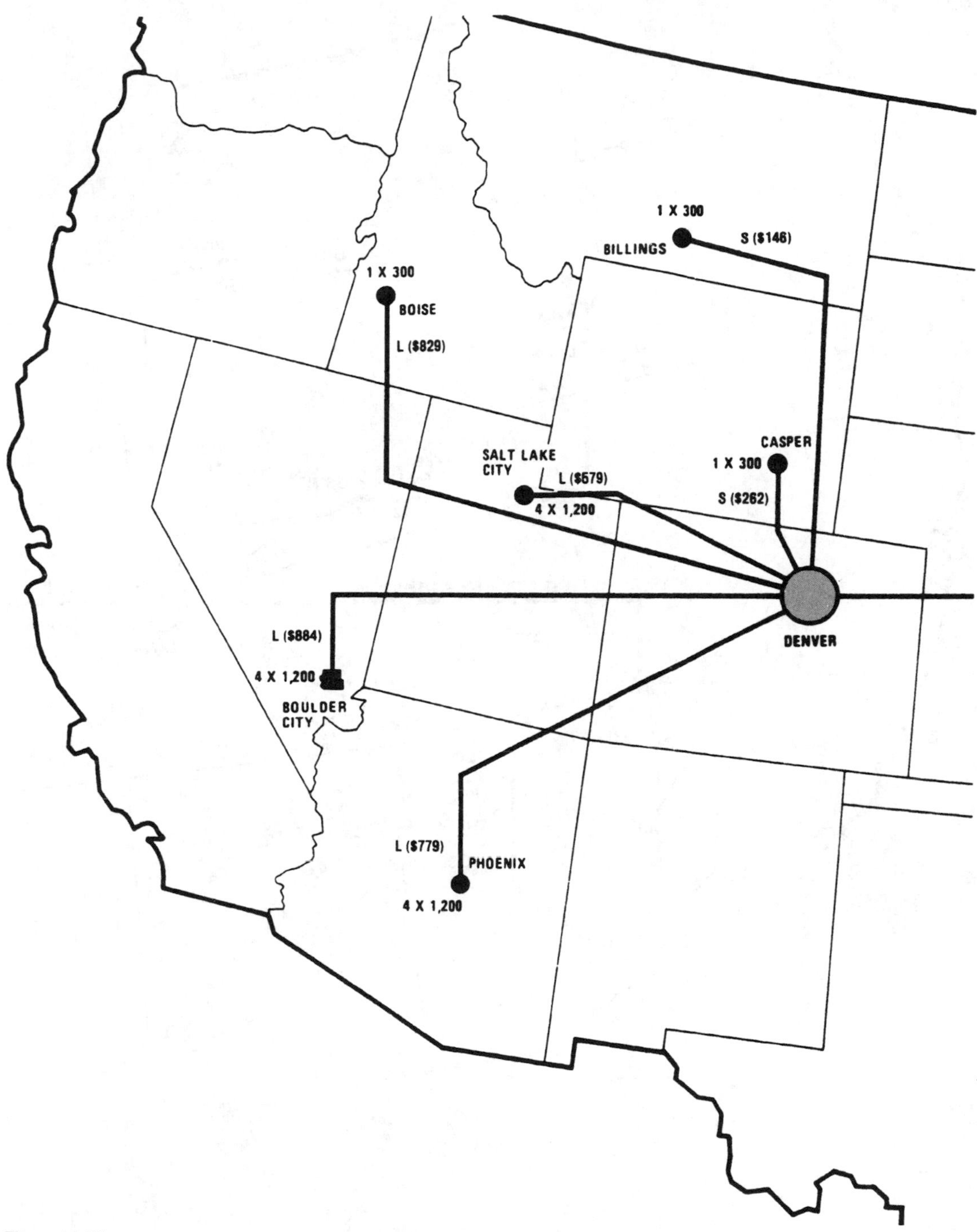
1 X 300
BILLINGS
S ($146)
1 X 300
BOISE
L ($829)
CASPER
1 X 300
S ($262)
SALT LAKE CITY
L ($579)
4 X 1,200
DENVER
L ($884)
4 X 1,200
BOULDER CITY
L ($779)
PHOENIX
4 X 1,200

Figure 14.4.2

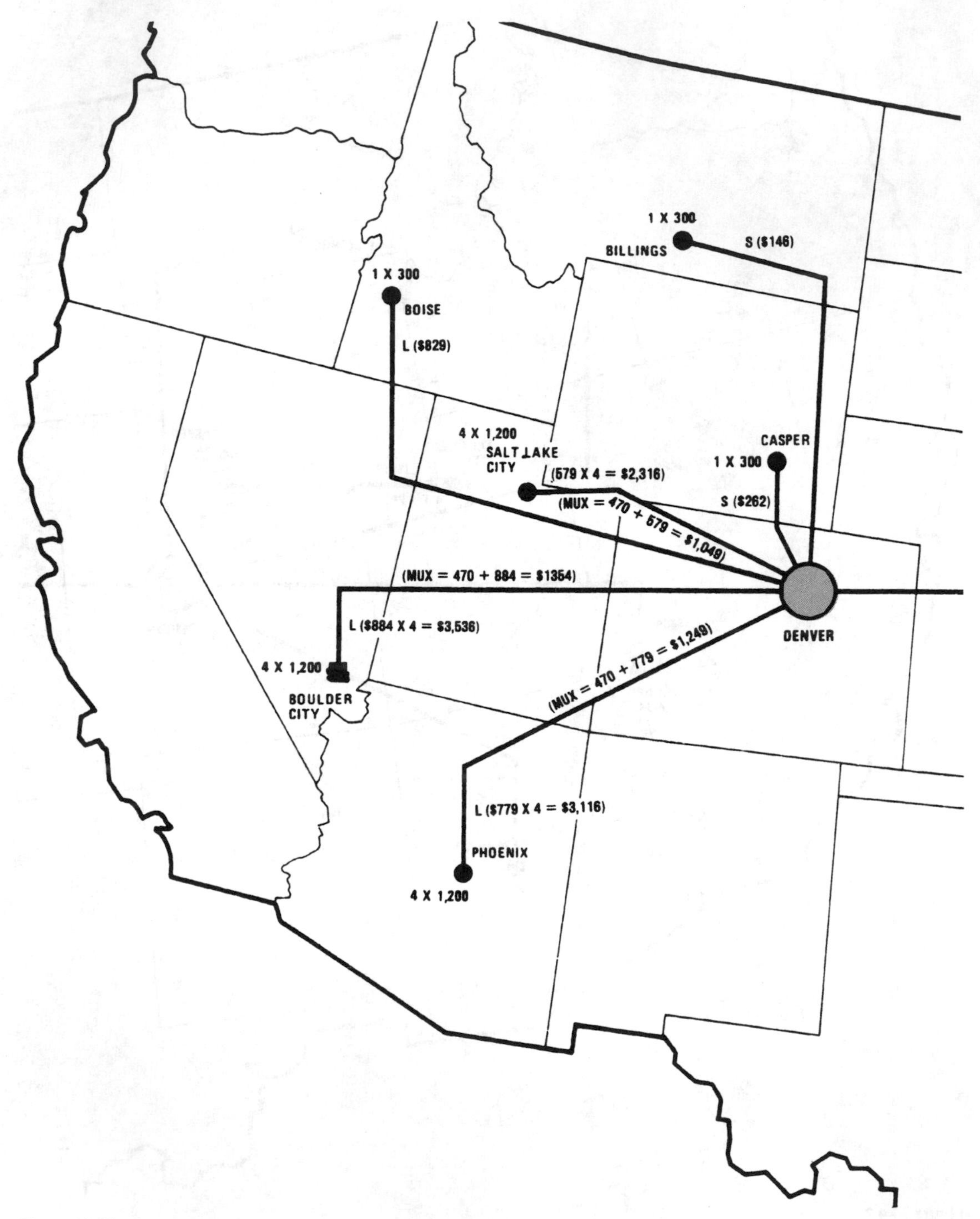
1 X 300
BILLINGS
S ($146)
1 X 300
BOISE
L ($829)
4 X 1,200
SALT LAKE CITY
(579 X 4 = $2,316)
(MUX = 470 + 579 = $1,049)
CASPER
1 X 300
S ($262)
(MUX = 470 + 884 = $1354)
L ($884 X 4 = $3,536)
4 X 1,200
BOULDER CITY
DENVER
(MUX = 470 + 779 = $1,249)
L ($779 X 4 = $3,116)
PHOENIX
4 X 1,200

Figure 14.4.3

Homing points

Using the initial multiplexer locations, next examine the cost trade-offs of using each location as a hub, connecting other remote terminals in the geographical area through the multiplexer to the computer center in Denver. From the layout in Figure 14.4.3, it is apparent that Boise is closer to Salt Lake City than to the other two multiplexer locations, and Billings and Casper are closer to Denver than to any multiplexer site. Thus, it would be impractical to connect either Billings or Casper terminals through a multiplexer. This would increase costs because switched service was originally computed to be more economical than a leased line, and a leased line from Billings or Casper to a multiplexer is longer and costlier than a leased line to Denver. Thus, we now concentrate on Boise, whose formerly computed most economical cost was $829 per month using a leased line to Denver. From an IXC mileage table, we obtain a distance of 297 miles between Boise and Salt Lake City. The cost of this circuit is $532.01 plus $72.10 for two station terminals for a monthly cost of $604.11—which is less than the $829-per-month cost of the leased line from Boise directly to Denver, representing a potential savings of $225 per month.

Because the Salt Lake City 4-channel multiplexer is operating at 4800 bit/s, higher-speed modems connecting Salt Lake City with Denver and a larger-capacity multiplexer will be required. Notice that the cost of modems for the line connecting Boise with Salt Lake City is not considered because a modem is required at Boise in any event, as is one at the computer site in Denver.

Suppose that a larger-capacity multiplexer rents for an extra $50 per month per device. Because two are required, this reduces our potential savings to $125 per month. Now comes the bad part. Upgrading the modem to the next higher synchronous data rate, for example, 7200 bit/s, costs an extra $75 per month per modem as well as $21.15 per month for line conditioning, converting the potential savings to an extra expenditure of $46.15 per month.

Instead of using an expanded conventional multiplexer and obtaining a higher-speed modem, suppose that a statistical multiplexer is installed at Salt Lake City and Denver to support the four 1200-bit/s channels at Salt Lake City and the one 300-bit/s channel from Boise. Suppose that the statistical multiplexer rents for $175 per month and has an asynchronous compression ratio of 4:1. This means that the total input data rate of (4 × 1200) + 300 or 5100 bit/s, which requires a modem capable of transmission at 1275 bit/s or more. Thus, although the statistical multiplexer costs $25 per month per device more than the conventional multiplexer and its expansion unit in this example, its use might result in considerable modem cost reductions.

The statistical multiplexer permits use of 2400-bit/s modems on the Salt Lake City-to-Denver circuit, which can be obtained for approximately $65 per device per month. Using the configuration in Figure 14.4.4, the monthly cost to support Boise and Salt Lake City terminals is listed in Table 14.4.5.

In comparing the cost shown in Table 14.4.5 with the costs listed on Figure 14.4.3, the Salt Lake City-to-Denver transmission cost has increased $10 per month, to $1059. This is because the extra cost of the statistical multiplexers ($175 vs. $100 per month) is not completely offset by the lower cost of the modems required ($65 vs. $135). However, the total cost for supporting both Boise and Salt Lake City ter-

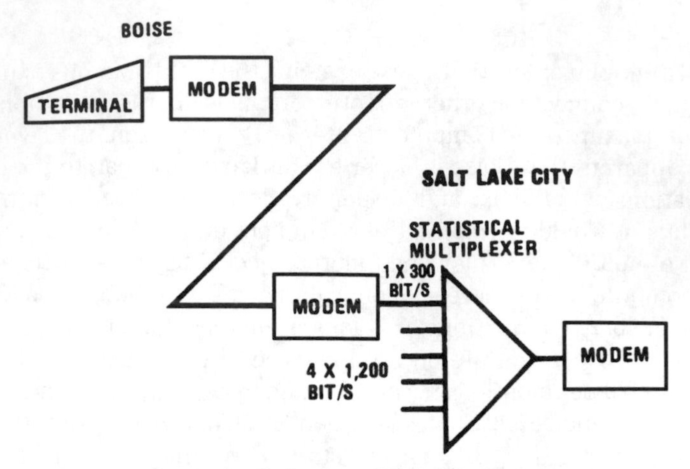

Figure 14.4.4 Supporting Boise through Salt Lake City.

TABLE 14.4.5

Salt Lake City—Denver	
Statistical multiplexers 2 @ $175	$ 350.00
Leased line cost	579.00
Modem cost 2 @ $65	130.00
Subtotal Salt Lake City to Denver	$1059.00
Boise to Salt Lake City	
Leased line and terminations	604.11
Total	$1663.11

minals has been reduced from $1878 ($829 + $1049) to $1633.11 per month. Thus, the new network configuration would be as illustrated in Figure 14.4.5.

Prior to finalizing the network layout, you can examine each of the three multiplexer locations to determine if additional cost reductions is realized. Consider transmitting data from one multiplexer location to another in order to eliminate or reduce the leased-line mileage charges for long-distance circuits. The distances between the three multiplexer locations and the computer site in Denver are illustrated in Figure 14.4.6.

In examining the distances between locations, connecting Boulder City to Denver via Phoenix would appear to offer the best possibility for potential savings. This is because the distance from Boulder City to Phoenix (236 miles) is less than any of the other interconnection distances. The cost of a pair of multiplexers and high-speed modems, as well as the leased line between Boulder City and Denver, was $1087 per month, as computed in Figure 14.4.3. If the line cost between Boulder City and Phoenix plus multiplexing equipment cost and any additional modem charges are less than that amount, multiplexing the data from the terminals in Boulder City through Phoenix will be more economical than directly to Denver.

The cost of a 236-mile circuit is $474.67 per month. $72.10 per month for two station terminals makes a total monthly cost of $546.77.

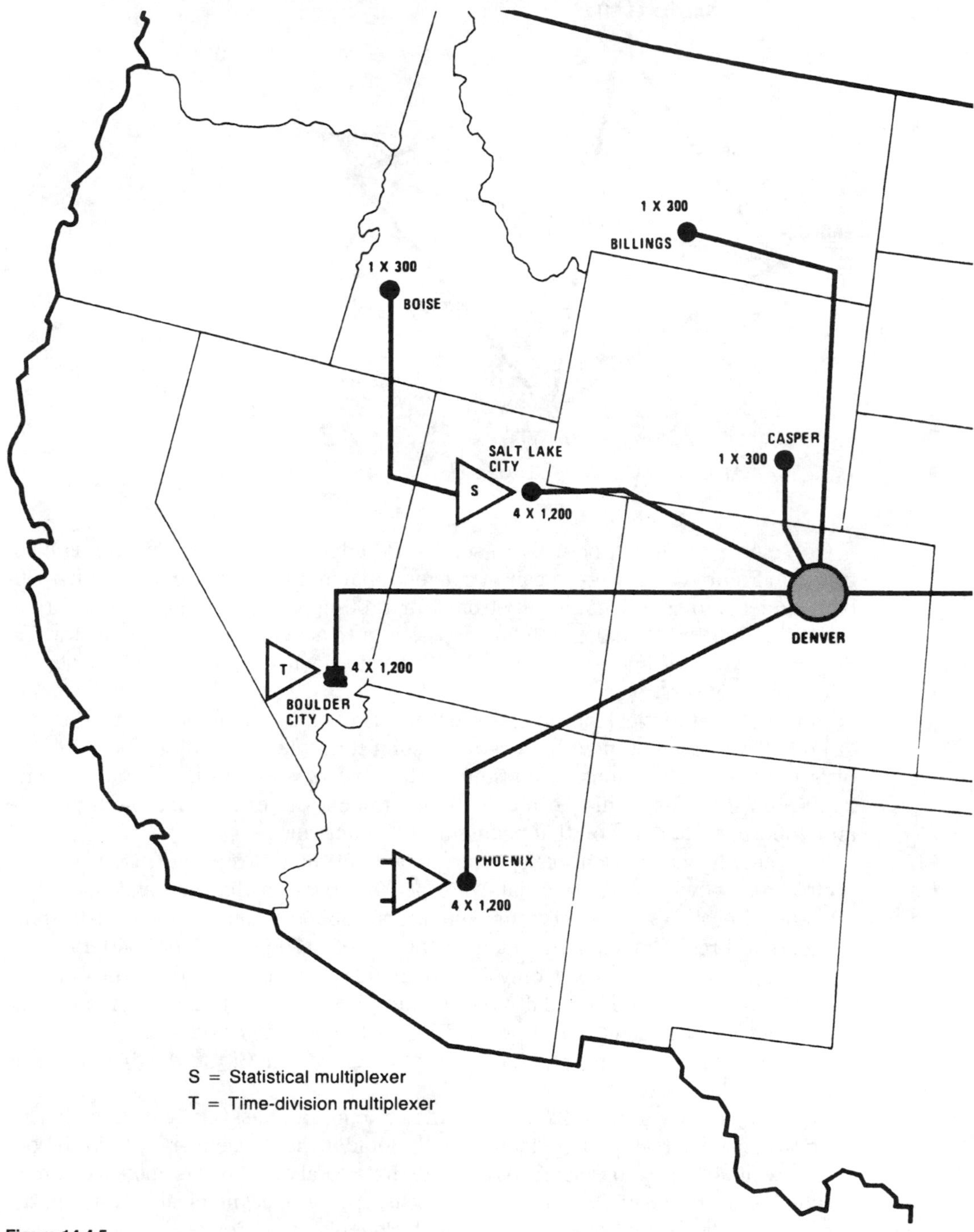
1 X 300
BILLINGS
1 X 300
BOISE
CASPER
1 X 300
SALT LAKE CITY
S
4 X 1,200
DENVER
T
4 X 1,200
BOULDER CITY
PHOENIX
T
4 X 1,200
S = Statistical multiplexer
T = Time-division multiplexer

Figure 14.4.5

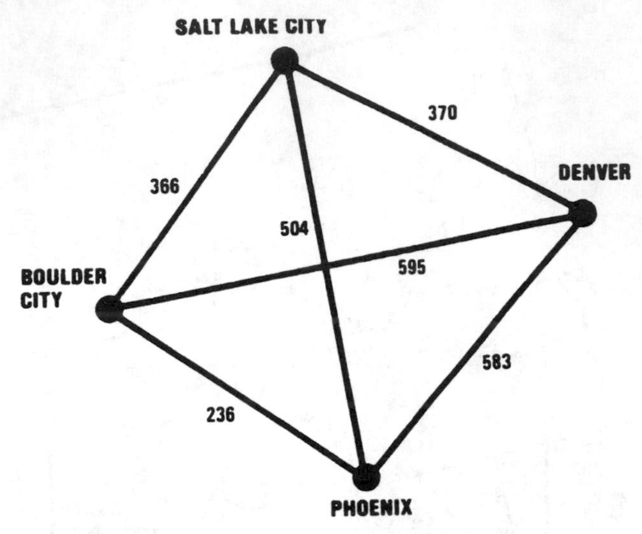

Figure 14.4.6 Multiplexer interconnection distances.

One method that can be used to service Boulder City through Phoenix is illustrated in Figure l4.4.7. Here, a pair of conventional multiplexers are used with 4800-bit/s modems to multiplex the data from four 1200-bit/s terminals in Boulder City to Phoenix. They are then demultiplexed, passed into a statistical multiplexer with an 8-channel capacity, and remultiplexed with local Phoenix terminal traffic for transmission to Denver.

Notice that statistical multiplexers are not used on the Boulder City-to-Phoenix circuit. If you assume monthly costs for such multiplexers and 2400- and 4800-bit/s modems as previously mentioned, the cost of statistical multiplexers and 2400-bit/s modems would be higher than the cost of conventional multiplexers and 4800-bit/s modems on that segment of the network.

For the Phoenix-to-Denver segment, statistical multiplexers are used because their absence would require a data rate of 9600 bit/s on that line and 9600 bit/s modems. Because a 9600 bit/s modem costs $235 per month, using 2400 bit/s modems will result in a monthly modem savings of $170 per end, or $340. Because a statistical multiplexer costs only $25 per month over an expanded conventional 4-channel multiplexer, using that type of multiplexer will cost $290 per month less ($340 to $50) than using conventional TDM's with 9600 bit/s modems.

Based on the preceding, the cost of servicing Boulder City through Phoenix is listed in Table 14.4.6.

For comparison to the $2275.77 cost of servicing Boulder City and Phoenix, the previous cost can be found in Figure 14.4.3. Boulder City to Denver costs $1354 per month while Phoenix to Denver costs $1249, for a total of $2603 per month on an individual basis. Based on the preceding analysis, it is more economical to transmit the multiplexed data from Boulder City through Phoenix on to Denver.

With this, the initial network can be finalized, as shown in Figure 14.4.8. In this illustration, the total monthly service cost per segment is shown in parentheses. Note

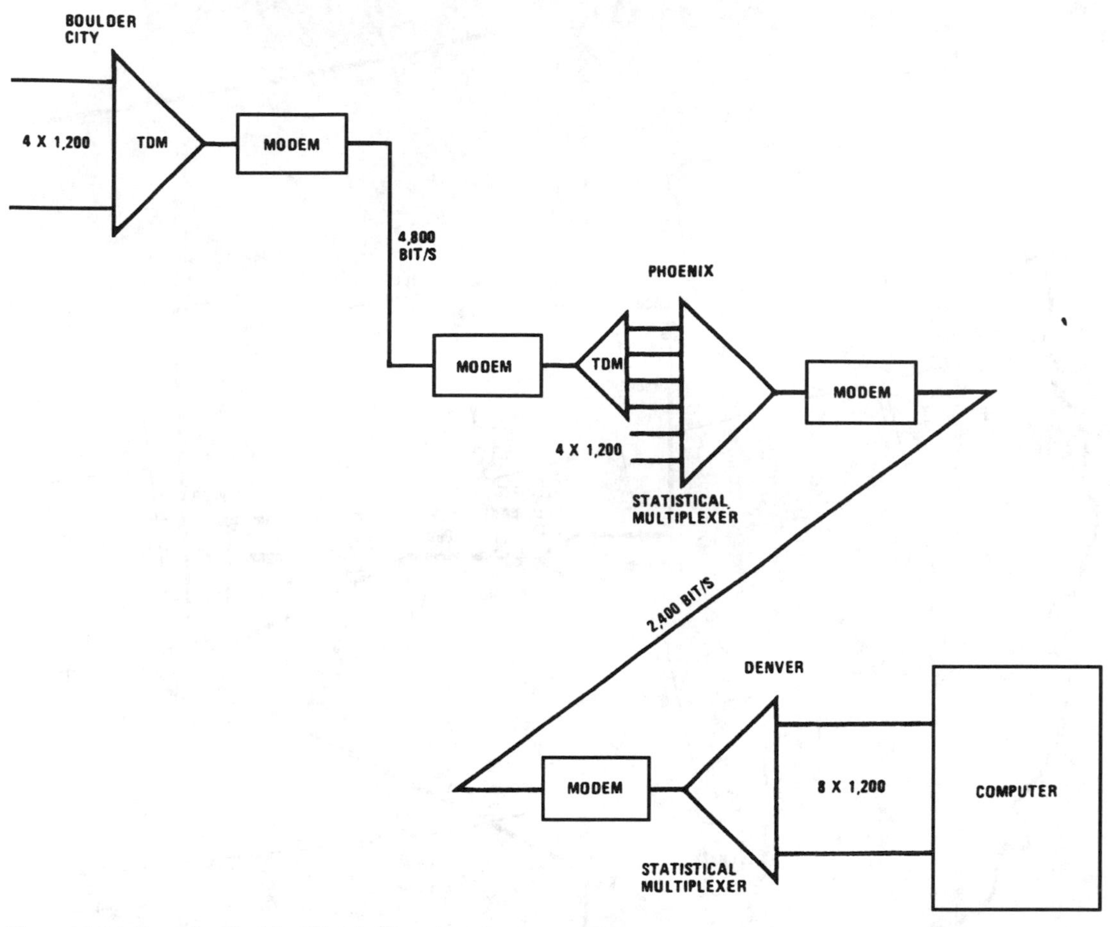

Figure 14.4.7 Servicing Boulder City via Phoenix.

TABLE 14.4.6 Boulder City—Phoenix—Denver Cost

Boulder City to Phoenix	
Multiplexers 2 @ $100	$ 200.00
4800-bit/s modems 2 @ $135	270.00
Leased line cost and terminations	546.77
Subtotal Boulder City to Phoenix	$1016.77
Phoenix to Denver	
Statistical multiplexers 2 @ $175	$ 350.00
2400-bit/s modems 2 @ $65	130.00
Leased line cost	779.00
Subtotal Phoenix to Denver	$1259.00
Total Cost	$2275.77

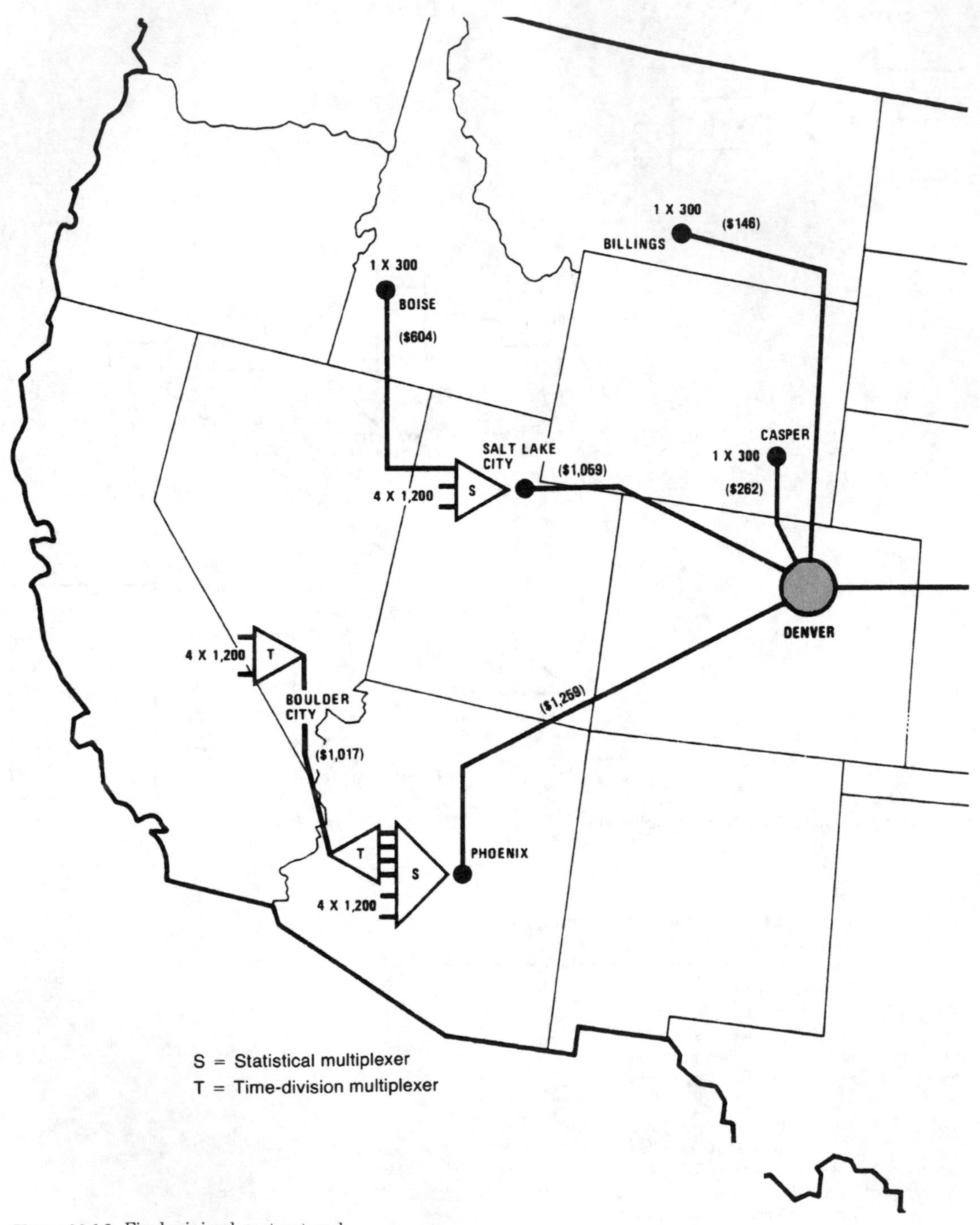

Figure 14.4.8 Final minimal-cost network.

that the monthly network cost is now $4347 per month. In comparison, the network illustrated in Figure 14.4.8 with the initial multiplexer placements would cost $4884 per month. The most economical method from Table 14.4.3, column 13, results in an expenditure of more than $10,000 per month because data concentration equipment had not yet been considered. Although the obvious use of multiplexers reduced the network cost from over $10,000 to $4889 per month, through an optimization process, the cost was reduced more than $500 per month.

Again, all costs are for illustrative purposes only.

14.5 Optimization Programs

The task of topology optimization can be long and tedious, even when not considering performance constraints—which might have large effects on the design of complex networks. The interconnection of network locations as economically as possible while also meeting performance constraints is one of the most complex tasks facing the network analyst.

In addition to initially selecting the types and locations of data concentration equipment, each terminal in the network must be examined to determine if it is economical to transmit data through a concentration device or directly to the host computer. Next, concentration points must be examined. The routing and data rates of the circuits from each of the concentration devices are routed to ther such devices to determine if transmission costs can be reduced while maintaining a desired level of performance.

While these tasks are being pursued, the network analyst must evaluate the tariff structures, not only of telephone companies, but also of specialized carriers, satellite carriers, and value-added carriers. For large networks containing several hundred terminals scattered over a large geographical area, the calculations required could result in a considerable effort, and a change in one or more tariffs could result in errors in the initial computations, making obsolete the initially designed minimum-cost network.

Fortunately for the network analyst, a number of data communications software network planning tools have been developed by several consulting organizations.

These tools can be purchased, leased, or used on a timesharing-agreement basis. They assist the network analyst in automatically creating least-cost network layouts, once-specified parameters are entered. They help analyze network performance under varying conditions, and create and maintain computerized network databases. These databases then form the foundation for network analysis under numerous changing environments. One of the leading network planning tools is MIND.

The MIND system

The MIND (modular interactive network designer) system was developed by Network Analysis Corp. (now Contel) of Great Neck, New York, for users with little or no computer experience who desire to implement the most modern and cost-effective network solutions.

Using any type of asynchronous terminal, you can access the MIND system through dialing a timeshared computer service organization at which the software and tariff databases reside. With the aid of a users' manual, reinforced by the online prompting of MIND, users describe their network requirements to MIND's network editor module, which is used as the database for later design and analysis. Through the editor, one can move, add, or delete terminals, concentrators, multiplexers, and lines, and change traffic patterns or hardware characteristics. MIND analyzes each set of requirements, tells the user how best to satisfy them with a least-cost line layout, and provides accurate data on operating cost, response time, reliability, and a host of other factors.

Optimization with MIND

To show how the MIND program works, assume that you wish to connect one terminal located in each of seven cities to our computer center in Macon, Georgia. Also assume that the terminals operate synchronously and can be connected via one or more multipoint lines because of the availability of appropriate poll and select software in the front-end processor.

From an analysis of anticipated operations, estimate the transmit and receive traffic in kbit/s, as illustrated in Table 14.5.1. This data is used to build the database required by the computer program for analysis and design of the network.

In Figure 14.5.1, a portion of the process required for building and verifying the database is shown.

As illustrated by the numeric 1 circled, all cities in the network are entered with identification by city and state abbreviation, including the computer center, which is entered as MACGA. Next, area codes and exchanges are entered for Macon and the other network locations. The numeric 2 circled indicates the data entry for area code and exchange to identify Macon.

Next, by the entry of the characters SN, denoted by the numeric 3 circled, the program is instructed to accept traffic data for the six remote cities. A portion of the data entry interaction between the terminal operator and the computer is illustrated in Figure 14.5.2 for Houston and Dallas.

Once traffic data, as well as the terminal area codes and exchange locations, have been entered for all cities, the terminal operator lists the previously entered data for verification, entering the symbols TN to the program. Next, by the use of the symbol *, which denotes "all," all node data is printed in tabular form, as illustrated in Fig-

TABLE 14.5.1

Location	Area code/ exchange	Traffic in kbit/s	
		Transmit	Receive
Atlanta, Ga.	404363	.05	.125
Houston, Texas	713220	.06	.250
Dallas, Texas	214321	.06	.125
Cincinnati, Ohio	513242	.10	.250
Cleveland, Ohio	216921	.06	.125
Columbus, Ohio	614221	.05	.125
Denver, Colo.	303292	.05	.125

```
   *** MIND II *** Version 8.1.1 *** MAR. 2, 1982
%
% 4/2/82 MIND CONTAINS REVISED ATT TARIFF RATES
%        REPRESENTING AN APPROX. 1.6% RATE INCREASE
%
% Type HELP for more info. End all input with a carriage return.

NDS198I: PAGE LENGTH IS          66

READY AT HOME !EDIT
.ENTER EDITING COMMAND OR ?.  !ET
READY AT EDIT/ET   !AN
.ENTER NODE ID OR ? . !MACGA,ATLGA,HOUTX,DALTX,CINOH,CLEOH,COLOH,DENCO,①
EDT304I: NODE MACGA HAS BEEN ADDED.
EDT304I: NODE ATLGA HAS BEEN ADDED.
EDT304I: NODE HOUTX HAS BEEN ADDED.
EDT304I: NODE DALTX HAS BEEN ADDED.
EDT304I: NODE CINOH HAS BEEN ADDED.
EDT304I: NODE CLEOH HAS BEEN ADDED.
EDT304I: NODE COLOH HAS BEEN ADDED.
EDT304I: NODE DENCO HAS BEEN ADDED.
READY AT EDIT/ET    !
.ENTER EDITING COMMAND OR ?.  !EP                            ⎫
READY AT EDIT/EP   !SN                                       ⎪
.ENTER NODE ID OR ? .     !MACGA                             ⎬ ②
ENTER KEYWORD OR ?  !ACEX                                    ⎪
ENTER Terminal Area Code and Exchange (AAAXXX)     !912744   ⎭
ENTER KEYWORD OR ?  !ROLE
ENTER Role of Terminal in Network  !CONC
ENTER KEYWORD OR ?  !
READY AT EDIT/EP   !SN ③
```

Figure 14.5.1

```
READY AT EDIT/EP   !SN
.ENTER NODE ID OR ? .     !HOUTX
ENTER KEYWORD OR ?  !ACEX
ENTER Terminal Area Code and Exchange (AAAXXX)     !713220
ENTER KEYWORD OR ?  !HOME
ENTER Homing Point of Terminal!MACGA
ENTER KEYWORD OR ?  !RC
ENTER Receive Traffic Value   !.250
ENTER KEYWORD OR ?  !TX
ENTER Transmit Traffic Value  !.06
ENTER KEYWORD OR ?  !
READY AT EDIT/EP   !SN
.ENTER NODE ID OR ? .     !DALTX
ENTER KEYWORD OR ?  !ACEX
ENTER Terminal Area Code and Exchange (AAAXXX)     !214321
ENTER KEYWORD OR ?  !HOME
ENTER Homing Point of Terminal!MACGA
ENTER KEYWORD OR ?  !RC
ENTER Receive Traffic Value   !.125
ENTER KEYWORD OR ?  !TX
ENTER Transmit Traffic Value  !.06
ENTER KEYWORD OR ?  !
```

Figure 14.5.2

```
READY AT EDIT/T      !TN
.ENTER NODE ID OR ? .      !*
NODE   ACEX      CITY       ST NDEV    RC     TX    AVAIL  ROLE HOME   TER
MACGA 912744 MACON        ,GA   1   0.000  0.000  .0000 CONC           0.00
ATLGA 404363 ATLANTA      ,GA   1   0.125  0.050  .0000 TERM MACGA   0.00
HOUTX 713220 HOUSTON      ,TX   1   0.250  0.060  .0000 TERM MACGA   0.00
DALTX 214321 DALLAS       ,TX   1   0.125  0.060  .0000 TERM MACGA   0.00
CINOH 513242 CINCINNATI,OH   1   0.250  0.100  .0000 TERM MACGA   0.00
CLEOH 216921 CLEVELAND ,OH   1   0.125  0.060  .0000 TERM MACGA   0.00
COLOH 614221 COLUMBUS   ,OH   1   0.125  0.050  .0000 TERM MACGA   0.00
DENCO 303292 DENVER       ,CO   1   0.125  0.050  .0000 TERM MACGA   0.00
```

Figure 14.5.3

ure 14.5.3. Here, the column TER denotes the monthly station terminal charge, which is zero for all locations because up to now data has been entered in the build mode of the program and the optimization portion has yet to be executed.

Once the database is completed, the network optimization program is executed. This program will not only provide routing and pricing information, but it is also used to graphically display a network map of the resulting multipoint circuits.

Figure 14.5.4 shows a portion of the output available from the execution of MIND: the Trunk Line Summary. From the Trunk Line Summary, the total monthly cost of the multipoint circuits is $3219.32. The drop-cost of $288.40 per month is actually the sum of the station terminal costs. The V and H coordinates listed and used for computing airline mileage between points were based on the area code and exchange location of each terminal that was previously entered. Thus, the program alleviates the necessity of obtaining VH coordinates by permitting users to enter the more commonly known area and exchange data elements. Again, referring to the interexchange channel details (Figure 14.5.4), the AB column heading denotes the rate category of the circuit. (All costs are illustrative only.)

Lastly, Figure 14.5.5 shows the line layout for the optimized network. As shown, one circuit links Cleveland, Columbus, Cincinnati, and Atlanta to Macon, and the second multipoint circuit connects Denver, Dallas, and Houston via Atlanta to Macon. One line connects Macon to Atlanta, which, through the use of split-stream modems, connects the two multipoint circuits on one line to Macon.

```
   TRUNK LINE SUMMARY:
  NO TRUNK LINES IN NETWORK

*** NAC'S MIND: A DESIGN TOOL OF NETWORK ANALYSIS CORPORATION ***
    TOPO: CIRCUIT REPORT                   Tue 18-May-1982 12:27PM
 AT CONCENTRATOR MACGA:
      CIRCUIT MACGA1        :
         STATION DETAILS
   TERID ---LOCATION--- ---CITY--- ST  TRANSMIT   RECEIVE NDV --COST--
   MACGA                MACON      GA     0.000     0.000   0     0.00
   DENCO                DENVER     CO     0.050     0.125   1    36.05
   DALTX                DALLAS     TX     0.060     0.125   1    36.05
   HOUTX                HOUSTON    TX     0.060     0.250   1    36.05
   CINOH                CINCINNATI OH     0.100     0.250   1    36.05
   COLOH                COLUMBUS   OH     0.050     0.125   1    36.05
   CLEOH                CLEVELAND  OH     0.060     0.125   1    36.05
   ATLGA                ATLANTA    GA     0.050     0.125   1    72.10

INTEREXCHANGE CHANNEL DETAILS
                          CITY       ST -ACEX- VVVV HHHH AB ALMI     COST
           FROM DENCO DENVER     CO 303292 7501 5899 A
           TO   DALTX DALLAS     TX 214321 8436 4034 A  660.    779.32
           FROM DALTX DALLAS     TX 214321 8436 4034 A
           TO   HOUTX HOUSTON    TX 713220 8938 3536 A  224.    369.48
           FROM HOUTX HOUSTON    TX 713220 8938 3536 A
           TO   ATLGA ATLANTA    GA 404363 7260 2083 A  702.    818.80
           FROM ATLGA ATLANTA    GA 404363 7260 2083 A
           TO   CINOH CINCINNATI OH 513242 6263 2679 A  368.    504.84
           FROM CINOH CINCINNATI OH 513242 6263 2679 A
           TO   COLOH COLUMBUS   OH 614221 5972 2555 A  101.    253.86
           FROM CLEOH CLEVELAND  OH 216921 5574 2543 A
           TO   COLOH COLUMBUS   OH 614221 5972 2555 A  126.    277.36
           FROM MACGA MACON      GA 912744 7364 1865 A
           TO   ATLGA ATLANTA    GA 404363 7260 2083 A   77.    215.66
  TOTALS FOR CIRCUIT MACGA1       :
LOCATIONS =  7  TRANSMIT=      0.430  DROP COST=$    288.40
DEVICES   =  7  RECEIVE =      1.125  LINE COST=$   3219.32
MILEAGE= 2258.  TOTAL   =      1.555  TOTAL    =$   3507.72
```

Figure 14.5.4

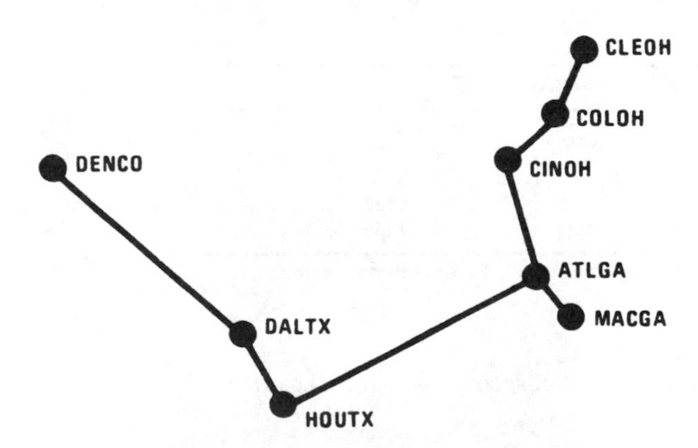

Figure 14.5.5

Questions

14.1 Design a minimum-cost network based upon the following information:

A.

	Terminal Data		Terminal Call Data	
Location	Data Rate	# of terminals	# of calls/day	connect time/call (min)
1	300	1	2	05
2	1,200	4	1	20
3	1,200	1	2	20
4	300	4	1	50

B.

Terminal Location	Distance To Computer	Rate Category
1	750	A-A
2	400	B-A
3	375	A-A
4	1,040	B-A

C. Interchange mileage between locations

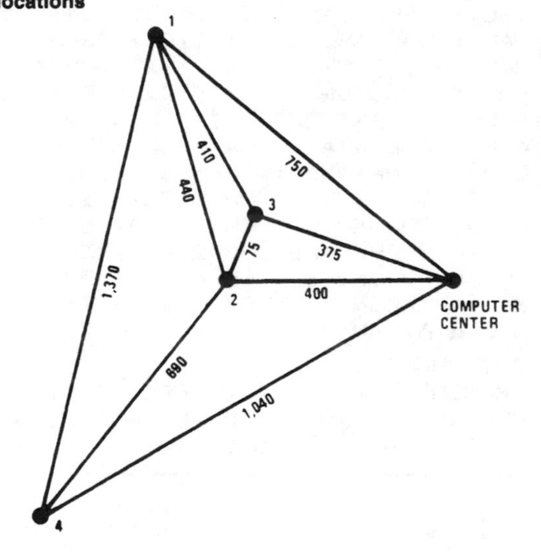

D.

Equipment rental charges/month	
4-channel multiplexer	$100
8-channel statistical multiplexer	$175
2,400-bit/s modem	$ 65
4,800-bit/s modem	$135
9,600-bit/s modem	$235

Note: Statistical multiplexer has a 4:1 compression ratio for asynchronous data.

14.2 In data communications, what is sizing?

14.3 How is the erlang defined? How is the call-second related to the erlang?

Chapter

15

Using Specialized Data Communications Components

15.1 Introduction

This chapter introduces you to the operating characteristics and uses of several specialized data communications components.

Once the descriptive material has been completed, you can turn to the four network integration problems, starting with Section 15.10. These problems can be solved by the use of the information contained in this chapter and material previously covered. All costs are for illustrative purposes only.

15.2 Limited-Distance Modems and Line Drivers

Savings can be realized by using hardwired, twisted-pair lines driven by limited distance modems (LDMs). However, choosing the most economical way to drive data pulses down these lines involves tradeoffs in cost, performance, and quality of transmission. Although LDMs lack capabilities found in conventional modems, they are often chosen because of their lower cost for localized configurations, such as office complexes, universities, and factories.

Typical point-to-point, multipoint, and repeater configurations are shown in Figure 15.2.1. Because such applications often have large numbers of terminals, the savings that can accrue by selection of the most economical transmission devices can amount to a substantial portion of the total network cost. However, in evaluating the possible savings, the cost of wire and its installation should not be overlooked because these expenses can be considerable. If wiring costs are prohibitive, the alternative is to use limited distance modems in conjunction with leased telephone lines, but there might be problems in selecting units that meet telephone company standards, which will be covered later.

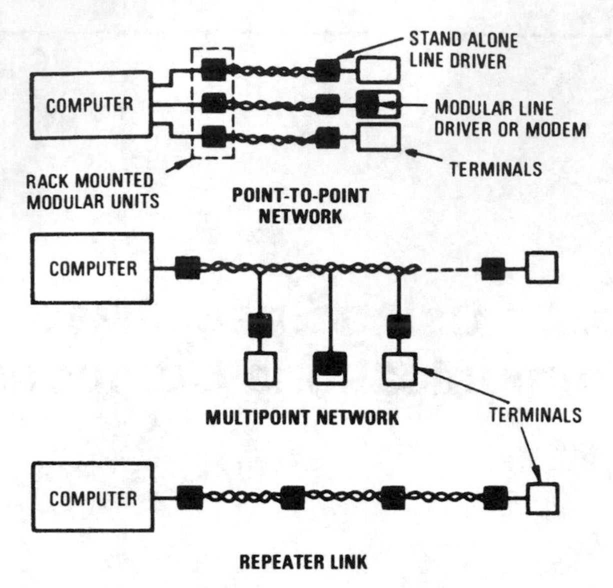

Figure 15.2.1 Link configurations.

The four methods of transmitting data, each more expensive than the preceding one, should be carefully evaluated:

- Direct connection by ordinary copper-wire pairs.
- Line drivers, which reshape distorted pulses.
- LDMs, which are simpler versions of conventional modems.
- Conventional telephone-line modems.

The range of distances that can be achieved by the various interconnection methods are shown in Figure 15.2.2. As might be expected, the greater the distance, the greater the cost.

The EIA-232 and CCITT V.24 standards limit direct connections to 50 feet for data rates to 20 kbit/s. For distances greater than 50 feet, first consideration should be given to line drivers.

A single line driver generally provides adequate performance for hundreds of feet (depending on the unit) at data rates up to 9.6 kbit/s. Beyond the distance normally specified for the line driver, signal attenuation and line distortion become significant, and tradeoffs between line drivers and LDMs must be considered. However, line drivers can be strung out as repeaters (Figure 15.2.1).

The distortion that results from long line lengths is caused by electrical characteristics of twisted-pair cables, which round off the leading edges of pulses and displace them in time. Pulse rounding, in turn, causes pulse sensing to be delayed, adding to existing delays. The effect of increasing attenuation as distance increases sometimes makes signal levels fall below the thresholds of pulse sensing circuits, thereby causing bits to be dropped.

The factors affecting the usable distances of line drivers and LDMs for long wire runs are transmission rates, distances between points, and wire types. By using line drivers, data can be transmitted at low rates over several miles of wire. For instance, at speeds up to 300 bit/s, data can be transmitted over 5000 feet of #22 twisted-pair wire without excessive distortion. But to minimize errors, high data rates require squarer pulses than low data rates. Therefore, data rates much higher than 300 bit/s suffer from reduced range. For example, 2.4 kbit/s is limited to several hundred feet between line drivers.

When to use LDMs

If line drivers cannot do the job, the next consideration should be LDMs. The basic elements of an LDM are shown in Figure 15.2.3. These devices are simpler and less

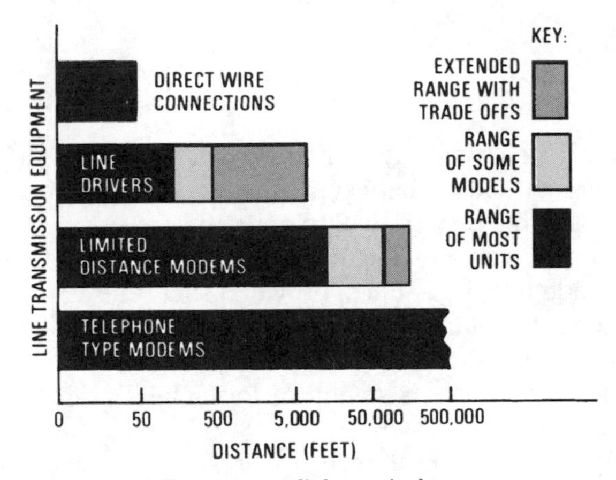

Figure 15.2.2 Four ways to link terminals.

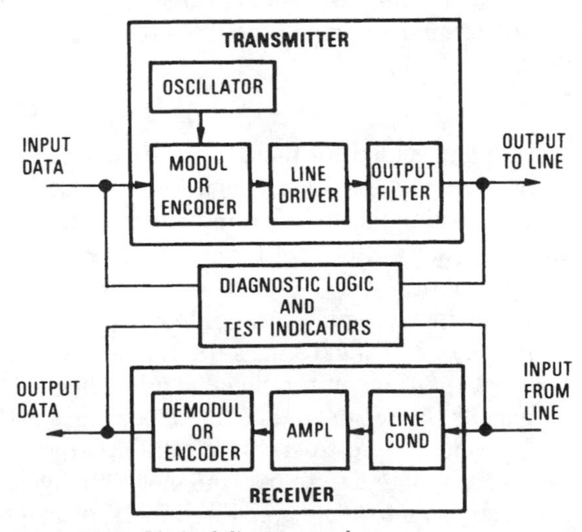

Figure 15.2.3 Limited distance modem.

expensive versions of conventional telephone-line modems. The cost advantage of an LDM increases with the data rate because three major functions performed by conventional modems can be eliminated or relaxed. At low data rates, the cost differences are not great. At high speeds, the differential can be thousands of dollars.

Among the unnecessary modem functions are multilevel modulation schemes, which compress high data rates into the narrow voice-grade channels supplied by telephone companies. Another function that can be eliminated in LDMs is immunity to frequency offset. This effect often occurs in the modulation and demodulation processes in long-distance telephone circuits. If the modulation and demodulation carriers are not precisely at the same frequency and phase, then the received signal is offset, causing demodulation errors.

A requirement that can be relaxed in LDMs is noise rejection. This factor often plays an important role in dictating the modulation technique of a conventional modem and tends to make it costly.

Settling for less

In return for low price, LDMs lack some capabilities, which could limit their usefulness. Therefore, in considering tradeoffs between LDMs and conventional modems, the buyer must decide whether his network will tolerate the deficiencies.

First, LDMs are used mostly for private line, hardwired links. Some also can operate over local-loop voice grade telephone lines, provided that: there are no loading coils between the two points; the total distance the signal must travel can be guaranteed to be within the range of the modem; the bandwidth is within the range of the telephone lines; and the power level meets standards for metallic circuits (Bell System Technical Reference 43401).

Some LDMs require a degree of technical expertise and certain test equipment to change speeds and make line-matching adjustments. Others must be returned to the factory for these jobs. However, a number of LDMs do facilitate these adjustments by means of front-panel controls or wire jumpers. As might be expected, the ease with which adjustments can be made often depends on the price of the unit.

What else to look for

At short distances, the error rates specified for most units are about n error in 10^8 bits, but these rates increase rapidly with distance. Diagnostic capabilities vary from one LDM model to another. Self-testing methods permit the user to find out if his unit is defective or if trouble exists elsewhere, with one or more lights to indicate equipment status and alarm conditions.

The self-testing features usually involve some form of loopback testing. The simplest diagnostic is a local loopback test, which returns the transmitter output to the receiver of the same LDM so that the transmitter signal can be checked for errors. Another test, dc busback, transmits the received data and clock back to the LDM at the other end of the line to provide an end-to-end test. Still another test, remote loopback, triggers dc busback at a remote site to test the elements bypassed in the preceding test. In all of these tests, an internally generated message can be used instead of existing data.

In a polled network, only one driving unit is needed at the primary site. For networks in which the wire runs exceed the limits of the line driving units, repeater stations might be needed.

Data rates of most LDMs range between 2.4 and 9.6 kbit/s. However, some units offer speeds down to 300 bit/s or less and others reach data rates beyond 1 Mbit/s.

Hooking up the network

An important consideration in designing a network containing LDMs is selection of the most economical form of wiring. If the equipment is wholly situated on private property, cables can be strung, or telephone company lines can be leased.

Wire sizes and types deserve special attention. Twisted-pair wire serves for relatively short distances, and coaxial cable might be needed for longer runs. For example, one vendor specifies that a data rate of 4.8 kbit/s can be transmitted over 15 miles of #19 twisted-pair wire, but only five miles for #24 wire, at speeds near 1 Mbit/s, these distances drop to 10 percent of the 4.8 kbit/s distances. If, instead, the units are interconnected by video-quality coaxial cables, the signals can be reliably transmitted as far as 17 miles at 4.8 kbit/s and 2.4 miles at 1 Mbit/s. Each manufacturer can provide a set of curves showing suggested bit rates for wire lengths of various sizes. An example is shown in Figure 15.2.4.

15.3 Port Sharing Units

Port sharing units are devices that sit between host computer and modem and control access to and from the host for up to about six terminals. In this way, port sharing units are able to cut down on the number of computer ports needed for these terminals.

The port sharing unit is versatile, comparatively inexpensive (about $500), and available from many modem and terminal manufacturers. It can save the cost of a relatively expensive multiplexer that does essentially the same job, but may have more capabilities than are needed. Port sharing units can also be used on local peripherals, and so expand the job that can be done by a single port on the host computer.

To put the concept of port sharing into perspective, be aware of related devices designed to cut networking costs. Modem-sharing units and line sharing units are

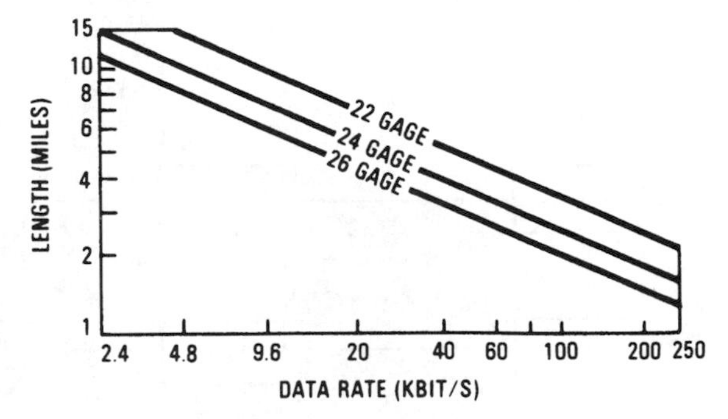

Figure 15.2.4 Heavier wire means more distance.

available to minimize modem and line costs at remote locations, but they do not deal with the problem of overloading the host computer's ports. Modem- and line-sharing units are a partial solution to the high cost of data communications networking, but they are limited as to the types of problems they can handle. They can, by themselves, complicate the life of the network designer.

One problem that surfaces when either modem- or line-sharing units are used by themselves is the distribution of polled terminals within the network. For either kind of sharing unit to be effective, the terminals should be placed so that several are grouped close together. Modem-sharing units with an EIA-232 interface option can be used to serve remote terminals. Although this interface allows the data communications network a slightly more flexible configuration, with a number of terminals remote from the sharing unit, the number of remote terminals that can be served by any one unit is usually limited to one or two.

Another disadvantage of this arrangement is that if either modem on the high-speed link between the computer and the modem-sharing unit should fail, or if the circuit itself goes down, all the remote terminals become inoperative. Multiplexed terminals can use the dial-up network to restore data communications if the dedicated line fails. Polled terminals, however, do not have this advantage because the host computer software is set up to seek and recognize the addresses of specific terminals in a certain order on the line. Thus, polled terminals must stay in their respective places, relative to each other, along the communications route. Any change in route necessitates changes in hardware at the terminals as well as software at the host computer.

Port sharing alone

Port sharing, then, is presented either as an alternative or as a supplement to modem and line sharing, in networks without multiplexers. A port sharing unit is connected to a computer port and can transmit and receive data to and from two to about six either synchronous or asynchronous modems (Figure 15.3.1). Data from the computer port is broadcast by the port-sharing unit, which passes the broadcast data from the port to the first modem that raises a receiver-carrier-detect (RCD) signal. Data for any other destination will be blocked by the unit until the first modem stops receiving. The port-sharing unit thus provides transmission by broadcast and reception by contention for the port connected to it. Like a modem-sharing unit, a port-

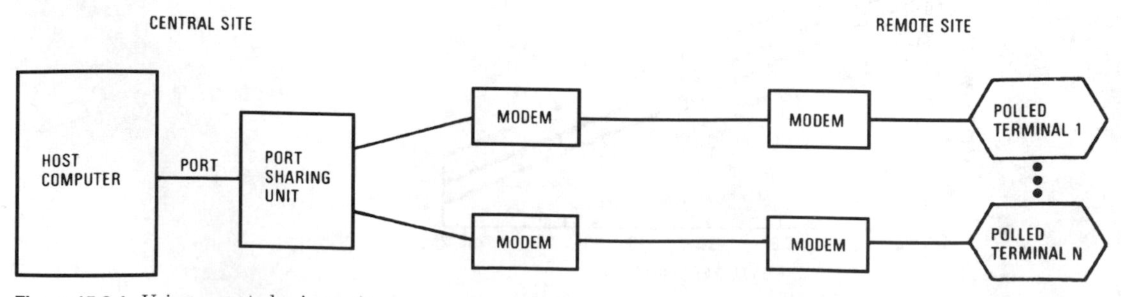

Figure 15.3.1 Using a port sharing unit.

sharing unit is transparent with respect to data transmission. Data rates are limited only by the capabilities of the terminal, modem, and computer port.

To gain the same results without a port-sharing unit would require a multidrop configuration. Both the port sharing unit and a multidrop network allow a large number of terminals to be served by one computer port, but in a multidrop network, the failure of any part of the circuit will put all terminals beyond the failure out of action. In the configuration in Figure 15.3.1, however, failure of a modem or an outage on the line will cut out only a terminal on that segment. Failure of a computer port or of the port sharing-unit would, of course, bring down the entire network, but these devices are stable and such failures are fairly unusual.

Port sharing as a supplement

Port-sharing units can also be used alongside modem sharing units. If modem sharing units alone are used, a situation can arise where there are not enough ports to serve the network, as in Figure 15.3.2A. If each modem-sharing unit serves its full complement of terminals, and all the computer ports are in use, expansion of the network, even by just one port, might require a second mainframe computer.

This problem can be dealt with by using a port-sharing unit at the central site, which, by cutting down the number of ports needed, allows a network to expand without using additional computer ports. Figure 15.3.2B shows how one port-sharing unit with a two-modem interface can free a computer port from the configuration shown in Figure 15.3.2A.

One versatile feature of port-sharing units is an option that allows the unit to accept a local interface instead of the normal EIA-232 interface so that two local terminals can be operated without modems at the central site.

Although both modem-sharing units and port-sharing units are similar in the way they are used, there is an important difference in the normal placing of their interfaces. Table 15.3.1 compares the characteristics of a port-sharing unit with those of modem- and line-sharing units. Table 15.3.2 compares the typical monthly cost of a four-terminal network in multiplexed and port-sharing configurations. The break-even point comes when the cost of each leased line (X) reaches $180 a month. This figure can be arrived at by taking the total known costs (for multiplexers, computer ports, modems, and terminals), and adding an unknown cost—for the leased line—which remains constant. Balanced against this is the somewhat smaller total rental amount for a single port-sharing unit, one computer port, four lower-speed modems, and four terminals.

More leased lines

The lower part of the table, however, shows an increase in the number of leased lines from one to four. The upper part of the table, therefore, gives a fixed cost of $1420 with a variable amount on a one-time basis for a single leased line, while the lower part shows a significantly lower equipment cost, but gives a variable cost for leased lines four times that for a single-line multiplexed configuration.

In this example, $1420 less $880 equals $540 or 3× (with × still representing the variable cost of the leased line). Dividing $540 by three gives a break-even point of

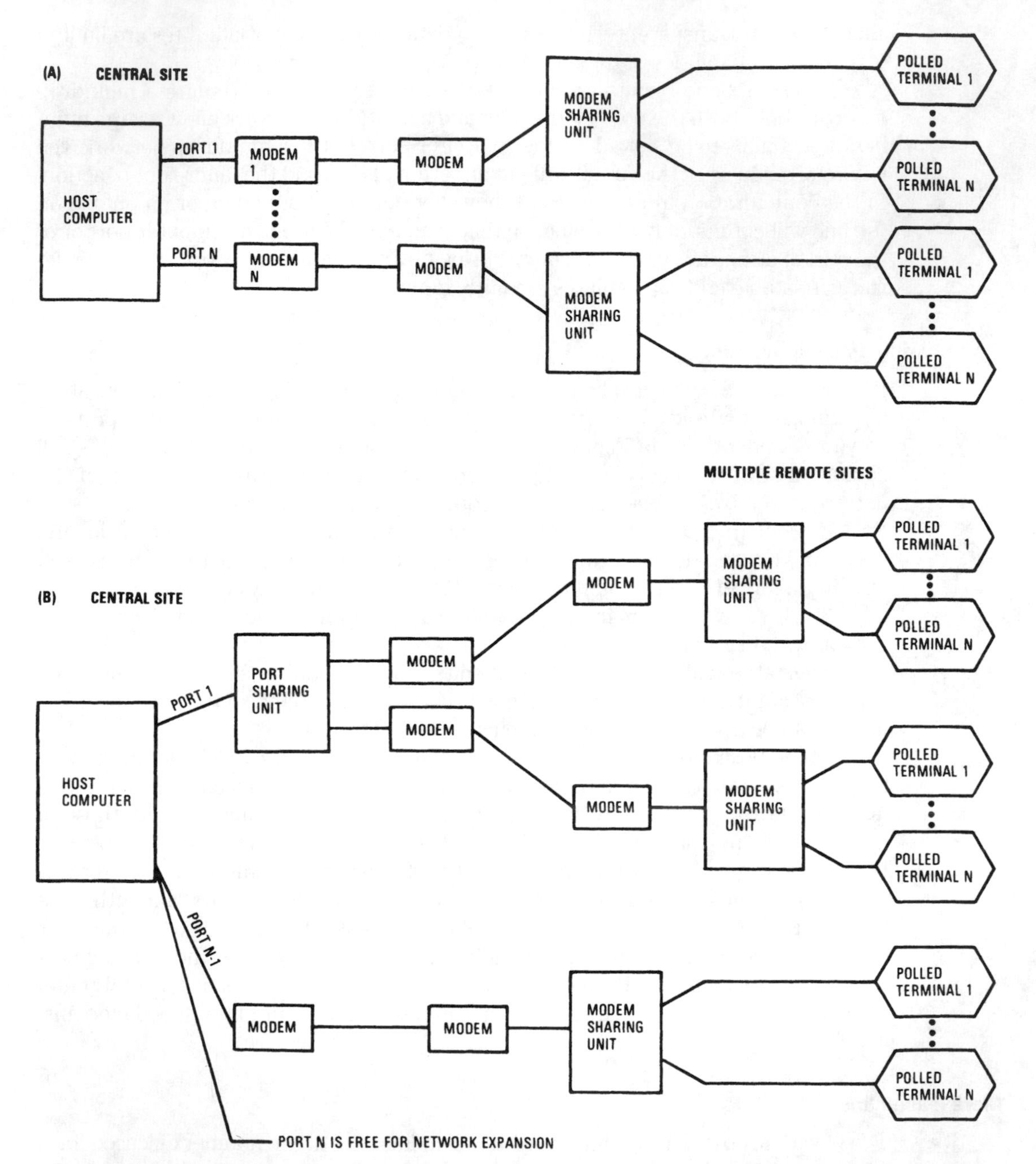

Figure 15.3.2 Two sharing techniques combined.

$180. Thus, until monthly leased-line charges total $180 for each line, the use of a port-sharing unit is more economical.

In addition, users can (without increasing network costs) add up to two more local terminals to a configuration based on a port-sharing unit because the table is

TABLE 15.3.1 Features of Sharing-Units

Feature	Modem sharing/ line sharing unit	Port sharing unit
Transmit mode	Broadcast	Broadcast
Receive mode	Contention	Contention
Number of modems interfaced	2 to 32	2 to 6
Terminals supported	Polled	Polled
Options	EIA-232 interface (MSU to modem)	Local interface (PSU to terminal)
Normal interface placement	Between modem and terminal	Between computer port and modem

TABLE 15.3.2 Comparison of Monthly Costs

Multiplexed configuration costs	
Two 4-channel TDM's @$120	$ 240
4 Computer ports @$35	140
2 9600 bit/s modems @$220	440
4 Terminals @$150	600
1 Leased line	X
Where X = monthly cost of a leased line	$1420 + X
Port sharing unit configuration costs	
1 Port sharing unit @$25	$ 25
1 Computer port @$35	35
4 2400 bit/s modems @$55	220
4 Terminals @$150	600
Leased lines	4X
Where X = monthly cost of a leased line	$880 + 4X

based on the costs related to four terminals, and the average port-sharing unit can support up to six. The only additional cost would be for the rental of the terminal units.

In order to carry two additional terminals in a multiplexed configuration, however, the user would have to pay substantially more. On the basis of the table, $150 would be required for each terminal, and an additional $35 each for two more computer ports. As explained earlier, the cost of adding two computer ports can be further aggravated if all ports are already in use on the host computer so that any extra load requires another entire processing unit.

The port-sharing unit, therefore, is most evidently a cost-saving tool when the user is already straining the CPU to its limits. Although saving money is a full-time preoccupation for all cost-conscious data communications managers, and port sharing should be considered in any polled-terminal situation where instantaneous response is not the most important network condition, there are times, such as when the CPU runs out of capacity, that the cost of any further network expansion takes a leap from a few hundred dollars to perhaps tens of thousands for another CPU.

15.4 Port Selectors

Even in a network in which terminals must contend for access to ports on the host computer, it is unlikely that all ports will be busy at the same time. So most networks can operate with more terminals than computer ports. But because each terminal is not connected to its own dedicated port, a switching device, called a *port selector* or *intelligent patch panel (IPP)*, is needed to connect incoming messages to available ports.

Port selectors allow terminals to be added to a network without increasing the number of computer ports. When a network is being designed, the selectors can reduce the cost of its host computer by cutting back on the number of ports needed.

These IPPs operate in a manner similar to telephone rotaries (stepping switches that sequentially search for available outgoing lines), except that the selectors provide appropriate interfaces between computers and terminals. Some selectors have extra features specifically applicable to data networks.

The difference between port selection units and port sharing devices is that port sharing devices are used in polled networks (where the host controls the traffic flow), while port selection units are in contention networks, in which terminals transmit to the host on a random basis. With a port selection device, access to any port is on a first-come, first-served basis.

The port selector looks for incoming data from terminals connected to its line side, and at the same time keeps track of any computer ports that are idle. Some port selectors can be arranged to form subgroups of contending terminals and ports, so that certain terminals can only be connected to an assigned group of ports. When one of these terminals requests access, the selector searches for an available port in its group and connects the terminal to that port. If all the designated ports are taken, the selector continues to scan until one becomes available, or until the request for access ceases. Another option is a "busy out" feature, whereby the port selector signals new callers that all available ports are in use.

Port selectors can be used at remote locations to make input terminals contend for a smaller number of communications lines or multiplexer ports. Consider a remote location with 32 terminals each communicating with the central computer at 300 bit/s. One economical method to move the data is to multiplex transmission from the 32 terminals over a single leased line, at an aggregate transmission speed of 9600 bit/s. This requires two 32-port multiplexers and 32 computer ports. Assume that 16 additional 300 bit/s terminals must be added to the remote site. Because the leased line is already operating at its 9600 bit/s limit, one alternative is to upgrade the multiplexers by adding 16 ports to each one, and to replace the modems and voice-grade communications link with wideband facilities. This method requires 16 additional computer ports to service the additional terminal traffic. But the added line capacity would be very expensive because wideband facilities are far more costly than voice-grade lines.

Another option would be to install a pair of 16-port multiplexers, one additional leased line between them, and two 4800-bit/s modems to transmit on the new line. But 16 additional computer ports would still be required to service the additional terminals.

Evaluating port selection

Increasing the number of terminals that can be handled by the central site by means of a port selection arrangement can indeed be less expensive than doing it by expanding the capacity of the multiplexing equipment. To find out if port selection is the practical answer to a particular situation, the user should first assess the demands the present terminals are making on the network, then estimate the additional demands that will arise because of the proposed new terminals. If, for example, it is found that even though 48 terminals are needed, the total number demanding access to the central computer at any one time is unlikely to be more than 32; and that this limit of 32 is going to be exceeded during a total of perhaps two hours a week; and that even in those two hours, the total demand will hardly ever mount to more than 35 terminals at any one time. Then, if the user is willing to risk the chance that three terminals could not immediately be able to reach the computer during about two hours in the week, a standard 32-port selector can be installed (Figure 15.4.1). It services 48 lines at 32 ports on a contention basis.

The port selector then offers a contention ratio of 48 to 32, or 3 to 2 (48 terminals connected to the line side of the selector and 32 outputs to the multiplexer ports). It eliminates the need to add 16 computer ports, to upgrade or add multiplexers, or to upgrade or add communications links. Although a telephone company rotary could be used for the port selection process, 80 modems would then be needed (48 for the terminals and 32 for the carrier lines). A port selector can thus help reduce the number of modems needed for port selection by rotaries, as well as the expense of 48 communications links. By directly connecting the terminals to the line side of the port selector, it becomes possible to do away with the low-speed modems. A port selector that can make 48 lines contend for 32 ports might cost about $15,000. But the rotary cost would equal that of a port selector in a little less than a year.

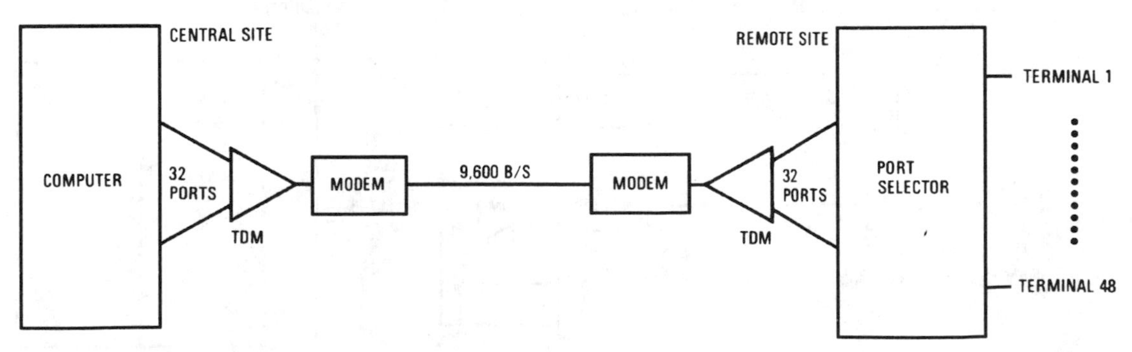

Figure 15.4.1 Serving 48 terminals with 32 ports.

A typical computer network, as in Figure 15.4.2A, illustrates what must be taken into account in coming to a decision about port selectors. In this example, 48 computer ports handle messages from two multiplexers and a local rotary. Each multiplexer provides 16 terminal-to-computer connections, as well as 16 dial-in ports at the computer site. If the network is distributed over several time zones, peaks in utilization of the terminals will occur at different times in different places (Figure 15.4.2B).

Load patterns

A typical profile of the number of users logged onto the network from each geographical area can be computed with statistical software packages provided by computer manufacturers. Although other networks might not reveal the same patterns, they might be similar because normal working habits are a significant factor in the fluctuations of network utilization.

At the start of working hours, use gradually builds as people arrive at the office and settle down to work. During the morning coffee break period, the number of users decreases temporarily, with the length of time and the degree of the dropoff varying from place to place. Morning peak use is followed by a drop in activity during the lunch period. Use then builds up until the afternoon break, and peaks again

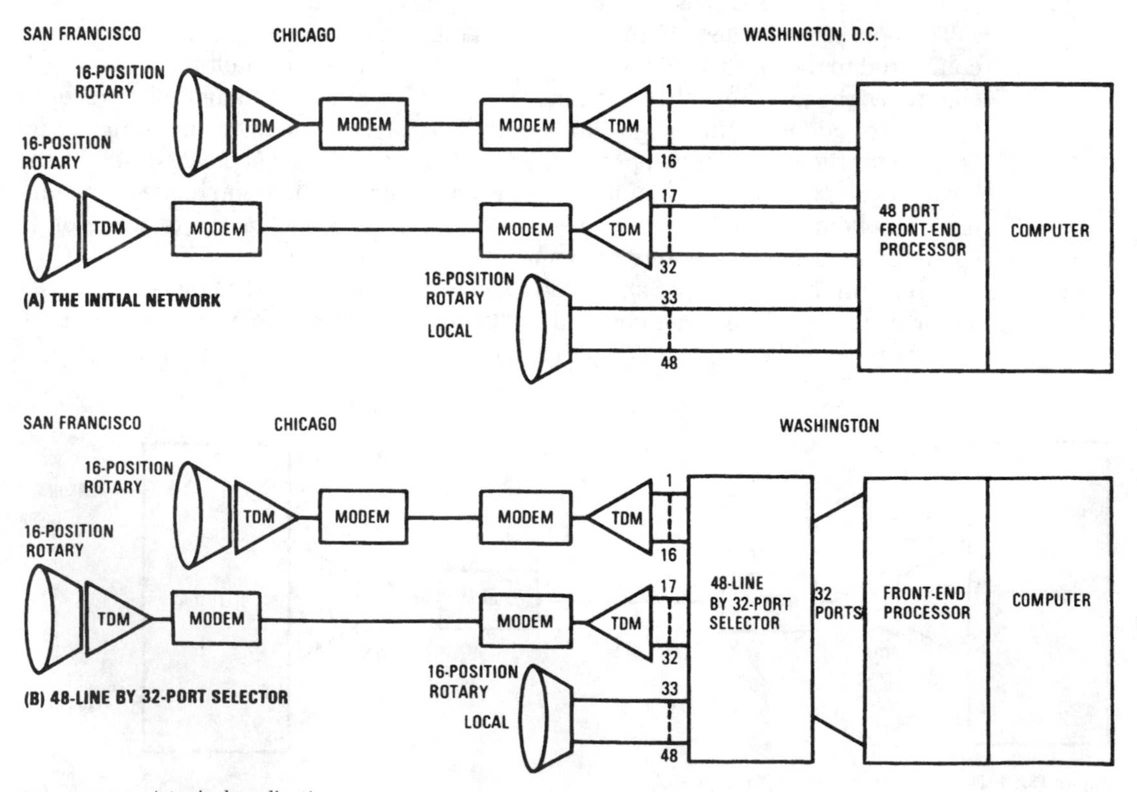

Figure 15.4.2 A typical application.

as people rush to complete the day's work. As close of business approaches activity tapers off until only a few terminals remain on line.

By combining profiles, an overall network profile can be developed which presents the total number of terminals on line. Assume that you find that 95 percent of the time, up to 37 terminals will be on-line or seeking access between 4:15 p.m. and 4:45 p.m., and 99 percent of the time there will be as many as 40 users between 4 p.m. and 5:30 p.m.

Returning to the network in Figure 15.4.2A, it is now apparent that 99 percent of the time eight or more of the 48 computer ports are not in use, and that 95 percent of the time, 11 or more ports will be idle. Thus, the use of a port selector becomes a question of economics against inconvenience. Is the cost of a number of mostly idle computer ports worth the seldom used advantage of being able to connect all terminals, simultaneously and without delay? A related question that the network designer must answer is whether the computer can process all messages rapidly enough when all terminals are on-line.

Assume (Figure 15.4.2A) that the 90th percentile of port usage is decided on. This would mean that a 32-port selector would be needed (Figure 15.4.2B). After calculating the cost saving possible with such a port selector, you can determine if the sacrifice of 16 continuously available ports is justifiable. Investigations might also be made into the savings possible with a 48-line by 37-port selector (95th percentile), or a 48-line by 40-port selector (99th percentile).

Although the saving might be considerable for a single computer installation, additional savings are possible for installations that have redundant computers because every excess computer port that can be eliminated on one front end can also be eliminated on the other. In addition, reductions can also be possible in the capital outlay for devices that switch between central processing units.

"Blue-sky" outlook

The intelligent patch panel evokes much wishful thinking because of its inherent flexibility. The three most frequently mentioned "wish-list" items are:

1. Speed, code, and protocol conversion to make the IPP a truly universal interface and allow its use with large mainframes
2. Preprocessing capability so that certain CPU functions could be off-loaded to the IPP
3. Demultiplexing of remote multiplexers to further enhance network cost savings (such as a remote eight-channel multiplexer connected to a single local IPP interface, with the IPP acting as the local multiplexer).

However, these desirable features would also increase the price and complexity of such equipment and push its function into the realm of the large, expensive front-end, or communications, processor.

The intelligent patch panel benefits both the user and the site manager. For the user, the IPP is a simple interface that affords keyboard selection of authorized computer resources and the feedback and control to minimize intervention by computer-site personnel. To the computer-site manager, the IPP gives the information and control necessary to dynamically change the entire network. This results in a cost-effective configuration, savings in computer ports and operator time, and increased user efficiency.

Finally, because of the nature of any microprocessor-based device, the IPP is inherently programmable, thus allowing for future enhancements and avoiding rapid obsolescence.

15.5 Data Communications Switches

At one time, data communications switches were found mainly in technical control rooms, where they help in on-line monitoring, fault diagnosis, and digital and analog testing. But now they are also being used to reroute data quickly and efficiently and to replace several dedicated backup units with just one, enabling a single terminal (say) to act is standby for several on-line terminals.

The four basic categories of switches are fallback, bypass, crossover, and matrix. Two or more of these, from the same or different categories, can be chained to serve still other data communications requirements. Furthermore, within each category there are two types of switches: the so-called telco switches, which transfer four-wire leased or two-wire dial-up telephone lines, and the EIA switches, which transfer all 24 leads of an EIA-232 interface.

Fallback switches

The fallback switch is a rapid and reliable means of switching network components from on-line to standby equipment. The EIA version selects either of a pair of 24-pin-connected components, which, as shown in Figure 15.5.1 can be terminals, modems, or channels on a front-end processor.

In the first example, two terminals share a single modem. This configuration might be required; for example, when terminals have the same transmission speed, but use different protocols so that each communicates with a different group of remote terminals or computers.

In the second example, (Figure 15.5.1 B), one terminal is provided with access to two modems, one of which is redundant but needed for uptime reliability. Alternatively, the first modem might enable the terminal to transmit to another terminal at 2000 bit/s during one portion of the day, and the second lets it "talk" to a central computer at 9600 bit/s during other periods of the day. Then, depending on operational requirements, one terminal with a fallback switch for modem selection could be more practical than installing two terminals.

In a third application, an EIA fallback switch (Figure 15.5.1C) permits a modem to be transferred between front-end processors. Although called a *line-transfer device* by some manufacturers, in effect it selects which front-end processor will service the modem.

A telco fallback switch similarly allows the user to select one of two sets of telephone lines. As shown in Figure 15.5.1D, it can select one line from among various combinations of dedicated and dial-up lines that have been installed to fit the needs of a particular application. For a critically large data-transfer application, for in-

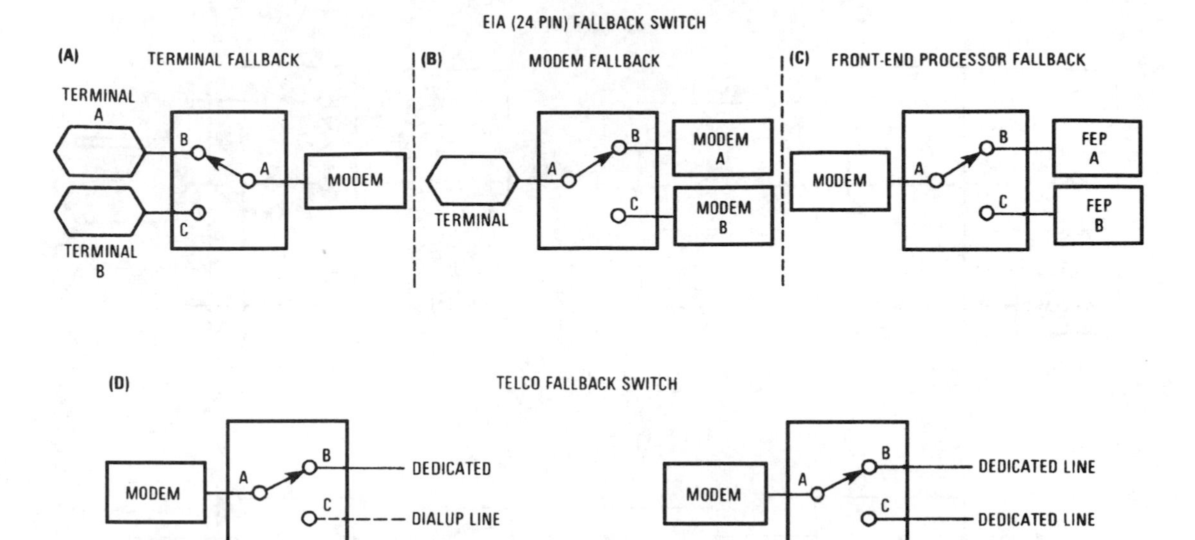

Figure 15.5.1 Fallback switches.

stance, it might be connected to a pair of leased lines, one a primary circuit and the other an alternate circuit.

Bypass switches

The EIA bypass switch connects several EIA interfaces of one type (for example, modems) to the same number plus a spare of another EIA interface type (for example, terminals) and can switch any member of the first group to the spare member of the second group. One application is at a computer installation (Figure 15.5.2A). Here, one front-end channel is reserved as a spare in case any of the existing channels, which normally service predetermined modems, should need to be connected quickly to a spare channel.

In another application (Figure 15.5.2B), the EIA bypass switch can substitute a standby spare terminal for a failed on-line terminal and do away with the need for a spare modem. Although seldom used for multiple-terminal access, a bypass switch can also enable many terminals to share a single modem and line.

A telco bypass switch transfers any one of a group of two- or four-wire telco lines to a spare communications component. For example, as shown in Figure 15.5.2C, if modem 1 should fail, line 1 can be switched to the spare modem. Conversely, a telco switch can transfer a spare line to an operational communications component like a modem (Figure 15.5.2D). Telco bypass switches can be used to switch leased or dial-up lines to modems, automatic dialers, or acoustic couplers.

Crossover and matrix switches

Crossover switches supply their users with an easy method of interchanging the data flows between two pairs of communications components. Each switching module

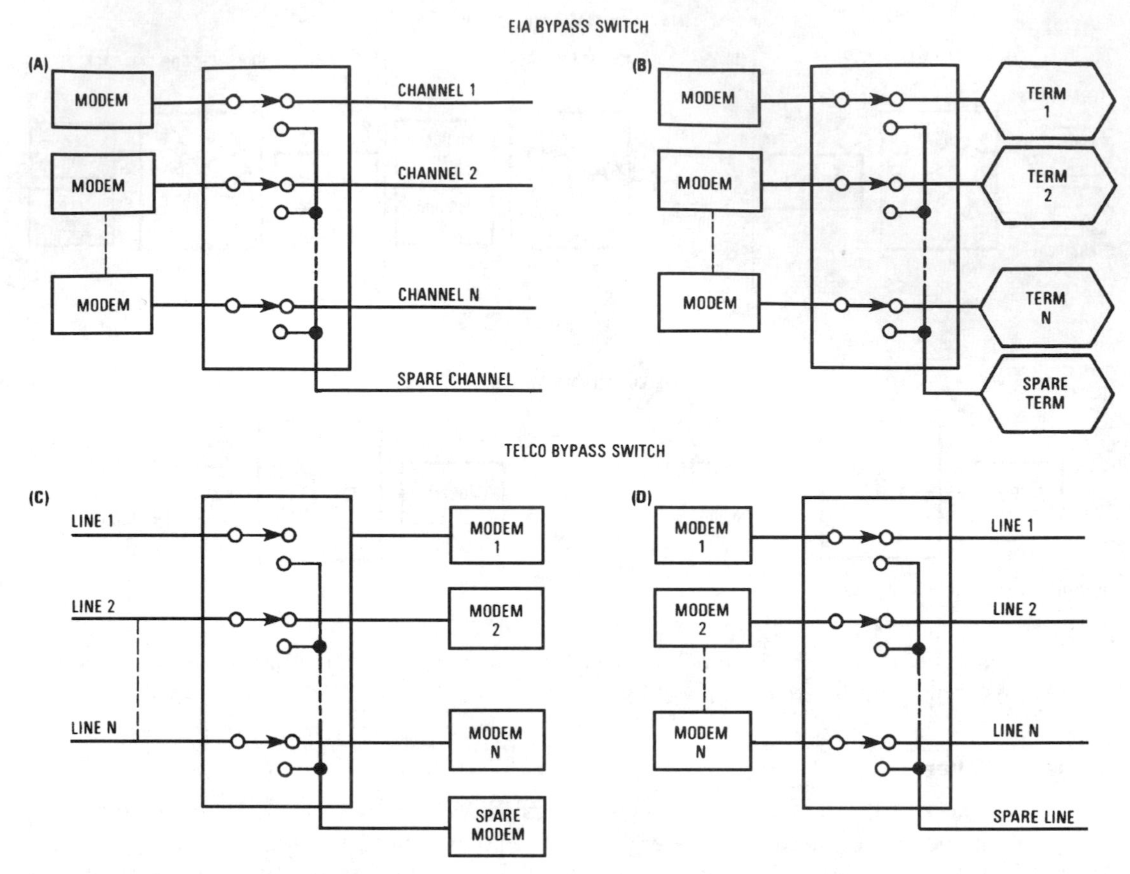

Figure 15.5.2 Bypass switches.

has four connectors, one for each of the two pairs of communications components connected to it. As shown in Figure 15.5.3A, an EIA crossover switch permits the data flow to be reversed between two pairs of EIA-interfaced components. Here, modem A (which is normally connected to the front-end channel A) and modem B (normally connected to front-end channel B) swap channels when the switch is moved by the network operator from the normal to the crossover mode of operation.

Similarly, a telco crossover switch permits the user to interchange the data flow between two telco lines and two modems (Figure 15.5.3B). Although two dedicated lines are shown connected to the crossover switch, the switch can also be used to cross over two dialup lines or one dedicated and one dialup line. In any of these cases, crossover line A (which is normally connected to modem A) becomes connected to modem B, and vice versa.

With a matrix switch the user can interconnect any combination of a group of incoming interfaces to any combination of a group of outgoing interfaces. Matrix switches are manufactured as *N*-by-*N*, with 4-by-4, 8-by-8, and 16-by-16 combinations typically available.

As shown in Figure 15.5.4A, an EIA 4-by-4 matrix switch is a quick and efficient way of connecting any combination of four modems to any combination of four front-end-processor channels. The circles represent the activated switch combinations so that, in this case, modem 1 serves front-end processor (FEP) channel 1, modem 2 serves FEP channel 3, modem 3 serves FEP channel 2, and modem 4 serves FEP channel 4. Of course, in any configuration the user is free to designate one or more modem or front-end-processor channels as spares, or a combination of modems and channels as spares.

The telco 4-by-4 matrix shown in Figure 15.5.4B similarly permits the transfer of any combination of four incoming lines to any combination of four outgoing lines. One type of application warranting use of telco matrix switches arises when remote terminals require access to two or more adjacent computers. If the terminals are used heavily enough to justify installing leased lines from the remote sites to the central computers, the telco matrix switch enables the user to switch the incoming leased lines to outgoing cables, which, via modems, are connected to different computers.

Derived functions

From the four categories of switches, a number of additional switching functions have been developed. For instance, a spare-component backup switch is basically a pair of fallback switches contained in one housing. As shown in Figure 15.5.5A, this switch permits a normal and a backup mode of operation. The normal mode permits

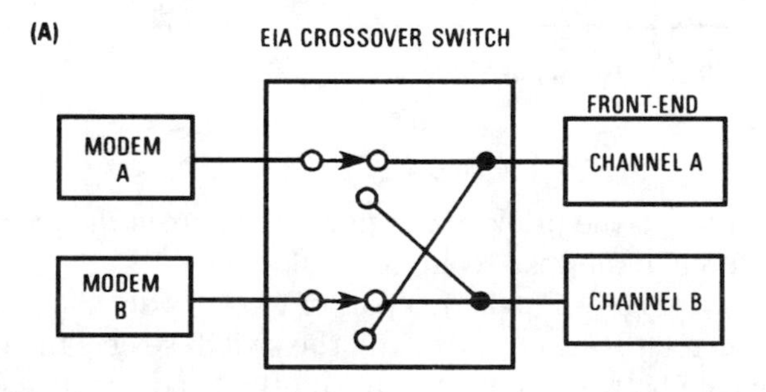

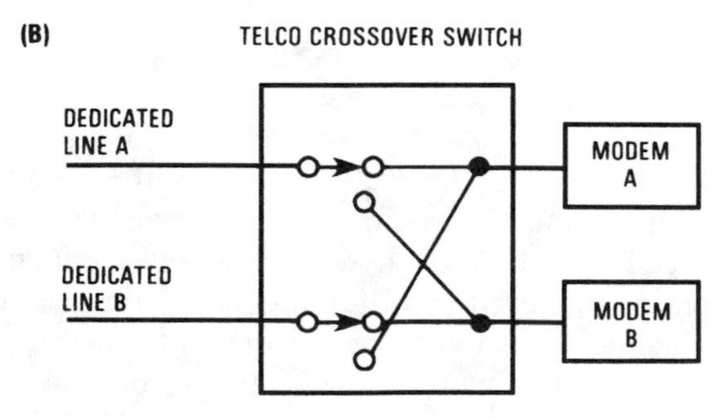

Figure 15.5.3 Crossover switches.

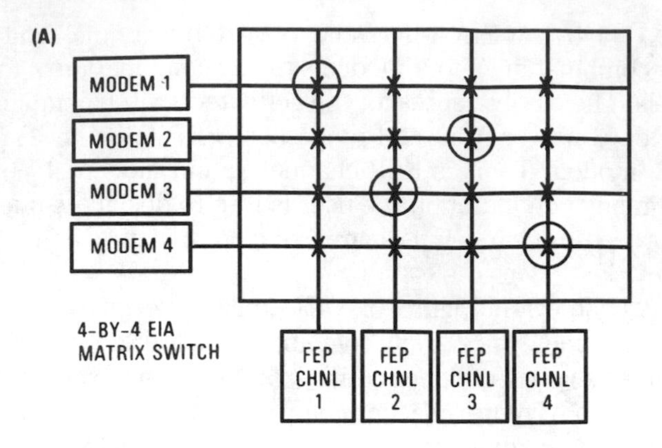

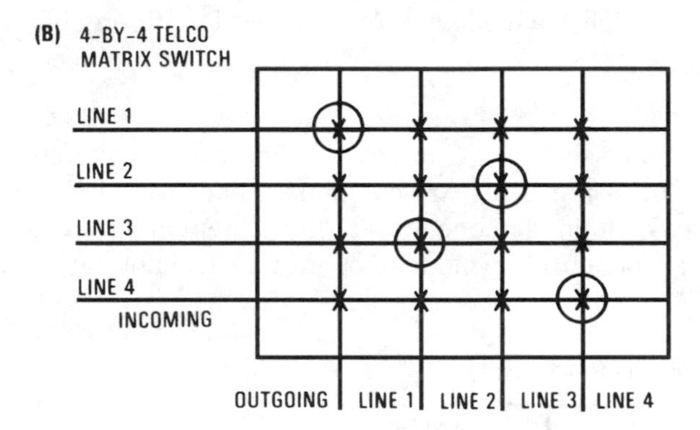

Figure 15.5.4 Matrix switches.

data to be transferred through the primary component, whereas the backup mode switches the data flow through the spare components.

In another configuration (Figure 15.5.5B), a pair of modems are the primary and spare components connected to one terminal, and the switch selects the modem to be used in transferring data between the terminal and the telco line. Because three EIA interfaces are involved, this configuration is called a *3-of-4 EIA interface bypass switch*.

In a 4-of-4 EIA interface (Figure 15.5.5C), four interface devices are connected to the switch. In this configuration, the switch selects one of the two encoders to encode terminal data for transmission through the attached modem.

A second common switch derivation is a multiple fallback switch. Besides the EIA and telco versions, this switch is manufactured in a 1-of-*N* version, with *N* being the number of possible selections. Figure 15.5.6A shows two possible configurations for a 1-of-4 EIA fallback switch. At the left, the switch allows the terminal to be connected to any one of four modems; at the right, any one of four terminals can be connected to a single modem. Similarly, Figure 15.5.6B shows how the 1-of-4 telco fallback switch allows either four modems to share a single line or four lines to share a single modem.

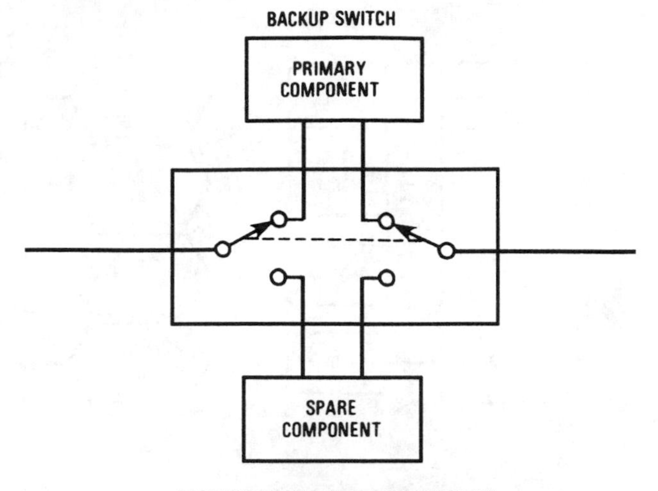

BACKUP SWITCH 3-OF-4 EIA INTERFACES

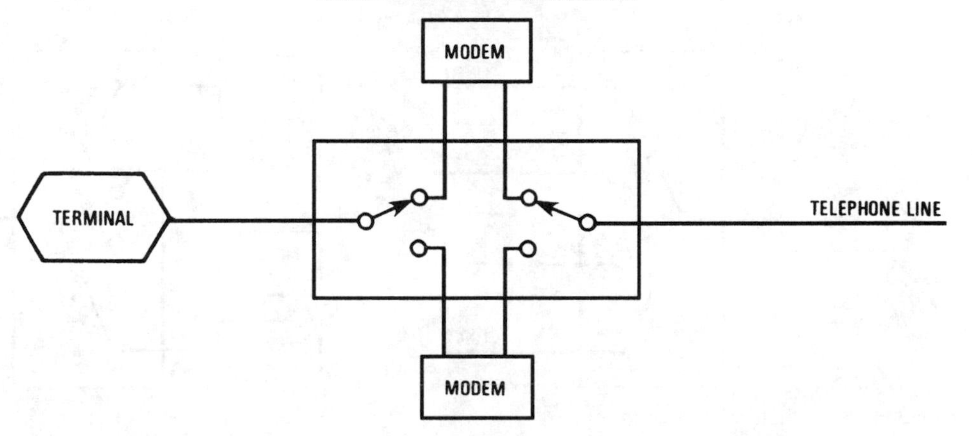

BACKUP SWITCH, 4-OF-4 EIA INTERFACES

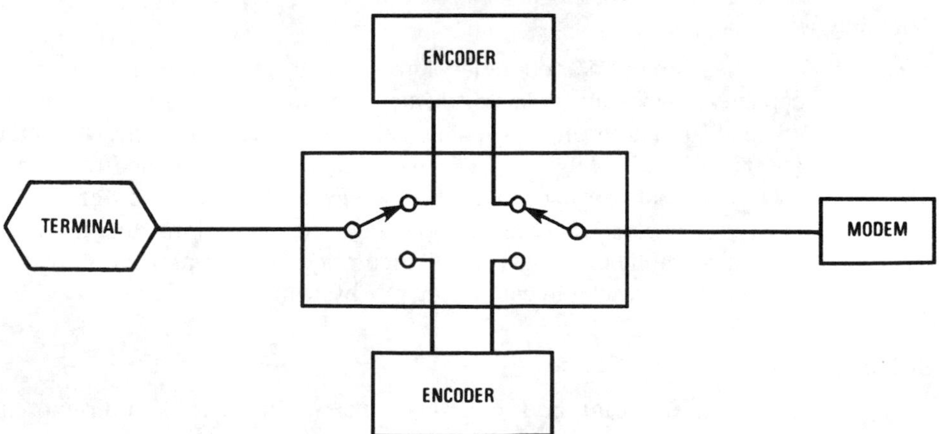

Figure 15.5.5 Backup-switch variations.

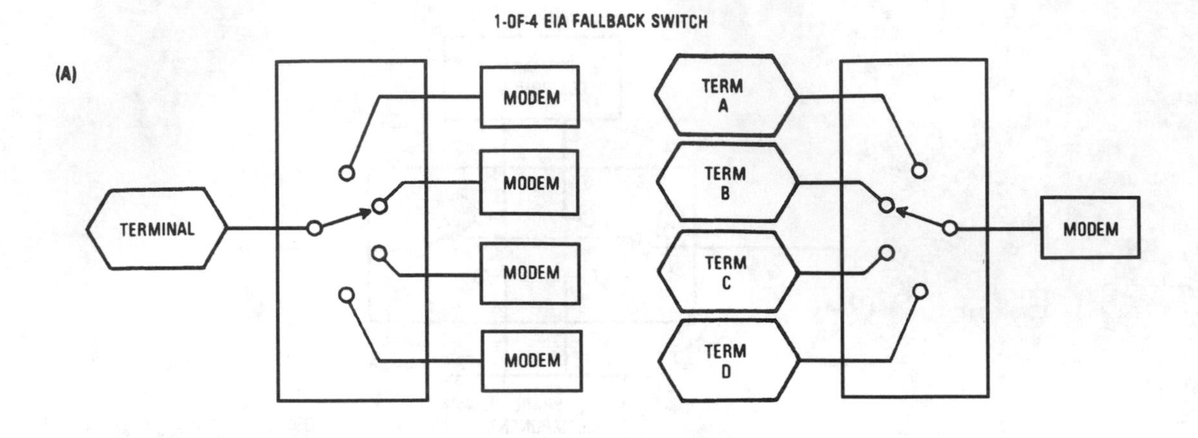

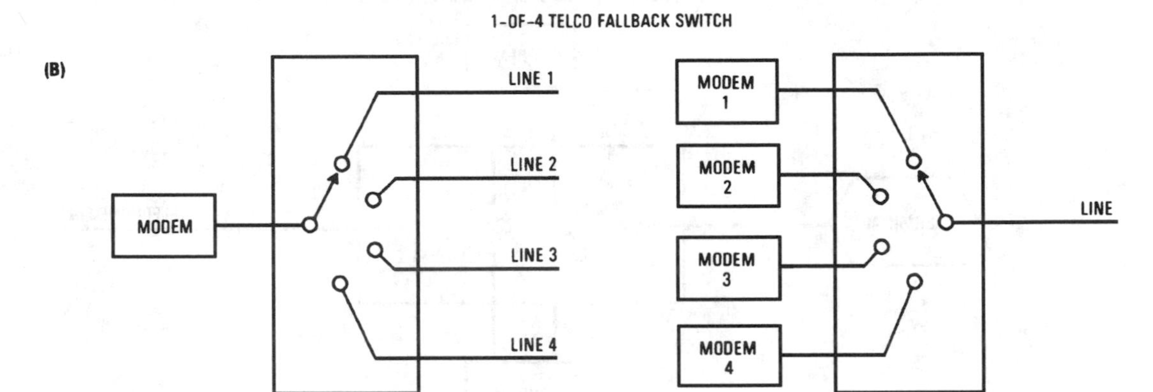

Figure 15.5.6 Multiple fallback switches.

Chaining switches

Switches can be chained to develop additional switching functions or to increase the capacity of existing network devices. Even more usefully, different categories of switches and different types of switches within the same category can be chained.

Figure 15.5.7 shows a 4-by-4 telco matrix switch chained to a 4-by-4 EIA matrix switch so that the user can interconnect any combination of lines, modems, and front-end-processor channels to arrange the desired information path. For this example, the number of possible configurations is increased to N^3 from the normal N^2 combinations available with a single N-by-N matrix switch.

Switch control

The commonest methods of activating a switch are local and remote manual, ASCII unattended remote, and via a business machine or central host computer.

For a remote manual switch a manufacturer produces a remote control panel equipped with a pushbutton and a cable connecting it to the remote switch. This setup also has the advantage that shorter cable lengths can be run from communications components to the switch.

The ASCII unattended remote control permits a switch to be controlled or monitored at any remote site at which a telephone line can be installed. An adapter interfaces the switch (or switches) to the telephone line and turns it on or off on receiving a coded message consisting of the switch number and the state to transfer to. The adapter then reports back the switch's new status. Also available is a query mode that allows the operator to check a remote switch's position.

When a business machine (computer) is involved, switching is controlled directly by the machine—normally through a 5-volt TTL circuit.

Asynchronous transfer mode

ATM's growing prominence is based largely on its capability to switch data at megabit- and gigabit-per-second rates. ATM also enables the merging of voice and data communications into a common format—notable because previous techniques were optimized for one or the other.

In asynchronous transfer mode, traffic is switched by means of an address contained in a cell (packet). An ATM cell's length is fixed at 53 bytes, five bytes of which is the header. The ATM standard does not specify transmission rates, framing, or means of transport. Therefore, LANs, public networks, and switches can readily adapt ATM to their operation.

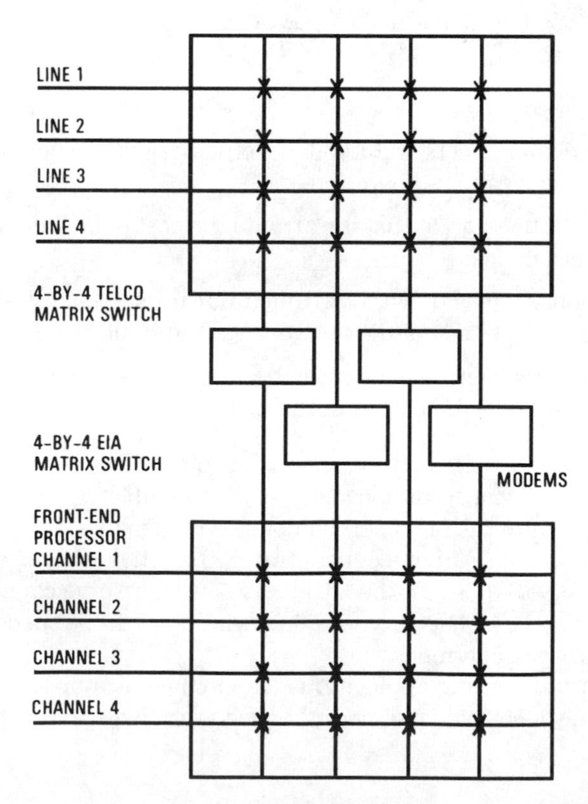

Figure 15.5.7 Chaining telco and EIA matrix switches.

The ATM technology is connection oriented: each transmitted cell travels over a route previously specified. However, ATM does not reserve its connection-oriented path for a particular user. Rather, if a previously reserved path is not in use, it becomes available to another user. Thus, the connection between network nodes is often called "virtual." By enabling numerous virtual connections to share the same physical facility, circuits can achieve efficiencies well beyond the previously achieved 40 percent for either voice or data.

ATM LANs will not use the shared medium of conventional LANs. Instead, users would have their own dedicated ports on a central ATM switch. Similar to a private branch exchange (PBX), the ATM switch controls the network, routes traffic, and imposes flow control. There is no port contention, and each transmission moves in one continuous unsegmented burst. Users operate over their own medium at their own data rate, with the switch handling any necessary conversions. And a separate port would enable connection to a WAN, if so desired.

Much of this ATM coverage could have been reserved for the final chapter, "Trends in Data Communications." But growing network throughput concerns might well overtake publication, compelling network administrators to consider adapting ATM solutions to their problems, rather than deem ATM as a "trend."

15.6 Switching Applications

The cost of providing communications switches can range from less than a few hundred dollars to well over $50,000. What makes the price vary so much rests on the answers to such questions as:

- What devices are most likely to fail?
- What tangible and intangible effects will a failed network device, such as a concentrator, have on the organization's operation?
- Would the operational loss be so great that it warrants the cost of including backup equipment and transmission lines?
- When a network component goes down, how much downtime—if any—is allowable to activate backup devices and get the network back into full-scale operation?
- To obtain speedy network recovery, what are the best types of switches for the application, and where should they be placed in the network?

This section centers around the ramifications of switching between dual collocated concentrators. Here, one concentrator can be assigned completely to back up the other unit, or each concentrator might be servicing its own terminals during normal operation. In either case, on failure of one concentrator, the other takes over all duties if it has enough capacity to do so. In the latter case, if the reserve capacity isn't available, then a secondary job, such as driving a line printer, might be suspended as long as concentrator downtime continues.

In the basic setup of the following applications, each concentrator location services a number of relatively local low- and medium-speed terminals so that each has a number of terminal-to-concentrator links. Each concentrator merges all traffic from its terminals and sends it on a high-speed line to a remote host computer.

The applications tend to become more complex and more expensive. The actual choice depends to some extent on network application and on the severity of the consequences of a device failure.

Hot-start configuration

The two main methods of integrating collocated concentrators to service remote terminals are commonly called "hot start" and "cold start." The hot-start approach (Figure 15.6.1) means that a backup computer is energized and fully programmed with a duplicate of the software in the primary concentrator, and can be continuously tracking the traffic in and out of the primary concentrator. When the primary computer fails, a computer-controlled switch can put the backup concentrator in control almost instantaneously.

Full effectiveness of such a hot-start arrangement requires the installation of an intercomputer (that is, interconcentrator) communications unit. When a failure, such as memory-parity errors or power loss occurs, the concentrator experiencing difficulty sends appropriate software commands through the communications unit. Additionally, an automatic command to a bank of computer-controlled telco fallback switches provides instantaneous transfer of the line from each terminal to the line-controller of the operating concentrator.

The near-instantaneous switching and the minimization of the loss of data are the important advantages of the hot-start configuration. However, there are significant hardware costs associated with the computer-controlled switches and the intercomputer communications unit. In addition, the necessary software modifications to permit the desired switching are complex, involving experienced personnel, much patience, and large amounts of machine time for testing the developed software. Overall, the cost for hot start might well reach over $100,000—not counting the cost of the concentrator itself. But it might be well worth the money to ensure that the network remains continuously operational and available.

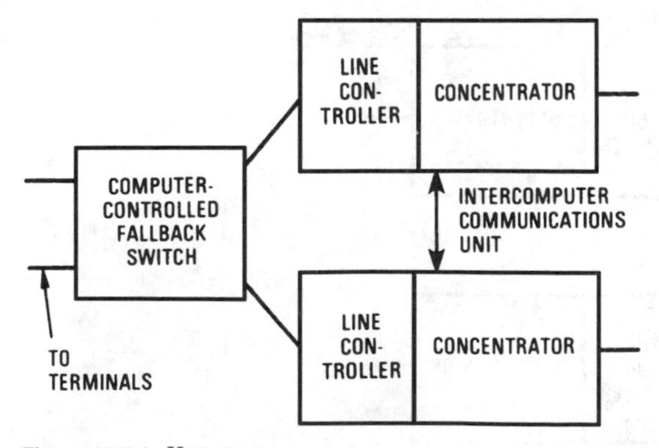

Figure 15.6.1 Hot-start.

Cold-start configuration

Telco fallback switches represent one method of providing an alternate path between the remote terminals and the two concentrators (Figure 15.6.2). Here, the occurrence of a concentrator failure or a concentrator-to-host link failure will require manual intervention. When one or both failures occur, it becomes necessary to switch the telco units to ensure that the remote terminals are connected to the operating concentrator. Furthermore, the standby concentrator must have its programs loaded from a high-speed storage unit, such as a disk. Thus, if the concentrators are initially sharing the terminal workload, the failure of one concentrator might require the other concentrator's software to be reconfigured to service the entire workload. This configuration can be completed in a few minutes by manually activating the switches and reading the backup programs from the disk into the operational concentrator's memory.

Some data being transmitted through the concentrator can be lost during the reconfiguration time. But the low cost of the cold-start configuration might justify the extra time associated with satisfying retransmission requests for lost messages.

Sharing a backup concentrator

The availability requirements of the network might be such that neither operating concentrator has the reserve to serve as backup for the other. But it might be possible to service both devices with a single backup concentrator (Figure 15.6.3). Here, telco fallback switches allow the terminals in Building 1 or Building 3 to be connected to the backup concentrator in Building 2. The number of modems interfaced to the telco switching units in Building 2 only needs to equal the maximum number of such devices in either Building 1 or Building 3. Thus, if the possibility of two concentrators failing at the same time is disregarded, the cost of the fallback switches is more than offset by the savings as a result of the lesser number of modems necessary at the backup concentrator.

Backup with EIA switches

An alternative approach to servicing the terminals in Building 1 and 3 by the backup concentrator in Building 2 can be obtained through the use of EIA fallback switches

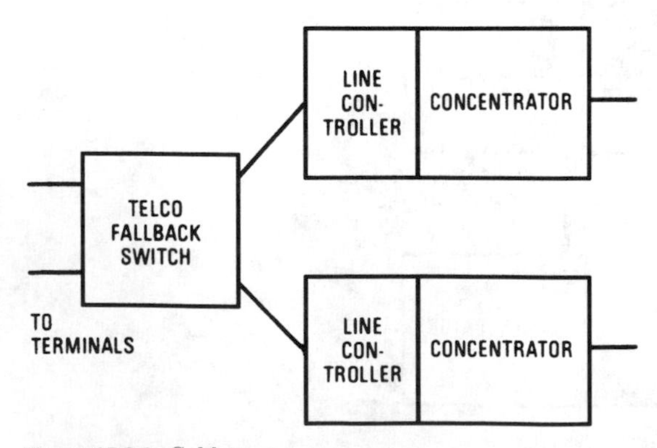

Figure 15.6.2 Cold-start.

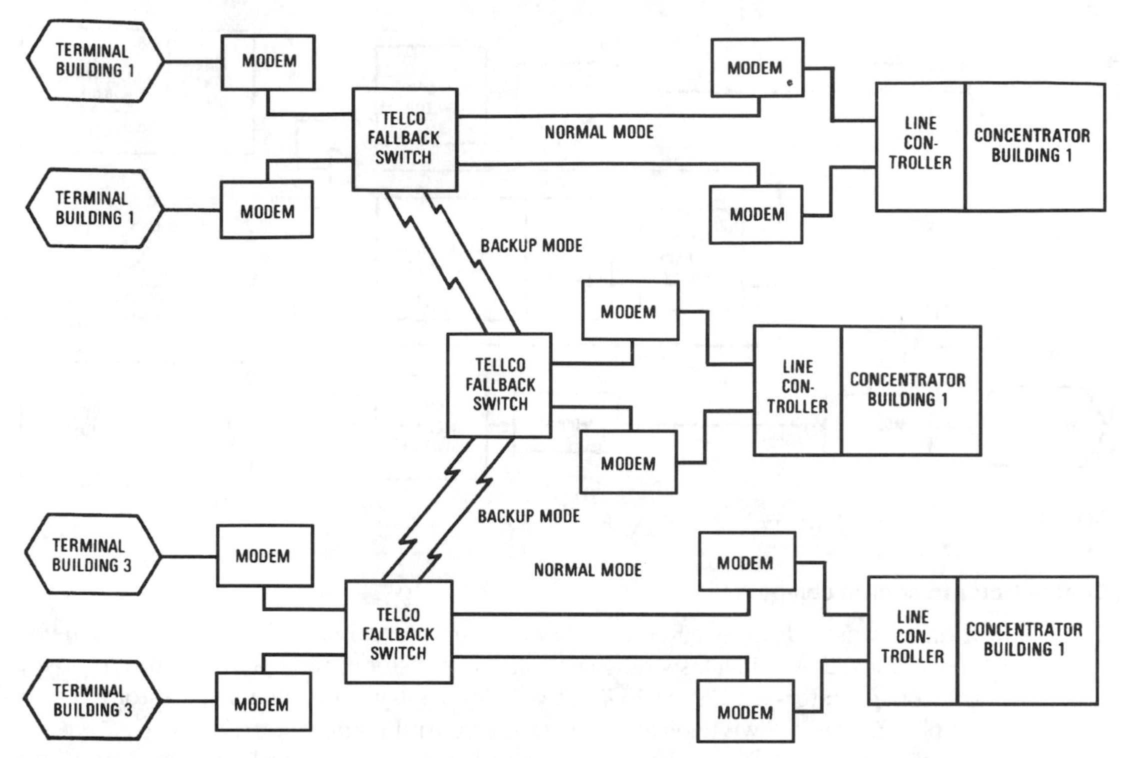

Figure 15.6.3 Sharing a backup concentrator.

(Figure 15.6.4). Instead of installation between the modems, as with the telco switches in the preceding application, the EIA switches are between the modem and the line controller of the concentrator. Depending on the distance between either primary concentrator and the backup concentrator, line drivers or modems become necessary to permit an undistorted output signal to reach the backup. Assuming relatively short distances that permit the use of lower-cost line drivers, rather than modems, a telco fallback switch will suffice in Building 2 for each pair of terminal-to-concentrator links in the other buildings.

In the normal mode of operation, the terminals in Building 1 or 3 communicate with their respective concentrator via a pair of modems and an EIA fallback switch. If either concentrator fails, the operator must position the fallback switch into its backup mode and position the telco switch in Building 2. Doing this provides a new set of circuits from the affected terminals to the concentrator in Building 2.

This and the previous application have a concentrator added to the basic configuration of two such devices. In either case, the user can set up a network of three primary concentrators that share the backup duties. As well as connecting terminals directly to the concentrator in Building 2, the user would have to install EIA or telco fallback switches to transfer the new data paths to either of the other two concentrators when backup service is needed.

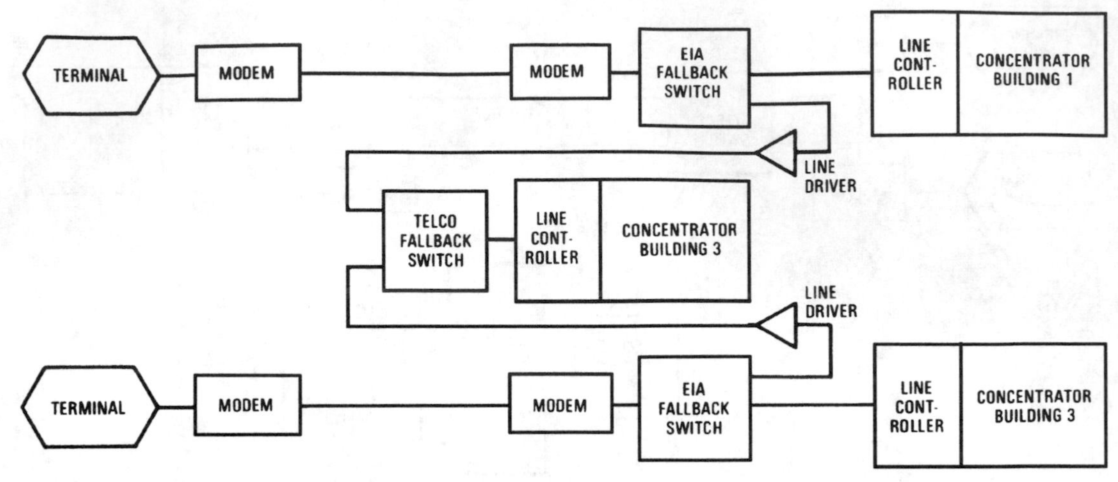

Figure 15.6.4 Backup with EIA switches.

Concentrator to central computer

If data transfer from each concentrator to the central computer is via a few high-speed lines, EIA fallback switches permit the transfer of modems and lines between the concentrators. In Figure 15.6.5, two switches permit each concentrator to communicate over its own dedicated link to the central-computer complex.

This type of configuration compensates for a concentrator failure by permitting the remaining concentrator to communicate with the host computer over its line and the line of the other device. However, the failure of either one of the dedicated lines or of a modem would require selection of one of the concentrators to use the remaining data communications link.

Adding a third EIA fallback switch

If the user wants to overcome the shortcomings of the preceding configuration, the inclusion of a third EIA fallback switch and another modem interfaced to the dual-concentrator configuration can either prevent or minimize the failure of a modem or of a dedicated line (Figure 15.6.6).

In the normal mode of operation, each concentrator communicates with the central computer via its own dedicated line. If a modem or dedicated line should fail, the proper positioning of two of the switches allows the concentrator to communicate with the central computer via the middle modem over an alternative path—either a dial-up line or another dedicated line. A disadvantage of this configuration is that each concentrator has access to only one line at a time, unlike the configuration in the previous application.

Adding more switchable lines

Access to more than one dedicated line at a time can be obtained by adding lines for each concentrator and reconfiguring the EIA fallback switches, as shown in Figure

15.6.7. If one concentrator should fail, the other can communicate over both dedicated lines, and it still has access to the backup line. In this manner, throughput degradation should be minimized.

Chaining adds options

Chaining two EIA fallback switches results in another way of providing an alternative central-computer link for a dual-concentrator installation (Figure 15.6.8). Only one channel is required for each concentrator. In normal operation, each switch interfaced to each concentrator channel remains in the primary-modem position. If the dedicated line or the primary modem of either concentrator fails, the associated switch is positioned so that a path is provided to the backup modem. As previously, this backup modem can use a dial-up or a dedicated line to communicate with the central computer.

This configuration requires only one concentrator channel to provide a link in the event of modem or dedicated-line failure. However, should a concentrator fail, the

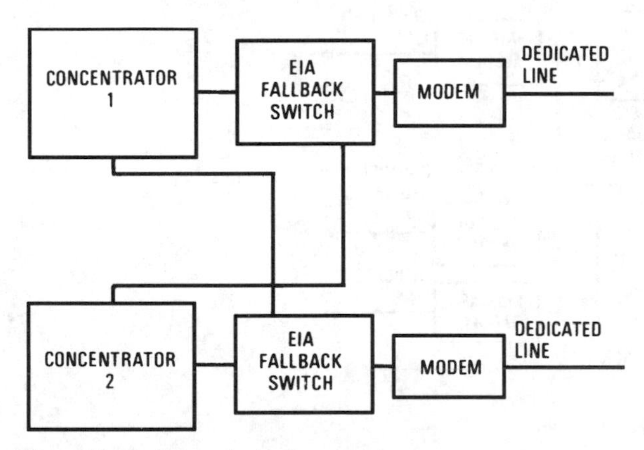

Figure 15.6.5 Concentrator to computer.

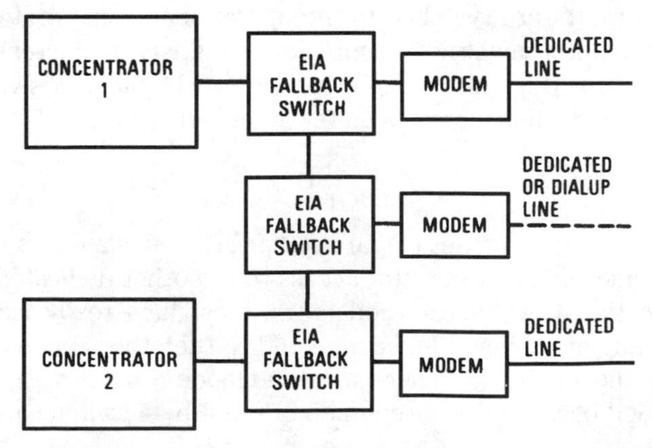

Figure 15.6.6 Adding a third EIA fallback switch.

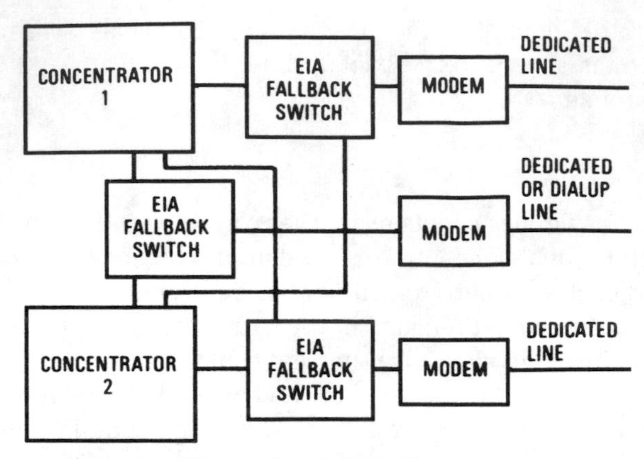

Figure 15.6.7 Adding more switchable lines.

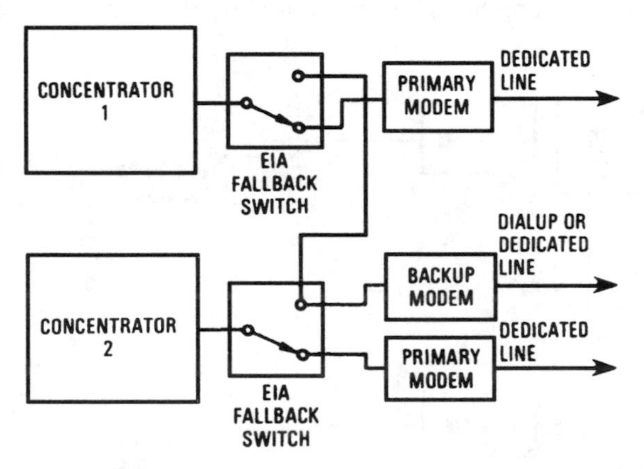

Figure 15.6.8 Chaining adds options.

other one is not provided with access to the failing device's line. Thus, if terminals from the failing concentrator are switched to the operational concentrator, the operational link to the computer might not be sufficient to satisfy the increased terminal traffic. Redundancy for this link through the use of EIA fallback switches can become rather complicated when more than a few lines require multiple access.

Access to other lines

Use of one or more EIA matrix switches (Figure 15.6.9), can alleviate switching complexity as well as provide each concentrator access to the other dedicated line. For example, with a single 4-by-4 switch, each concentrator can have ready access to the spare modem and to any modem and line connected to the other concentrator. Although only one spare modem is shown here, a second modem and its associated line facilities could be added because the output side of the 4-by-4 switch can interface one additional device.

As shown by the circles, concentrator 1 normally transmits data through modem 2 to the central computer, and concentrator 2 via modem 1. Either concentrator can be connected to the spare modem and associated line, should its primary modem or line fail.

If a concentrator fails, the other one can be connected to the failing device's primary modem and line, thus ensuring the continuation of full throughput to the central computer. If each concentrator communicates with the central computer through more than one link, the use of an 8-by-8 or a 16-by-16 switch, or the chaining of more than one matrix switch, should be explored.

Integrating switches into both links

The numbers of concentrators and of communications lines from each concentrator to the host computer depend on such factors as the number of terminals serviced by each concentrator in both primary and backup modes of operation, terminal-traffic patterns, and line-protocol overheads. The configuration in Figure 15.6.10 represents one way of integrating switches into both the terminal-to-concentrator and the concentrator-to-host-computer links. It provides an alternative path for both links when dual concentrators are within about 50 feet of each other.

Here, it is assumed that the remote terminals are in two buildings. However, because the distances between each terminal and the concentrators preclude direct attachment or the use of line drivers, modems are necessary. An equipment study establishes each terminal's need for access to a second concentrator in order to maintain the desired level of backup. At the same time, to maintain throughput at the full transmission speed after the failure of one concentrator, it is necessary to have the capability to switch the links of the failing device to the other one. Furthermore, if any modem or line of the concentrator link to the computer becomes inoperative, switching to a spare modem communicating with the host computer via the dial-up network is desirable.

If the equipment study shows that each concentrator requires one channel for communicating with the host computer, then two channels become necessary on

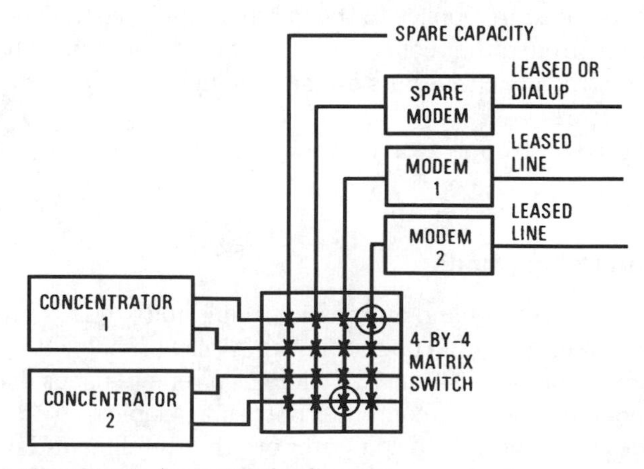

Figure 15.6.9 Access to other lines.

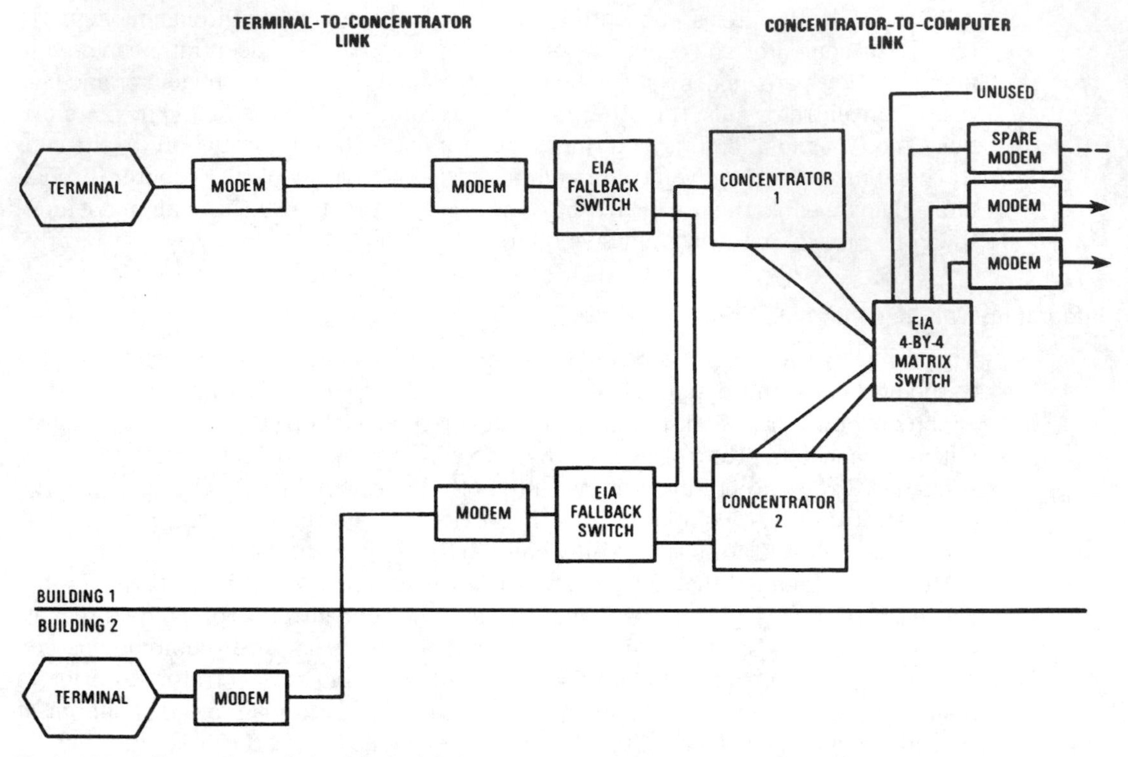

Figure 15.6.10 Integrating switches into both links.

each device in order for each to use the other's link, as well as its own at the same time. Thus, as shown here, the failure of concentrator 1 can be compensated for by positioning the EIA fallback switches on the terminal-to-concentrator link so that the terminals in Building 1 connect to the second concentrator. In addition, each concentrator link to the computer is connected via a 4-by-4 EIA matrix switch to the other concentrator. The same procedure applies to the failure of the second concentrator.

Should a modem or line from either concentrator link to the computer become inoperative, the 4-by-4 matrix switch permits a ready configuration to the spare modem and the dial-up network.

The procedures covered here apply to a network with any number of terminal and host links.

15.7 Automatic Answering and Calling Units

Automatic answering units (AAUs) and automatic calling units (ACUs) reduce the necessity of network intervention by an operator. An AAU is normally integrated into a modem that interfaces a computer. If the AAU detects a remote user's dial-up ring, it automatically connects the user to a computer port (if available).

An ACU, on the other hand, permits a computer or other intelligent device to automatically dial other devices. Dial-up numbers, calling times, and redial quantity and sequence (if necessary) can be programmed in the computer.

AAUs are normally in use in computer-timesharing applications, where many subscribers randomly call several computer dial-up lines. ACUs are cost-effective where transmissions are of short duration. The ACU, normally at a central location, enables automatic dial-up polling of remote devices during preprogrammed time intervals.

Speed and code converters

Lower speed, low-level-code terminals can communicate with faster, higher-level devices on the same network, thanks to comparatively low-cost converters for speed and code. The converters might be temporary or permanent network additions.

They make otherwise incompatible terminals compatible with the network. Compatibility brings the chance to reduce the number of expensive multipoint lines and computer ports, and to standardize on one modem speed.

Just how and where speed and code converters can be applied in the network is covered in the accompanying illustrations. First, though, it's important that speed and code converters are by no means standardized interchangeable products. All convert speed, but not all convert codes. They also differ in speeds and codes handled.

The network of Figure 15.7.1 has a mixture of 110- and 150-bit/s terminals connected to the computer. This requires at least one computer port and one multidrop circuit to service each type of remote terminal. Furthermore two different modems are required at the computer site (one 110-bit/s and one 150-bit/s) and at any remote site with both types of terminals.

If the user wishes to install 300 bit/s terminals at a few of her remote locations because of workload growth, she normally would have to add one more multidrop circuit similar to the setup in Figure 15.7.2. She would need a computer port and enough 300 bit/s modems to service the addition of the 300-bit/s terminals. However, through the installation of speed converters for each of the 110- and 150-bit/s terminals (Figure 15.7.2), all terminals could be serviced via the single 300-bit/s multidrop

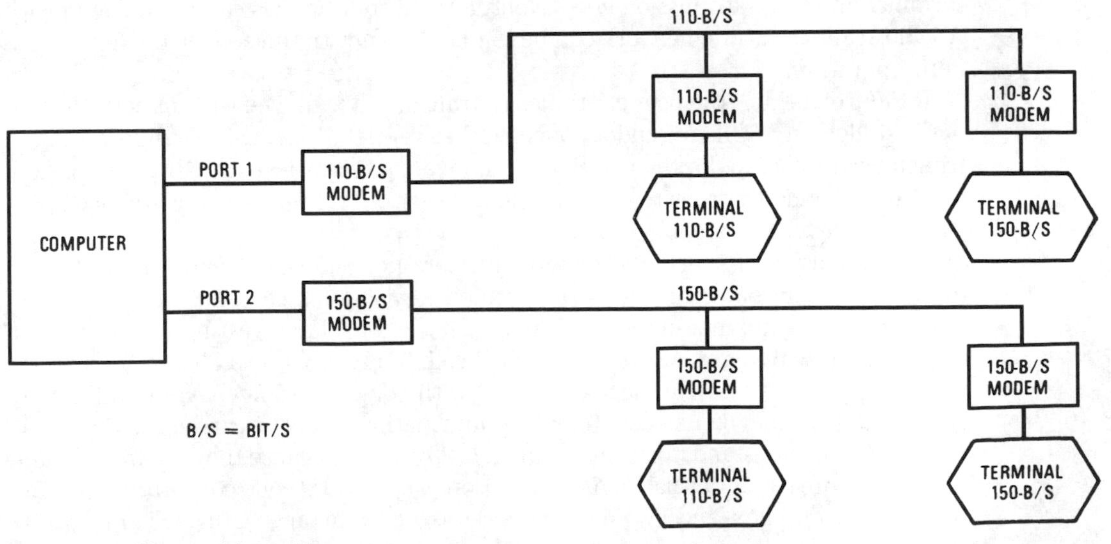

Figure 15.7.1 Nature of the problem.

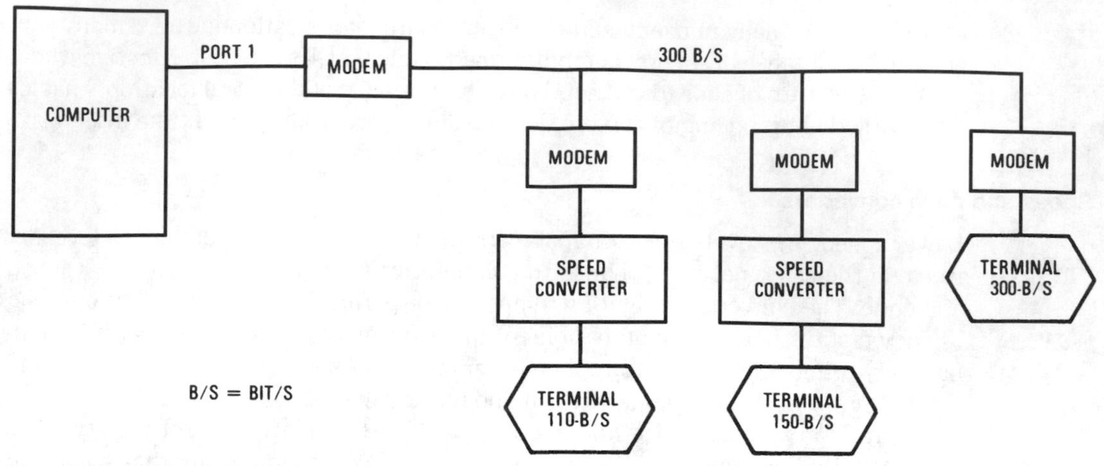

Figure 15.7.2 Adding higher-speed terminals.

line shown. This reduces the number of lines needed and requires only one modem and one computer port at the central site. Also, because all modems in the network can now operate at the same speed, the number of spare modems is reduced.

Like speed converters, code converters permit different types of terminals to be mixed within a multidrop line of the same network. They can be very useful during network upgrading, when a user gradually replaces a large number of older Baudot terminals with more modern ASCII or EBCDIC ones. Depending on the mix of new and old terminals and the conversion schedule, the user can go one of two ways. The existing network (Figure 15.7.3A) operates by the computer transmitting five-level Baudot code on a multidrop line to the five-level Baudot terminals interfaced to that line.

In Figure 15.7.3B, the user has converted transmission to eight-level ASCII code by installing Baudot-to-ASCII code converters between each five-level Baudot code terminal and the transmission line. Eventually these terminals can be replaced with ASCII terminals, or the network can be left as shown with the converters for permanent operation.

In Figure 15.7.3C, the user continues to transmit in five-level Baudot code, and so has installed an ASCII-to-Baudot converter between the new ASCII terminal and the transmission line. If she then decides to convert to eight-level ASCII transmission, she must decide whether to convert the older five-level terminals or replace them with eight-level terminals.

A combination of speed-and-code converters permits the user to satisfy two or more dissimilar communications requirements with a single terminal, thereby reducing both the number of terminals necessary and capital investment. Consider an organization with many ASCII-code eight-level Teletype models 33 and 35 used on the TWX network and in timesharing, and with some older, five-level Baudot Teletype models 28 and 32 used on Telex and international-record-carrier networks.

It is desirable to standardize on the models 33 and 35 because they are more widely used and more readily available. A combination of speed and code converters will make them usable on Telex- and TWX-like networks as well as in timesharing (Figure 15.7.4).

The ASCII even-parity serial code of 11 bits (100 words per minute) is converted to a Baudot serial code of 7.5 bits (66 words per minute) and vice versa. Because most ASCII terminals can be interfaced by a converter, CRT terminals can also be used to send messages through the Telex- or TWX-like networks.

15.8 Digitizing Voice

In the past, few users required, or even seriously entertained, exclusive use of end-to-end digitized voice transmission, mainly because of the high costs associated with analog-to-digital (A/D) and digital-to-analog (D/A) conversion devices. However, the cost of both digital transmission service and A/D and D/A conversion hardware is declining. As a result, digitized voice is emerging as a serious alternative to analog service. Moreover, given the continuing proliferation of digital transmission media, and many users' desires to integrate both voice and data in one network, digitized speech has become an increasingly important traffic component in the planning of future networks.

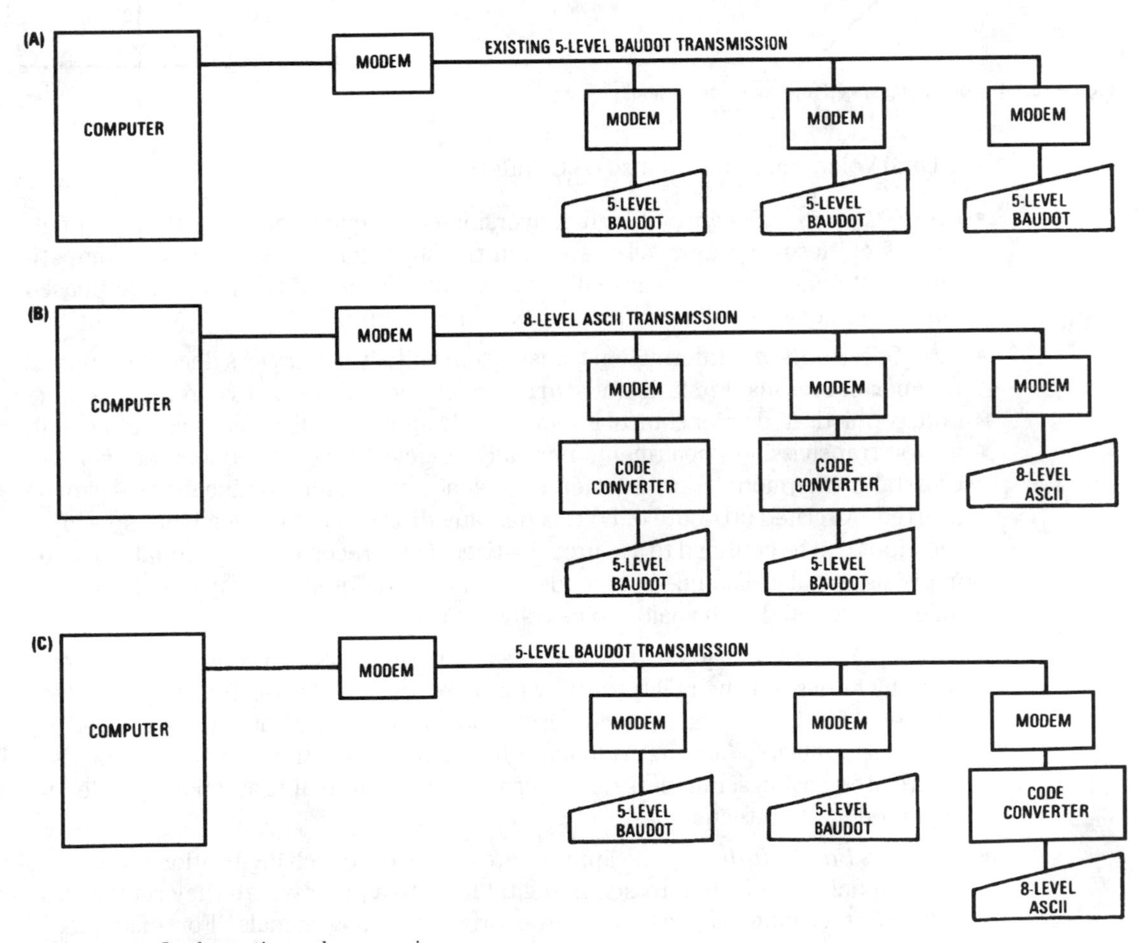

Figure 15.7.3 Implementing code conversion.

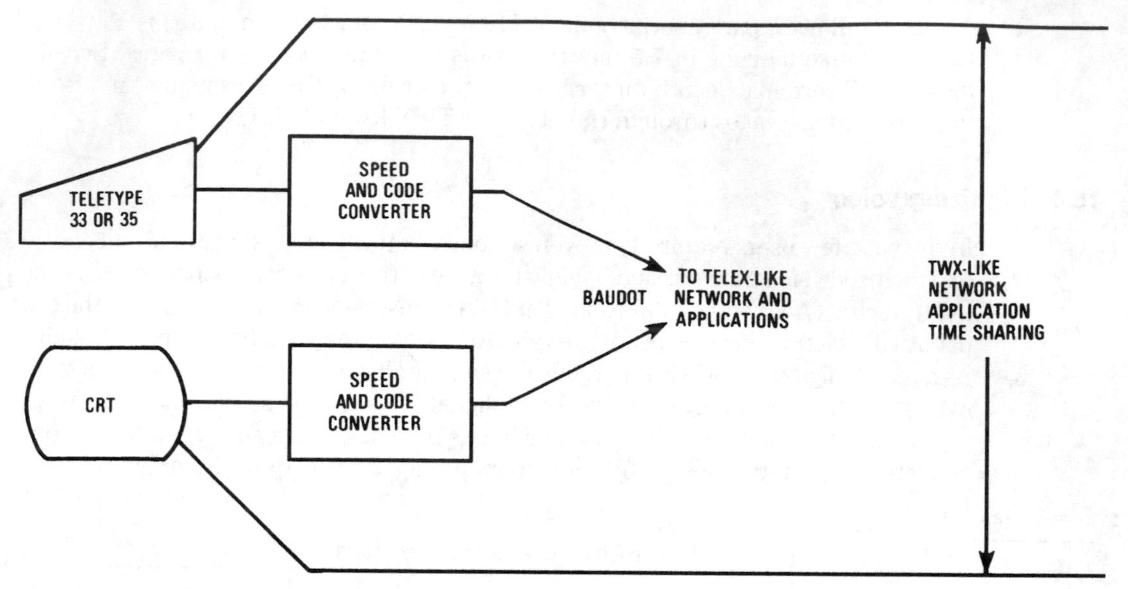

Figure 15.7.4 Combining speed and code conversion.

Digital voice communications (DVC) offers:

- *Compatibility.* Because of high conversion costs, compatibility with digital network facilities is not always beneficial in the short term, but eventually compatibility will be a necessity. This will come about because of the long-range phased elimination of analog-based media by common carriers.
- *Less Degradation.* Information transmitted digitally suffers less degradation for a number of reasons. First, digital signals are easily regenerated via repeaters. Second, sophisticated error control is more simply applied to digital signals. As a result, analog transmission impairments normally associated with telephone networks—crosstalk, intermodulation interference, echo, and filter nonlinearities—can be averted. A related attribute of DVC is that any distortion introduced into speech at the input can be confined to a source digitizer. Thus, received speech quality can be made essentially distance-independent, a major difference compared to speech quality associated with analog voice transmission.
- *Secure Communications.* For many years, the military and some large corporations have desired the ability to transmit conversational speech in a secure (encrypted) fashion. A modicum of protection from eavesdroppers is offered by analog scramblers. Although adequate for many applications, even the most sophisticated analog scramblers do not provide the degree of protection afforded by digital encryption techniques.
- *Reduced Bandwidth.* By the appropriate choice of speech digitization technique, voice signals can be compressed in digital form to a point where they require less equivalent channel capacity than the original analog signals. For example, if speech is digitized at 2.4 kbit/s, four simultaneous full-duplex conversations can

be multiplexed over a single voice-grade line driven by 9.6-kbit/s modems or 9.6-kbit/s digital trunks. This effectively quadruples communications capacity. With the inevitable decline of A/D and D/A conversion costs, these configurations become increasingly attractive.

- *Voice/Data Integration.* Once voice has been digitized, it can be freely intermixed with (digital) data traffic. Such flexibility relieves network planners of the burden of separate facilities management for individual dedicated networks. This also enables users to take advantage of inherent economies of scale resulting from integration.
- *Compatibility with Computers.* Speech in digital form can be readily processed, transformed, and stored by computers. Obviously, with the appropriate processing capabilities, communications networks provide voice-related services, including automatic speaker authentication, speaker identification, speech recognition, and computer-generated voice answerback in response to keyed inquiries.

Trade-offs will favor DVC

Many of these advantages are long-term in nature and are highly technology dependent and cost sensitive. The trade-offs will ultimately favor DVC.

The first trade-off to consider is cost versus voice digitization rate (VDR). The techniques are well understood and the complexity of the conversion hardware is minimal. For useful DVC, however, high VDR has limited practical appeal caused by the high transmission cost associated with even a single conversation. From the standpoint of bandwidth efficiency, the lowest possible VDR is always desirable. However, the cost of low VDR conversion devices is still quite high. Therefore, trade-offs between conversion device costs and transmission bandwidth must be made in order to arrive at the most cost-effective strategy.

To analyze this trade-off, an optimum voice digitization rate must be determined. Combining transmission costs and hardware conversion costs, as shown in Figure 15.8.1, a total-cost curve can be computed. From this graphic view, the "optimum"

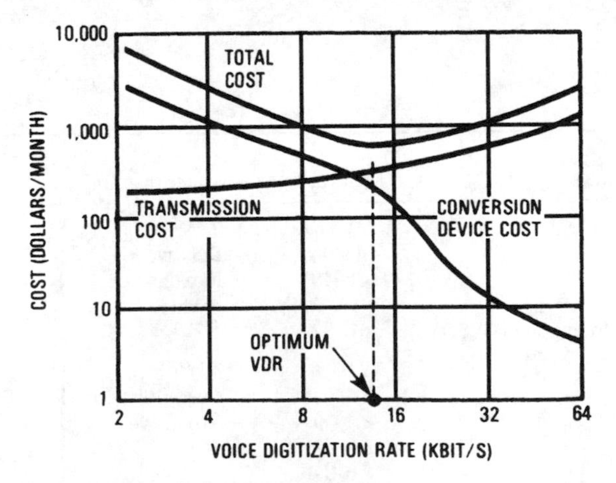

Figure 15.8.1 Trade-off speed and cost.

VDR is found at the point of least cost. This is only a representative analysis of a DVC cost picture, and any actual cost trade-off must take into account such real factors as current specific tariff charges, conversion device costs, and network topology.

Voice "quality" must also be considered. Voice quality is both difficult to define and difficult to quantify. Yet no single characteristic of DVC is more crucial to the user. Perceptual voice quality encompasses several different characteristics. The barest requirement is intelligibility or, stated simply, the listener's ability to understand what the speaker is saying. Several devices that operate at low VDR can distort specific phrases and impair intelligibility. Other digitization techniques provide adequate intelligibility, but sound synthetic or machine-like. In many instances, therefore, voice naturalness is also required.

At higher VDRs, additional speaker-related attributes are easily preserved, such as a listener's ability to recognize the speaker, and even discern his emotional state. In order to reduce VDR, some speech characteristics are compromised. This degradation depends, however, not only on the VDR, but also on the digitization technique used.

Digitizing speech

All human languages consist of certain basic sounds called *phonemes*. English, for example, has approximately 40 phonemes. In normal conversational speech, at most 10 phonemes per second are uttered. Therefore, if six bits are used to encode each phoneme, a bandwidth of 60 bit/s is required to transmit human speech. Table 15.8.1 lists several major voice digitization techniques, and illustrates the typical range of associated voice digitization rates at which "acceptable" quality speech can be obtained.

Caution must be exercised in order to reconcile the apparently large differences between the VDR which is theoretically required and that which is realized in practice because a great deal more information than only speech is normally conveyed. For example, phrasing, stress, articulation, and emotional content are all important parameters of conversational speech. Indeed, purely computer-driven synthesis techniques yield a singularly monotone output, yet have achieved rates as low as 75 bit/s. But, these devices are not applicable for conversion of user-generated speech in real-time applications.

TABLE 15.8.1 DVC Techniques

Digitization technique	VDR (kbit/s)	Cost characterization	Digitization methods
Linear Pulse Code Modulation (PCM)	90–110	↑	
Log PCM	48–64	Relatively	Waveform—
Differential Pulse Code Modulation (DPCM)	32–48	inexpensive devices	reconstruction
Continuously Variable Slope Delta Modulation (CVSD)	16–32	↓	
Sub-band Coder (SBC)	7.2–24	↑	
Linear Predictive Vocoder	2.4–9.6	Currently	
Channel Vocoder	2.4–4.8	expensive	Analysis
Cepstrum Vocoder	2.4–4.8	↓	synthesis
Format Vocoder	0.6–2.4		

The indicated VDR associated with each technique in Table 15.8.1 represents the typical bit rate required for acceptable voice quality. Each technique's VDR can be increased or decreased, but either change will have an effect on quality. For instance, in many of the strategies, the voice quality associated with different techniques is not comparable. Table 15.8.1 contains both commonly used digitizing techniques and techniques still considered to be experimental.

The most widely used method of digitizing voice is pulse-code modulation (PCM). It is used by carriers worldwide, and, in fact, implementations of PCM have been standardized by CCITT for voice digitization. With PCM, analog inputs are low-pass filtered and sampled at a fixed rate. The sampled waveform is quantized to a discreet level and then encoded. The VDR in PCM depends on the sampling rate.

A technique known as *companding* is often used to reduce the analog signal's dynamic range. Companding takes advantage of the human ear's operation, which is logarithmic—with higher sensitivity at lower amplitudes. Two standard companding techniques—μ255 and A-law—are used most often. Both techniques use fixed digitization rates.

Differential pulse-code modulation. (DPCM) differs from PCM in that it uses the difference between sample amplitudes and not the actual values as in PCM. Delta modulation (DM) is a special case of DPCM, which approximates a speech waveform with a "staircase" function. By using automatic gain techniques, the staircase can be made to adaptively track the analog waveform. A special form of adaptive DM is called *continuously variable slope delta modulation (CVSD)*. CVSD varies the staircase step size as a function of its average signal power. LSI-based CVSD devices have become cost competitive with PCM at equivalent performance levels.

Another voice digitization technique, sub-band coding, is still largely experimental. Sub-band coding techniques divide speech into continuous sub-bands, and each sub-band is quantized independently. Experiments indicate that sub-band coding provides greater control of quantization errors and can improve signal quality.

What's available

The preceding are all waveform reconstruction techniques. Another digitizing scheme, analysis-synthesis or source coding, uses vocoders, which exploit certain intrinsic properties of human voice. By digitizing only these parameters, a significant reduction in VDR can be achieved. As Table 15.8.1 points out, a number of different vocoder types are used in reconstructing waveforms. The choice of a speech digitization technique is a nontrivial affair; there exists a variety of competing strategies, each differing in device cost, complexity, VDR, quality of synthesized speech, and performance in the presence of errors.

The information on hybrid digitizing schemes is adapted from "Voice Compression Techniques," a paper by Dr. David L. Lyons, president of Pacific Communication Sciences, Inc. A third digitizing scheme is a hybrid of aspects of the other two. Three such techniques came into extensive use in the mid-1980s: Residual Excited Linear Prediction (RELP), Adaptive Predictive Coding (APC), and Multipulse Coding. Called *time domain hybrid coders*, their processing is performed on real-time samples of the speech signal. Other hybrid schemes are Time Domain Harmonic Scaling

(TDHS) with Subband Coding (SBC) and Adaptive Transform Coding (ATC). A more-recent time-domain technique is Code Excited Linear Prediction (CELP), with digitizing rates below 8 kbit/s.

Once speech has been encoded in digital form, it can be handled in a fashion similar to data traffic. Speech packets could be formed and transmitted in a store-and-forward manner through a packet network. Although new protocols would be required for accommodating speech packets (because of their different error and delay performance requirements compared to data), there exists a potential for saving of transmission bandwidth by a combination of packet switching and speech-actuated vocoding. By not transmitting information or utilizing communications facilities during periods of silence, approximately a 50 percent reduction in bandwidth requirements can be realized.

The future reliance on DVC will also spur the growth of man-machine communications applications on a network-wide basis. Speech is man's most natural form of communications; thus, ample motivation exists for "communicating" with computers in this manner. The widespread existence of digitized voice using quantizers and vocoders will facilitate such applications and, with additional research, will transfer them from the laboratory environment to on-line services.

Synthesize answerback

Computer-generated answerback already exists in the analog world, and is used extensively in banking and order-entry applications. A synthesized voice could be used in response to user-entered queries via standard keyboards or dual-tone multi-frequency keyboards (such as AT&T's Touch-Tone pad). Both speaker identification ("Who is this individual?") and speaker authentication ("Is this person who she says she is?") could be supported under DVC. Vocoders intrinsically generate certain unique parameters of voice, such as pitch. These parameters can then be used to verify or determine the caller's identity. The potential for erroneous identification is much higher when analog signals are used for this as a result of noise. In DVC using vocoders, however, the vocoder supplies the raw speech parameters required for identification.

Finally, the support of speech recognition and speech understanding devices under which users can orally input commands and/or data to a computer (although still a major research topic at many universities and laboratories around the world), is facilitated in a digitized environment.

15.9 Network Integration

Current facilities

The V.S. Cracker Co. operates a large computer installation at its corporate headquarters in Greensboro, N.C. Until recently, all corporate accounts were in New England. As a result of this, the company's network consisted of a number of leased lines from Greensboro to several branch offices in the northeast United States.

As a result of a high demand for crackers on the West Coast, a branch office is scheduled to open in San Francisco. Based on a recently conducted feasibility study, the San Francisco office will require the installation of one remote batch terminal (RBT) operating at 4800 bit/s and ten 1200-bit/s interactive Teletype-compatible terminals.

The prime-time activity factor of the RBT is expected to be unity, meaning that the device will be connected to the network continuously during the prime shift, although it might not necessarily be transmitting or receiving data during that time. For the interactive terminals, during prime time an average of four terminals are expected to be connected to the network. For 90 percent of the time, six or less terminals will be connected.

At San Francisco, all interactive terminals will be located on the same floor of one building, geographically distributed in several work areas. A room is reserved for the RBT and any required communications equipment. The average distance from any interactive terminal to the inside of the RBT room is approximately 275 feet. In discussing terminal operations with San Francisco personnel, it was ascertained that all terminals will be in fixed locations.

Because terminal operators also serve as order entry clerks, separate telephones must be installed for each terminal to ensure that no customer encounters a period of busy signals when a terminal is in use.

At the central computer site, 30 slots are available on the front-end processor for network expansion. Each slot can service a dual-capacity asynchronous/synchronous channel module, which is available from the computer manufacturer.

Requirement

Design an equipment configuration to service the terminals to be installed in San Francisco so that the remote- and central-site costs are minimized. For the interactive terminals, analysis shows that six terminals would not result in any measurable loss of productivity, and the servicing of 10 terminals might actually result in a gain to the company of $100 per month. Assume that the servicing of four such terminals would result in a loss of productivity worth $400 per month. The following equipment should be considered at the denoted monthly lease rates:

Equipment	Monthly cost
Acoustic coupler	$ 30
Auto-answer modem, 1200 bit/s	40
Telco rotary	30
Telephone	15
Dial-in-line	15
Line driver	10
TDM (4-channel)	90
TDM (8-channel)	120
Statistical TDM (12-channel)	275
9600 bit/s modem	200
9600 bit/s multiport medium	200
Front-end processor channel	35
Cable, per foot	0.10
Leased line, San Francisco to Greensboro	1873

15.10 Network Integration

Current facilities

Jiffy Jewelers presently has a corporate data processing center in Dallas, Texas, that is connected on line to a number of regional offices located throughout the United States. At the Chicago regional office, a remote batch terminal (RBT) is connected via a leased line to the computer in Dallas. This RBT internally performs code compression and operates throughout the prime shift, transmitting to Dallas at 9600 bit/s.

Expansion requirements

Because of the use of new marketing techniques, sales at the Chicago office have been increasing faster than the national average. The manager of the Chicago office believes sales can increase even more if his marketing staff has access to small portable terminals operating at 300 bit/s that could be carried to customer sites. Such terminals would enable the sales force to verify current prices and delivery schedules, as well as directly enter orders for goods whose prices fluctuate according to the price of silver and gold.

During a meeting with the regional manager, he mentioned that although at any time up to 20 salesmen could be visiting customers, the maximum number of simultaneous terminals in operation should not exceed eight. With sales increasing, the RBT would either require a partial second shift operation or an increase in the transmission rate to 14,400 bit /s. Responding to your question, he estimated that the cost of a second shift operation would be $1080/month.

Right before the meeting adjourned, the regional manager mentioned that he would like a CRT with an attached printer installed in the Chicago office. This terminal would enable his staff to enter orders called in from customers, check delivery schedules, and access the new MIS that provides information on sales trends and profit forecasting by region. For this application, a transmission rate of 2400 bit/s would be acceptable.

Equipment and operating cost

In examining alternative communications equipment configurations, you compiled a list of available equipment and their monthly costs, as well as the monthly second shift operating cost and leased line costs. The costs of such equipment and facilities are tabulated:

Equipment	Monthly cost
Voice-grade leased line	$ 910
Wideband circuit	4075
TDM (8-channel sync/async input)	120
TDM (12-channel sync/async)	200

Wideband multiplexer (4-channel sync input)	250
Inverse multiplexer	300
9600-bit/s modem	200
9600-bit/s multiport modem	220
4800-bit/s modem	135
2400-bit/s modem	50
300-bit/s auto-answer modem	20
Dial-in line	15
Telco rotary	30
Second shift RBT	1080

Requirement

Design a minimum-cost network configuration to service the remote data processing requirements of the Chicago office. Assume that the CRT will be co-located with the RBT and any required additional communications components.

15.11 Network Integration

Current facilities

Your organization operates a large multidimensional (batch, remote-batch, and time-sharing) computer in Chicago. Currently, two dual-function (asynchronous/synchronous) ports are unused and available for network expansion on your existing front-end processor. An additional front-end processor can be obtained with a minimum of 16 dual-function ports and leased for $1500 per month.

Expansion requirements

Management has decided to open a branch office in St. Louis and a customer-inquiry office in Little Rock. At the branch office, five CRTs and one remote batch terminal are required. At the Little Rock office, one additional CRT is required. If each terminal is connected to an individual port, asynchronous 1200-bit/s CRTs can be used. If CRTs are clustered, 4800-bit/s synchronous devices must be obtained. The remote batch terminal must operate at 4800 bit/s:

From	To	Monthly cost
St. Louis	Chicago	$404
Little Rock	Chicago	675
Little Rock	St. Louis	432

Equipment cost

The following equipment should be considered at the indicated monthly cost:

Equipment	Monthly cost
4800-bit/s modem	$ 120
9600-bit/s modem	200
9600-bit/s multiport modem	220
8-channel TDM	160
Modem-sharing unit	35
1200-bit/s terminal	100
4800-bit/s terminal	125
Computer port	35
Front-end processor	1500
Remote batch terminal	500

Requirement

Assuming that existing facilities cannot be modified, what network configurations should be considered and what are the implications of those configurations?

15.12 Network Integration

Current facilities

Presently, your organization has five warehouses in major cities throughout the southeastern United States, with a distribution center in Atlanta. Each regional office has data communications equipment and facilities, as illustrated in Figure 15.12.1.

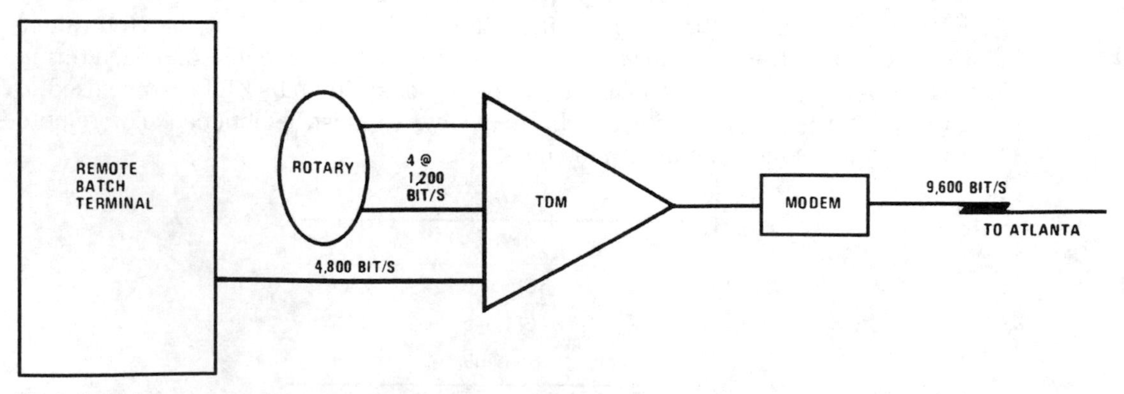

Figure 15.13.1

Expansion requirements

Currently, each warehouse terminal communicates with only the distribution center because goods are shipped from the distribution center to each warehouse directly. The low-speed terminals at each warehouse interrogate the company master database in Atlanta to determine the status of items. The remote batch terminal transmits "ticket" information of goods, leaving the warehouse en route to various distribution centers.

Management would like to transship goods between warehouses. To do so, the low-speed interactive terminals would continue to access the database in Atlanta. However, the remote batch terminal at each warehouse would now require direct communications with the RBTs located at other warehouses. Based on a feasibility study, it was estimated that each RBT would average 12 warehouse-to-warehouse calls per month for transmission of data at an average cost of $5.50 per call.

Upgrading the present network

At the central computer site, a message-switching software package could be added to the front-end processor. This package costs $900 per month and would require an additional memory module, at a cost of $750 per month.

Examining other options

The hardware costs which should be investigated for alternative communications are:

	Monthly cost
4800-bit/s modems	$120
Fallback switch	10
Matrix	200

If the switching is not automatic, as can be obtained from using appropriate software, an additional 50 cents per call will be required for coordination.

Problem

In addition to message switching, what other options are available, and what is the economic impact of such operations?

[illegible]

Chapter

16

Videoconferencing

Some of the material for this chapter is adapted from reports supplied by the Reston Consulting Group, Reston, VA, and from The Insight Research Corp., Livingston, NJ.

16.1 Introduction

This chapter explores the evolution of videoconferencing from its primitive AT&T PicturePhone origins to its luxury stage to its current growing desktop and multipoint popularity. Along the way, the prime movers to its current "affordable" status—codec and standards developments—are examined. Also covered is the shift from rented videoconference rooms to user-owned equipment (such as desktop video) to multipoint networked video.

16.2 Terminology and Definitions

- *Video teleconferencing*, now shortened to *videoconferencing*, is the real-time, two-way transmission of digitized video images, with sound, between two or more locations. Transmitted images can be freeze-frame (where the image is "repainted" every few seconds) or full motion, where the latter usually requires about double the bandwidth of the former.
- Multipoint videoconferencing is just what it indicates: a videoconference among more than two sites. It might require the use of a multipoint control unit (MCU), which is a rather complicated and expensive bit of video switching equipment.
- The coder/decoder (codec) converts the analog video signals to compressed digital. At the receiving screen, a codec reverses the process for viewing. Video compression algorithms have enabled fair-quality transmissions at data rates as low as 56 kbps.

16.3 Evolution

At the 1964 New York City World's Fair, AT&T demonstrated the feasibility of video communications with its PicturePhone. The device was an enhanced telephone that was meant to be a person-to-person visual communicator using its small black-and-white screen. By today's standards, the picture quality was poor.

Until the early 1980s, corporate users found videoconferencing service quality deficient and the cost was exorbitant. This situation was caused in great part to inadequate codecs and a lack of cost-effective transmission services, such as on-demand circuits operating at 1.544 Mbps (T1). Early implementations required special costly videoconference rooms that resembled movie studios.

The initial, sparse implementations of full-motion video used dedicated 6-MHz point-to-point satellite circuits between two sites. When not being used for a videoconference, the satellite circuits remained largely idle, except possibly to service other internal corporate communications. Some of these satellite-driven implementations were used by universities for instruction and to communicate with research organizations and other universities. Because of the higher utilizations, these applications proved to be a good value.

Boardroom videoconferencing was born in 1983, driven by the availability of T1 service from the public networks for point-to-point transmission, which spurred development of codecs that compressed and transmitted video signals at T1 rates. Typical was the T1 service provided by AT&T with its Accunet 1.5 Reserved offering, which contrasted with the company's Accunet 1.5 full-period leased-line service. The reserved offering was specially targeted to supply a more cost-effective videoconferencing service; it was provided in increments of 30-minutes duration.

The costs of Accunet 1.5 Reserved service depend on mileage, time of day, and usage time. As for the equipment, two vendors initially cornered the market with codecs that compressed and transmitted video at T1 rates. The companies were Compression Labs Inc. (CLI) and NEC. The combination of the availability of reserved T1 service and of T1 codecs was a major step that made videoconferencing practicable.

However, early implementations were poor replacements for face-to-face meetings. Broadcast quality, both video and audio, were poor. And there were no controls to zoom in on individuals or objects. Demand for improvements soon resulted in the availability of controllers that provided up to four or five picture-screen partitions, and controls that activated and manipulated cameras using voice detection. Such implementations were packaged as fixed or as roll-about with hand-held remote controls.

Improvements in video codecs, enabling a fair-quality picture at 56 kbps, have resulted in incorporating these codecs into workstations. On the transmission side, the workstations can operate with the switched digital rate of 56 kbps that is provided by interexchange carriers. Better quality is achieved by using two 56/64 kbps circuits for a total bandwidth of 112/128 kbps. A poorer picture quality is provided by installations that operate at 2.4 to 9.6 kbps.

The costs of various videoconferencing arrangements vary. In the most expensive boardroom implementation, considerable costs are incurred for the environment, such as for lighting, acoustics, and furniture. The codec and the controller are usually the most important and expensive electronic elements in a boardroom application.

Business-quality implementations—those that operate at 56 kbps and above—were introduced about 1987. For a 100-mile circuit, the usually more-costly full-period DDS 56 service breaks even with the Switched 56 service at about 42 hours per month usage, after which DDS 56 is less expensive. For a 1000-mile circuit, the break-even point is at about 95 hours.

16.4 Standards

Videoconferencing technology and standards consist of three parts: displays, codecs, and transmission media. There are three worldwide display standards: the National Television Standard Code (NTSC), which is used mostly in the USA; Phase Alternating Line (PAL), used in most of Europe, Asia, Africa, and Australia; and SECAM, used by France. All three standards use analog technology and are not compatible with each other.

The dominant videoconferencing standard developed by the ITU-T (formerly CCITT) Group 15 is H.320, which describes videoconferencing terminals. However, H.320 has also come to represent a suite of specifications for enabling compatible videoconferencing sessions.

The compression portion of H.320 is the video codec H.261 standard. It designates a series of compression algorithms for $P \times 64$ digital channels. (P is an integer from one to 32, resulting in rates from 64 kbps to 2.048 Mbps.)

The data encryption methods that H.320 supports is specified by H.233. And the electronic management of encryption keys is standardized by H.KEY.

MCUs are covered by two standards: H.231 specifies how three or more H.320-compatible videoconferencing installations are linked. H.243 describes the MCU protocol standard. (Multipoint videoconferencing is covered in more depth in a later section.)

16.5 Desktop Applications

A wide range of videoconferencing products have become available. They include Unix network software packages; Windows-based kits that provide videoconferencing over ISDN; and software that converts a Windows PC into a videoconferencing station on a network.

One resultant application is "whiteboarding"—also known as *document conferencing*—where people work jointly on an electronic document. Another application enables participants to work on projects using the same word processor or virtual spreadsheet. These applications do not require the quality of videoconferenced meetings, where human images have to be transmitted. Thus, lower-priced installations can often be implemented.

Although there are no published standards for videoconferencing over LANs at this writing, the ITU-T is developing them. In addition, an IEEE videoconferencing standard for isochronous Ethernet is in the works. Presently, some proprietary packages are available to enable LAN videoconferencing, but operation is not interoperable among different LAN-product vendors.

With the advent of ATM, with its 155 Mbps to beyond 2 Gpbs rates, videoconferencing should be more readily enabled, not only on a LAN, but also from LAN to WAN. And LAN-based products lend themselves more smoothly to multipoint conferencing than do ISDN-based products.

Currently, desktop videoconferencing operation readily achieves a freeze-frame rate of 15 frames per second (fps). Some reach the conventionally defined full-motion rate of 30 fps, with that much more network traffic generated, of course.

16.6 Multipoint Videoconferencing

Many large corporations that once rented videoconference rooms and equipment now possess their own. Service providers are increasingly addressing implementations between companies, especially of those that are not immediately compatible. Differing codecs and transmission rates have to be reconciled. And service between those that have their own setups to those that do not is an area for the provider's expertise. A rented, fully equipped public room is usually part of this arrangement. Large users with their own facilities often rent out their videoconference rooms, when idle, to another user. This has given rise to the service provider that specializes in being a videoconference-room broker.

Videoconferencing has come of age with the advent of the multipoint application. When a multipoint hookup is required, the user usually turns to the service provider for installing and operating the requisite MCUs. Once the equipment is in place, the parties contact the service provider when they are ready to establish a multipoint call. Besides transmission costs, users usually pay a connection fee.

The major piece of equipment, the digital MCU, exists in both proprietary form and based on the ITU-T's H.243 standard. An MCU that complies with H.243 will be compatible with H.320 codecs for data rate, compression, and audio encoding. For those using proprietary equipment, service providers offer data rate and codec conversions to enable pseudo-compatible operation.

An interim measure for those not yet compliant with standards-based digital MCUs is the analog bridge. Here, a digital image and its accompanying audio are converted to analog via a codec, then sent over a bridge to another codec, which digitizes the signals for its receiving device. The plus of this approach is interoperability with just about any type of codec. However, the signals do tend to degrade. And analog bridges do not facilitate displays of high-resolution documents and multiple windows.

One unexpected problem that multipoint users might frequently encounter is incompatibilities between long-distance carriers. Quite often, interoperability between such carriers is achievable only with specially designed gateways—and then, only if such devices are sanctioned by the carriers. The good news is that gateways are under development so that different carriers will become increasingly interoperable.

Other multipoint problems include scheduling. Service providers will usually not commit to establishing a multipoint videoconference unless they are given weeks (sometimes months) advance notice, especially for first-time users. And interoperability is lacking when proprietary solutions are applied to such operations as application sharing, imaging, and file exchanges.

The major cost element in a multipoint application is the MCU, close to $100,000 at this writing. Experimental solutions to non-MCU multipoint operation involve ISDN interfaces, which enable linking of up to four sites. Short of ISDN, the service provider with the MCU remains the facilitating factor. One bright spot on the horizon is the ITU-T's upcoming T.120 standards suite. These standards are intended to facilitate videoconferencing interoperability.

Questions

16.1 Define videoconferencing.
16.2 What distinguished AT&T's Accunet 1.5 Reserved offering from its other Accunet 1.5 offering?
16.3 What factors determine Accunet 1.5 Reserved costs?
16.4 What data rate is considered as offering business-quality videoconferencing?
16.5 What is the dominant videoconferencing standard? What does it describe?
16.6 What is an MCU? What is its application?
16.7 What is whiteboarding?
16.8 What frame rate is considered full-motion?
16.9 What is an interim alternative to the digital MCU? Describe its operation.

Chapter

17

Network Diagnostic and Monitoring Considerations

Once a data communications network becomes operational, a number of transmission problems are likely to occur, mostly on a random basis. As a result of the evolution of networks with computers, modems, circuits, and terminals supplied by many different vendors, a method to pinpoint transmission problems to specific devices must be available to the user. Ideally, only the responsible vendor is notified of the fault in the equipment or on the circuit.

To assist in fault isolation, a number of devices have built-in test-character generators and circuitry to compare data transmitted with data received during a loop-back test. Although such test features can resolve certain fault-isolation problems, they will not normally denote such factors as the sources of transmission distortion. To obtain additional information, line monitors and other test equipment must be used.

This chapter first focuses on wide area networks by reviewing some of the sources of transmission errors and how line monitors can be used to trap and display data to provide a precise picture of line activity. Because many computer installations justify spare equipment to maintain operations continuity if a device or circuit should fail at the central site, the design and construction factors involved in a technical control center will be covered. Personnel at a well-designed center not only can monitor and determine the causes of transmission problems, but it can also be able to patch users around those problem devices and circuits while remedial action is taken.

After the section on WANs is information on LANs. You can study the data flow on a token ring network through the use of a software network-monitoring product. This product will enable you to examine many LAN-specific communications parameters.

17.1 Sources of WAN Transmission-Line Distortion

Transmission-line testing is a well-developed subset of electronic knowledge. An understanding of how lines are tested, the parameters utilized as variables, and the interpretation of the results will aid any network designer or user in the understanding of a network. Consideration of these testing procedures will entail giving attention to the following points:

- Key parameters relevant to transmission-line testing
- Common-practice measurement techniques
- References to specifications and procedures

Knowledge of these factors gives an understanding of the parameters that affect the data transmission capabilities of transmission lines. Transmission lines, as covered here, are the voice-bandwidth private-line data circuits provided by a communications carrier.

Basic measurements

As detailed in Section 5.2, the decibel (dB) is a unit that is defined as the ratio of output signal power to input signal power.

$$dB = 10 \log_{10} \frac{\textit{output power}}{\textit{input power}}$$

Logarithms are used because a signal level in dB can be easily added and subtracted, and because the ear responds to signal levels in an approximately logarithmic manner. Notice that if the output power is less than the input power, the result is negative and the channel is said to have a dB loss. Decibels are simply the measure of a power ratio. Measurements made in this way are expressed in decibels relative to one milliwatt (dBm), where

$$dBm = 10 \log_{10} \frac{\textit{signal power in milliwatts}}{\text{1 milliwatt}}$$

Therefore, 0 dBm means 1 milliwatt, and absolute power levels can be expressed as so many dBm.

For reference, common carriers establish a zero transmission level point (TLP) as some measurable point in a channel at which a specific test-tone level (0 dBm) is expected. A +7 TLP, for example, is a point at which the test-tone level should be +7 dBm. For voice circuits, the test tone used is a 1004-Hz, 0-dBm tone at the zero TLP. This is normally referred to as a 1-kHz test tone, or simply as a test tone.

Other tones can be used for testing. These tones are all 1 kHz, but are at different levels. To reference them back to test-tone level, another unit of measurements is used. Decibels referred to one milliwatt with respect to test-tone level (dBm0) are defined as:

$$dBm0 = \textit{signal level in dBm} - \textit{test tone level in dBm}$$

Data is normally transmitted at a level 13 dB below test-tone level. Therefore, data level is –13 dBm0. At a zero TLP, data level would be –13 dBm; at a +7 TLP, data level would be –6 dBm. It is important to note that data transmission-line measurements are made with a 1-kHz tone—called a *holding tone*—at data level, not at test-tone level.

The noise factor

In theory, noise is the limiting factor in data transmission. In actuality, noise alone rarely is the cause of errors in low- and medium-speed transmission. For high-speed transmission, the effects of other transmission impairments reduce the noise margin—the signal above noise level—that makes noise the figure of erit in comparing different modulation schemes (as is covered under modems).

The audible hiss on a telephone line is noise. This uniformly distributed, random background noise is measured in several ways. Noise measurement units are decibels above reference noise, where

$$0 \text{ dBrn} = -90 \text{ dBm}$$

is the reference noise level. This arbitrary level represents the lowest noise level an "average" listener can hear on a telephone. A C-message filter is used as a "front end" to a level meter to give a representative noise measurement. Although this is a voice telephone measurement, it has carried over into data transmission. When a C-message filter is used in measurement, the units are called *decibels above reference noise C-message weighted (dBrnc)*.

Noise on a channel is a function of the telephone channel and telephone equipment, so an increase in signal level will bring about an increase in noise distortion. To measure noise under "actual" conditions, a 1-kHz holding tone (data level) is transmitted. At the receiving end, a second filter is used to remove, or "notch out," the tone. This is strictly defined as a C-notched noise measurement.

Figure 17.1.1 shows the response of a C-message filter. Notice how it severely attenuates power-line harmonics (60, 120, 180 Hz, etc.). A second filter, 3-kHz flat, is used as a check for power-line harmonics. Both C-message and 3-kHz-flat noise measurements are made, if power-line harmonics are suspected.

As with the dBm, an absolute reference level for noise is often specified, based on a transmission-level point. Decibels above reference noise, C-message weighted, with respect to test-tone level (dBrn0), are dBrn0 = noise level in dBrnc—test-tone level in dBm. Figure 17.1.2 illustrates the relationships between these measurements.

In practice, a telephone channel will transmit frequencies from 300 to 3.3 kHz. The amplitude loss at each frequency might be different. Curves of level vs. frequency, referenced to the loss at 1 kHz, define the amplitude response—or attenuation distortion—of a transmission channel.

Another source of distortion is from different frequencies propagating through a channel with different velocities. This means that if a number of different frequency tones are transmitted simultaneously into a channel, they will be received at different times. This is significant because a data signal is composed of many different fre-

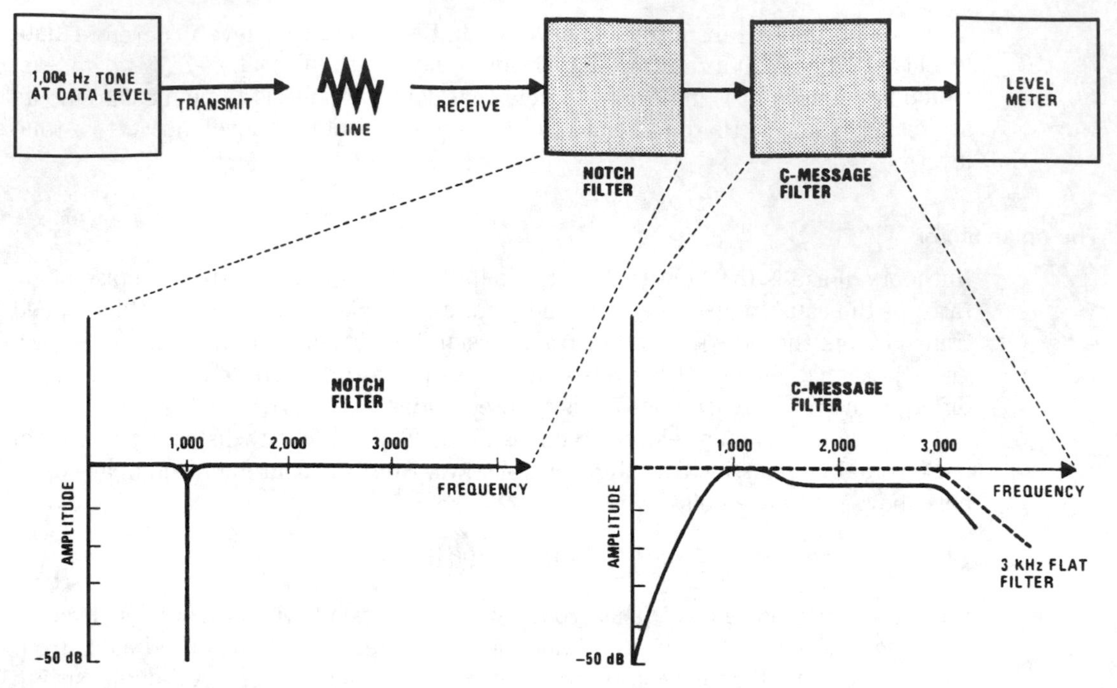

Figure 17.1.1 Measuring the noise.

quencies. Figure 17.1.3 shows that there will be a difference in arrival times (or in phase) between two tones of different frequencies.

This difference can be expressed in terms of microseconds of delay. Absolute delay is not really measurable because a frame of reference (when did the signal start?) cannot be established. The relative difference between two tones can be measured, however. Envelope delay distortion at a given frequency is, then, measured as the difference in microseconds when compared with the delay experienced by a reference tone at 1.8 kHz. Notice that a 1-kHz tone is not used for reference. Sophisticated methods are used to actually measure envelope delay distortion.

Distortion caused by power-line harmonics, in multiplex configurations for example, creates "jittering" of zero crossings, or of the instantaneous phase of a signal, as shown in Figure 17.1.4. This phase jitter is measured by looking at zero crossings of a holding tone. Noise, which can affect zero crossings as shown, can strongly influence phase jitter measurements. To make accurate measurements, a data-level holding tone should be used because the ratio of signal level to noise level is a factor. Notched noise measurements should be made in conjunction with phase jitter measurements. This should help to determine what is actually being experienced: true phase jitter or just the effects of a high noise level.

At low to medium speeds, phase jitter is rarely noticeable compared with the effects of amplitude-response distortion, envelope-delay distortion, and line hits (covered later). For high-speed transmission, phase jitter is more crucial, as will be covered under modems.

Nonlinear distortion

A channel has many nonlinear components. These components distort a data signal by generating unwanted harmonics that add to the signal in a detrimental manner. As illustrated in Figure 17.1.5, for a single tone, the effects can be pronounced. The example is a simplification, but the "clipping" effect on the wave (Figure 17.1.5C) is

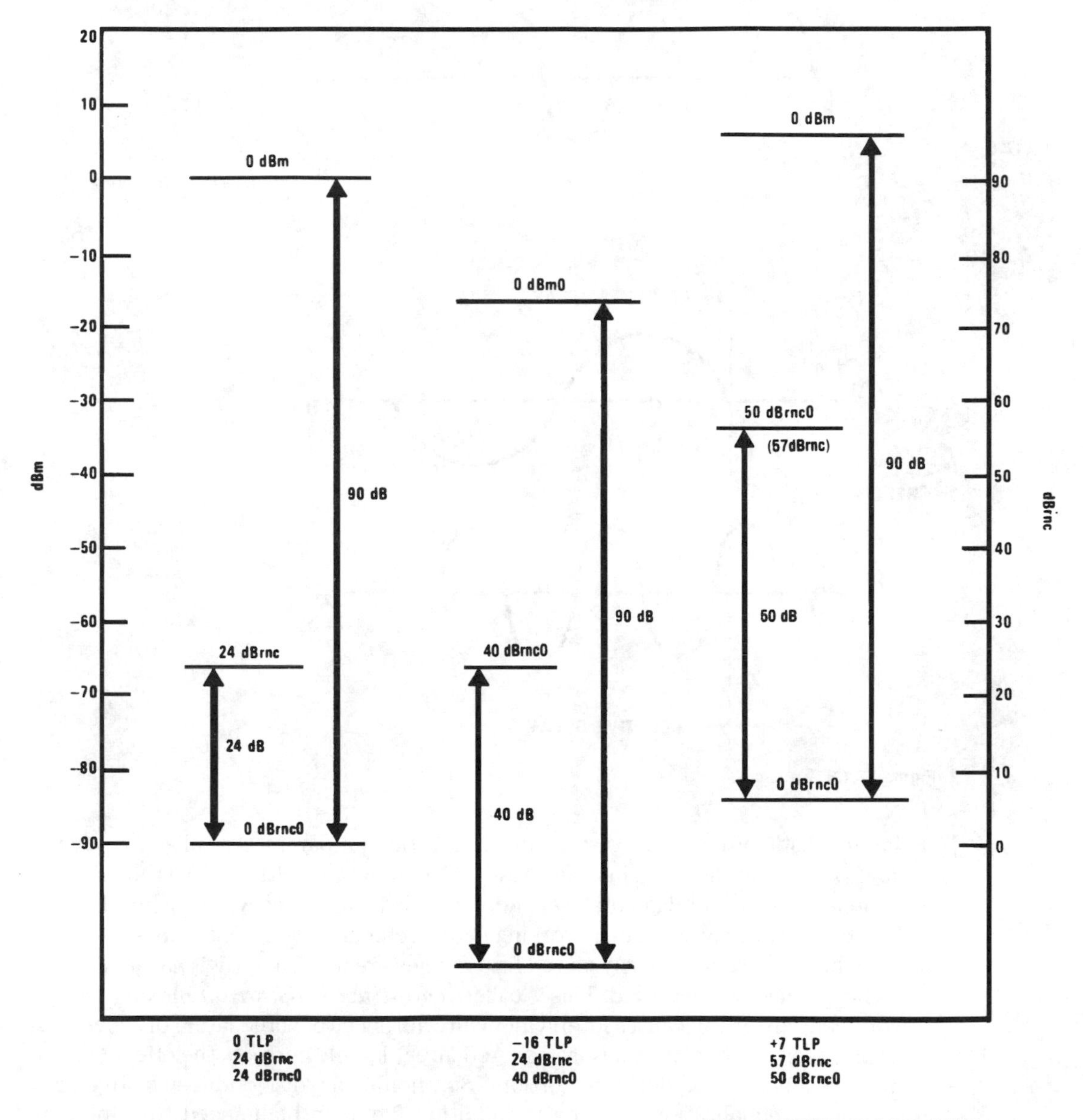

Figure 17.1.2 Relationships between measurements.

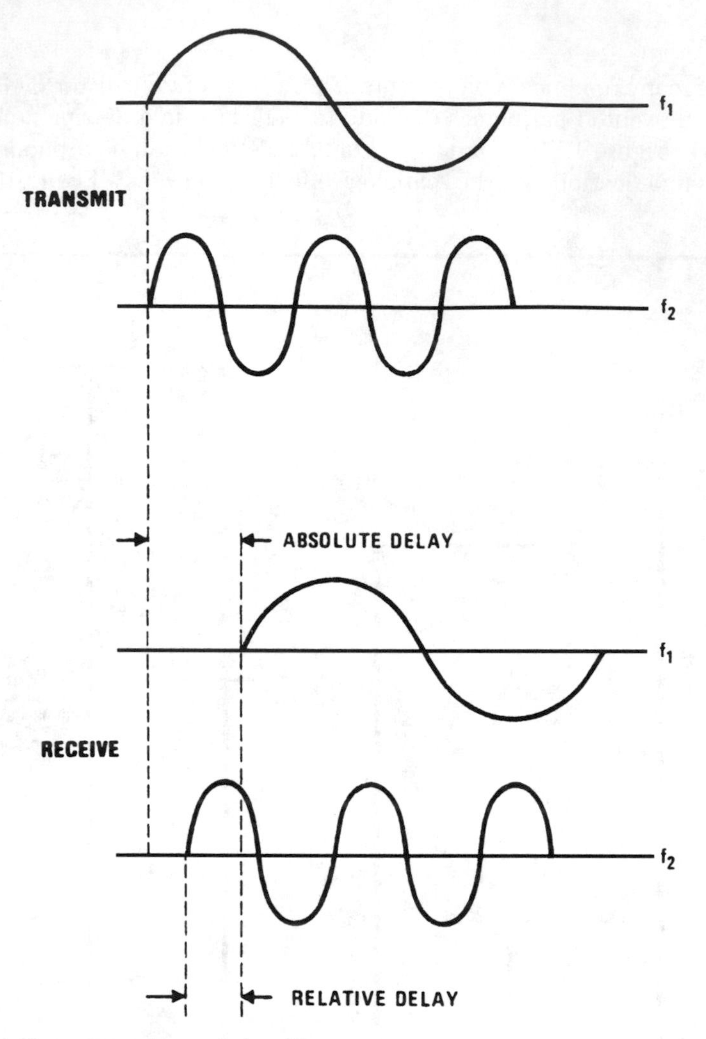

Figure 17.1.3 Transmission delay.

a characteristic of harmonic distortion. At one time, a parameter called *harmonic distortion* was measured in just this way. A tone was transmitted down the channel, and the level of the fundamental, second, and third harmonics was measured.

To more accurately simulate a working data circuit, a newer technique is in use today in which two sets of closely paired frequencies are used, and their harmonics and cross harmonics are measured. This is called a *nonlinear-distortion measurement.*

Nonlinear distortion causes intersymbol interference (covered later) of a low magnitude. However, equalization (also covered later) largely corrects the effects of amplitude response and delay distortion. So, nonlinear distortion—which is not corrected by equalization—can be the limiting factor in high-speed transmission. The presence of nonlinear distortion is the reason for "D"-type conditioning provided by the telephone company for high-speed lines.

Another source of transmission errors is called a "line hit." It is the primary source of error in low- and medium-speed transmission. Any hit of this type can render unrecognizable any data transmitted for the duration of the hit, and to the end of the block in progress. Line hits can be classified into specific types, such as:

- *Dropouts.* Sudden, large reductions in signal level that last more than several milliseconds.
- *Phase hits.* Sudden, uncontrolled changes in phase of the received signal.
- *Gain hits.* Sudden, uncontrolled increases in the received signal level.
- *Impulse noise.* Sudden "spikes" of noise of very short duration. In telephone reception, those clicks that can be heard on occasion.

Four-wire, full-duplex private lines have independent transmit and receive "sides" or channels. Transmission test instruments normally have independent transmit and receive sections. The transmit section is usually capable of generating a 1-kHz tone and variable-frequency tones at adjustable levels. The receive section can simply be a level meter. In addition, the test instrument could contain a frequency counter, noise measurement filters, as well as phase-jitter, envelope-delay, impulse-noise, and

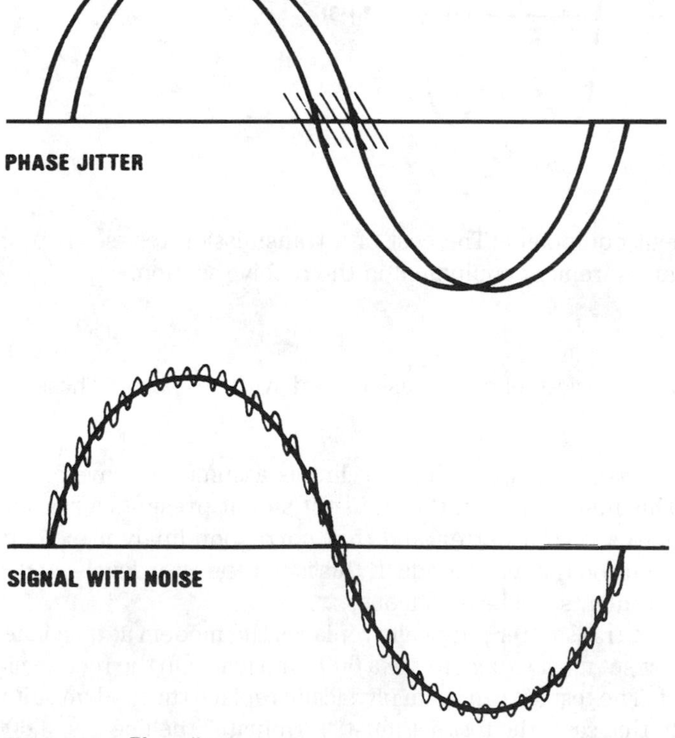

Figure 17.1.4 Phase jitter.

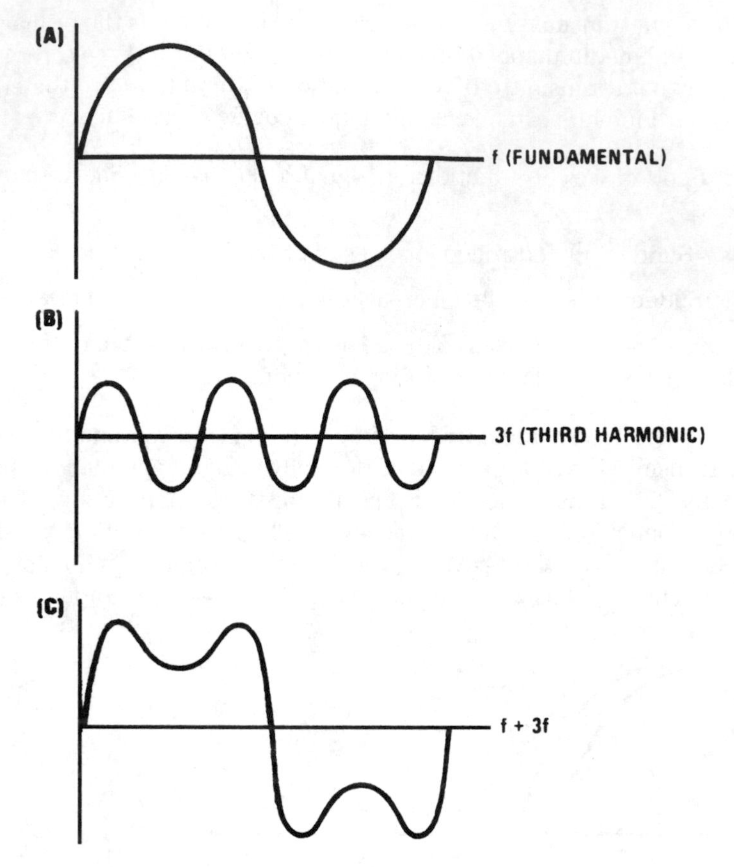

Figure 17.1.5 Harmonic distortion.

hit-measurement equipment. The cost of a transmission test set is proportional to the extent of measurement equipment in the receive section.

Measurement methods

There are several basic techniques associated with test sets. These involve some items to be careful of.

- *Bridge vs. terminate.* A transmission line is assumed to have an impedance of 600 ohms. This means that on the transmit side it presents a resistance or load of 600 ohms to the transmitter, and that correspondingly it expects to "see" a load of 600 ohms on the receive side. If this is not the case, level—and level-dependent—measurements will be erroneous.

  The test-set transmitter physically replaces the modem at its connection to the line, so the test set must expect to see a 600-ohm load. On the receive side, two possibilities exist. The test set can again physically replace the modem at its connection to the line. In this case, the test set must "terminate" the line into a 600-ohm load. Or, the test set can be attached to the line without replacing the modem. In this case, the test set "bridges" across the line and must present a high impedance to the line so that the line does not see both its load and that of the modem. Level, and all level-dependent, measurements will be wrong if a line is bridged when it should be terminated, or vice versa. Most test receivers can switch between these modes.

- *Setting levels.* The correct transmit and receive levels must be established. Common practice is to have a +13 TLP on transmit, and a –3 TLP on receive. This means that the transmit level should be 0 dBm and that the receive level should be –16 dBm because the data level is –13 dBm0 (Figure 17.1.2). All measurements with tone should use a data-level tone at 1004 Hz, except for envelope-delay distortion, which uses a 1.8-kHz tone.
- *Circuit loss.* With the proper level tone being transmitted, the user should verify that the proper tone level is received. Care should be taken about whether a bridged or terminated measurement is being made. The received level should be within 1 dB of the expected level. If it is not within this tolerance, circuit-level adjustments should be made to correct the error. It is not good practice to attempt to compensate for the difference with the test set because all level-dependent measurements will then be incorrect.
- *Noise measurement.* C-notched noise is normally what is measured—using both a notch filter and a C-message filter (Figure 16.1.1). Care should be taken to use a data-level holding tone, and to know what the test set is actually measuring (dBrn, dBrnc, or dBrnc0). A 3-kHz flat measurement should be made to check for power-line interference.

  If the test set has an audio monitor, it might be worthwhile to measure idle channel noise and listen to the circuit. This is done by leaving the test set receiver in a C-message noise measurement mode and putting a "quiet" termination—the characteristic impedance—on the transmit end. Most test sets have this termination, which might simply be a 600-ohm resistor placed across the transmit line with the transmitter disconnected. Listening on the receiver, you might hear muffled conversations from other lines (crosstalk), or tone interference (a tone). There is not a specified measurement technique for single-tone interference, so the ear might be the only practical detection device.
- *Envelope delay.* The test set manual should be checked for arrangements to measure forward (transmit)- or reverse (receive)-channel delay. A 1.8-kHz tone is used as the zero millisecond reference for the delay measured at other frequencies.
- *Phase jitter.* A data-level tone—the 1-kHz holding tone—should be used. Phase jitter measurements should be distinguished from the measurement of high noise levels.
- *Distortion.* Distinction should be made between the measurement of harmonic or of nonlinear distortion. Noise affects distortion measurements so that a part of the measurement process is to check the noise contribution. The measurement can then be corrected by a table-lookup procedure.
- *Hit measurements.* Again, a data-level tone should be used. Test readings are normally made for a duration of 15 minutes. Correlation should not be made with the results for tests of other time durations because of random phenomena.

It is often possible to mistake dropouts, gain and phase hits, and impulse noise for one another. To avoid this misidentification, hits are normally measured simultaneously in a prescribed order. The hit-count order is: first, a dropout locks out all other hit counters. Second, a gain or phase hit preempts any impulse-noise count.

Improving the line

Consider a line with ideal phase and amplitude response. Assume that a pulse with a baseband frequency response, as shown in Figure 17.1.6A is modulated, passed through the line, and demodulated. The resultant time response of the pulse at the receiver would be as shown in Figure 17.1.6B. Suppose that at time T, a second pulse is transmitted. At the receiver, the result would be the addition of the two pulses shown in Figure 17.1.6C. Note that if the received signal is being sampled at a rate of T times a second, the received pulses can be accurately interpreted because there is no interference between the received pulses at time T. (One pulse is at its maximum amplitude, and the other is at zero.) This conclusion means that under ideal circumstances, it is possible to signal at an unmodulated rate of double the fl pulse rate through a line that is band limited to fl Hz. This rule is a simplification of a Nyquist criterion of communications.

For an actual line with a given amplitude and phase response, the situation changes. First, because the amplitude and delay response are so bad at the edges of the voice band, only the center portion of the band is used to transmit data. Normally, only 2.4 kHz (1.6 kHz in severe cases) of the band centered around 1.7 or 1.8 kHz is used, as shown in Figure 17.1.6D. Next, a method is used to compensate for the amplitude and phase response to approximate an ideal line more closely.

To understand the need for this compensation, consider this example of the effects of delay distortion. Suppose that a pulse like one of those discussed previously is transmitted on an actual line. The received pulse is distorted, as shown in Figure 17.1.6E. Notice that now at intervals of time T, there is some contribution of amplitude of other pulses, which could be transmitted at multiples of time T. This effect is known as *intersymbol interference*. Compensation is, then, used to minimize this type of interference.

The compensation process is called *equalization*. Equalization can be fixed, manual, automatic, or adaptive. Equalization is covered in detail in Section 5.4.

For speeds up to 1.8 kbit/s, the standard modulation method is frequency-shift keying (FSK). This type of modulation is very simple to implement, costs little, and is quite rugged for this low-speed transmission. For 2.4-kbit/s transmission, differential phase-shift keying (DPSK) has become the standard. In DPSK, the reference phase required for signal detection "rides" with phase jitter and other impairments of the line. That is, to correctly sample and interpret a symbol, all that is required is the previous symbol; whatever distorted the current symbol probably affected the previous symbol the same way so that a type of immunity to signal impairments is inherent. DPSK gives reasonable efficiency at a moderate cost.

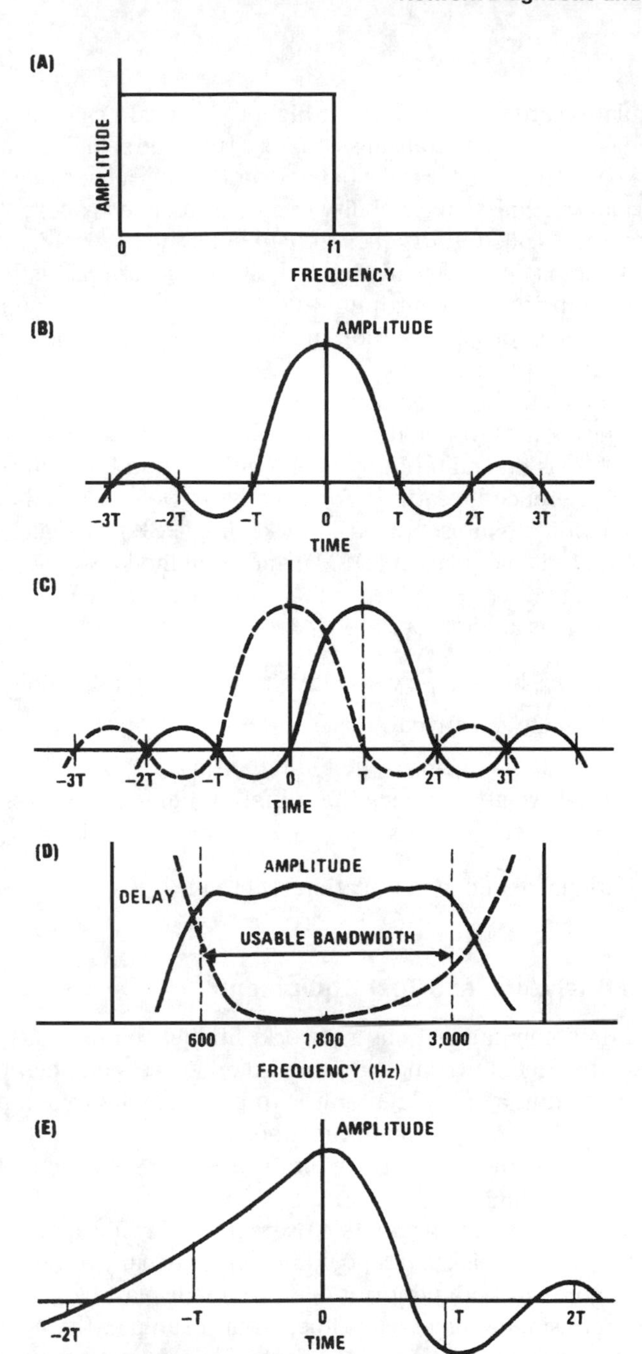

Figure 17.1.6 Why line equalization is needed.

The higher-speed modem

The choices of modulation are not as simple for higher speeds. Recall that about 2.4 kHz of the voice band is used for data transmission. As a rule, one signaling interval (or bit per second) can be transmitted per cycle of bandwidth. This means that 2.4 kbit/s is the maximum normal transmission capability of a line. At speeds above 2.4 kbit/s, therefore, another method, called *multilevel coding*, must be used.

The number of bits transmitted per modulated signaling intervals is limited by noise. All transmission impairments contribute to errors, and can be related back to noise. Therefore, the primary design criterion in modems becomes one of maximizing the margin against noise (signal-to-noise ratio).

Different types of modulation methods on lines with identical bit-error rates can be compared for various coding levels and signal-to-noise ratios. Modem-vendor data can be useful for conditions encountered by users with certain line conditions not listed in the available reference literature. For speeds between 4.8 and 9.6 kbit/s, several types of modulation are in use: phase-shift keying (PSK), vestigial sideband (VSB), quadrature amplitude modulation (QAM), and combined PSK and amplitude modulation (AM).

Some general observations are appropriate:

- VSB schemes are efficient, but are very sensitive to phase jitter and noise.
- PSK schemes are extremely sensitive to noise and phase jitter.
- Amplitude-phase-modulation schemes—including QAM—appear superior in the presence of noise and phase jitter. Notice that different implementations will vary in their noise and phase jitter immunity.

Section 5.3 has additional detail on modulation techniques.

17.2 Line Monitors, Protocol Analyzers, and Test Equipment

When trouble hits a data communications network, finding the fault fast is paramount. That's why an investment in a line monitor and other test equipment is a wise decision. Such devices permit a capable operator to isolate faults to a component, such as a modem, terminal, line, or a communications processor. In the hands of an experienced technician, these devices can pinpoint problem areas within a defective unit to expedite troubleshooting.

Line monitors are passive devices that enable users to display line activity as it occurs, or after the fact because of disk and tape units built into many devices. Because control characters are normally both nonprintable and nondisplayable, line monitors decode such characters using mnemonics. Thus, as an example, the acknowledgment character might be displayed as $^{A}{}_{K}$ in the form of a superscript and subscript on the monitor's screen, which occupies one character position. This function is more formally called *decoding* and it varies considerably between line monitors. Some line monitors decode ASCII and EBCDIC control characters, and others decode binary data streams into their appropriate fields in an SDLC, HDLC, or X.25 frame.

Another device that provides a line monitoring capability, but which is much more sophisticated, is a protocol analyzer. The additional features built into protocol analyzers can include abilities to trap data, to emulate or simulate predefined activity, or to perform a statistical analysis of line activity.

Protocol analyzers

Protocol analyzers trap and display data and control characters to provide precise pictures of line activity, eliminating the difficulty in deciphering pulses viewed on an oscilloscope. In addition, the analyzer might permit control signals on EIA-232 connections to be checked when there are problems in establishing and maintaining message exchanges.

In a multivendor installation, a protocol analyzer can pinpoint a problem so that the proper service organization can be called to make repairs. Downtime and servicing costs can be excessive if the user calls the wrong service organization.

Conventional troubleshooting requires the measuring of voltage and current levels, reading phase jitter, and viewing bit streams with oscilloscopes. The proper application of these techniques requires more expertise than is ordinarily available in many communications installations. An equally valuable role of protocol analyzers is frequent on-line testing to check for degradation of service, which often signals incipient failures. Problems found during such routine tests can be corrected before they cause costly, unexpected breakdowns.

Protocol analyzers provide readouts for data and control characters. Information is displayed in one of four ways:

- Light-emitting diodes (LEDS) that indicate the states of individual bits. This readout must be converted into the appropriate code
- Labeled indicator lights
- Alphanumeric displays
- Alphanumeric on cathode-ray tubes

Most units permit data or control characters to be trapped and displayed on a selective basis. Trapping can be initiated on detection of errors, or by decoding specified characters, such as a synchronization character. The ability to read the states of the EIA-232 control leads allows the user to check the "handshaking" sequence if there are problems in establishing and maintaining message exchanges. Many protocol analyzers also permit these leads to be "clamped" at desired voltage levels, or programmed to simulate any aspect of operation.

Some units also provide test messages for polling and answerback, which are used in end-to-end network tests. Parity indications and counts, also common features, are used both in checking out specific problems and in determining line quality. Advanced data link controls, such as IBM's SDLC and Digital Equipment Corp's DDCMP, can be accommodated, so long as the message lengths do not exceed the buffer size of the analyzer.

The compactness of integrated circuits permits protocol analyzers to be built into small enclosures. The makers of units that have binary readouts squeeze them into

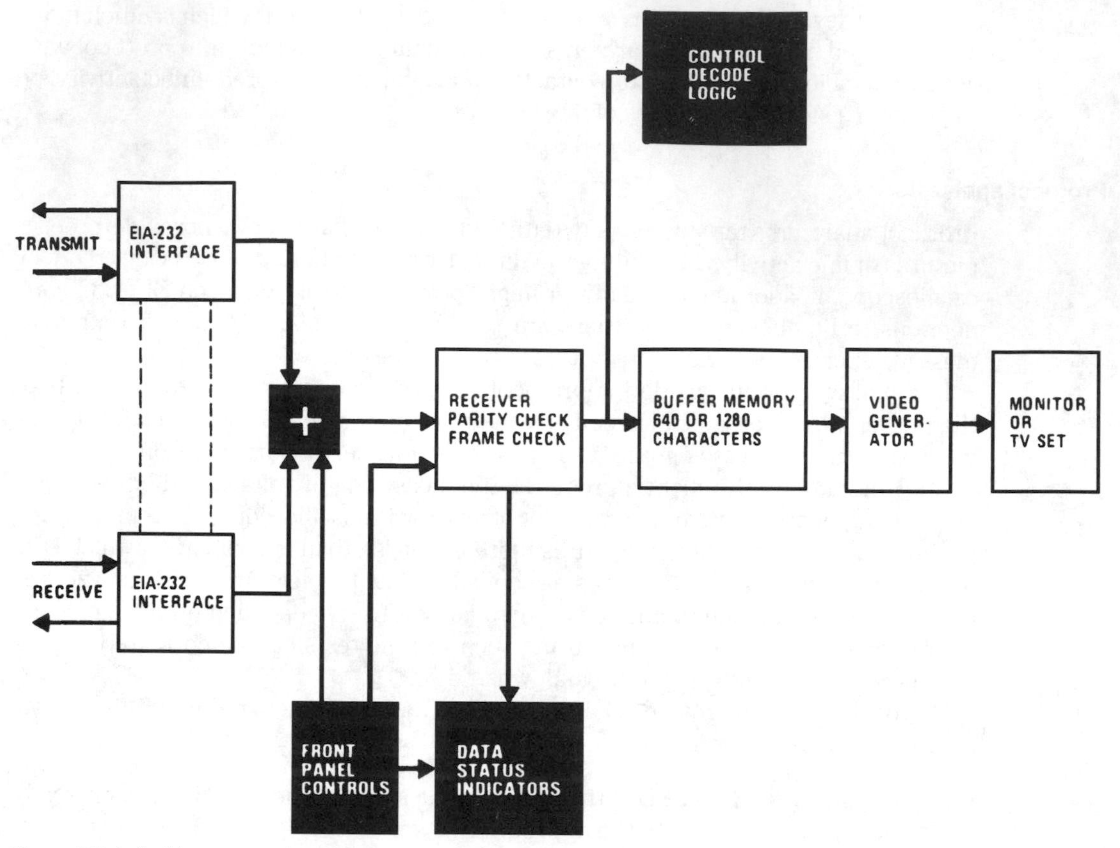

Figure 17.2.1 Inside a protocol analyzer.

suitcases small enough to fit under an airplane seat. Devices with cathode ray tubes are bulkier, but, nevertheless, they are still fairly small.

Most protocol analyzers are built around serial buffer memories that store the data selected for trapping, as illustrated in Figure 17.2.1. Operation of the buffer is indicated by logic circuits, which route the trapped characters to the memory and direct the readout into the selected mode. Analyzers interface the line with connectors compatible with EIA-232 standards.

Protocol analyzers are used in two ways: to check out control signals at the EIA-232 interface (establishing and maintaining a proper connection); and to check for the correctness and proper sequencing of control characters and data.

If the message exchange cannot be initiated or maintained, the EIA-232 lines should be checked with the line analyzer at both ends of the data link. If there is difficulty in transmitting, then, for most modems, the analyzer should check the data-terminal ready, data-set-ready, and request-to-send states. Each manufacturer provides for the display of different combinations of EIA-232 lines. The prospective buyer should ensure that the unit she is considering includes those tests deemed necessary for her troubleshooting requirements.

The next level of troubleshooting consists of testing data-link control characters and data. Some protocol analyzers trap only control characters, and others also capture data. This should also be considered in selecting a unit.

Some protocol analyzers provide complete, self-contained testing capabilities, which include polling, answerback, and generating test messages. Other types can participate in these tests only if additional equipment is used to generate signals and responses. Examples of tests using protocol analyzers are shown in Figures 17.2.2 through 17.2.7.

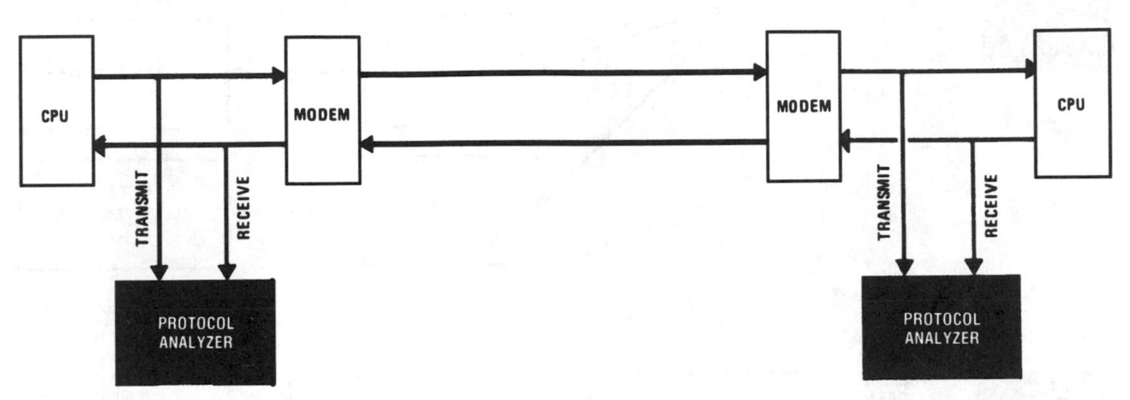

Figure 17.2.2 Real-time monitoring.

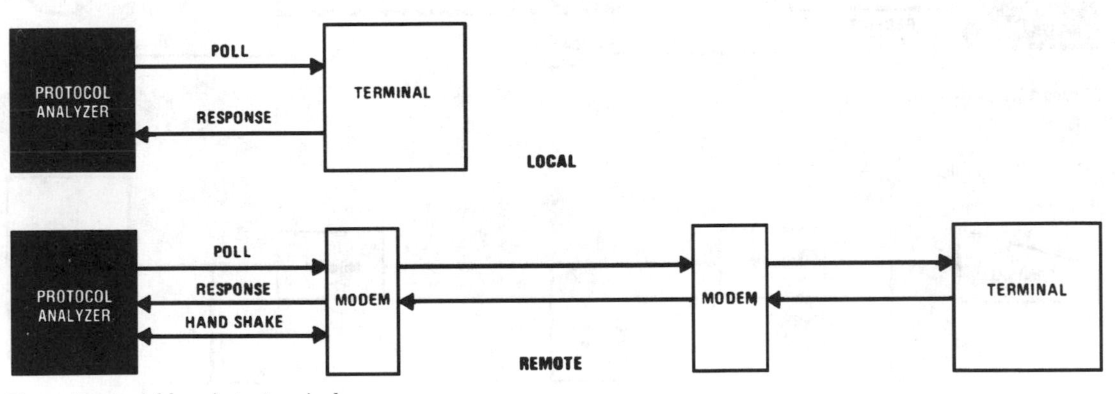

Figure 17.2.3 Addressing a terminal.

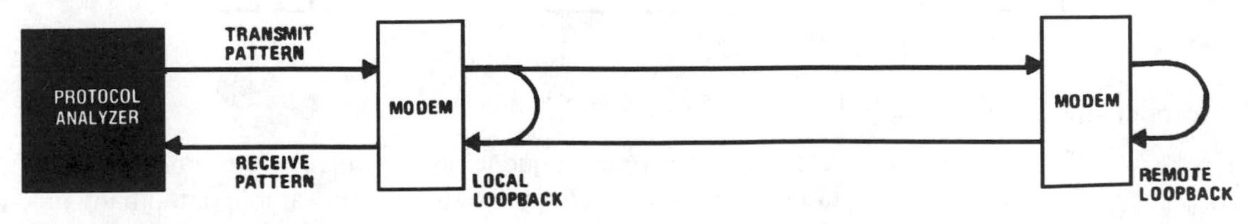

Figure 17.2.4 Testing modems.

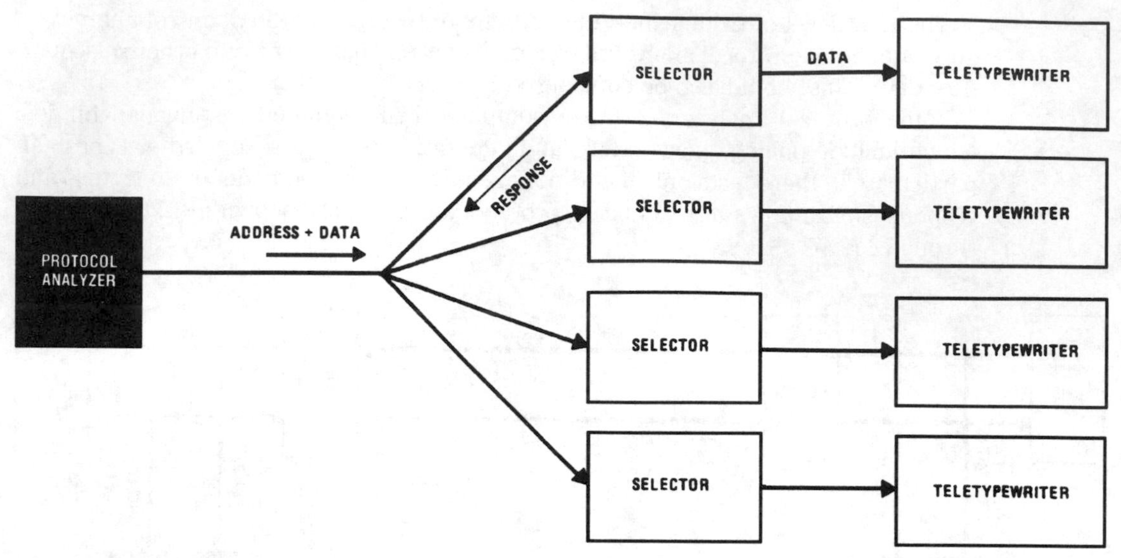

Figure 17.2.5 Polling selectors.

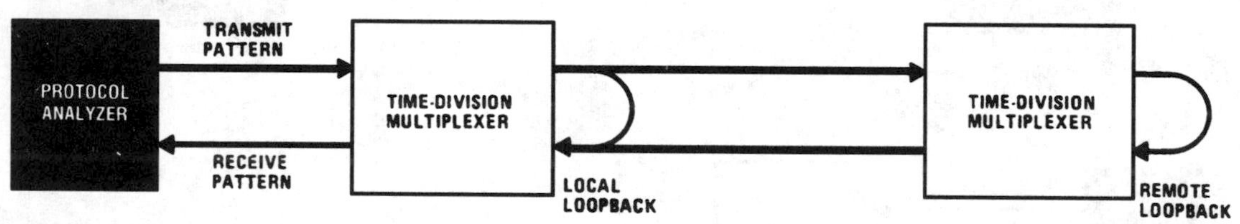

Figure 17.2.6 Testing TDMs.

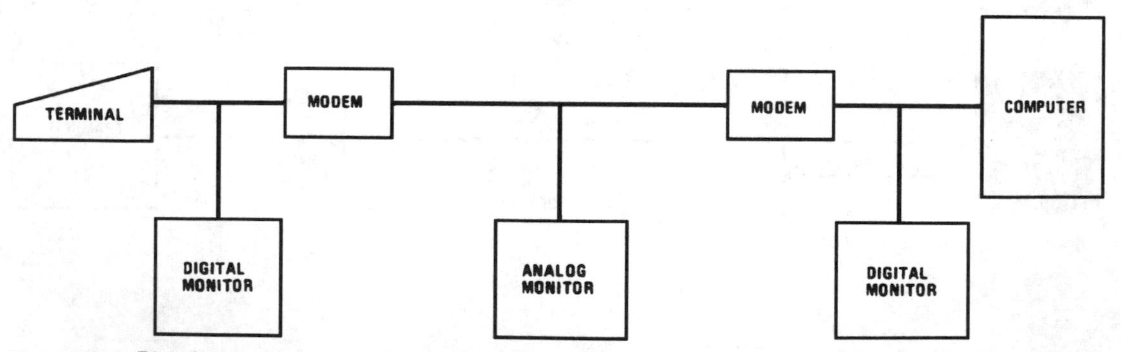

Figure 17.2.7 Digital and analog monitoring.

Protocol analyzer features

To check the correctness and proper sequencing of control characters and data without having to obtain a terminal and operator to transmit sample data, many protocol analyzers contain special features that can generate the desired test data (Table 17.2.1).

TABLE 17.2.1 Protocol Analyzer Features

Code-level selection	Modulation rate
Stop-element length selection	Random-word generator
SDLC	Dot-pattern generator
Stored-message generator	Error count
Send RY/U*	Nonvolatile program memory
Character-error count	Selectable character trap
Distortion analysis	Programmable sync characters

- *Code level.* The level of a code is defined as the number of information bits used to encode the characters in an alphabet, generally 5, 6, 7 or 8 bits, Baudot, ASCII, or EBCDIC.
- *Stop-element length.* For asynchronous formats, the stop element consists of a pulse with 1-, 1.42-, 1.5-, or 2-bit intervals, depending on the code and equipment used.
- *SDLC.* Several sets offer automatic SDLC (IBM's synchronous data link control) capability as an option. With SDLC, block-check-character treatment is important. Although it is not difficult to calculate and manually enter the BCC (block check character) for BSC (binary synchronous communications) protocols, it is cumbersome to do so for SDLC formats.

  In such cases, block-check-character entry involves additional manipulations for zero-insertion and zero-complementing NRZI (nonreturn to zero, inverted). The inverse of these manipulations must be performed on the received message for it to be intelligible. Unlike the BSC protocol with its negative-acknowledgment response, an absent or wrong BCC and zero insertion means a polled station will not reply.
- *Stored-message generator.* Stored messages—generally in a programmable read-only memory—are available in such codes as Baudot (International Telegraph Alphabet No. 2), ASCII, and EBCDIC. The stored message might be a standard "fox" (the quick brown fox jumps over a lazy dog's back, 1234567890) or some other combination of characters.
- *Send RY/U*.* Alternating 1s and 0s can be sent instead of information bits in a string of characters by using the character pairs RY for 5-level code and U* for 8-level code.
- *Character-error count.* For error-rate determination, parity-error count is generally regarded as being equivalent to character-error count. The latter is obtained by bit comparison of received characters with the generated message. On the average, 80% of the character errors are found by parity checks.
- *Peak- or bias-distortion analysis.* Peak distortion refers to peak individual distortion when operating in the synchronous mode (EIA Standard 334 explains synchronous distortion), and to peak "gross stop-start distortion," or "telegraph distortion," during operation in the asynchronous mode. This type of distortion is

explained in EIA Standard 404 and Bell System Technical Reference PUB 41003. Peak-distortion analysis refers to both synchronous and asynchronous kinds. Bias distortion refers to the elongation of either the mark intervals (marking bias) or the space intervals (spacing bias).

- *Modulation rate.* The distortion display on some test sets can indicate overspeed; that is, the difference between the modulation rate of the received signal and a standard speed provided by the internal clock. If the internal clock has a fine enough adjustment, it can be used to measure the actual modulation rate of the received signal.
- *Random-word generator.* Up to six pseudo-random bit sequences can be stored in the test set. An example is the 511-bit pattern in CCITT recommendation V.52.
- *Dot-pattern generator.* A dot pattern—that is, a square wave, also known as *reversal* or *1:1*—is obtained by generating alternate mark and space levels for analysis.
- *Error count.* Error count and error-rate capability can be used as interchangeable terms by some manufacturers. Some vendors use the acronym BERT (bit error rate test). Normally, the line monitor will provide a direct readout of the error rate and a readout of the total count and time interval associated with an error count.
- *Nonvolatile-program memory.* Devices with this feature permit retention of programs fed into memory via the keyboard or bit switches, even though the power is disconnected. The retention time varies, depending on the equipment.
- *Selectable character trap.* With some devices, incoming data can be compared with one or more "trap" characters that the operator has programmed into a storage register. When a sequence of incoming characters matches the programmed characters, the operator can turn on a trap indicator that will initiate some function, such as transmitting an acknowledgment of a poll message to the station where the test set is located. For simplicity, the selected trap character(s) could be the station's address.
- *Programmable sync characters.* For transmitting text over synchronous data links, one or more synchronization characters are required to bring the receiver and transmitter into the correct character-phase relationship. If the device has this feature, its receiver circuit is programmed with the correct character(s).

Analog measurements

Up to now, this section has been limited to digital monitoring between the transmission device and the terminal equipment, as illustrated in Figure 17.2.7. Although such monitoring is important to ascertain such information as why a specific terminal does not respond to a poll—using a line monitor can show that the line address generated by the computer is incorrect, or perhaps an end-of-transmission character inadvertently follows every poll because of a software error—digital monitors will not provide information on dropouts, phase hits, gain hits, impulse noise, envelope delay, and other circuit data technicians can use to determine the causes of problems on the line. For such measurements, an analog line monitor, capable of measuring the previously covered information, as well as such additional data as signal strength, amplitude modulation, and harmonic distortion, must be used. Because the control of a circuit normally rests with the communications carrier, only a few organizations use analog test equipment. Most organizations merely report an outage to the communications carrier by informing the carrier that a loopback test generates lots of errors, whereas a local loopback test on both ends of the circuit in question shows both transmission devices to be working properly.

17.3 Diagnostic Center Components

Any ad hoc approach to troubleshooting in a data network can run into problems and duplication almost as soon as it can be classified as a network; i.e., when there are two locations. With large networks, these and other considerations make a centralized and systematized troubleshooting operation practically essential. Commonly called a *tech control center*, such a diagnostic facility often pays off additionally by pinpointing and pruning unnecessarily complex line routings, thereby keeping the network neat and easy to maintain.

The first step in designing a tech control center is to define the network clearly. Up-to-date information might be readily available in a new network, but in an older network it often is not updated frequently enough. The information on the central computer site should include the computer's line address for each channel, identity of the digital line that connects the front end to the modem, the modem type, the party responsible for the modem, the modem's telephone or leased-line number, the modem's data-transmission rate, type of data-access arrangement, if applicable, routing of remote lines, and the party responsible for remote lines. Similar information is needed at terminal sites, including locations and terminal types.

Once this data is mapped out, the next step involves adjustments to include future network elements so that they can be accommodated in the tech control center.

The mapping and documentation indicate the number and types of channels that must be monitored. The types are categorized first in terms of transmission medium (cable, satellite, line-of-sight), then by whether the medium is dial-up or leased, digital or analog, synchronous or asynchronous, and two-wire or four-wire. This information indicates the number of channels that are needed in the patching and test equipment, and the types of lines that must be handled by this equipment.

Obviously, not all channels need to be monitored simultaneously; thus the actual number of parallel channels in the tech center might be substantially less than the total. But knowing the total of present and future channels provides the basis for an intelligent estimate of the channel capacity.

The basic equipment must include a patchfield for interconnection and a bit-error-rate test set (BERT). The patchfield permits network components to be connected or bypassed, provides for test equipment to be connected into signal lines without disturbing them, and provides for loopback tests. A BERT generates bit sequences. These can be applied to various portions of the network, and checks made to see if the bits are properly transmitted and received.

Other important devices are a set of lights to indicate the status of the interface cable's control leads, a voltmeter and oscilloscope to monitor signal levels and pulse shapes, and, if the carrier cannot be relied on for adequate or prompt line testing, a variety of analog line-monitoring test instruments. Error detection and control-lead monitoring often are included in line-monitoring equipment, which also provides for readout and trapping of characters, and operation in specific line protocols.

In the following diagnostic center component sections, examples are based on synchronous transmission on a four-wire (full-duplex) circuit.

Patchfield

A patchfield provides a way for monitoring lines, injecting signals, and reconfiguring a network. If it is well designed, a patchfield also provides access to any local equipment interface.

As illustrated in Figure 17.3.1, the equipment and line jacks are normally connected together internally. If a plug is inserted in the equipment jack, the V-shaped metal prong is lifted and the internal contact is broken, disconnecting the line and connecting the equipment at the opposite end of the plug. Notice that the circuit shown is for a two-wire connection, with ground not shown. A four-wire connection requires two sets of jacks.

In the diagram, the modem is connected to the front end. But if a BERT is used to simulate the front-end output to the modem, then the test set is plugged into the modem jack, so the actual front end is automatically disconnected. The same is true for the computer jack, except that the test equipment is connected to the front end to substitute for the modem. The patchfield can be used in a similar manner to switch out a defective unit and replace it with a spare piece of equipment.

The monitor jack is used to provide access to a line without disturbing it. Plugging into the monitor jack does not break the circuit, but only provides a point of contact for connecting test equipment.

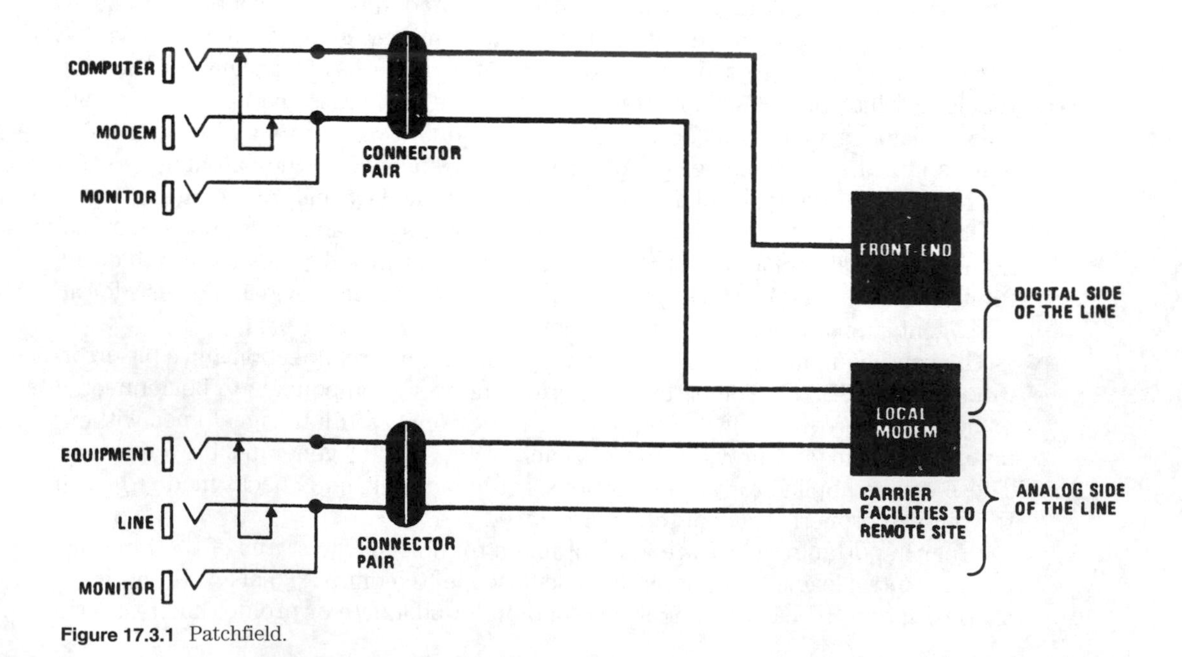

Figure 17.3.1 Patchfield.

LEAD	DESCRIPTION	COMMON GROUND	DATA		CLOCK		CONTROL	
			TO MODEM	FROM MODEM	TO MODEM	FROM MODEM	TO MODEM	FROM MODEM
1	PROTECTIVE GROUND (PG)	X						
2	TRANSMIT DATA (TD)		X					
3	RECEIVE DATA (RD)			X				
4	REQUEST TO SEND (RTS)						X	
5	CLEAR TO SEND (CTS)							X
6	DATA SET READY (DSR)							X
7	SIGNAL GROUND (SG)	X						
8	CARRIER DETECT (CD)							X
9	DATA SET TEST (DST)							
10	DATA SET TEST (DST)							
11	UNASSIGNED							
12	SECONDARY CARRIER DETECT (SCD)							X
13	SECONDARY CLEAR TO SEND (SCS)							X
14	SECONDARY TRANSMIT DATA (STD)		X					
15	TRANSMIT CLOCK (TC)					X		
16	SECONDARY RECEIVE DATA (SRD)			X				
17	RECEIVE CLOCK (RC)					X		
18	UNASSIGNED							
19	SECONDARY REQUEST TO SEND (SRS)						X	
20	DATA TERMINAL READY (DTR)						X	
21	SIGNAL QUALITY (SQD)							X
22	RING INDICATOR (RI)							X
23	DATA RATE SELECTOR (DRS)						X	X
24	EXT TRANSMIT CLOCK (XTC)				X			
25	UNASSIGNED							

Figure 17.3.2 EIA-232 lead assignments.

The diagram is simplified for the purpose of illustration in Figure 17.3.1. For an actual setup, such as one that conforms to EIA-232 specifications, there are a total of 25 data, clock, and control lines in the interface cable between the front end and the modem. Their signals are combined in the modem into each two-wire channel. Therefore, the tech center should have provision to patch many leads simultaneously, but not necessarily all equipment in the network.

EIA-232 lead assignments

The EIA-232 lead assignments (Figure 17.3.2) that must be patched (assuming the protective ground is common) are:

- Transmit data (TD)
- Receive data (RD)

- Transmit clock (TC)
- Receive clock (RC)
- Data terminal ready (DTR)
- Data set ready (DSR)
- Request to send (RTS)
- Clear to send (CTR)
- Carrier detect (CD)
- Signal ground (SG)

Normally, at least one or more of the following leads must also be patched:

- Ring indicator (RI)
- Signal quality detector (SQD)
- External transmit clock (XTC)
- Data-rate selector (DRS)

Note that provision for the variety and combinations of leads that can be patched must be based on operating conditions—both present and future. (Refer to Chapter 9 for details on EIA-232, EIA-449, and V.24. Table 9.3.2 gives the equivalents.)

Using BERT

The most common application of the bit-error-rate test set is in connection with loopback tests. The bit pattern produced by the BERT is passed through one or more elements of the network. The output is returned to the test set and compared to the original test pattern. Each time a mismatch is sensed, an error counter on the BERT is incremented. This count is an indication of the channel's error rate (errors per unit of time).

Loopback testing is generally performed in successive stages, encompassing a greater proportion of the network at each stage. In the first stage, the test set is looped back on itself to determine that it is operating properly. Then, the BERT is connected to the local modem, the output is looped back, and the error rate is checked. The signal is turned back at the modem output before it enters the transmission line. Connections are made through the patchfield, as shown in Figure 17.3.3.

Next, the test is repeated, this time including the front end and modem. The signal is turned back just before it enters the remote modem so that the line is also checked out. Then, the remote modem is included, with the signal being turned around at the digital side. The test sequence continues in this manner until all elements of the network have been added to the chain. At each stage, any added error rate indicates the degradation of signal quality contributed by the last unit connected.

In looping back the data from any point on the digital side of the remote modem, some changes in clock synchronization are needed. The transmit- and receive-clock signals in a modem are not normally synchronized with respect to each other. The transmit clock is produced by an internal oscillator, and the receive-clock signal is

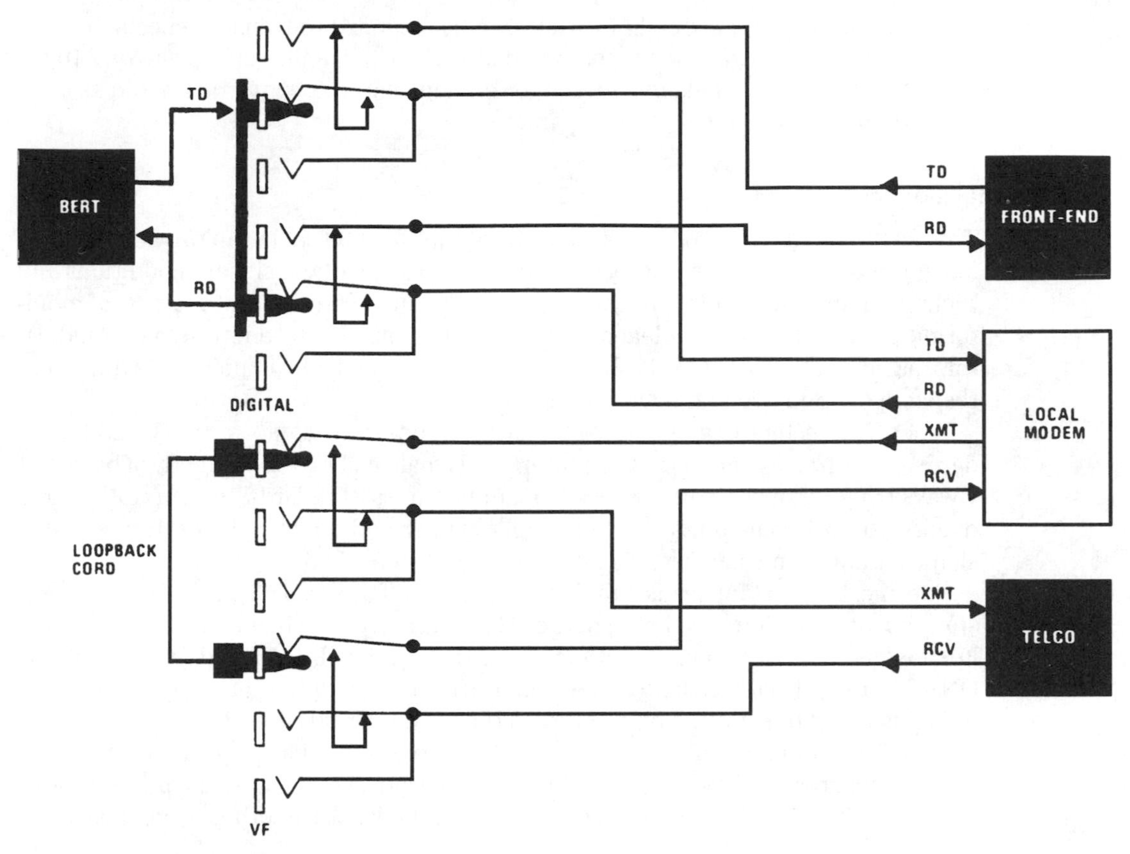

Figure 17.3.3 Bit-error-rate test setup.

derived by phase-locking on the received data stream. The receive clock is thus synchronized to the transmit clock at the remote modem. But when the data is looped back, the received data must be synchronized more accurately so that it can be properly sampled. To accomplish this, the transmit clock at the local site becomes the master clock, and both the transmit and receive clocks at the remote site must be slaved to it for synchronization.

This is achieved at the remote modem by strapping it for operation with an external transmit clock, connecting pins 15, 17, and 24 so that all three clocking signals are synchronized. In some modems, the receive clock is connected automatically to the external transmit clock in the loopback mode. To achieve loopback of the data itself, the transmit data and receive data pins (2 and 3, respectively) are connected. Some modems also require the presence of request-to-send (pin 4) and/or data-terminal-ready (pin 20) signals to operate in a loopback mode. The 12-volt dc supply needed to simulate these signals in most modems can be derived by connecting a lead from the data set test line (pin 9) to pins 4 and/or 20.

A return path is not available on two-wire channels. Therefore, a BERT is needed at both ends—one to transmit the test signal and another to receive and count the errors.

Line monitors used instead of BERTs can be arranged to transmit special bit patterns, such as protocol-control characters, to check the ability of the network to establish and maintain a link, in addition to sending character patterns treated as data. Some monitors also trap selected characters.

Digital signal monitoring

The activity on the control and clock leads of the EIA-232 cable can provide valuable insight into the setup's operating sequence. Knowledge of the signals' conditions and their sequence often points to the source of a failure. The simplest method of monitoring the signals on these leads is to connect them to light-emitting-diode indicators, as illustrated in Figure 17.3.4. The LEDs are individually illuminated when their respective inputs are in the ON state.

When connecting new equipment that is supposed to conform to EIA-232 standards, it is often assumed that the interface signals will match perfectly. This is not always true. Because of differences in manufacturers' standards, interpretation, and quality-control, some units will work well together and others will not. Hence waveshapes, timing, and signal levels must be carefully checked.

The most important leads are TD, RD, RTS, CTS, CD, DSR, and DTR. Understanding the function and timing of these leads is extremely helpful in fault isolation. For instance, if the terminal and modem are operational, DTR and DSR should be ON. If control signals are being exchanged, RTS, CTS, and CD should be active. And if data is being transmitted or received, TD and RD should be active.

Two other useful monitoring devices are a voltmeter and a oscilloscope. The voltmeter measures the dc voltages on the interface leads, while the scope permits viewing of activity and waveshapes on data and clock lines. Still another useful feature of

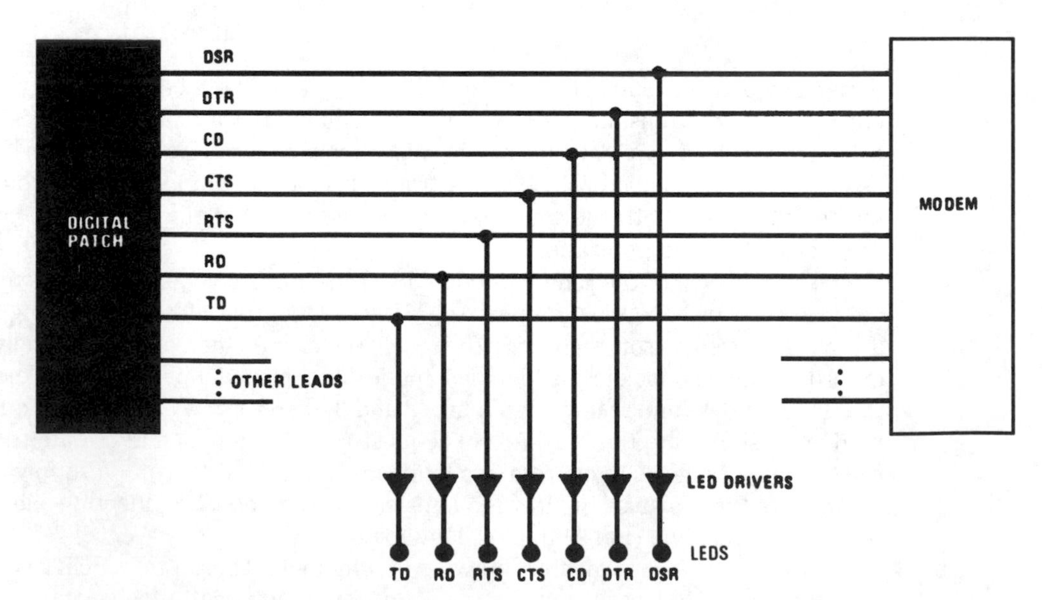

Figure 17.3.4 Digital signal monitoring.

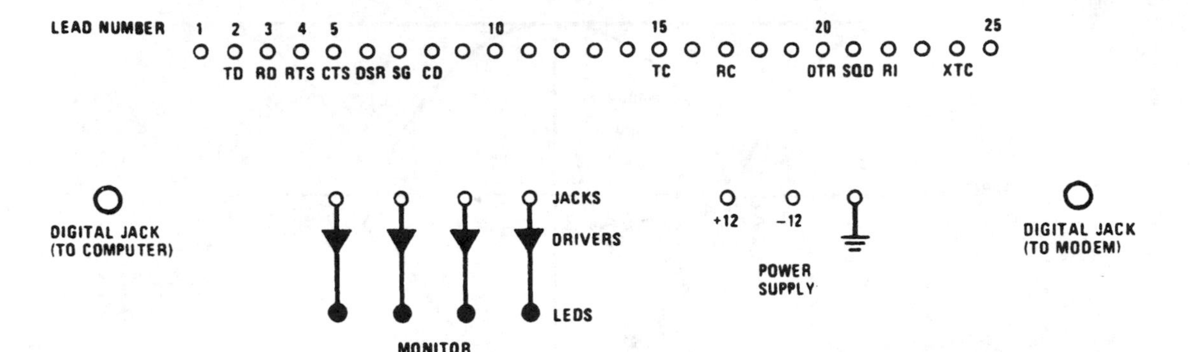

Figure 17.3.5 Monitoring panel.

a tech center is a 12-volt dc source to simulate EIA-232 signals for controlling the operating modes of the equipment under test, as described for loopback testing. An example of a monitoring panel with jacks, control leads, LEDs, and power-supply leads appears in Figure 17.3.5.

Analog signal monitoring

In addition to the basic digital test equipment, a tech control center can contain equipment to measure carrier-line characteristics. Among the parameters measured are circuit loss, noise, attenuation distortion (frequency response), delay distortion, phase jitter, frequency shift, phase hits, gain hits, dropouts, and nonlinear (harmonic) distortion. Of these, the simplest ones to measure are circuit loss, noise, attenuation distortion, and frequency shift. These can be measured with a single instrument, called a *level/noise/frequency test set*.

Other integrated instruments measure both phase jitter and impulse noise. Delay-distortion test sets can be valuable for testing conditioned lines.

Several instruments that measure all the aforementioned parameters are now available. One even provides a visual display of frequency response and delay distortion simultaneously. In the absence of these instruments, however, an oscilloscope can be used to obtain a qualitative visual indication of the important transmission characteristics. The resulting display is called an *eye pattern* because of its characteristic shape, as shown in Figure 17.3.6.

For an eye pattern, two signals are needed from the modem. One is the received baseband data signal (analog) and the other is the symbol timing wave, a binary signal. On low-speed modems, one bit is transmitted for each one-bit sampling interval. In high-speed modems, which use multilevel coding, each symbol bit represents more than one data bit. In most modems, this signal is not available at the outside of the modem, but instead must be tapped from an internal test point.

The eye pattern is produced by applying the demodulated data input to the vertical oscilloscope channel, and triggering the oscilloscope with the symbol timing wave. Positive-going signals form the upper half of the eye; negative-going signals, the lower half.

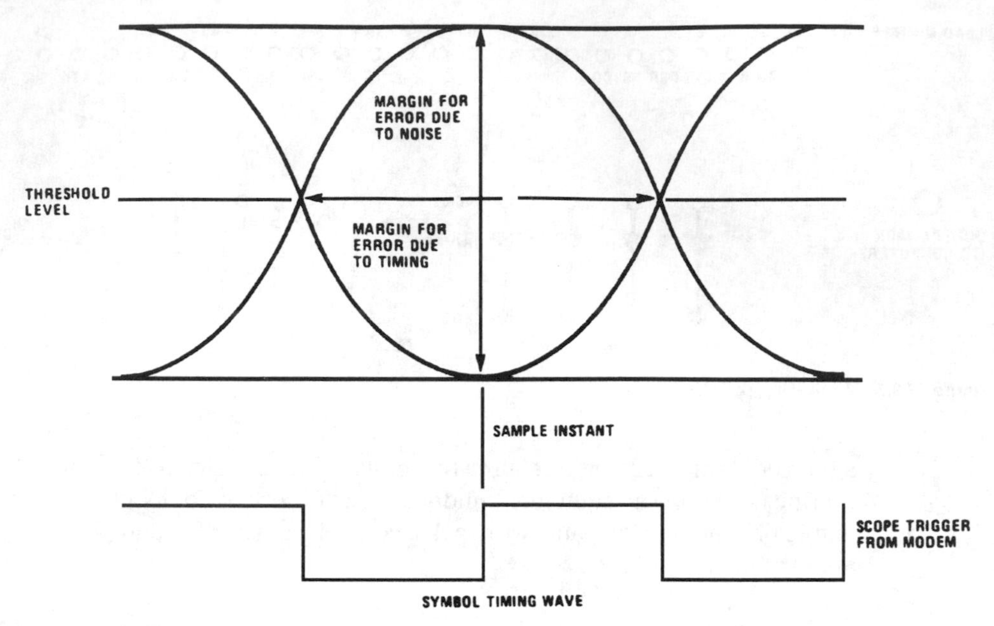

Figure 17.3.6 Eye pattern display.

The waveform is illustrated in Figure 17.3.6. The pattern shown is for a modem with one-level coding. With multilevel coding, modems will show several "eyes," stacked one above another at each coding interval. Also of significance is the shape of the eyes. A pulse gated at the sampling instant is determined positive or negative by whether it is above or below the threshold level. In order for the pulses to be correctly recognized, the eye must be open, and the sampling instant and threshold level should be in the center.

The vertical amplitude is affected by noise, intersymbol interference, echoes, and gain variations. Excess amounts of any of these factors, as well as timing degradations, will tend to narrow the eye. A properly shaped eye, therefore, indicates the excellent "health" of a number of transmission characteristics. Even a nontechnical viewer can be relied on to check it. If the shape is incorrect, further testing is required.

Line testing is a gray area of responsibility. Assuming that the lines are not privately owned, testing must be done with the consent of the carrier, and it should not interfere with the carrier's normal operations. Line-test equipment should therefore be fully compatible with the line. In the event of cooperative testing, it should be compatible with the equipment used by the carrier.

17.4 Building a Diagnostic Center

Generally a diagnostic test center (commonly called a *tech control center*) must, in addition to providing sufficient space for the equipment, allow access to connecting and test points, use only as much cable as necessary, and provide adequate ventilation. Taken one at a time, these considerations can be fairly obvious, but to overlook any one of them is to invite problems later on.

Human factors also deserve attention. The effectiveness of a tech control center is determined to a great extent by the willingness of people to use it. And if its operation involves excessive bending, stretching, or other difficulty in taking readings or making connections, users might try to work around it.

Section 3 of this chapter dealt with selecting and applying the test equipment and patch panels for a tech center. This section concentrates on the layout and installation of the previously covered components.

The most important consideration in selecting a location is to minimize cable lengths, bearing in mind that to avoid signal degradation, RS-232-C cables should not exceed 50 feet. The optimum location from this point of view is between the modems and the front end. Moreover, to facilitate access to modems for line and other network tests, the modems are often housed right in the tech center.

For distances exceeding 50 feet, cable drivers can extend the usable cabling distances, sometimes by hundreds of feet, depending on the data rate. But be careful. Many equipment manufacturers will not support setups using cable drivers. And when they are used, be sure to test the cables with the tech center's test equipment, and the data processing and communications equipment, before the cables are placed in normal operation. This is to be sure that any signal degradation caused by the drivers and long cables is acceptable.

In designing a tech center for a network not yet installed, the cable routings must be carefully planned and their lengths accurately computed. All obstructions that interfere with direct routing should be carefully taken into account because 50 feet of cable can easily be consumed, even when the straight-line distance between interconnected units is only a fraction of the 50 feet.

Computer-type cabinets provide the best housing for tech center equipment. Rack-mountable units usually conform to their standard widths and can be ordered or fitted with appropriate mounting hardware. Typical cabinets are 6 feet high, 2 feet wide, and 2 to 3 feet in depth. The deeper cabinets are recommended because space is needed behind the equipment for cables and air flow. Rails are usually provided at the front and rear of the cabinets for mounting the equipment. The standard spacing between rails is 19 inches, the most common width of rack-mounted equipment.

Other desirable features are roller slides and rear doors to facilitate access to equipment, and open cabinet bottoms to permit installation of under-floor cabling. For additional cooling, the top of the cabinet should be open, as well. The cabinets should have hardware for distributing ac power and "loops" or "ladders" for routing cables.

A word about space for cables and connectors—design parameters that are easily underestimated. Providing access to each unit entails the ability to interconnect through a patch panel, also called a *patchfield*. A full-duplex channel—both channel types can be housed in a single cable—usually requires four lines: two RS-232-C cables for connecting the terminal or computer to the modem via the patch panel, and two telephone cables to tie the modem to the line, also through the patch panel. Therefore, as the number of channels increases, the number of cables and connectors needed at the patch panel and the amount of space needed to house them increase fourfold. Still more lines would be required if such items as multiplexers, switching equipment, and spares are also to be tied through the network's tech center.

To ensure that there is adequate room behind the cabinet for service workspace, the rear of the cabinet should be at least 3 feet from the nearest obstruction. Therefore, the total floor space required for each cabinet is at least 5 to 6 feet deep and 2 feet wide.

An alternative to cabinets is the inexpensive, open "radio-relay" rack. These racks are less attractive than computer cabinets, but they are easier to access, and their openness provides good ventilation.

An attractive, although more expensive, method of providing access to the patch panel, controls, and indicator lights is to build a wraparound console. But consoles also consume a great deal of floor space. So, unless the tech center is visited constantly, a wraparound console might not be advisable.

Care must be taken in positioning test equipment, patch panels, and control and indicator panels in the cabinets so that they are at convenient heights. Test equipment, and control and indicator panels, should be as close as possible to eye level, which is 4 to 6 feet above the floor. Patch panels can be somewhat lower, from 3 to 5 feet. Modems and other equipment not frequently accessed can occupy the space in the lower 3 feet.

A list of recommended distances above the floor for various types of equipment is given in Table 17.4.1. Notice that the table specifies equipment heights in RMS units, which here stand for *rack-mounting space*. A term frequently used by cabinet sup-

TABLE 17.4.1 Equipment Mounting Heights

Equipment type	Equipment height (RMS)	Mounting range (feet)
Cabinet	40	½ to 6
Patchfield (analog and digital for 12 to 16 modems)	4	3 to 5
Status monitoring panel (for 12 to 16 modems)	2	5 to 6
Test equipment	2 to 4	4 to 6
Miscellaneous patchfields (analog, digital, spare modems, etc.)	2	3 to 5
Asynchronous modems (typically eight per mount)	4	½ to 4
Synchronous modems, 2400 bit/s or less (typically two per second)	4	½ to 4
Synchronous modems, 4800 bit/s or more	4	½ to 4
Drawers (for patch cables, accessories)	2 to 4	3 to 4
Writing surfaces	1 to 2	3 to 4

NOTE: Distances are from the floor (feet) or lowest mounting point (RMS) to the bottom of the mounted equipment. (1 RMS = 1¾ inches)

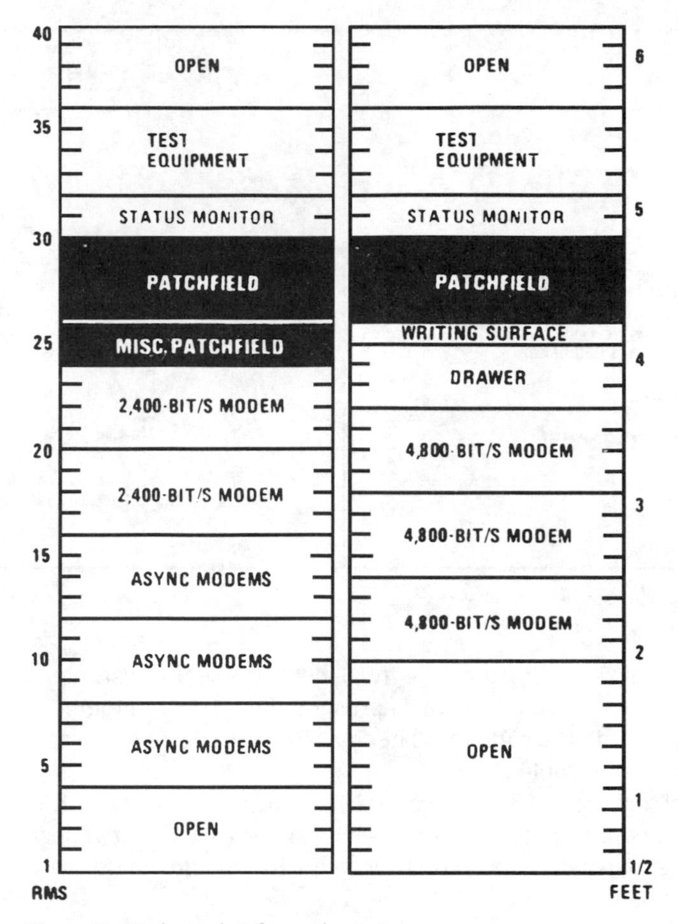

Figure 17.4.1 Arranging the equipment.

pliers, one RMS unit is equal to a linear distance of 1¾ inches. So 2 RMS equals 3½ inches, and so on.

Patch panels can present some problems in finding enough room to locate all the jacks at convenient heights. Commercial patch panels are designed for 12 to 16 modems, and each modem requires two EIA-232 connectors. Experience indicates that the recommended height range provides room for a set of three panels, which can accommodate 36 to 48 modems.

Equipment layout can be simplified by using worksheets, as shown in Figure 17.4.1, for arranging the equipment, calibrated both in feet and RMS units. Notice that the RMS measurements begin at the lowest possible mounting location, generally 6 to 8 inches above the floor, but the distance in feet is measured directly from the floor. The reason for the discrepancy is that RMS is used to measure usable mounting space, whereas the distance in feet—the actual distance from the floor—is used for measuring cable lengths.

EQUIPMENT	HEIGHT (RMS)
PATCHFIELD (MOST FOR 12 OR 16 MODEMS; TWO FOR 24 MODEMS)	2 X 4 = 8
STATUS MONITOR (TWO REQUIRED)	2 X 2 = 4
MOUNT FOR ASYNCHRONOUS MODEMS (IF THERE ARE EIGHT PER MOUNT, THREE MOUNTS REQUIRED)	3 X 4 = 12
MOUNTS FOR 2,400-BIT/S MODEMS (IF THERE ARE TWO PER MOUNT, TWO MOUNTS REQUIRED)	2 X 4 = 8
MOUNTS FOR 4,800-BIT/S MODEMS	3 X 4 = 12
MOUNT FOR TWO PIECES OF TEST EQUIPMENT	2 X 4 = 8
	TOTAL EQUIPMENT HEIGHT = 52 RMS

Figure 17.4.2 Estimating total mounting space.

The procedure for designing a tech control center is first to estimate the amount of front mounting space needed to house present and future equipment, and then to lay out the equipment in the cabinets. The final step is to arrange the cabinets on the floor to minimize digital cable lengths.

The balance of this section illustrates this procedure for a setup with 12 low-speed modems, two of medium speed, and two of high speed. Assume that space must also be provided for expansion by 50% to 18 low-speed modems and three each of the medium- and high-speed modems.

How many cabinets?

After choosing the test equipment, obtain the rack-mounting dimension in RMS units. For the setup described, assume that the test equipment consists of two patch panels, two status monitor panels, and a bit-error-rate tester. Also assume that the modems will be housed in the tech center, so that their mounting dimensions must also be determined. As indicated in Figure 17.4.2, Estimating total mounting space, the total height of the equipment is 52 RMS units. Computer-type cabinets that are 6 feet high provide about 40 RMS of mounting space. Taller cabinets are available, but the extra height makes it difficult to reach the equipment on top. As a result, the 52 RMS of height should be built into a double-width cabinet or two cabinets bolted together. Another advantage of two cabinets is that the equipment can be spread out to ensure sufficient cabling space and ventilation. Also, the loose arrangement provides adequate space to mount additional test equipment, should the need ever arise.

Arranging the equipment

Once the total mounting space has been estimated, and the required number of mounting racks determined, the equipment can be laid out in the cabinets, with the locations selected according to the recommended equipment heights.

In Figure 17.4.1, a drawer and a writing surface have been added. The drawer holds cables, documentation, and other accessories. A "miscellaneous" patch panel also is provided for access to backup modems, special test points, or to expand the patching capability to other units in the network. The need for an extra patch panel often develops as troubleshooting and maintenance techniques evolve.

Minimizing cable runs

Because the patch panel is connected between the modem and the digital equipment, the cable run should be determined by summing the lengths of the cable between the modem and the patch panel, and the cable between the patch panel and the digital device. From Figures 17.4.3 and 17.4.4, you can determine, for example, that a 24-foot straight-line distance between a modem and computer (indicated as underfloor distance) might require a 60-foot cable.

The problem now is how to reduce the cable length to 50 feet to meet RS-232-C standards. If the digital device and tech center cannot be moved closer to each other, then one way to save a few feet is to use a cabinet without a partition or to drill holes in an existing partition so that the cables connected across the cabinet can be routed directly, instead of around the partition. Another solution is to move the modems and patch panels closer to the floor.

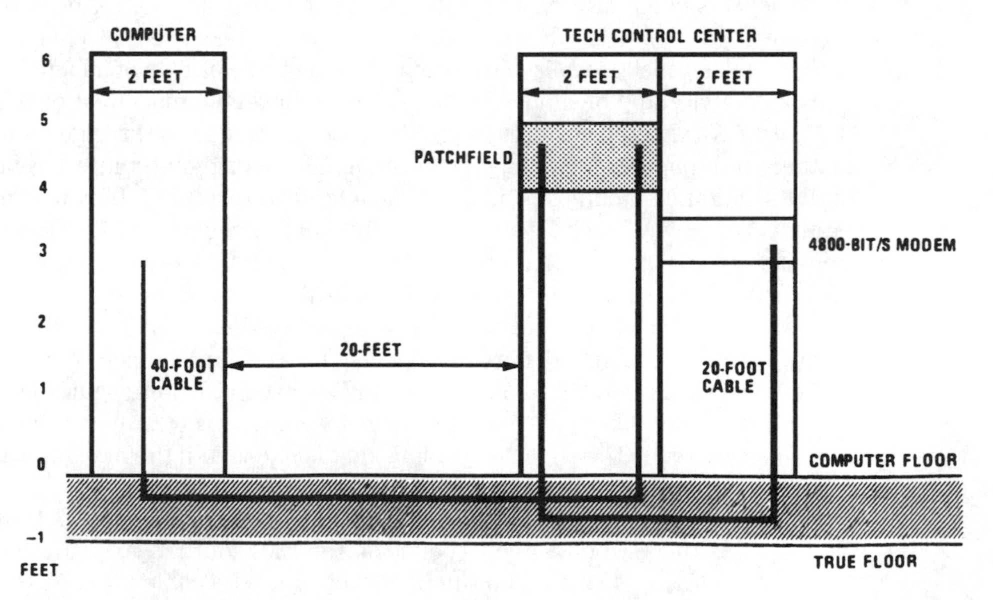

Figure 17.4.3 Cable runs.

CABLE LENGTH FROM MODEM TO PATCHFIELD:	
MODEM TO FLOOR (ASSUMING 1 FOOT OF CABLE IN THE CABINET)	1 + 4 + 1 = 6 FEET
UNDER BOTH CABINETS	2 + 2 = 4 FEET
FLOOR TO PATCHFIELD (ASSUMING 1 FOOT OF CABLE IN THE CABINET)	1 + 5 + 1 = 7 FEET
CABLES USUALLY MADE IN MULTIPLES OF 5 FEET, SO A 20-FOOT CABLE WOULD PROBABLY BE USED	**TOTAL LENGTH = 17 FEET**
CABLE LENGTH FROM PATCHFIELD TO THE COMPUTER:	
PATCHFIELD TO FLOOR (ASSUMING 1 FOOT OF CABLE IN CABINET)	1 + 5 + 1 = 7 FEET
UNDER FLOOR (AND BOTH TECH CONTROL AND COMPUTER CABINET)	2 + 20 + 2 = 24 FEET
FLOOR TO COMPUTER (ASSUMING 1 FOOT OF CABLE IN CABINET)	1 + 3 + 1 = 5 FEET
	TOTAL LENGTH = 36 FEET
THEREFORE, A 40-FOOT CABLE WOULD PROBABLY BE USED	

Figure 17.4.4 Estimating cable length.

17.5 Monitoring LANs

Many aspects of monitoring a LAN are similar to those of a WAN. And several transmission impairments are common to both. However, because the cabling associated with a LAN is usually localized to within a building, the effects of inclement weather, sunspot activity, and of similar problem sources have minimal effect on a LAN.

Most LAN-related transmission problems can be traced to three general areas: violations of equipment cabling specifications, electromagnetic interference (EMI), and utilization. A cabling specification violation can result in a LAN becoming inoperative. Trace why: each LAN type transmits data using a modified form of digital unipolar signaling. On Ethernet LANs, this signaling takes the form of Manchester encoding; on token ring networks, differential Manchester encoding.

As the digital signal propagates on a cable, the cable's resistance, inductance, and capacitance increasingly distort the signal. At a certain distance from the sending device, the signal becomes so distorted that it becomes unrecognizable. That distance is normally slightly beyond the cable specifications for a particular LAN. This explains why, when the specified cabling limit is exceeded, the network can become inoperative.

EMI results in the generation of noise, which adversely affects data transmission. How great is this EMI effect? It depends on the EMI source's strength, the distance between the EMI generator and the LAN cable, and the cable type. As to the latter, unshielded cable is obviously more susceptible to EMI-induced transmission problems than is shielded cable.

One problem that EMI can cause is jitter, which is noted by the displacement of a signal from its intended location by a small amount of time. Too much jitter, caused by an excess of EMI, results in a high error rate that can play havoc with a LAN's operation. Fortunately, there are several types of monitors that provide jitter information by transmitting test patterns and measuring the received signal for a variance in data position. When jitter occurs, rectification of the problem can be accomplished by relocating a cable away from machinery or from a fluorescent-light ballast.

The third area that can adversely affect a LAN's performance is its utilization level. As utilization increases beyond 50 percent for Ethernet/IEEE 802.3 networks and 60 percent for token ring networks, users will experience throughput delays. Although such delays might not be perceived when performing file transfers, they become more noticeable when interactive client-server operations are performed.

There is no universal diagnostic monitor that provides an insight into all LAN problems or their possible solutions. However, you can obtain an appreciation for the operation of a LAN monitoring product by examining one. The one examined here is TokenVision, a software product developed by Triticom, designed for use on token ring networks.

Figure 17.5.1 illustrates TokenVision's main menu. This menu provides the user with the ability to monitor network traffic, set network alarms, set workstation options, enable and disable logging of monitored information, generate a variety of reports, or exit the program and return to DOS. The best way to become familiar with the operational capabilities of a program is through its use.

When you select the Monitor Traffic option in the TokenVision Main Menu, you are prompted to select the station address to be monitored—either source or destination. TokenVision supports a number of certain adapters (manufactured by different vendors), one of which must be installed in the workstation in which it operates. The

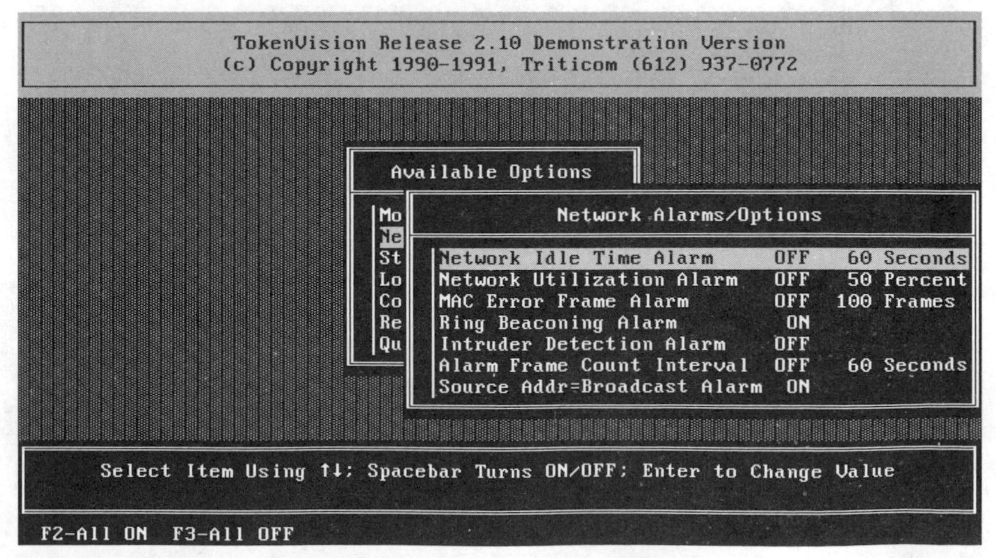

Figure 17.5.1 TokenVision main menu.

reason a specific vendor's adapter is required is that the program must be able to read IEEE 802.5 frame fields passing through the adapter card. To do so, the developers had to write software to access registers and memory locations within the adapter. Because many adapter cards differ in their internal design, TokenVision currently supports the use of only those cards for which it is capable of reading explicit locations to obtain required information. Once you select the station address to be monitored, TokenVision begins to accumulate statistics based on that address while monitoring frames passing through the station's adapter card.

Figure 17.5.2 illustrates the TokenVision monitoring screen display. As the frames are read, the program identifies new station addresses as well as incrementing the frame count of previously encountered station addresses. When Figure 17.5.2 was printed, a total of 17 stations were identified and are listed in the left column, along with their frame counts.

In examining the entries in Figure 17.5.2, notice that source address 0000C900000D has generated by far the greatest number of frames (14,119) for only one minute and 14 seconds ("Elapsed" block in lower right corner of screen) of monitoring. If this heavy traffic persists for a prolonged period, it would indicate that the station user at that address might be hogging LAN resources, perhaps as a result of an excessive amount of file transfer activity.

The reverse video bar across the monitoring screen under the station address column provides additional information about the highlighted source address. In this example, the first station address is highlighted. The inverse video bar identifies a previously associated network user ("Aristotle") with the hex address of the adapter, the vendor or manufacturer of the adapter, number of frames and bytes transmitted, percentage of LAN capacity used by the highlighted address, average transmission in bytes, and any transmission errors attributable to that address.

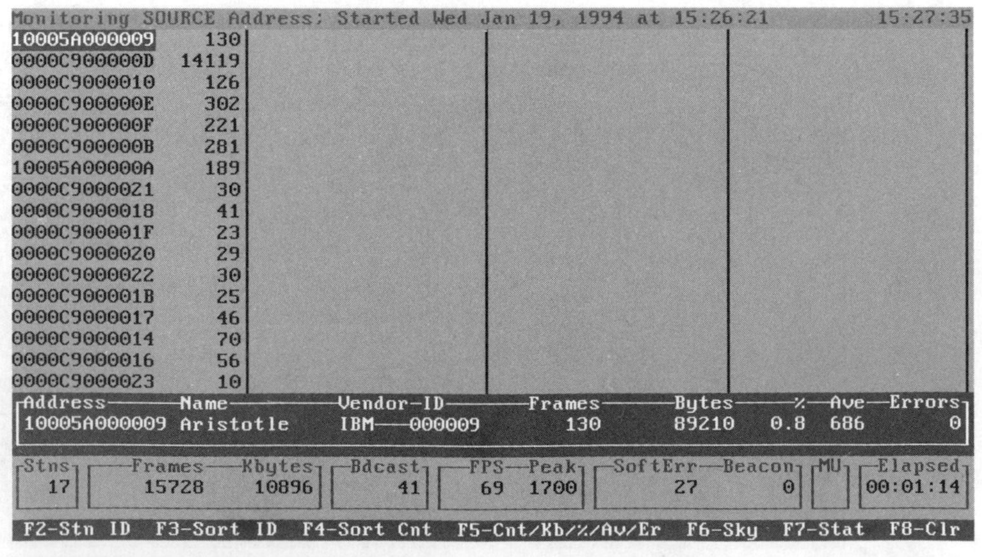

Figure 17.5.2 Monitoring based upon the source address.

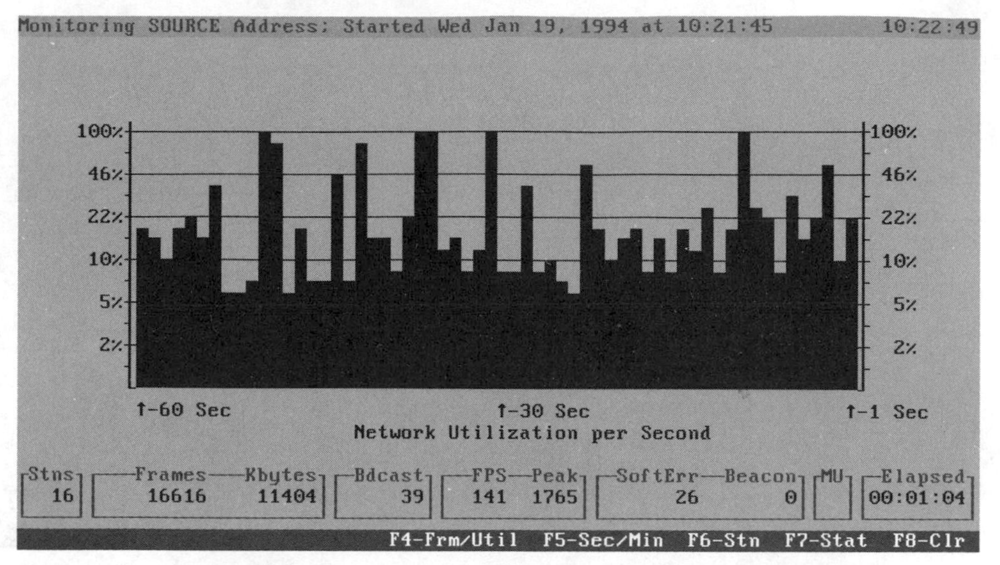

Figure 17.5.3 Skyline display.

Notice that the next horizontal bar (shaded in gray) provides summary information for all the LAN stations. The bottom (inverse video) bar indicates the assignment associated with different function keys.

In the summary information (gray) bar, the soft error ("SoftErr") entry refers to a frame sent by a station at a set interval (usually a two-second default) when it detects a transmission error or experiences a physical error. This frame includes information about the station's nearest addressable upstream neighbor (NAUN), the error type, and the number of errors of that type. Because TokenVision can read all frames, the program can decode information contained in all Report Soft Error Media Access Control (MAC) frames.

The Beacon entry in the gray bar (Figure 17.5.2) indicates the number of Beacon frames detected. A station transmits a Beacon frame when it detects a hard error. The frame identifies the station that detected the error and its NAUN. If the Beacon frame returns to the originating station, logic on the adapter card causes the station to remove itself from the ring and perform an internal diagnostic test to determine if reinsertion into the ring is possible. If reinsertion is not possible, the Beacon frame enables a station with defective wiring or a faulty connection to remain removed from the ring.

One of the more interesting TokenVision screen displays is its skyline illustration. You can generate this display from the monitoring screen shown in Figure 17.5.2 by pressing the F6 key. Figure 17.5.3 illustrates the resulting skyline display. It provides a visual indication of network utilization at one-second intervals, which you can change to one-minute intervals by pressing the F5 key.

TokenVision's Statistics display (Figure 17.5.4) is accessed from the skyline display by pressing the F7 key. Although both the skyline and statistics display indicate utilization, the value of the latter display is its classifying of the frame distribution by size. Half of all interactive client-server frames are normally less than 64 bytes long.

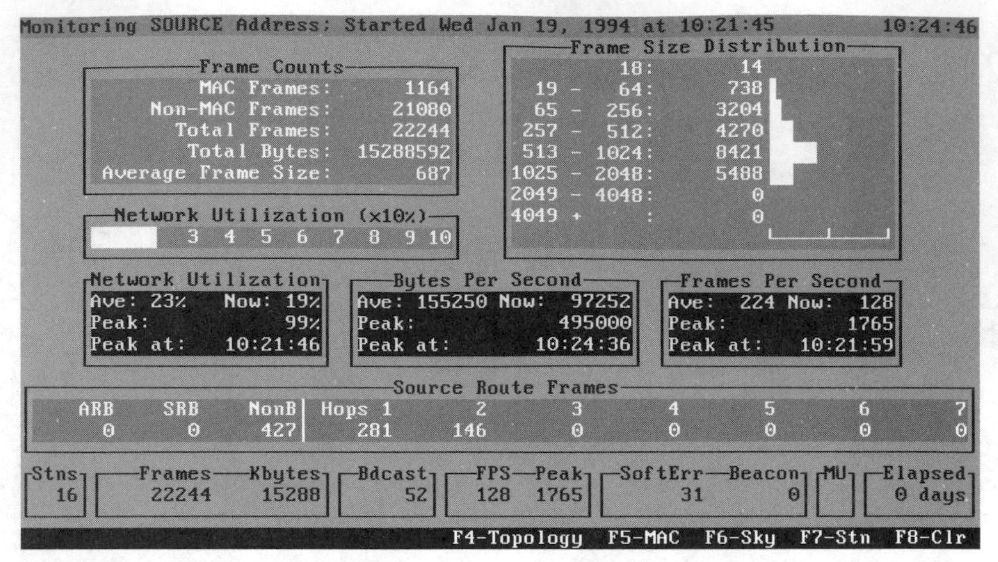

Figure 17.5.4 TokenVision statistics display.

Almost all file transfer frames are the maximum allowable token ring length: 4500 bytes for 4 Mbps; 18,000 for 16 MBPS. Therefore, you can use the frame size distribution information to determine the type and distribution of network usage. Doing so might provide insight to the reason for a high utilization and yield a starting point for actions to reduce it.

The key to the use of many types of hardware and software diagnostic tools is their ability to set alarm thresholds. When a threshold is exceeded, the diagnostic device generates audible or visible warnings and/or logs the incident to a printer or disk. Similar to many WAN diagnostic devices, TokenVision provides users with the ability to set different types of alarm thresholds, then perform other tasks. When a threshold condition is exceeded, TokenVision generates an alarm and can be configured to log alarm information to a printer or disk.

Figure 17.5.5 illustrates the TokenVision Network Alarms/Options screen display. From this screen, you can toggle alarms on and off and, if a value is associated with an alarm, you can change it.

Once an alarm is triggered, its occurrence is displayed on the top line of the screen during traffic monitoring. If you also enabled the optional audio alarm, a warbling sound occurs to gain your attention.

The idle time alarm (Figure 17.5.5) is reserved for the period during which TokenVision does not sense traffic. This alarm can be used to check certain conditions, such as that of file servers operating under Novell software. They periodically transmit frames to make them aware of each other. The absence of traffic could indicate that part of your network is not functioning properly.

The network utilization alarm permits you to monitor potential network saturation conditions. The MAC error frame alarm counts the total of several types of error frames. The ring beaconing alarm occurs only when a Beacon frame is encountered. The intruder detection alarm is triggered if a new station is detected that was not previously defined by having a name assigned to its adapter address. The alarm frame count interval defines the time interval for the count in the MAC error frame to cause an alarm; it represents an interval for generating an alarm rather than the alarm itself. The last alarm shown in Figure 17.5.5, "Source Addr = Broadcast Alarm," generates an alarm if a source address has its broadcast bit set; this indicates multiple source addresses, even though all source addresses must be unique.

To sum up TokenView's capabilities: You can monitor LAN traffic, determine network utilization on both an individual and composite basis, set alarms, and perform other network-related insightful activities. Other vendors, including Triticom, provide similar software tools for Ethernet and ARCnet networks, as well as products that measure bridge performance. Such products do not provide network planning information regarding cabling standards. But their use does provide you with the capability to examine your network's health and to consider methods for improving network performance.

Questions

17.1 What are the gains or losses in decibels of the following signal power measurements?

A. Output power 10 mW, input power 100 mW

B. Output power 100 mW, input power 1 mW

17.2 Discuss the differences between dropouts, phase hits, gain hits, and impulse noise.

17.3 Discuss two methods that can be used to improve line transmission.

17.4 What is the key difference between digital and analog monitoring?

17.5 Discuss the use of five protocol analyzer/digital test set features should be considered.

17.6 What elements are the specific concern of analog monitoring, and why is analog monitoring less frequently used than digital monitoring?

17.7 What is a patchfield and how can it be used?

17.8 What impairments that normally affect a WAN do not typically affect a LAN? Why not?

17.9 What are the three general areas to which LAN-related problems can be traced?

17.10 What are three typical functions of a LAN-monitoring software tool?

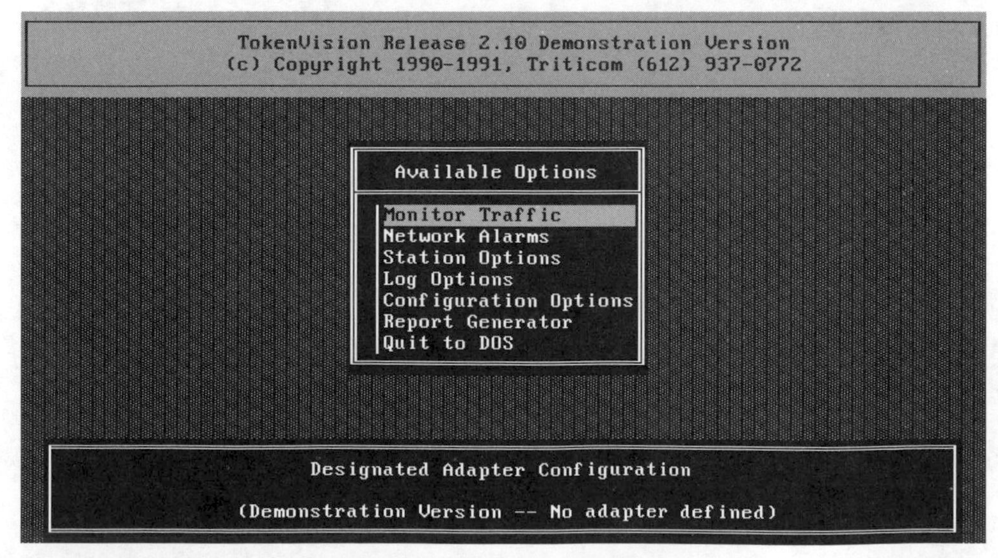

Figure 17.5.5 TokenVision network alarms/options display.

Chapter

18

Network Planning and Design Alternatives

Preceding every efficiently designed network is proper planning. The network plan can be considered a roadmap that will result in the network analyst achieving an effective network that meets the requirements of his organizational users.

This chapter features some of the methods used to collect information, the typical data elements that should be collected and their use, and the types of operational requirements that a supervisor should query network users about. Next, current and anticipated resources are examined. Some of the typical types of hardware limitations, and methods that can be used to alleviate those limitations, is covered. Finally, coverage of carrier and equipment selection, installation, and operation show the relationship of those elements to the network plan.

18.1 Information Collection

In formulating a network plan, the most difficult element might be obtaining the requirements of current and anticipated network users. If these requirements are stated incorrectly or glossed over as a result of misunderstandings, then the most methodical network plan will not be successful.

Several methods can be used in the information collection process, ranging from personal interviews to surveys, forms, and checklists. Although an interview can be costly and time consuming, especially if current or anticipated users are scattered over a wide geographical area, often this method of information collection results in the discovery of items or classification of certain factors that no amount of survey forms or checklists can accomplish. When sitting down with the network user, the analyst could discover, for instance, that lower seatbelt production for the next six months will result in the termination of second and third shifts at a plant during that period. The analyst might then recognize that the factory on-line data collection network will be inoperative during half the year for all, but the first shift. Therefore, the circuit connecting the factory with the company's data processing center could be used to support a remote batch terminal that is now using the public switched telephone network for transmission.

Although checklists have been used on occasion, they are very difficult to use. They can result in user frustration because they normally do not permit a deviation from listed elements. When using the checklist approach, the network planner prepares a menu of all possible responses to the information, permitting the network user only to check the relevant block. Thus, if the analyst wants to know the planned operating speed of terminals to be acquired during the next year, she might have blocks denoting 300, 1200, 2400, 4800, and 9600 bit/s for the user to check. If the user required 110-bit/s terminals and another block were available, she could check that block and write in his requirement. In doing this, the checklist has been modified. By the use of these types of modifications, she could wind up with a survey.

The survey

Although a survey might not be able to ferret out items of information that an interview could obtain, this method provides much more flexibility than a checklist. It is the most popular technique used for obtaining information. (Notice that the Teletype and CRT devices mentioned in this survey example can be readily replaced by PCs. And the data rates could well be much greater, depending primarily on the modems and media used.)

Figure 18.1.1 contains a terminal and data communications survey form, and one possible set of instructions for completing the survey is listed in Table 18.1.1. Because, in this case, data traffic that must be supported by a communications network originates from remote terminals, the survey is called a *Terminal* and *Data Communications Survey*. For the particular requirements of the network analyst, only 12 items have been placed on the survey. However, several additional items are covered later in this section, which could be included.

In this particular survey, the first item to be completed denotes the location of the terminal. This will be important when the analyst prepares a network layout in preparation for developing a minimal-cost network design. Although in Table 18.1.1, the survey respondent is not asked to indicate the street address and room number of terminals, this could be incorporated into the survey. The analyst might believe that remote terminals could be directly connected to communications equipment at one or more remote locations, or might wish to explore the use of line drivers and limited-distance modems to connect terminals to data concentration equipment. For such cases, the analyst might request a diagram showing terminal locations by building floor and the layout of conduits on each floor.

The second item on the survey assumes that only three types of terminals will be considered. For the less-knowledgeable individual, a definition of each type could be included in the survey instructions.

In the third item, the user is to indicate the number of terminals required over a three-year period. Although the analyst could use the first year's information to develop a network if none already exists, one or two additional years of projected in-

1. Location: ______________________________
2. Type of terminal: Type I — Teleprinter __________ Type II — A/N Display __________
   Type III — Batch __________
3. Quantity of terminals required (#): 1995 __________ 1996 __________ 1997 __________
4. Transmission speed required (bit/s): a. 110 ______ b. 300 ______ c. 1,200 ______
   d. 2,400 ______ e. 4,800 ______ f. 9,600 ______ g. other (explain) __________
   ____
5. Terminal options required: ______________________________
   ____
   ____
6. Normal operating hours (EST) per day: __________ To __________
7. Peak hours (EST):
   To host: From ______ To ______ From host: From ______ To ______
8. Calls per day per terminal (#): To host ______
9. Connect time (hours and minutes) per call: Average ______ Peak ______
10. Messages per call per terminal (#):
    To host: Average ______ Peak ______ From host: Average ______ Peak ______
11. Average characters per message per terminal (#):
    a. To host ______ b. From host ______
12. Comments: ______________________________
    ____
    ____
    ____
    ____
    ____
    ____

Figure 18.1.1 Terminal and data communications survey.

formation might be desired for planning purposes. In the instructions contained in Table 18.1.1, the survey respondent is instructed to complete different forms when terminal usage conditions change.

The fourth item, transmission speed required, permits the user to indicate the operating data rate desired. In some organizations, the respondent might only be asked to furnish expected operational statistics. The network analyst will review them and then inform the requestor what the terminal operating speed should be, based on the user's expected operational data transfers.

Item 5, terminal options required, could be expanded by the network analyst to obtain additional information. As an example, the analyst might inquire about the operating speed and usage of printers for CRTs. Such printers could operate at one-half to one-fourth the data rate of CRTs. Extensive printer usage could permit more CRTs to be placed on a polled circuit than if the CRTs were extensively used for job output scanning that requires a fast look at extensive amounts of information by the terminal operator at the full data transfer rate of the CRT.

In items 6 through 11, operational communications data factors are requested to enable the network analyst to examine several methods of connecting each terminal to the host computer. If the computer is located in the Eastern Standard Time (EST) zone, then items 6 and 7 will request the normal and peak operating hours with respect to EST. This permits the analyst to examine the effect of terminal operations in many time zones on the host computer and the network. Thus, if peak usage oc-

TABLE 18.1.1 Instructions for Completing Terminal and Data Communications Survey

The following explanation and examples are given to assist you in completing the survey. The survey form may be copied if more are needed.

For each of the twelve items on the survey, indicate the required information as follows:

Item #1–*Location*

A separate survey form *must* be completed for each different location where a terminal will be installed. Indicate *city*, *state*, and denote if *Region* or *Area Office*. *Specific location must be indicated.*

Item #2–*Type of terminal*

A separate survey form *must* be completed for each of the three categories of terminals. Indicate type of terminal by placing a *check mark* in the appropriate space.

Item #3–*Quantity of terminals required*

A separate survey form may be required for each year if all other conditions are not the same. For example, if three teleprinters are required in 1995 and three more are required in 1996, but their usage is different, two separate forms should be completed. Indicate *number* of terminals required in the appropriate space.

Item #4–*Transmission speed required*

Indicate transmission speed required by placing a *check mark* in the appropriate space.

Item #5–*Terminal options required*

Indicate any *optional terminals features* you may require. Examples: (1) printers for CRT, (2)special printing features, (3) special keyboard requirements, (4) portability, (5) graphics, (6) protocol.

Item #6–*Normal operating hours per day*

Indicate the hours (EST) during which the terminal(s) will normally be operating.

Item #7–*Peak hours*

Indicate the *peak hours (EST)* during which you will be transmitting to and receive from the host computer. More than one group of peak hours is possible.

Item #8–*Calls per day per terminal*

Indicate the number of calls to the host computer per day per terminal.

Item #9–*Connect time per call*

Indicate the connect *time (hours and minutes)* per call (average and peak).

Item #10–*Messages per call per terminal*

Indicate the *number* of messages per call per terminal going to the host computer (average and peak).

Item #11–*Average characters per message per terminal*

Indicate average number of characters per message per terminal going to and being received from the host computer. A carriage return constitutes the end of a one-line message to the host, and the end of a print line constitutes the end of a one-line message from the host.

Item #12–*Comments*

Indicate any additional information which you feel may be pertinent to the survey.

curs at the same relative time across the country, the difference in time zones might permit the analyst to reduce the number of front-end processor ports more than if time zone differences were not considered.

The call-per-day and connect-time-per-call data requested in items 8 and 9 can be used to compare transmission cost over the switched network with, for example, leased-line utilization. When items 10 and 11 are included, the analysts can cost out value-added carrier utilization and then compare that cost with that of telephone company and perhaps satellite carrier service. Lastly, item 12 permits any additional information to be added to the survey form. One of the problems with such a catch-all item is that personnel completing the form might not be knowledgeable enough to state other factors that are not asked for on the survey. A modified form of item 12 is illustrated in Figure 18.1.2.

As shown in Figure 18.1.2, supplemented by relevant instructions, users can now express special operational requirements, as well as add comments. In part A of item 12, response-time requirements can be denoted. Applications of a critical nature that might require special communications backup and equipment redundancy can be listed in part B. If the application is of such a nature that data encryption might be warranted, this fact can be expressed in part C. For organizations that do not centrally fund communications, but charge back costs to users, a billing or chargeback portion of item 12 should be considered. A variety of data, from a user's charge number to a dollar limit for the expenditure of funds on communications, could be filled in part D of item 12. Lastly, the catch-all "other" phrase can be used. However, the person completing this modified form now has four examples to think about before entering comments.

Survey analysis

Assume that the Chicago customer service center of an organization completed one Terminal and Data Communications Survey form, as illustrated in Figure 18.1.3. From this completed form, the network analyst can plan the method of integrating the communications requirements of the Chicago customer service center with the other network requirements of the organization. From items 2, 3, and 4 of the survey, the service center will be using 300-bit/s teleprinter terminals and has a requirement for 10 such devices in 1995, seven more for 1996, and an additional four in 1997. Thus, while the analyst's immediate concern is with supporting 10 terminals in Chicago, plans to support 11 additional devices over the next two years should also be made. In items 6 and 7 it appears that there are no peak hours of operation; thus calls will normally occur on a random basis throughout the normal business day.

12. Special Operating Requirements and Comments: ________________

A. Response Time

B. Backup/Redundancy

C. Security

D. Billing/Chargeback

E. Other

Figure 18.1.2 Modified portion of survey form.

1. Location: CHICAGO CUSTOMER SERVICE CENTER
2. Type of terminal: Type I – Teleprinter ✓ Type II – A/N Display ____
   Type III – Batch ____
3. Quantity of terminals required (#): 1995 10 1996 7 1997 4
4. Transmission speed required (bit/s): a. 110 ____ b. 300 ✓ c. 1,200 ____
   d. 2,400 ____ e. 4,800 ____ f. 9,600 ____ g. other (explain) ____
5. Terminal options required: Portable and Lightweight will be moved throughout building due to frequent organizational changes.
6. Normal operating hours (EST) per day: 8AM To 5PM
7. Peak hours (EST):
   To host: From ____ To ____ From host: From ____ To ____
8. Calls per day per terminal (#): To host 6
9. Connect time (hours and minutes) per call: Average 30 min Peak 45 min
10. Messages per call per terminal (#):
    To host: Average 40 Peak ____ From host: Average 300 Peak ____
11. Average characters per message per terminal (#):
    a. To host 25 b. From host 60
12. Special Operating Requirements and Comments: ____
    A. Response Time
    B. Backup/Redundancy
    C. Security
    D. Billing/Chargeback
    E. Other

Figure 18.1.3 Completed form sample—terminal and data communications survey.

Using the data furnished in items 8 and 9, the average and peak connect time per terminal can be computed by multiplying the calls per day by the average and peak connect times per call. Thus each terminal in Chicago will be on line an average of 180 minutes per day, and for high-volume days have a peak connect time of 270 minutes. This information can be used to determine the optimum size of data concentration equipment, or to determine the cost of switched-telephone-network and private-line service, depending on where the computer center is located.

Using the information in items 8 and 9 in conjunction with the data in items 10 and 11, the network analyst determines the cost of using a value-added carrier to support the Chicago service center's transmission requirements. From item 10, an average of 40 messages per call per terminal will be transmitted to the host computer. In item 11, the average characters per message per terminal to the host is 25. This results in a total of 1000 characters on the average transmitted to the computer during one call. Similarly, an average of 18,000 characters will be transmitted from the host computer to the terminal during a call, for a total of 19,000 characters transmitted and received.

Multiplied by the six calls per day, this results in a daily average of 114,000 characters transmitted and received by each terminal, or a daily total of 1.14 million characters for the 10 terminals. For the value-added carrier that has a charge element based on packets and not characters, the information in items 10 and 11 can be used to determine the number of packets transmitted and received. As long as the number of average characters per message, as indicated in item 11 is less than the maximum number of characters that can be contained in a packet, the messages per call per terminal in item 10 are equal to packets. If the average characters per message (item 11) exceeds the number of characters per packet, each of the messages in the messages per call (item 10) represents more than one packet. This normally occurs when terminals transmit synchronously and a large number of characters is grouped into a block for transmission. If the value-added carrier under consideration has a packet size of 128 characters and the average per message transmitted is 160, then each message per call (item 10) would represent two packets, one containing 128 characters, the second containing 32 characters.

Resource examination

In addition to collecting information on user requirements, the network planner must consider the available resources and their expansion potential. These resources range from existing circuits and data concentration equipment to the central-site hardware. Especially included are existing front-end processors and the availability of additional ports to service additional terminal connections to the processor.

One method that can be used by the network analyst to understand the capability and expansion potential of existing equipment and circuits is the use of an "Expansion Table." Such a table might contain a list of present equipment, current monthly cost, expansion potential of equipment, and the anticipated cost of the expansion. In Table 18.2.1, a portion of the expansion table is illustrated for the Chicago segment of a network. Here, an 8-channel time-division multiplexer and an 8-position rotary connecting eight dial-in lines are listed, and their expansion potential and cost of expansion indicated. As illustrated, you can indicate the channel data rates on the TDM, which can be used to determine how many of the expansion channels at various data rates can be used before the capacity of the high-speed modem or leased line is exceeded.

Although a telephone company rotary can step through a series of perhaps 1000 numbers, when such devices are installed normally only a small group of contiguous telephone numbers will be left available on user request to serve as expansion potential. In the expansion table, you might list the quantity of numbers reserved and their extensions. Information that can be added to the table for remote sites includes

TABLE 18.2.1 Potential Equipment Expansion

Location: Chicago

Equipment	Monthly rent	Expansion capability	Projected monthly cost
TDM, 8 chnl (4 × 300, 4 × 1,200)	$150	8 channel	$100
Rotary (8 dial-in)	$ 35	10 contiguous numbers XXX-444	Installation only
Modem: Bell 209	6,000 bit/s effective transmission on GD4726 circuit		

the type of transmission equipment used and transmission medium information, if applicable. In Table 18.2.1, a Bell 209 modem, which is capable of transmitting data at 9600 bit/s, is listed with the notation "6000-bit/s effective transmission on GD4726 circuit." The GD4726 is the telephone company designation for a particular leased line that connects the customer's Chicago location with the computer facility. The notation about 6000-bit/s effective transmission indicates that an additional 3600 bit/s can be transmitted on that medium.

Central-site resources

Although the host computer site can have a very large number of communications components and circuits, a good starting place for resource investigation is the computer's front-end processor. This is because the available resources and expansion potential of that device are the governing factors behind the design, establishment, and growth of a communications network.

Software limitations

If the front-end processor cannot support multipoint lines as a result of the unavailability of poll and select software, then the network designer might not be able to consider that method during the network configuration process. If multipoint lines can significantly reduce potential communications costs, then the analyst should explore the cost of modifying existing software, purchasing software from a vendor that will enable the front-end processor to poll and select terminals, or obtaining and using software from the computer manufacturer that might require some modifications.

Similarly, the network analyst should investigate the protocol software-support modules available from the computer manufacturer as well as from software houses. The analyst will then know what other types of terminal support is available and what the cost might be. In Table 18.2.2, some of the more common terminal software protocols are listed.

Even when a terminal protocol is available at no cost or for a nominal fee, care must be taken to examine the effect of adding another software module to the front-

TABLE 18.2.2 Common Terminal Software Protocols

Teletype	110 bit/s ASCII
Teletype	135.5 bit/s EBCDIC
Teletype	300 bit/s ASCII
Teletype	1200 bit/s ASCII
CRT synchronous block mode	
CRT poll and select cluster	
IBM 2780/3780	
IBM 3270	

end processor. This is because each software module must be core resident at all times if terminals requiring that specific protocol support are to be permitted to connect to the network at any time during the day. Because the processor's memory size is fixed, the size of the protocol module must be examined to determine if that module will "fit" in the processor, or if an expansion memory module is required.

Obviously, if the processor is fully configured and not enough memory is available for the required module, then another front-end processor would be required. This would be very expensive, for example, if the analyst were faced with the situation where only one remote user wanted to obtain a terminal whose protocol was not previously supported; if the organization's front-end processor were fully configured; and all memory were utilized by existing software modules.

Here, even if the analyst can obtain the required terminal protocol support module at no cost, the cost of an additional processor to support one terminal might not be justified. The analyst informs the user that the specific terminal desired cannot be supported for a reasonable cost and that the user should select another terminal whose protocol is already supported by the front-end processor.

Hardware constraints

In conjunction with software limitations, hardware constraints of front-end processors must be examined. Although no techniques exist for fitting software modules onto hardware once all memory is used, other than obtaining additional memory or another processor, a variety of techniques can be used to alleviate hardware constraints. These techniques range from front-end processor channel reconfiguration to obtaining devices to "front-end" the front-end processor.

To understand some of the options available to the network analyst for alleviating front-end processor hardware constraints, let us examine a typical hardware configuration, illustrated in Figure 18.2.1. In this configuration, sixteen 300-bit/s and sixteen 1200-bit/s teleprinter channels are used to support the output of two 16-channel multiplexers. Four 9600-bit/s batch terminals and four CRTs, each on an individual circuit, are supported by a total of eight processor channels. Eight channels are available for network expansion.

Assume that this 48-channel front-end processor cannot be expanded to accommodate additional channels in excess of 48. Also, because of the growth of the organization, there is a requirement for an additional 16-channel multiplexer at a third remote site to support 300-bit/s terminal operations and 14 additional CRTs at various locations.

For the first expansion requirement, the utilization of a port selector can be an optimum solution, depending on the cost of that device in comparison to the cost of an additional front-end processor. If you use such a device and estimate that no more than 32 users will require simultaneous access 95% of the time, then the time-division multiplexers feeding the front-end processor via a port selector would appear as illustrated in Figure 18.2.2. In this illustration, a port selector capable of cross-connecting 48 input connections to 32 output lines is used. During a typical day, 95% of the time users will not encounter a busy condition. When such a condition does occur, the user can reestablish the connection and obtain access to the network if an-

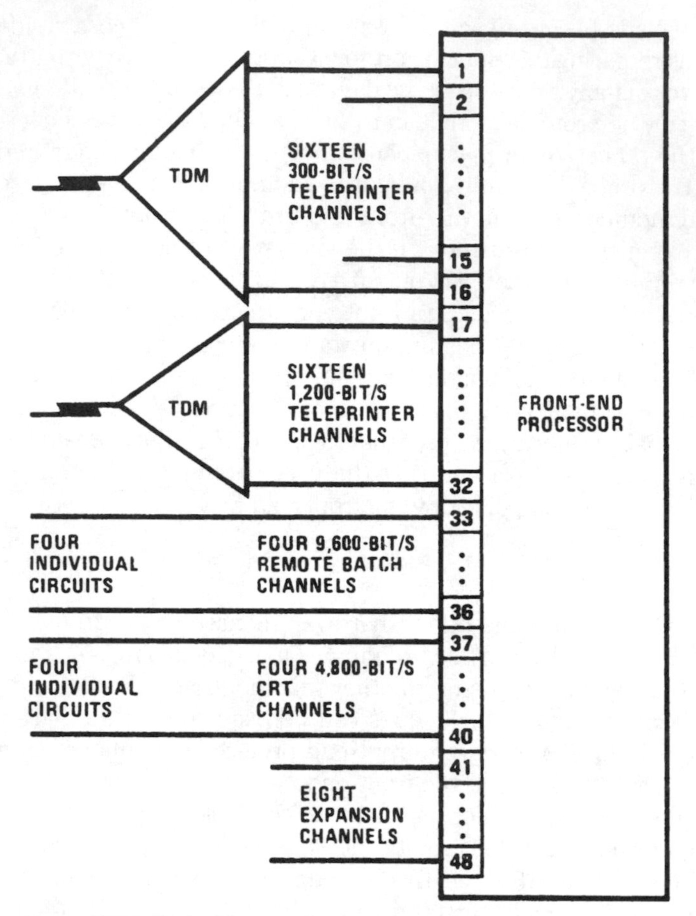

Figure 18.2.1 Typical front-end processor configuration.

other user has completed processing activity in the interim. In this particular application, the port selector permits 16 additional teleprinter channel connections with no increase in the number of front-end processor channels.

The CRT expansion requirements depend on the relationship of the location of the additional CRTs with respect to existing terminals of that type. Several approaches can be considered to alleviate processor hardware constraints.

If the CRTs are co-located at a few geographical areas, modem-sharing units could be used to service those devices, as illustrated in Figure 18.2.3, part A. Modem-sharing units permit from four to 32 terminals to share a common modem. Thus, one channel on the front-end processor services the CRTs connected via this configuration.

When the CRTs are geographically distributed, as illustrated in Figure 18.2.3B, one or more multipoint circuits connecting a number of terminals via a common circuit to the computer center can be used. Again, for each multipoint circuit only one front-end processor channel is required, regardless of the number of terminals connected by that line, within the front-end's software limits.

For partially colocated terminals, one or more modem-sharing units with DCE options can be used. Figure 18.2.3C shows colocated CRTs, directly serviced by the

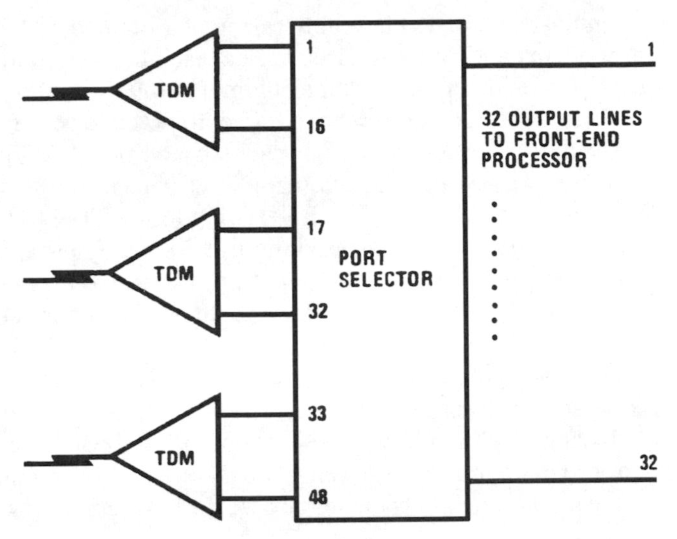

Figure 18.2.2 Employing a port selector.

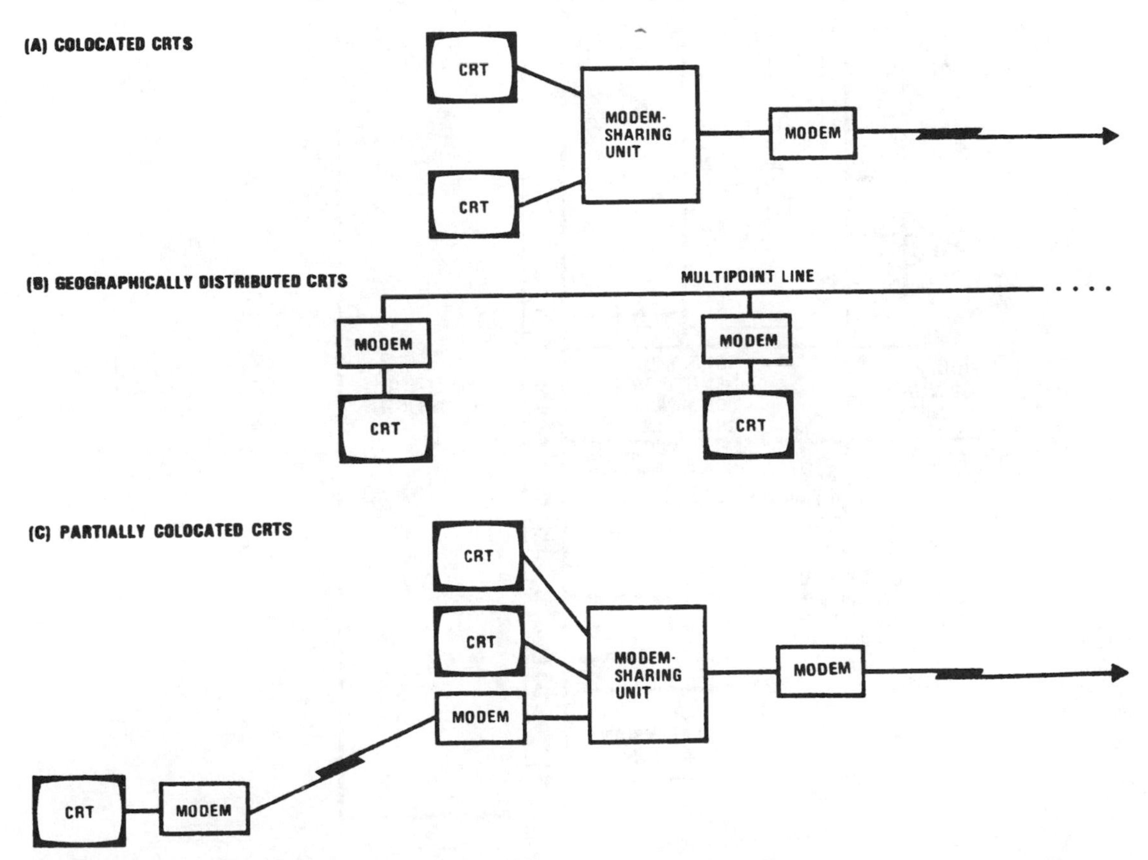

Figure 18.2.3 Alternative approaches for servicing CRTs.

sharing unit at the terminal location. Each remote terminal is connected to the sharing unit via a pair of modems and an extra circuit. Because all CRTs connected via the sharing unit use a common medium, only one channel at the front-end processor is required to support the colocated and remote CRTs connected in this manner.

Returning to the expansion requirement to service 14 additional CRTs, assume that an optimum use of equipment and circuits results in the use of three multiport lines and two modem-sharing units to support the existing four CRTs and the 14 additional devices. The final front-end processor configuration could now appear as in Figure 18.2.4. Here, after the use of three multipoint circuits and four modem-sharing units, four or five front-end processor channels are still available for network expansion.

Even if all channels are used, other hardware can be considered to free a channel on the front-end processor for other use. One such device the network analyst can investigate is a port-sharing unit. To realize the potential of this device, consider the partial network illustrated in Figure 18.2.5A. Here, four modem-sharing units, one at each of four remote locations, provide transmission to the host computer from nu-

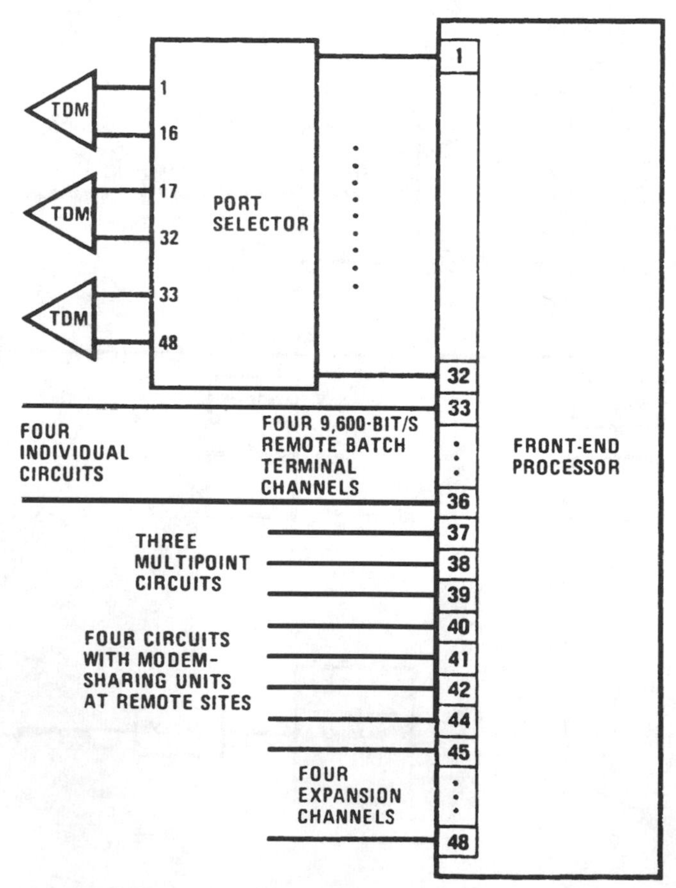

Figure 18.2.4 Front-end processor after network expansion.

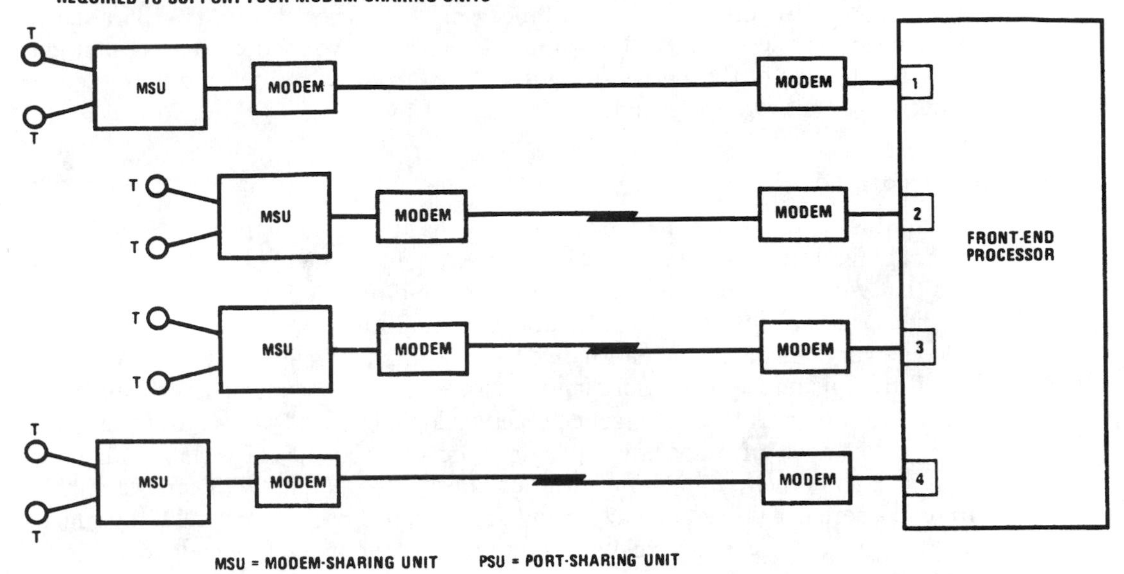

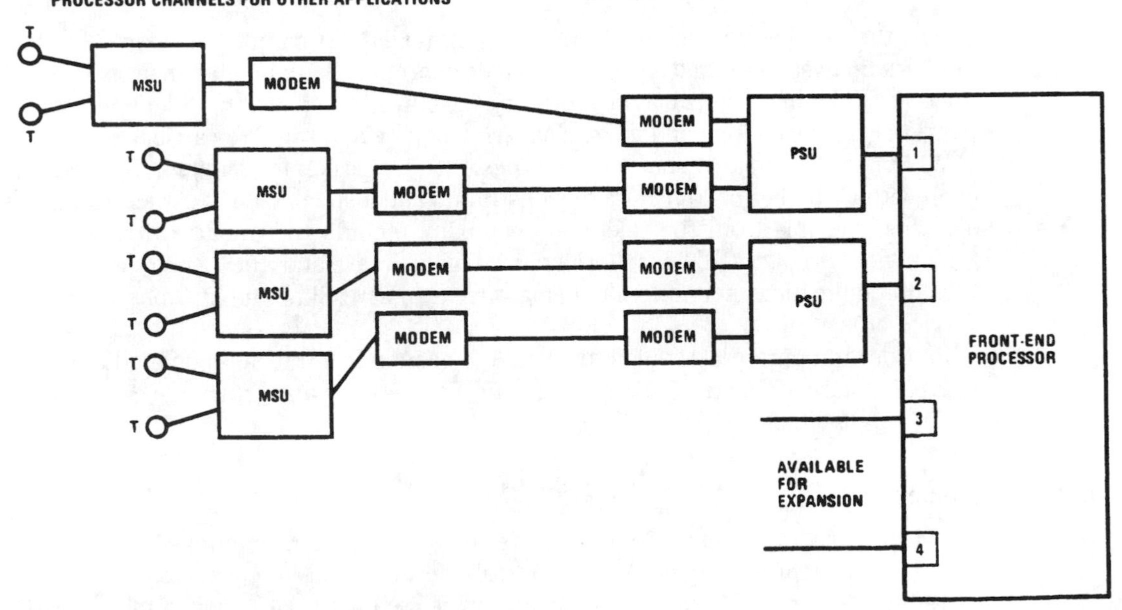

Figure 18.2.5 Employing port-sharing units.

merous clustered terminals at each site. The use of this configuration requires four channels or ports on the front-end processor. In Figure 18.2.5B, two port-sharing units have been installed, with each unit supporting two modem-sharing unit clusters. By the use of the port-sharing units, two front-end processor ports become available for network expansion.

Throughput considerations

So far, this coverage of resource examination has been restricted to alternative network configurations without being concerned about throughput. Although it is obvious that you cannot continue to service more and more terminals by continuing to modify, add, and alter equipment and lines, computing throughput constraints can become very complex. One alternative to computing a large series of equations can result from using the processor manufacturer's reference manual. Many vendors offer series of curves that represent processor loading based on the number of connected terminals of a specific type, the type of circuit used, terminal data rate, protocol, and other parameters. With this information, the network analyst only has to read a series of utilization factors and accumulate a utilization sum. If the sum exceeds unity, the processor will be saturated under the projected loading conditions, and some service must be dropped or moved.

18.3 Transmission Medium Alternatives

Today, the number and scope of transmission media that can be considered by the network analyst can result in confusion unless proper planning and analysis is conducted. Communications carriers' offerings are available from the telephone company, packet-switching vendors, specialized carriers, and satellite carriers.

Within each company, one or more types of transmission media might be offered, each with a number of transmission constraints and a specific tariff. As an example, services available from the telephone company include the public switched telephone network, leased lines, foreign exchange lines, and the use of wide-area telecommunications service—all analog offerings, as well as the Dataphone digital service offerings.

In order to appropriately plan for the most economical transmission medium that is capable of servicing the anticipated traffic, the network analyst should construct a traffic-volume table.

Traffic-volume table

Using the data from Figure 18.1.1, the Terminal and Data Communications Survey (or an equivalent mechanism), a traffic-volume table can be constructed, as illustrated in Figure 18.3.1. Notice that each row is completed on a per-terminal basis, with the "Total" row reflecting the expected traffic at that node. The Total columns with checkmarks indicate the specific Total columns the network analyst will normally use to determine the type of medium that appears most economical, and at the same time be capable of servicing the required traffic volume.

NODAL POINT ________

										AVERAGE			
	CALL/ DAY	CONNECT TIME/CALL		TOTAL TRANSMIT TIME		MESSAGES/DAY TO HOST		MESSAGES/DAY FROM HOST		CHARACTERS/ MESSAGE		CHARACTERS/DAY	
		AVG	PEAK	AVG	PEAK	AVG	PEAK	AVG	PEAK	TO HOST	FROM HOST	AVG	PEAK
TOTAL				✓	✓							✓	✓

Figure 18.3.1 Traffic-volume table.

Once the table (Figure 18.3.1) is completed, the analyst computes the number of bits/second to be transmitted from a specific nodal point over a common data link. As an example, assume that a peak of 15 million characters/day was computed as the total number of characters to be transmitted in an 8-hour working day. If each character is composed of 10 bits, then 150 million bits per day will be transmitted and received by the terminals at the specified nodal point. Over an 8-hour day, this represents 18,750,000 bits per hour, or 5208 bits per second. Depending on the number of terminals, their operating data rate, and connect time per day, if data should be concentrated at the nodal point for transmission, then a medium capable of supporting a peak data rate in excess of 5208 bits per second must be utilized.

Nodal points

The preceding paragraphs intimated that a nodal point is a location where you can consider the concentration or multiplexing of data. However, the geographical boundaries of the nodal point have not been defined. Concerning its geographical boundary, a nodal point can be as small an area as an office building or as wide an area as perhaps the United States west of the Mississippi. The size of the nodal point will be governed by the number, type, and operational characteristics of terminals that will have their data transmitted and received from a central computer via a data concentrator. The selection of nodal points was covered in Chapter 14.

18.4 Installation and Operation

When planning the installation of a communications network, you must look beyond simply allocating sufficient floor space and appropriate power sources, and allow for the installation of air conditioning, new flooring, and electrical power receptacles, as well as fire-detection equipment, partitions, and environmental control systems. Also, unless you integrated such items as communications lines, modems, terminals, and data concentration equipment into the installation plan, there could be quite a time gap between the network's successful installation and the start of its useful operating life.

With turnkey systems, the vendor generally supplies not only all the hardware and software components, but also takes responsibility for combining them into an operational installation. Here, one key advantage for the user is that, for any maintenance or expansion problems, his point of contact is limited to a single source. Although this is a valuable time-saver when maintenance problems arise, the user ultimately pays for the service. Most turnkey vendors use a uniform profit markup. It's not uncommon for a user to pay 50 to 100 percent more for terminals, modems, multiplexers, and communications lines, than if the user assembled the network with components from different suppliers. This latter approach results in a more cost-effective network because the middleman is eliminated. But if each vendor maintains only its own equipment, the user must know how to best select the required equipment, and also to plan ahead to avoid "fingerpointing" when network problems arise.

Although the exact sequence of events to follow in communications-site preparation and planning varies from installation to installation, Table 18.4.1 serves as a useful guide for developing a pre-installation schedule. About three months prior to installation, the user should be at the stage of having defined the necessary communications lines and ancillary equipment required for the network, and is ready to place orders with the appropriate vendors. With the exception of local telephone companies, which can install a standard telephone circuit usually within a week after order, almost all other communications equipment is quoted on a 90-day after-receipt-of-order (ARO) basis. This means that, unless sufficient time is left for contingency purposes, a critical network component could just be leaving the vendor's factory on the date that the network is scheduled to become operational.

TABLE 18.4.1 Pre-Installation Schedule

90 days before installation

- Define needs for and order long-distance telecommunications lines
- Define needs for and order local telecommunications lines
- Define needs for and order ancillary equipment—modems, multiplexers, rotary, data access arrangements, cable, and rack-mounting equipment

60 days before installation

- Define and install electrical power requirements according to specifications
- Complete environmental specifications
- Install convenience outlets
- Develop maintenance plan

30 days before installation

- Install all telecommunications lines and equipment
- Establish and check out the functioning of modems, remote terminals, local terminals, and communications lines
- Obtain test equipment

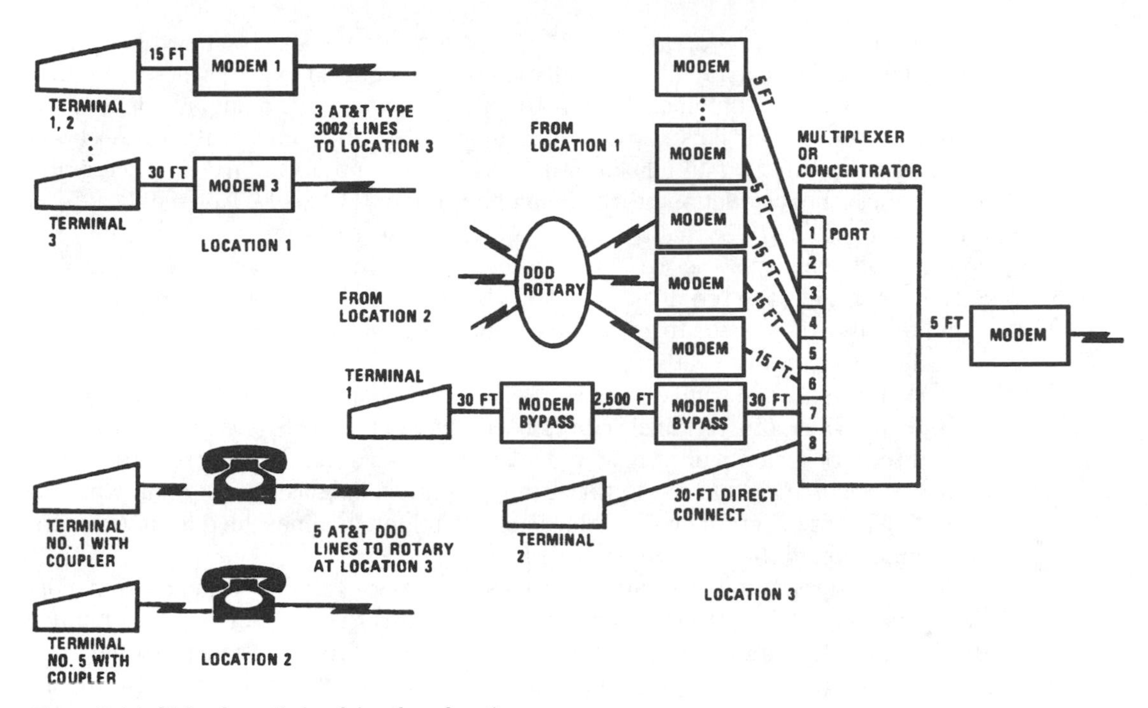

Figure 18.4.1 Link schematic involving three locations.

Consider modem options

In defining and ordering the required modems, check with vendors to investigate all the available options. Some of these might be of a convenience nature, but unless you consider all options carefully, it's possible that a modem of the properly rated transmission speed might not be usable with other equipment in the network. The magnitude of customer options varies among modem manufacturers as well as among the types of modems (bit rate) under consideration. One of the most common modems, the Bell 103A, provides the user with 32 option possibilities, ranging from auto-call unit to mark or space hold, and from auto-answer permanently wired or key controlled to the terminal responding or not responding to disconnect and initiating or not initiating disconnect.

Although some options might appear trivial, they can save the user a considerable amount of money. For example, consider a network where the central computer automatically dials numerous remote locations during each evening and requests transmission of the day's activities. If the modem options are selected so that the terminal will respond to a disconnect signal, then, in effect, each remote terminal could be powered down after completion of the data interchange instead of drawing power throughout the night.

Link schematic

One way to help define and order the required equipment is to prepare a link schematic. In the hypothetical situation represented in Figure 18.4.1, suppose that locations 2 and 3 are in the same state. (If the two sites are not within the same local telephone company area, the user will have to contact at least two local telephone companies to arrange for installation of the required modems, assuming that such modems can be obtained from the telephone company and not an independent manufacturer.) In this illustration, it is assumed that the terminals at location 1 will communicate with the multiplexer or data concentrator at location 3 via voice-grade leased lines, the installation of which can be arranged with AT&T or the local telephone company. (Notice that the use of "AT&T" in these examples is strictly for convenience. It could just as well be another long-distance carrier, such as MCI or Sprint.) Assuming the terminals at location 2 are selected with built-in acoustic couplers, the user only needs to contact the local telephone company for installation of five telephones at this location.

At location 3, it's assumed the user desires to install two terminals close to the multiplexer. Here, one terminal is directly connected to port 8, and the other is connected to port 7 via a pair of modem bypass units—sometimes referred to as *data transmit-receive extenders*. In this arrangement, as long as the user limits his cable laying to his own property and does not use telephone lines, he does not require any connection to the telephone network.

Here, the three-number rotary is assumed to service the five terminals transmitting from location 2. If users obtain non-Bell modems that are not registered under the Federal Communications Commission Equipment Registration program, they must contact the local telephone company or independent manufacturer for the installation of data access arrangements for transmission over the switched network.

Although the link schematic is a simplified one, it is interesting to note that the user could have conceivably chosen terminals from 10 different vendors, the modems from 10 suppliers, the rotary, DAAs, and telephones from three local operating companies, the modem backup units and cables from two additional vendors, and the leased lines from AT&T or other interchange carrier.

To alleviate the problems that can arise from selecting modems from several vendors, you should ensure that either the modems are paired from the same vendor for each data link, or that the modem of one vendor can communicate with the modem of another.

Equipment interface

One important point that is often neglected is to denote the type of interface for each item of equipment, as well as to define the differences between equipment when preparing a link schematic diagram. The two primary methods of interfacing equipment are based on voltage and current. For voltage interfacing, the EIA-232 is an industry standard specification that dictates, for example, what the modem output must be for matching with telephone lines. With EIA-232 equipment, data is bipolar, voltage serial.

Current-loop interfacing replaces the mark and space levels by current, rather than voltage. Some vendors offer three current-loop interfaces, with the difference being in the current level provided. Although current-loop interfacing can be more complex than voltage interfacing, its principal advantage is that it can eliminate the use of modems or acoustic couplers because a current-loop circuit can drive a device at distances up to one mile, depending on the amount of current and the gauge of wire used.

If the equipment uses a voltage interface, there might be a 50-foot cable limit because of the high driver impedance associated with this technique. If equipment spacing exceeds the cable limit, a user can rearrange the equipment, make provisions for obtaining a remote connection unit that allows an extension of line-driving capability up to 300 feet, obtain modem bypass units that allow communications at distances up to several miles, investigate the feasibility of using current-loop equipment, or use modems.

Most multiplexers, concentrators, and terminals can be selected with a current interface option, which permits the user to eliminate modems or couplers when her terminals and multiplexers are located within her building or at a site where she can lay her own cable. When using current-loop interfaces, the user should consult the vendor's literature to determine the type of cable necessary. Usually, the vendor will specify the wire resistance and current-carrying capacity. Although 18 or 20 AWG (American Wire Gauge) copper conductor wire is commonly used, smaller-gauge (i.e., higher-number AWG) wire can be installed if distances are under 2000 feet.

When transferring information from the link schematic, supplement the material with data from applicable vendor literature to complete a link worksheet. This worksheet not only serves as a guide in planning electrical outlet requirements, but it can be furnished to selected vendors to help them meet your installation requirements. Also, you can perform simple crosschecks to ensure that such taken-for-granted factors as cable-length limitations and modem compatibility are met to anticipate installation problems.

About two months before the installation date, all electrical power requirements should be defined at each site location, and arrangements made for installation according to the specifications developed for the network. Unless terminals are designed to operate from a fixed location, the cost of adding convenience outlets before installation will be less than trying to incorporate a degree of flexibility later.

Existing environmental (i.e., heating, cooling) tables should be updated to include all applicable communications equipment. Although it's doubtful if the addition of communications equipment will change air conditioner requirements, adding the dimensions, weight, temperature, and humidity characteristics to an environment table permits more complete analysis of the site.

For locations where you plan to install a large number of modems, multiplexers, or combination of ancillary equipment, consider the utilization of cabinets, enclosures, or rack mounts. Besides centralizing equipment and eliminating cluttered configurations, this will greatly facilitate test and maintenance procedures.

Maintenance contact

For each site location, prepare a maintenance-contact sheet in addition to a master maintenance sheet, which should reside at the centralized location. Each sheet should list the point of contact and negotiated responsibilities for the equipment of the appropriate vendors.

If the required equipment is ordered 90 days prior to installation, and the work plan is such that 30 days are allowed for installing and checking out the equipment, this latter period can be considered as contingency time. It can be utilized to establish and check out the functioning of communications lines and equipment. During this time, the user can become familiar with equipment maintenance requirements and diagnostic routines. Because most modems and multiplexers have on- and off-line diagnostics to help determine the operational status of each piece of equipment, the user should learn to pinpoint the source of problems to reduce fingerpointing once the network is functioning.

Besides relying on the built-in diagnostic capabilities of the equipment, the user should purchase test instrumentation with a high degree of flexibility. Portable test equipment is useful when installations have the operating equipment distributed over a wide area so that the test equipment can be taken to the work by the technical staff.

Cutover

Cutover is the point at which the new component replaces the old component, or the new setup replaces the old one. Several methods of cutover should be explored by the network analyst to determine the optimum method to use based on operating personnel, budget constraints, and other limitations.

The two primary methods of cutover are serial and parallel, with combinations of these two methods possible in an infinite variety of configurations. In the serial cutover method, as illustrated in Figure 18.4.2A, the old method of communications ceases when the new method is installed. This is initially the least costly cutover method. However, if the new method does not work as designed, then operational problems could result that could force the reinstallation of the old method while "bugs" are eliminated from the new one.

In Figure 18.4.2B, a parallel cutover method is illustrated. Here, the initial cost is higher as a result of operating two parallel networks for a period of time. But total costs can be reduced if—during the period bugs are being resolved—employees have the old network as fallback.

(A) SERIAL CUTOVER

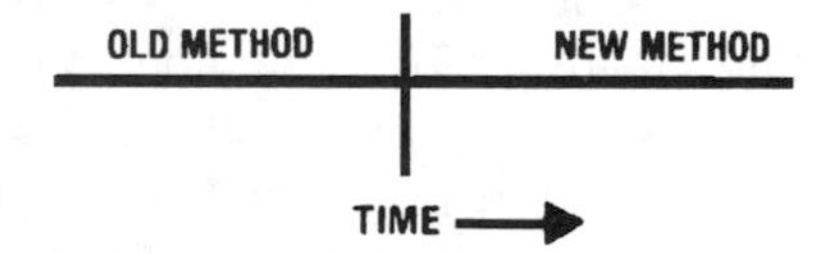

(B) PARALLEL CUTOVER

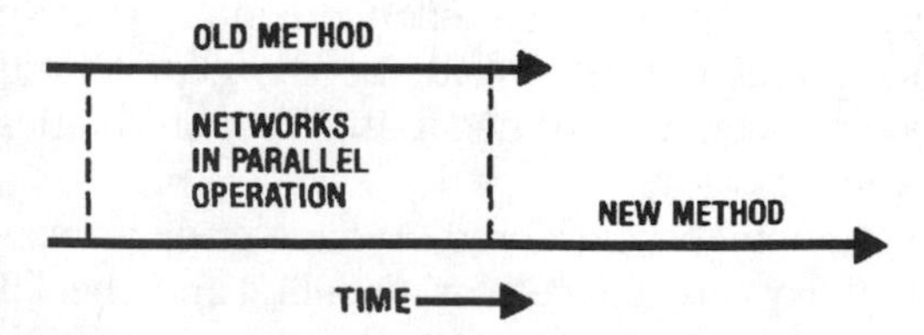

Figure 18.4.2 Cutover methods.

MASTER STATION LOG TECHNICAL CONTROL CENTER			DATE	TIME PERIOD	
				FROM	TO
			5-15-83	0700	2400

CHANNEL OR CIRCUIT	TIME	OPERATOR INITIAL	ACTION/EVENT
GD 4720	0630	SN	OPERATOR S. BROWN AT WASH REPORTS CIRCUIT GD 4720 INOP, UNABLE TO GET INTO SYSTEM.
GD 4720	0645	SN	CHECKED MODEMS BTB AND LINE WITH S. BELL; ALL APPEAR GOOD AND NO ERRORS.
GD 4720	0655	SN	RAN MODEM CHECKS AND LINE CHECKS WITH OPR BROWN FROM WASHINGTON TERMINAL; ALL NORMAL AND NO ERRORS.
GD 4720	0905	SN	FOUND LOOSE CONNECTION ON FRONT-END. TURNED CKT BACK TO OPS AND NOTIFIED BROWN CKT RESTORED.

Figure 18.4.3 Master station log.

Monitoring and reporting

A number of forms should be completed for both remote and central-computer sites, to facilitate network management and control.

At the central-computer site, a master station log can be completed by technical-control-center personnel and routed to network management the next business day. A sample completed log with four entries is illustrated in Figure 18.4.3. The purpose of this log is to record all outages and other significant occurrences pertaining to the organization's data communications network and its associated equipment. By examining the network log, managers obtain a better understanding of equipment and circuit operations, and problems experienced and reported by field personnel.

Another form that should be considered is a data transmission log, as illustrated in Figure 18.4.4. If communications funding is not centralized, then the network manager might not be able to obtain local and long-distance telephone bills and reevaluate portions of the network based on new usage conditions. By submitting the data transmission log to terminal operators (who at the present time dial long distance to a computer), the network manager can periodically reanalyze the usage of the particular terminals in question. Then, she can determine if a new transmission medium should be used, based on a change in transmission activity for a sustained period of time.

LOCATION ______ QUARTER ENDING ______ TERMINAL ______

DATE	NO. OF CALLS	TOTAL TRANSMISSION TIME (MINUTES)	DATE	NO. OF CALLS	TOTAL TRANSMISSION TIME (MINUTES)

FORWARD NLT 15 DAYS AFTER END OF QUARTER TO:

CHIEF, DATA COMMUNICATIONS

Figure 18.4.4 Data transmission log.

Although there are no firm rules to follow in planning a network installation, these guidelines should help alleviate the more common problems that arise. If you devote careful consideration to the plan at an early enough date, it will go far in starting network operation at the scheduled time.

Questions

18.1 Name and discuss three methods used in the information-collection process.

18.2 What problems can occur when a checklist is distributed to users?

18.3 What are some of the advantages and disadvantages of interviews to obtain information?

18.4 Discuss the function performed by an expansion table.

18.5 Why is it important to know all protocols supported by a front-end processor when the processor is required to support one terminal type?

18.6 Discuss the possible effects of adding software modules to a front-end processor.

18.7 Name and discuss two techniques to alleviate front-end-processor port constraints.

18.8 What are the two primary methods of cutover? Discuss the advantages and disadvantages of each method.

Chapter

19

Network Management and Administration

With the evolution of data communications as a professional field of specialization, numerous techniques have been tested for their applicability to network management and administration. Although some of these techniques were originally developed for use in other fields, certain techniques have been specially developed to assist the network analyst in the management and administration of a network.

In this chapter, two of the more important techniques that the network manager can use for controlling cost, obtaining required equipment, and distributing the cost of operating the network is covered.

The first method, a procurement action, can be used with a high degree of success when a firm has a requirement for many devices, or one or more expensive pieces of equipment. By having many vendors compete for business, the competitive nature of organizations can result in a lower cost for the required product.

One problem now facing many organizations is how to distribute the cost of data communications usage in a fair and equitable manner. Until a few years ago, communications costs were an insignificant portion of an organization's total information processing budget. They were centrally funded and assessed to all company departments as an overhead expense. As a result of the growth of remote processing, communications expenditures have increased dramatically to the point where they might not be suitable as an overhead expense. This is because the department that is located at the computer center feels, and rightfully so, that it should not pay for the data communications costs of other departments located in other cities that want to use the company computer for new applications. A trend toward charging back of communications costs has evolved. This is covered further in this chapter.

19.1 Procurement Actions

A procurement action can be considered a formal approach for obtaining required goods and services. Such an action can range from contacting a vendor directly and placing an order, to writing a series of specifications for the items required, issuing the specifications and a list of conditions to several vendors, and evaluating their responses from both a technical and cost viewpoint. For the latter case, this type of action is considered a competitive procurement because more than one vendor is given the opportunity to compete for the business of the organization. For the former case, this type of procurement action is known as *sole-source procurement* because only one vendor is considered.

In general, sole-source procurements can be used to obtain required equipment in a much shorter time frame than a competitive procurement. Although a competitive procurement takes longer because a number of vendors must be contacted and their equipment and prices evaluated and compared, such procurement actions usually result in a reduction of cost to the organization.

Requirement specifications

Prior to contacting vendors, the network analyst should prepare a list of requirement specifications. The formality, use, and transmission of such specifications to vendors will usually be a function of the rules of the organization's purchasing office.

For some organizations, the procurement of any item with an expected dollar amount over a certain limit must have the requirement specifications issued in a formal procurement document. The network analyst is responsible for preparing the requirement specifications that lay out the technical characteristics or operating parameters of required components. The purchasing office becomes responsible for including company-standard clauses concerning alternative proposals, contract forms, shipping costs, and other factors of a general nature common to most company procurements.

When writing the specifications of the devices required, the network analyst should divide the features of every component into two categories: mandatory and desirable. To be responsive, a vendor must meet all the mandatory requirements. Thus, mandatory features can be considered as items or performance specifications that, if not met, would degrade the overall performance of the required device to an unacceptable level.

A desirable feature can be considered an item or performance specification that would be nice to have, but without which the component would still be able to perform as required. Consider, for example, a multiport modem capable of operating at 9600 bit/s. If the network analyst desires to multiplex two 4800-bit/s data streams, then a two-port multiplexer, combining two 4800-bit/s data streams, is a mandatory feature of the multiport modem. The analyst might believe that at some later date it is conceivable that, instead of two 4800-bit/s data streams, one data stream could be lowered to 2400 bit/s and the support of the number of such devices operating at 2400 bit/s increased to two. Then, a multiport modem that can be reconfigured would be a desirable feature.

Feature evaluation

In general, mandatory features are evaluated on a pass-fail basis. If the device has the item or items listed as mandatory, or meets the mandatory performance specifications, it is said to be *in compliance* with all mandatory features. If the device fails one or more mandatory features, then the device is excluded from further consideration.

A monetary value is normally assigned to each desirable feature. This dollar worth represents the typical cost to the user to procure the feature, or the value of the feature. The price that the vendor bids for a component is then adjusted by the value assigned to all the desirable features that the vendor's product includes. This cost value becomes the evaluated cost of the product for comparison with other vendor bids. This proposal evaluation method is known as the *cost/value technique* because the value of desirable features is subtracted from the cost of the equipment.

Returning to the multiport modem example, consider the two devices in Table 19.1.1. Here, the modems of both vendors have been judged as meeting all mandatory requirements. However, vendor A's modem also has a desirable feature that, it is felt, could be used at a later date. A value of $500 is assigned to that feature. After subtracting the value of the desirable feature from the cost of the modem, vendor A's product has the lower evaluated price.

Award consideration

Although price is an important factor, many additional items must be considered prior to awarding a contract. These considerations range from the equipment availability to maintenance and training support and equipment space and power requirements, as well as to the methods available to obtain the desired equipment. Methods include lease, purchase, and lease with option to purchase.

For an organization that believes its requirements could change in the near future but requires certain items of equipment for current operations, leasing could be considered. Although the user can send back the equipment after giving appropriate notice, certain financial benefits that one obtains with equipment ownership cannot be used by the leaser. These financial considerations include the use of the investment tax credit and depreciation of equipment. Normally, users request vendors to price their products with respect to both purchase and lease. This way, the user can evaluate both options and select the one most advantageous based on requirements and financial circumstances.

TABLE 19.1.1 Cost/Value Techniques

	Vendor A	Vendor B
Modem cost	$5750	$5580
Desirable feature	$ 500	---
Cost value	$5250	$5580

19.2 Network Cost Distribution

Providing adequate service at the lowest cost per user is one goal frequently sought in the design or modification of a data communications network. A problem occurs when the data communications staff receives and analyzes communications requests from other departments. Usually, these requests are formulated as a data communications service requirement for a particular activity or series of activities.

Although a low price tag is often emphasized as an important part of these requests, the answer to the question of "How much is enough?" in the way of equipment and service is normally not left up to the data communications staff. The reason for this is that most organizations erroneously believe that the department that requests the service is the most knowledgeable about its data communications requirements. The primary goal of the data communications staff is believed to be to design and modify the network, leaving the requesting department responsible for determining its requirements.

One method that can be effectively used to control costs despite this attitude is the implementation of communications chargeback. Through the utilization of communications chargeback, the cost of communications must be shared by all the users. Realizing that they will have to pay for the resources associated with the support of their requirements, departments requesting data communications support tend to become more concerned with cost. They tend to work more closely with the data communications staff, and usually start to eliminate frills. Thus, chargeback can be viewed as an effective method for bringing users "down to earth" by letting them know the dollar value of the services they want versus the services that they really need.

A series of chargebacks projected over a period of time can be used by individual departments to generate the data communications portion of their operating budget. When properly formulated and defined, chargeback information gives all levels of management a realistic understanding of the nature and cost of data communications.

In developing a chargeback method, a distinction should be made between the recovery of direct costs and indirect costs. Direct costs relate to modems, multiplexers, and communications lines, and are clearly visible as expenses incurred in providing data communications support. Indirect costs for test equipment, patch panels, and equipment used in a technical control center (as well as the communications staff's salaries) should normally be omitted from chargeback, because of the problems you can encounter in trying to distribute these expenses equitably.

Although the cost of test equipment could be assigned to all departments equally, such equipment might not be used on an equal basis. Even when test equipment is used by everybody, some departmental managers find it hard to justify their share of the payment when the equipment is often used to pinpoint a telephone line problem (for example), thus placing the onus of repair cost on the vendor. As a result of these circumstances, most organizations centrally fund their communications staffs and their technical control facilities.

Chargeback development

Two charges are associated with such communications components as lines, modems, multiplexers, and ancillary devices: nonrecurring and recurring. Nonrecurring charges include site preparation and equipment installation. If equipment is purchased outright, the cost of such equipment could be considered a one-time charge. However, a number of organizations centrally fund equipment purchases, and amortize the cost of the equipment to the user departments as a recurring depreciation charge.

Recurring charges are usually associated with the monthly rental cost of equipment. One recurring charge associated with purchased equipment, aside from depreciation, is maintenance. Although included in the monthly rental charge of leased equipment, this cost is usually separate for purchased equipment.

In computing depreciation for purchased equipment, several alternative methods are available for consideration. The most common method is called *straight-line depreciation.* This method assumes an equal write-off of the asset by dividing its cost by the length of its estimated life. The formula for straight-line depreciation is

$$D = \frac{C - S}{P}$$

where D = dollars/month depreciation
C = the cost of the asset
S = the estimated salvage value of the asset at the end of its life
P = the months of use

For a 9.6-kbit/s modem purchased for $7600, with an estimated life of three years, and an estimated salvage value of $1000, straight-line depreciation is

$$D = \frac{\$7600 - \$1000}{36}$$
$$= \$183.33\text{/month, or } \$2200\text{/year}$$

Other depreciation plans include the "sum-of-the-year's-digits" and the "declining-balance" methods.

A point-to-point, single-use line is the easiest type of communications asset to charge to the user. This circuit is dedicated to a single department, and all communications costs are billed to it.

Sample chargeback methods

With M as the monthly cost per modem and L as the monthly cost of the leased line, the monthly charge to the user for a point-to-point circuit is

$$CB = 2M + L$$

For point-to-point, multiple-use lines, chargeback becomes more complicated. Consider the user of multiport modems with integral synchronous multiplexers that permit several streams of data to be transmitted over one line.

If TB is used to denote the total bandwidth of a line in bit/s, and BU denotes the portion of the bandwidth used by one cost center, then

$$CB = (2M + L)\,\frac{BU}{TB}$$

Each cost center is charged for its share of the line's bandwidth and for the use of the modems. If terminal 1 associated with cost center 1 operates at 2.4 kbit/s, and terminal N assigned to cost center N operates at 7.2 kbit/s, then the chargeback to each cost center would be as follows:

$$CB\text{ (cost center 1)} = (2M + L)\,\frac{2400}{9600}$$

$$CB\text{ (cost center } N) = (2M + L)\,\frac{7200}{9600}$$

What is specified is that because the first center utilizes 25 percent of the bandwidth, it should pay for one-quarter of the data communications cost, and the second cost center should pay the remainder.

The chargeback of communications facilities on multidrop lines can sometimes present a problem because some configurations might not appear fair and equitable to users receiving such chargebacks. Figure 19.2.1 shows a simple multidrop configuration and allows the examination of techniques that reduce communications costs and chargeback problems.

For this multidrop line, all of the cities are on a horizontal plane and one cost center is associated with each city serviced by the multidrop line.

If N is the number of remote modems, L_i denotes the line cost for segment i, MC represents the modem cost at the central site, and M_i is the modem cost at each remote site, the chargeback debited to each cost center becomes

$$CB_i = L_i\,M_i + \frac{MC}{N}$$

Simply stated, this formula charges to each cost center the expense of the line segment from the preceding city, its own modem cost, and the prorated share of the cost of the central-site modem—which is distributed to each cost center based on the number of users on the line.

However, because most multidrop lines are not configured on horizontal planes, certain network optimization routings can cause knowledgeable users to complain that their shares of the cost are not equitable. One such network is illustrated in Figure 19.2.2.

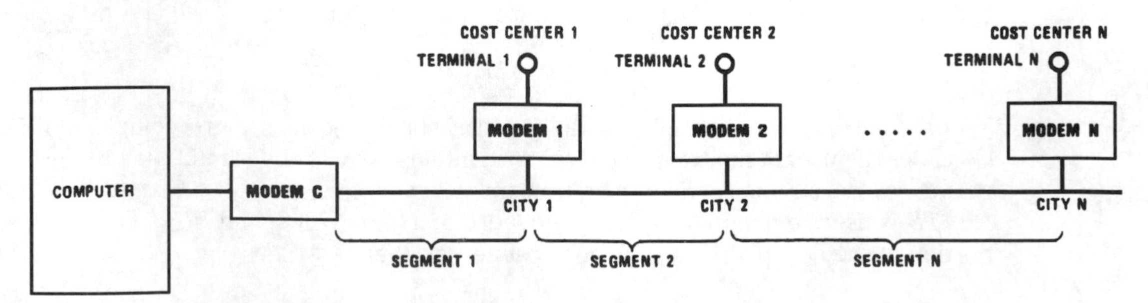

Figure 19.2.1 Multidrop line

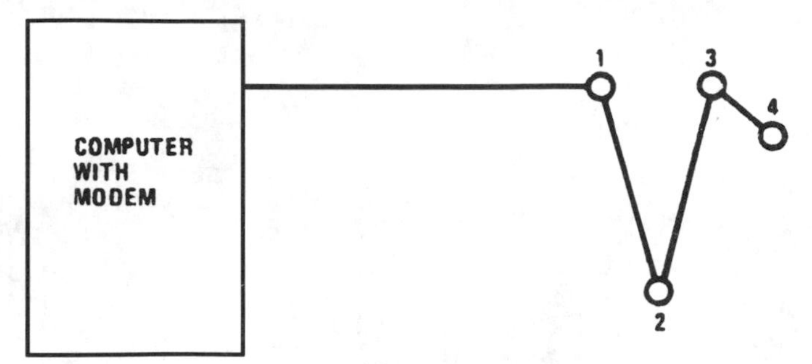

Figure 19.2.2 Optimized multidrop line.

Here, because of geographical locations, a "V" routing is used from cities 1 to 2 and 2 to 3. Although this routing reduces the total line cost of all locations, it increases the length and cost of the segment between cities 2 and 3 because connecting city 3 to city 1 is shorter than connecting city 3 to city 2. If one cost center is allocated to each city, cost center 3 could argue that it should not be forced to absorb higher individual costs to reduce the total cost to the whole organization.

Although this argument has merit, without a corporate network, cost center 3 would have to pay for the entire line segment from the computer to its location, as well as the full cost of the modem located at the computer. The optimization of a network can produce certain inequities. However, very seldom do these inequities justify an independent approach in which cost centers establish their own arrangements.

Prorated payment

Developing a chargeback methodology for contention equipment can also present problems. This is because the means of obtaining the most equitable determination of charges is the hardest to implement. Variable charges to different departmental cost centers are produced, and such charges are available only after the end of any given billing period.

One simple contention device is illustrated in Figure 19.2.3. Here, the contention equipment is a time-division multiplexer with two four-position telephone company rotaries connected to it at a remote location. One rotary connects four 300-bit/s auto-answer modems to the multiplexer, and the other rotary connects four 1.2-kbit/s auto-answer modems. With an aggregate throughput of 6 kbit/s, the multiplexer is connected to a 7.2-kbit/s modem, transferring data to the computer on the distant end at that speed. Assume that only two cost centers use the contention facility, with cost center 1 having six 300-bit/s terminals and four 1.2-kbit/s terminals, and cost center 2 has four 300-bit/s terminals and six 1.2-kbit/s terminals.

Several methods can be used to develop an equitable chargeback to both cost centers. One way to keep track of usage would be through terminal logs, with the total connect time of each terminal accumulated at the end of each billing period. The results of these logs might be as shown in Table 19.2.1 for one such period.

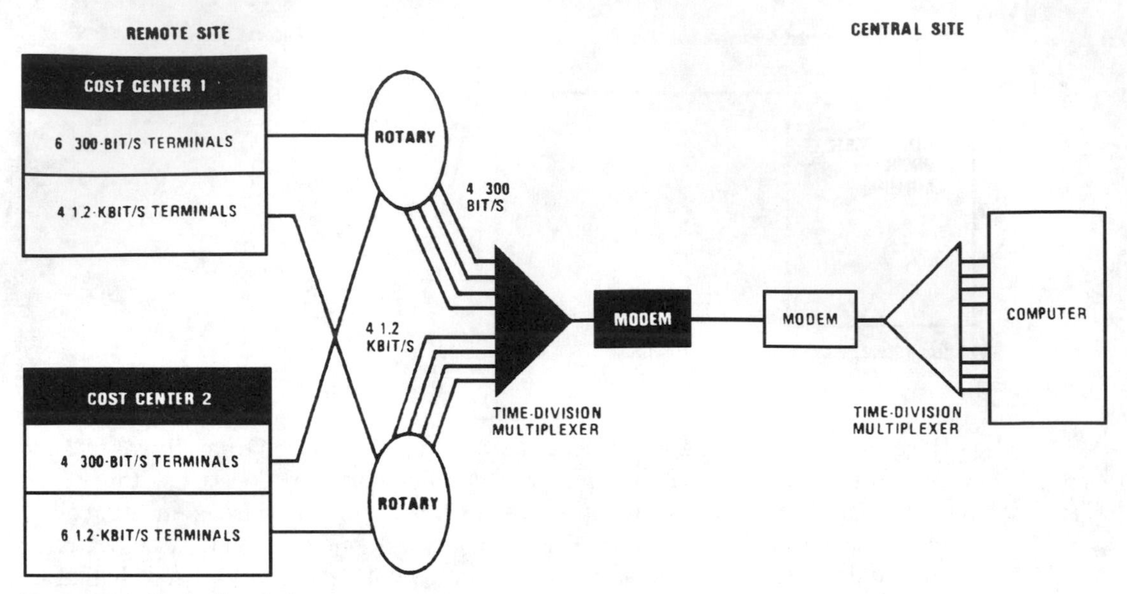

Figure 19.2.3 Contention device.

TABLE 19.2.1 Connect Time (Hours)

Modem speeds	300 bit/s	1.2 kbit/s
Cost center 1	35	87
Cost center 2	120	210
Total	155	297

For simplicity's sake, assume that the monthly cost of the remote-site low-speed modems, the telephone rotaries, both TDMs, the high-speed modems, and the leased line connecting the remote site to the central site is \$2000 per month. On a prorated usage basis, the cost debited to cost center 1 (chargeback) can be computed as follows:

$$CB1 = \frac{35}{155} \times \frac{1200}{6000} \times \$2000$$

$$+ \frac{87}{297} \times \frac{4800}{6000} \times 2000 = \$559$$

For these equations, 35/155 represents the prorated cost center 1 usage of the 300-bit/s port, and 1200/6000 represents the proportions of the four 300-bit/s ports to the total bandwidth of the network. Similarly, 87/297 is cost center 1's prorated usage of 1.2-kbit/s modems, with 4800/6000 representing the proportion of 1.2-kbit/s modems to the total bandwidth of the network.

Variations in connect times each month will cause the chargeback amount to change for each center sharing the contention equipment. One major problem in attempting to charge users on this basis is that logs are time consuming—both to complete and to analyze. In addition, users might neglect to record their usage, making such logs inaccurate.

Another method that can be used to generate chargeback costs for contention networks is by programming the corporate computer to do the required usage computations. A problem with this approach is that terminals are not necessarily fixed at a given location, and allowing a user to access the computer through different contention facilities would require a complex set of programs to produce the communications chargeback bill for each cost center.

Because contention facilities are normally accessed by timesharing or message-switching equipment, an easier way to use chargeback is to add a surcharge to the billing algorithm of each user. This surcharge is designed to recover the cost of providing communications for all contention facility users and is reviewed and modified as required to keep the total dollar surcharge as close to the communications cost as possible.

Returning to Figure 19.2.3, there is another, less sophisticated method to prorating the cost of a contention network. This method is based on the number of terminals used by each cost center. Using this approach, the chargeback for cost center 1 becomes

$$CB1 = \frac{6}{10} \times \frac{1200}{6000} \times \$2000$$

$$+ \frac{4}{10} \times \frac{4800}{6000} \times \$2000 = \$880$$

As shown, chargeback based on the number of terminals can produce results different from computation based on usage. One argument advanced for chargeback using terminal quantities is that the most important part of the initial design of many networks is the number of user terminals and the anticipated hours of their operation. Although actual operation undoubtedly deviates from initial projections, the information initially furnished to develop the communications network forms the basis for equipment procurement. Cost centers should be charged on the basis of the information furnished.

Although this section has presented a sample of chargeback techniques, users should recognize that the wide variety of network configurations precludes the development of textbook solutions. Even under the Federal Communications Commission's ruling that permits nonaffiliated users to share a leased-line network (joint-use provision), the only guideline is that communications costs must be shared equitably. Thus, the present state of communications chargeback is best defined by the words fair and equitable.

Questions

19.1 What are the differences between sole-source and competitive procurements with respect to price and time?

19.2 Discuss the difference between a mandatory and a desirable feature.

19.3 What is the difference between mandatory and desirable features with respect to their evaluation?

19.4 If vendors A, B, and C respond to a procurement action with bids of $8750, $8827, and $8905, and the value of their desirable features are $350, $500, and $150, respectively, who should be selected and why?

19.5 Two departments share the use of common communications. Total costs are $4500 per month. Each department's terminal usage for the month is:

	Terminal Hours	
	300 bit/s	1200 bit/s
Department 1	450	750
Department 2	850	375

How much should each department be charged on a prorated usage basis?

Chapter

20

Trends in Data Communications

20.1 Introduction

One overwhelming trend in data communications is the impressive acceptance, dominance, and continuing evolution of local area networks. The principal types remain Ethernet, token ring, and FDDI. At this writing, the proposed 100BASE-T (100 Mbit/s) or "fast" Ethernet standard is proceeding through its formal acceptance processes. Also, work on the interactive-traffic-handling FDDI-II proposed standard has been suspended in view of ATM's growing acceptance. Besides its other attributes, one of ATM's major strengths is its ability to readily integrate LANs and WANs.

Other trends that are examined in this chapter are what has happened to the promise of ISDN and to that of multimedia: the former does not appear to be making much headway, but the latter is becoming all the rage.

Another trend is the increasing use of the Internet as an information resource, a means of communication, and as a business platform. The Internet is no longer proprietary to academia or to the government, but is a fast-growing international public network of networks. A prime business adjunct is that of providing access to the Internet. Another indication of the Internet's pervasiveness is the proliferation of books and other publications on the subject, to which the reader is referred (see the Overview).

20.2 FDDI

The Fiber Distributed Data Interface (FDDI) has been with us since the mid-1980s. Until final endorsement of the proposed 100BASE-T specifications, FDDI remains the only fully approved 100-Mbit/s standard (see Chapter 10). Besides operating over its original optical-fiber medium, FDDI can function over shielded or unshielded twisted-pair copper. The ANSI standard, a TP-PMD (twisted pair-physical media dependent), replaced the proprietary methods that were being used to operate FDDI over copper.

Another development would tend to solidify FDDI's position further: the Digital Equipment Corp. (DEC) is licensing its FDDI Full-Duplex Technology (FFDT) that enables simultaneous send/receive operation, functionally doubling the throughput to 200 Mbit/s. Some DEC-supplied hardware and software modifications are required. Until ATM fully takes over as the high-speed networking technology—starting with the 25-Mbit/s products under development—and until 100BASE-T starts competing, FDDI should remain as the fully accepted 100-Mbit/s networking technology of choice.

20.3 100 VG-AnyLAN

Besides 100BASE-T, the IEEE has proposed 100VG-AnyLAN. This technology is said to be unique in enabling a user of either 10-Mbit/s Ethernet or 16-Mbit/s token ring to migrate to it. The allowable networking distances are similar to those of 10BASE-T: up to 3000 meters. Any LAN will operate over either single-mode or multimode fiber. Connection to a WAN requires a router.

20.4 ISDN

Progress in implementing the Integrated Services Digital Network (ISDN) in the USA has slowed perceptibly. But that is reportedly changing. The USA local exchange carriers (LECs) are filing tariffs that undercut switched 56/64-kbit/s and dedicated-line services. And the growth of the videoconferencing market (see Chapter 16) bodes well as a logical application for ISDN.

According to CCMI (Rockville, MD), availability of ISDN is upwards of 75 percent on the LECs' lines, ranging from about 50 percent at NYNEX to about 86 percent at Pacific Telesis. And for the deprived user, tariffed services—called *Virtual ISDN* and *ISDN Anywhere*—from the LECs can provide access to the nearest wire center that does offer ISDN.

ISDN architecture

At this juncture, a review of ISDN basics appears warranted. Under the ISDN architecture, access to this digital network will result from one of two major connection methods: basic access and primary access.

Basic access

Basic access defines a multiple channel connection, derived by multiplexing data on twisted-pair wiring. This multiple channel connection is between an end-user terminal device and a telephone company office or a local private (automated) branch exchange (PBX). The ISDN basic access channel format is illustrated in Figure 20.4.1.

As indicated in Figure 20.4.1, basic access consists of two bearer (B) channels and a data (D) channel (2B + D) that are multiplexed by time onto a common twisted-pair wiring medium. Each bearer channel can carry one pulse code modulation (PCM) voice conversation or data at a transmission rate of 64 kbit/s, thus enabling

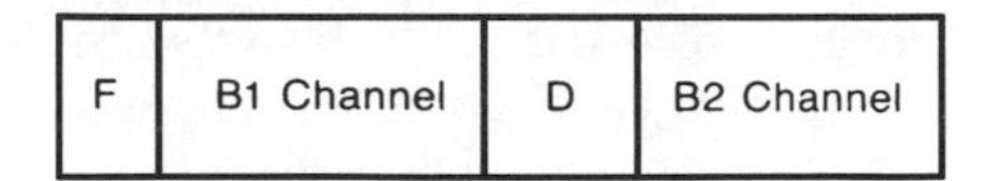

- B channels are 64 kbit/s each.
- D channel is 16 kbit/s.
- 2B + D service is a 144 kbit/s data stream.
- When framing (F) and multiplexing overhead is considered, the actual data rate of a basic access channel is 192 kbit/s.

Figure 20.4.1 ISDN basic access channel format.

end users to simultaneously transmit data and conduct a voice conversation on one telephone, converse with one person and receive a second call, or place one person on hold and answer the second call.

The D channel was designed both for controlling the B channels through shared network signaling functions on this channel and for transmitting packet-switched data. With the ability to transmit packet-switched data, the D channel facilitates a number of such new applications as the monitoring of home alarms and the reading of utility meters on demand. Because these types of applications have minimum data transmission requirements, the D channel can be used for a variety of applications, in addition to providing the signaling required to set up calls on the B channels.

Primary access

Primary access can be considered a multiplexing arrangement whereby a group of basic access users shares a common line facility. Primary access is usually used to directly connect a PBX to the ISDN network. This access method eliminates the need to provide individual basic access lines when a group of terminal devices shares a common PBX that could be directly connected to an ISDN network via a single high-speed line. Because of the different types of T1 network facilities in North America and Europe, two primary access standards have been developed.

In North America, primary access consists of a grouping of 23 B channels and one D channel (23B + D) to produce a 1.544 Mbit/s composite data rate, which is the standard T1 carrier data rate. In Europe, primary access consists of a grouping of 30 B channels plus one D channel to produce a 2.048 Mbit/s data rate, which is the T1 carrier transmission rate in Europe.

Network characteristics

The capabilities of an ISDN network offer a strong argument for its implementation by communications carriers. An ISDN can:

- Integrate voice, data, and video services.
- Offer high transmission quality.
- Improve and expand services as a result of B and D channel data rates.
- Promote greater efficiency and productivity with its ability to handle simultaneous calls on one line.

Because ISDN's digital nature combines voice, data, and video services, end-users only need to access one facility for each service. In addition, its end-to-end digital transmission can readily generate new pulses to replace distorted ones. Digital transmission has a lower error rate and provides a higher-quality transmission signal than an equivalent analog transmission facility. Although the amplifiers used by analog transmission facilities boost the strength of transmission signals, they also increase any impairments in the signal.

Because basic access provides three signal paths on a common line, ISDN can expand its services to the end-user. Under ISDN, each B channel can support a 64 kbit/s transmission rate, and the D channel will operate at 16 kbit/s. In fact, if both B channels and the D channel were in simultaneous operation, a data rate of 144 kbit/s can be obtained on a basic access ISDN circuit, which would exceed current analog rates.

Because each basic access channel consists of three multiplexed channels, an end-user could handle several operations simultaneously without using separate multiplexing equipment. For example, the end-user could receive a call from one person, transmit data to a computer, and have a utility company read the electric meter—all at once. The ability to conduct up to three simultaneous operations on one ISDN line should result in greater efficiency and productivity.

Terminal equipment and network interface

One of the key elements of ISDN is a small set of compatible, multipurpose, user-network interfaces that was developed to support a wide range of applications.

These network interfaces are based on the concept of a series of reference points for different user terminal arrangements, which is then used to define these interfaces. Figure 20.4.2 illustrates the relationship between ISDN reference points and network interfaces.

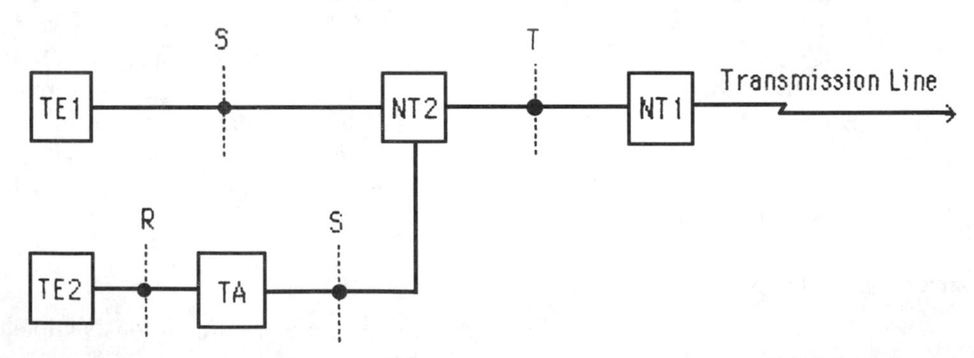

TE1 Terminal Equipment 1. These devices comply with the ISDN network interface.

TE2 Terminal Equipment 2. These devices do not have an ISDN interface and must be connected through a terminal adapter (TA) functional grouping.

NT2 Network Termination 2. Includes switching and concentration equipment, which performs functions equivalent to Layers 1 through 3 of the OSI reference model.

NT1 Network Termination 1. Includes functions equivalent to Layer 1 of the OSI reference model.

Figure 20.4.2 ISDN reference points and network interfaces.

The ISDN reference configuration consists of functional groupings and reference points at which physical interfaces exist. The functional groupings are sets of functions that might be required at an interface, while reference points are used to divide the functional groups into distinct entities.

The terminal equipment (TE) functional grouping is comprised of TE1- and TE2-type equipment. Examples of TE equipment include digital telephones, conventional data terminals, and integrated voice/data workstations. TE1-type equipment complies with the ISDN user-network interface and permits such equipment to be directly connected to an ISDN S-type interface, which supports multiple B and D channels. TE2-type equipment, which lacks an ISDN interface (e.g., EIA-232 or the ITU-T X- or V-series interfaces), must be connected through a terminal adapter (TA) functional grouping. The TA converts a non-ISDN interface (R) into an ISDN interface (S) by performing a physical interface conversion, protocol conversion, and rate adaptation (if necessary) to permit a TE2 terminal to operate on ISDN.

The network termination 2 (NT2) functional group includes devices that perform switching and data concentration functions equivalent to the first three layers of the OSI reference model. Typical NT2 equipment includes PBXs, LANs, terminal controllers, concentrators, and multiplexers.

The network termination 1 (NT1) functional group is the ISDN digital interface point and is equivalent to Layer one of the OSI reference model. Functions of NT1 include the physical and electrical termination of the loop, line monitoring, timing, and bit multiplexing. In Europe, where most communications carriers are government-owned monopolies, NT1 and NT2 functions can be combined into a common device, such as a PBX. In such situations, the equipment serves as an NT12 functional group. In the U.S., however, the communications carrier might provide only the NT1, and third-party vendors can provide NT2 equipment. In such situations, the third-party equipment would connect to the communications carrier equipment at the T interface.

20.5 Multimedia

Multimedia is the interactive use on a PC of video, audio, and animation, along with traditional text and computer graphics. *Networked multimedia* refers to sending all these types of information as digital data over computer networks. Networking enables connected users to interactively access and share the information. One tangible multimedia product is a compound electronic "document" that is formed from the mix of speech, text, animation, graphics, video, and still images.

Multimedia applications can be classified as

- image retrieval
- whiteboarding (document conferencing)
- video (with audio) playback
- real-time videoconferencing

The image-retrieval application appears to work well with conventional token ring or Ethernet networks. Whiteboarding is still developing, but some acceptable versions are currently available. Both video playback quality and the possible number of simultaneous users need improvement. As for the videoconferencing application, the quality achieved in using ISDN links is currently considerably ahead of that over LANs; here, ATM might be the answer.

The accelerating growth of multimedia owes much to the development of the videoserver, which permits integrating video traffic with that of existing LANs. The videoserver functions include compressing and framing digital video signals, converting analog TV signals to compressed digital video, linking to public videoconferencing and circuit-switched services, and storing video "clips."

Appendix

Answers to Questions and Problems

Answers to Questions and Problems in Chapter 1

1.1 To communicate, a transmitter, a receiver, and a transmission medium are necessary. For a telegraph system, the wire conductor is the medium, and the telegraph key serves as both the transmitter and receiver. When you listen to a radio broadcast, the station you are tuned to serves as the transmitter, the atmosphere is the transmission medium, and your radio set becomes the receiver.

1.2 In Morse code, different characters have a different number of elements to denote the series of dots and dashes that define the character. The lack of uniform character length made it difficult to develop a machine to interpret characters that varied in length. In addition, the lack of a prescribed time interval between characters and of a method to denote the beginning and ending of characters resulted in the absence of a method to synchronize automatic sending and receiving units.

1.3 The Baudot code provided a constant length for each character in that code's character set. The addition of start and stop elements to each character by Howard Krum permitted transmitting and receiving devices to be synchronized.

1.4 Satellite transmission is well suited for transmitting information both between two locations and from one location to multiple locations. In comparison, the high data rate of fiber-optic transmission is well suited for use by telephone companies in combining a large number of digitized voice and data calls on one fiber cable. These calls are routed between cities or in a local area network, where many computers and terminals require the ability to communicate with one another.

Answers to Questions and Problems in Chapter 2

2.1 Babbage's analytical engine and modern-day digital computers are designed to include a storage or memory area, components to perform calculations (arithmetic unit), and a means of controlling the operation (control unit).

2.2 Not only were vacuum tubes bulky, but they consumed large amounts of power, produced large amounts of heat, and—as a result of their relatively low mean time between failure in comparison to solid-state devices—at any given time a large computer consisting of vacuum tubes would have several such devices inoperative.

2.3 **A.** 182 **C.** 183
B. 107 **D.** 255

2.4 **A.** 11011 **C.** 110011010
B. 10111010 **D.** 11001110

2.5 **A.** 377 **C.** 305
B. 253 **D.** 214

2.6 **A.** FF **C.** C5
B. AB **D.** 8C

2.7

Decimal	Binary	Octal	Hexadecimal
17	10001	21	11
18	10010	22	12
19	10011	23	13
20	10100	24	14
21	10101	25	15
22	10110	26	16
23	10111	27	17
24	11000	30	18
25	11001	31	19
26	11010	32	1A

2.8 **A.** 100111 ones complement is 011000
Twos complement is 011001
Thus, subtraction becomes addition with twos complements or:

```
  101110        101110
– 100111    = + 011001
               1)000111
```

B.

```
  1101110        1101110
– 1111001    = + 0000111
                 1110101 (–10000000; or, in decimal: –128 + 117 = –11)
```

C.

```
  100011        100011
– 011010    = + 100110
               1)001001
```

D.

```
  011111       011111
– 100000     + 100000
               111111 (–1000000; or, in decimal: –64 + 63 = –1)
```

2.9 Data should be represented octally in an 18-bit computer as follows

X	X	X	X	X	X	X	X	X	X	X	X	X	X	X	X	X	X
17	16	15	14	13	12	11	10	9	8	7	6	5	4	3	2	1	0

For a 24-bit computer

X	X	X	X	X	X	X	X	X	X	X	X	X	X	X	X	X	X	X	X	X	X	X	X
23	22	21	20	19	18	17	16	15	14	13	12	11	10	9	8	7	6	5	4	3	2	1	0

For a 32-bit computer

X	X	X	X	X	X	X	X	X	X	X	X	X	X	X	X	X	X	X	X	X	X	X	X	X	X	X	X	X	X	X	X
31	30	29	28	27	26	25	24	23	22	21	20	19	18	17	16	15	14	13	12	11	10	9	8	7	6	5	4	3	2	1	0

2.10

	Procedure-oriented language	Assembly language	Machine language
Program machine dependency	Yes	Yes	Yes
Relative storage size for program	Larger	Smaller	Smaller
Execution time	Larger	Smaller	Smaller
Documentation	Easier	Harder	Harder
Debug time	Shorter	Larger	Larger
Programming complexity	Easier	Harder	Hardest
Program maintenance	Easier	Harder	Hardest

2.11 Several reasons exist that make assembly language the primary method of communications programming. First, no higher-level language has been developed exclusively for data communications. Such a language would require access to bit and byte manipulation, which is normally available only by the use of assembly language. Secondly, procedure-oriented languages are inefficient in object code and run time when compared to programs written in assembly language. Because communications programs are normally time dependent (i.e., gathering bits into a character before the first bit of the next character is received), time constraints might preclude the use of procedure-oriented languages for some types of communications applications.

2.12 A byte is the lowest addressable unit and can be used to group a series of bits to represent a character. Therefore, by manipulating bytes, you can manipulate characters; whereas by manipulating bits, you can manipulate only portions of a character.

2.13 Single and multiline controllers contain clocking and buffering circuitry and serve as an interface mechanism between a computer and communications lines. In the transmit mode, they convert a character, comprised of a number of bits, from a parallel data stream of the computer into a serial data stream for transmission. In the receive mode, they build characters by converting the received serial bit stream into a parallel group of bits that represents a character.

2.14 Using a front-end processor can reduce or eliminate the burden of communications processing from the host computer. Such functions as character generation or building by sampling incoming bit streams, code conversion, speed detection, and other communications functions can be moved to the front end, thereby permitting the host computer to concentrate on information processing.

2.15 The personal computer user could first access the other computer by making the sending PC function as a terminal. Once access is obtained to the other computer, the PC user could manipulate a database and extract the information required. Then, using the file transfer capability of communications software, the PC user could transfer the information to a file on the PC and use one or more applications on the computer to further manipulate the extracted information.

2.16 With appropriate hardware and software, a personal computer can be used to emulate a specific type of terminal that communicates with a specific mainframe computer. By obtaining an additional communications software program and, possibly, additional hardware, the personal computer might be able to emulate a second type of terminal, permitting communications access to another type of mainframe computer.

Answers to Questions and Problems in Chapter 3

3.1 In the context of the data communications field, a terminal is a device that permits information to be transmitted on or received from a data communications network.

3.2 Basic communications links include:

- Terminal-to-computer
- Terminal-to-terminal
- Computer-to-computer

These links can be modified by the inclusion of specialized data communications components to obtain such additional links as:

- Terminal-to-concentrator
- Concentrator-to-concentrator
- Concentrator-to-host

3.3 Automatic send and receive (ASR) terminals have local storage capability, usually paper tape or cassette, permitting data to be generated off line and transmitted automatically, or received and recorded automatically. When a keyboard send and receive (KSR) terminal is used, as data is entered manually, a hard copy is produced on the terminal's printer; however, no local storage capability is available to permit automatic operations. A receive only (RO) terminal does not have a keyboard because it can only receive information transmitted to it and cannot be used to respond to the transmitted information. Thus, an RO terminal operates as a simplex device. (See Section 4.3)

3.4 The cursor on the CRT terminal is used to indicate the position on the display where the operator can enter, modify, or delete data. Via appropriate cursor movements, the operator can display characters at any point on the CRT screen.

3.5 The key advantage of a multipoint circuit is that it permits a number of terminals to share a single transmission path. When one company conducted business with a large number of other companies, it became physically impossible to connect large numbers of teleprinters to a single party line, resulting in the development of message-switching and circuit-switching networks.

3.6 Examples of terminals used for order-entry applications:

A. Optical scanner connected to point-of-sale terminal consisting of an electronic cash register
B. Credit card reader that reads magnetic encoding on card.

For both applications, special equipment required includes the optical scanner, electronic cash register, and the credit card reader. To transmit the data to a computer off line, an off-line storage mechanism is required. If transmission is to be on line, specialized transmission devices, covered in later chapters, are required.

3.7 Both teleprinters and remote batch terminals normally transmit and receive data. Received data is usually printed at the terminal or on a storage mechanism attached to the device. Normally, a remote batch terminal contains a small processor that permits limited local processing in addition to transmission and reception of data to and from the host computer. Usually, a remote batch terminal has a much higher transmission volume than a teleprinter; thus, it operates at a much higher data rate.

3.8 A terminal emulator program translates or converts one set of coded signals that governs the operation of a terminal into a second code supported by the device the program is executed on. The coded signals that are converted include cursor positioning information and such video attributes as highlighting, underlining, blinking, and the color of a character on a color display. Also included could be the internal codes used to represent special purpose keys on the keyboard.

3.9 Typical inquiry and response systems include:

A. *Hotel reservations*. Requires a terminal that can produce hard copy confirmation slips to be mailed on customer request.
B. *Police vehicle identification*. Requires a terminal that can produce hard copy on request and whose transmission can be scrambled so that unauthorized persons cannot intercept data.

3.10 By connecting an optical scanner to a terminal, information can be entered into a terminal more rapidly than if data is entered manually via a keyboard.

3.11 Terminal characteristics that should be evaluated prior to selecting a specific device include:

A. Input/output media
B. Operating rate
C. Off-line capability
D. Data codes and character set
E. Operator convenience
F. Cost
G. Security

H. Control of user errors
I. Error detection and correction

3.12 Code conversion can occur in the terminal, the front-end processor, the host computer, or in a specialized device used between the computer and the terminal.

3.13 Parity is the addition of a redundant bit to make the number of bits that compose a character either odd or even. Through the use of parity checking, an error in transmission can be detected. As an example, consider the seven-bit character: 1011001

For the preceding character, four bits were set (to 1). If you use odd parity, the parity bit would be set; whereas if you use even parity, the parity bit would be reset on zero because the four bits set are an even number. Thus, the character to include the parity bit would appear as:

Odd parity	10110011
Even parity	10110010

3.14 In asynchronous transmission, each character contains a start bit and one or more stop bits that provide synchronization between the transmitter and receiver. In synchronous transmission, groups of characters are blocked together for transmission, and special synchronization characters are used to synchronize the transmitter and the receiver. Because data is blocked, and start and stop bits are omitted, synchronous transmission enables data to be transmitted at higher data rates than asynchronous transmission.

3.15 **A.** 10111010
B. 11101001
C. 11101111
D. 01110011

3.16 Synchronous terminals normally cost more than asynchronous terminals because they must buffer entered characters into data blocks for transmission. In addition to a buffer or memory area, synchronous terminals have more complex circuitry to include error-checking circuitry, as well as circuitry to perform special functions, such as insetting synchronization characters into messages to be transmitted and stripping such characters from received data blocks.

3.17 An intelligent terminal has the ability to modify its operating characteristics, but a dumb terminal lacks this capability. The capability to modify its operating characteristics permits an intelligent terminal to emulate the characteristics of other terminals, enabling it to access computers that support different types of terminals, if you obtain hardware and/or software required to modify the terminal's operating characteristics.

3.18 One parity check problem that can occur is a double bit error, which cannot be detected by that error-checking technique. Consider the seven bit character:

Bit position	1 2 3 4 5 6 7
Bit setting	1 0 1 1 0 1 0

With even parity, the parity bit would be zero, and the 8-bit character would be

1 0 1 1 0 1 0 0

If during transmission bits 1 and 3 are received in error, the result would be

0 0 0 1 0 1 0 0

which, from a parity-check view, would not indicate that an error had occurred.

3.19 Factors that should be considered during the terminal justification process include:
A. Time until information is required or it becomes obsolete.
B. Operational efficiency that can be gained through the use of terminals.
C. Management control through the use of terminals to enter data that becomes part of a database for a Management Information System.
D. Increased productivity and/or cost savings.

Answers to Questions and Problems in Chapter 4

4.1 A bit is the lowest level of information representation and signifies the presence or absence of a state or condition. A baud represents a unit of signaling speed equal to the number of discrete conditions or signal events per second.

When one bit is used as a signal unit, baud rate and bit/s are equivalent.

4.2 A computer word is fixed in size and can contain different size characters as long as the character size is less than the word size.

4.3 The three modes of transmission are simplex, half-duplex, and full-duplex. In the simplex mode, transmission occurs only in one direction. Half-duplex transmission can be in either direction, but only in one direction at a time. Transmission can occur in both directions at the same time in the full-duplex transmission mode.

4.4

	Asynchronous	Synchronous
Synchronization	Start/stop bits	Sync characters
Data rate	Under 2000 bit/s	≥ 2000 bit/s
Idle time between characters	Yes	No
Data formed into blocks	No	Yes

4.5 In serial transmission, the bits that compose the character to be transmitted are sent in serial sequence over one line. In parallel transmission, the characters are transmitted serially, but the bits that represent each character are transmitted in parallel—over several parallel lines. Parallel transmission takes one bit time, whereas serial transmission requires N bits, which comprise the character.

4.6 The three basic types of circuits are dedicated, switched, and leased. A dedicated line connects a device to a computer directly over a circuit owned by the user. A leased line is obtained from a common communications carrier. Dedicated lines normally connect local terminals to computers. Switched facilities normally connect terminals requiring low usage to a computer. Leased lines normally connect high-volume terminals to a computer.

4.7 The two basic types of line structure are point to point and multipoint. A comparison is shown on the next page:

	Point to point	Multipoint
Many terminals can share line	Yes, if collocated	Yes
Special poll and select software required	Yes	No
Data throughput	Number of terminals is key constraint	Terminal operating speeds is main constraint

4.8 Control characters control data transfer by the use of such controlling functions as

Acknowledgment	ACK
Negative acknowledgment	NAK
End of transmission	EOT
Enquiry	ENQ
Start of heading	SOH

4.9 A transmission code creates a correspondence between the bit encoding of data for transmission or for internal device representation and printed symbols.

4.10 Because the number of dots and dashes that form a Morse character vary, depending on character, this code is not practical for utilization in a computer communications environment that normally requires fixed-length characters for operation. In addition, Morse code uses the operator's ear to denote pauses between characters, but automatic data transmission requires a method of exact synchronization to indicate the start and ending of characters.

4.11 2^{10} or 1024 characters

4.12 0.69 millisecond for one bit, or 0.69×12 or 8.28 milliseconds for a 12-bit character.

4.13 A protocol is a method used to manage a data link. Protocol control tasks include

- Connection establishment
- Connection verification
- Connection disengagement
- Transmission sequence
- Data sequence
- Error control procedures

4.14

BSC	HDLC
Half-duplex	Full-duplex
Character oriented	Bit oriented
Pseudo-transparent	Naturally transparent
Fixed-length data blocks	Variable-length data message frames

4.15 In data transfer operations, a transparency mode of operation permits control characters to be transmitted without the expected operation (based on the control character) taking place. In BSC a two-control-character group, of which one is a data link escape (DLE) character, is used to obtain data transparency. Because HDLC is naturally transparent, no control characters are required to obtain a transparency mode of operation. But a 0 is inserted after five consecutive 1s to prevent a flag character from appearing in the information field. The receiver deletes it.

4.16 Analog service was developed for voice transmission, but digital service was developed for digital transmission. To transmit digital signals over an analog medium, data must be modulated into an analog or continuous signal, whereas data is transferred digitally end to end over a digital medium.

4.17 A leased line connects two or more points, or nodes, on the line to each other via a permanent routing. A WATS line permits users to use the switched network, at special tariffed costs, to call a number or receive calls from many telephones within a geographical area. An FX line is a hybrid combination of a leased line and the public switched telephone network.

4.18 An analog extension is a leased line that connects a location that does not have Dataphone digital service to a location that does. Analog extensions can be used to extend transmission to the digital network from nondigital network locations.

Answers to Questions and Problems in Chapter 5

5.1 Factors affecting shape of digital signals include the transmission distance, the transmission medium (type of wire gauge, low-capacitance shielded cable, etc.), and the pulse rise and fall time, which is proportional to the transmission speed.

5.2 The signal-to-noise ratio is the signal power divided by the noise power and can be used to characterize the quality of a circuit.

5.3 Power ratio $P = 10 \log_{10} \dfrac{\textit{Power transmitted}}{\textit{Power received}}$ dB

A. $P = 10 \log_{10} \dfrac{1}{1} = 10 \log_{10} 10 = 10 \text{ dB}$

B. $P = 10 \log_{10} \dfrac{0.2}{0.005} = \log_{10} 40 = 16 \text{ dB}$

C. $P = 10 \log_{10} \frac{0.25}{0.00025} = \log_{10} 200 = 23 \text{ dB}$

5.4 A line driver is used on a direct-connect circuit. It samples, amplifies, and regenerates data pulses and then retransmits such pulses, which now have their structures reformatted back into their original shapes.

5.5 The key difference between amplitude, frequency, and phase modulation is how the carrier signal is varied by the data signal. In amplitude modulation, the amplitude can be varied so that the carrier signal is either present at a certain amplitude or is absent as a result of zero amplitude. In frequency modulation, the frequency can be shifted according to the bit pattern. For phase modulation, the phase can be changed to represent a particular bit pattern.

5.6 Basically, the larger the channel's bandwidth, the greater its capacity for information transfer. The Nyquist relationship between bandwidth and the baud rate on a circuit is

$B = 2W$, where:
B = baud rate
W = bandwidth in Hertz

5.7 If one pulse represents one bit, then one baud equals one bit/s. If one pulse represents two bits, then one baud represents two bit/s.

5.8 Multilevel coding requires more complex logic for coding and decoding. In addition, it is more susceptible to distortion.

5.9 Nyquist's relationship between bandwidth (W) and baud rate (B) is given by:

$$B = 2W$$

Shannon calculated the theoretical maximum capacity of a channel of bandwidth W as:

$$C = W \log_2 \left(1 + \frac{S}{N}\right)$$

where S is the power of the transmitter and N is the power of thermal noise on the circuit. Shannon's formula defines the maximum bit rate (C) which, when divided by the baud rate (B), indicates the number of bits that must be represented by one signal element

$$N = \frac{C}{B}$$

5.10 A modem converts digital pulses into analog tones capable of being transferred on telephone circuits, and reconverts those tones back into digital pulses.

5.11 In asynchronous modems, the timing for synchronization is supplied by the transmitted character, whereas a timing signal is provided by an internal clock for synchronization of synchronous modems.

5.12 The four line-servicing groups and operating speeds of modems are

- Subvoice lines $\leq$ 300 bit/s
- Voice-grade lines $\leq$ 9.6 kbit/s

- Wideband lines ≥ 19.2 kbit/s
- Dedicated lines ≤ 1.5 Mbit/s

5.13 The Bell System 103/113 type modem uses a different set of frequencies to represent a mark and space from that used by the CCITT V.21 modem. Because the V.21 standard is followed in most foreign countries, the Bell System device could not be used at 300 bit/s to obtain transmission compatibility.

5.14 A V.22bis modem manufactured in the United States might follow the operation of a Bell System 212A device for fallback data rates of 1200 and 300 bit/s. A V.22bis modem manufactured in Europe may follow the operational characteristics of a CCITT V.22 device for fallback data rates of 1200 and 600 bit/s. Although a Bell System 212A and CCITT V.22 modem are compatible with one another at 1200 bit/s, they are not compatible with each other at their lower data rates. As a result, V.22bis modems might be incompatible with one another if placed into operation at their lower data rates.

5.15 Two key causes of signal distortion are signal attenuation (dB) and signal delay (millisecond or microsecond).

5.16 Because attenuation is greater at higher frequencies than lower frequencies, a reference frequency permits comparisons of attenuation between circuits.

5.17 Attenuation and delay equalizers and amplifiers can be used to condition leased lines. The attenuation equalizer adds a degree of signal loss to the lower frequencies of a modulated signal so that the loss throughout the transmitted frequency band is almost the same at all transmitted frequencies. To compensate for envelope delay, the delay equalizer introduces delay to some of the transmitted signals so that a uniform delay permits the entire signal to reach the receiver at the same time. Lastly, the amplifier is used to restore the transmitted signal back to its original shape.

5.18 Because leased lines have fixed transmission paths, they can be permanently conditioned. Switched circuits can have transmission routed via an infinite number of paths between the source and transmission receiver, and thus cannot be conditioned to satisfy all possible path characteristics. Modems use equalizers, usually when designed for data rates of 2400 bit/s and above. The equalizer is basically an inverse filter used to correct amplitude and delay distortions.

5.19 Through the utilization of equipment conforming to the RS-232-C standard, many types of equipment produced by a variety of equipment vendors can be interfaced to one another with minimal effort.

5.20 Modem handshaking is the exchange of control signals required to establish and maintain a connection between a modem and a business machine, and to disconnect the call at its conclusion.

5.21 Acoustic couplers permit terminal portability, and free a telephone line for other use when not coupled.

5.22 A serial unipolar signal uses a positive voltage to represent a mark and no voltage to represent a space. In bipolar signaling following a return-to-zero pattern, a binary zero is transmitted as zero volts and a binary one as either a positive or negative pulse—opposite in polarity to the previous binary one. Modified, or violated, bipolar signaling is a bipolar signal changed to incorporate network control information.

After an initial "handshake" between the calling and called modem, the fax transmission using modified half-duplex versions of the V.29 modem is in one direction, eliminating the effect of turnaround delays on throughput.

5.23

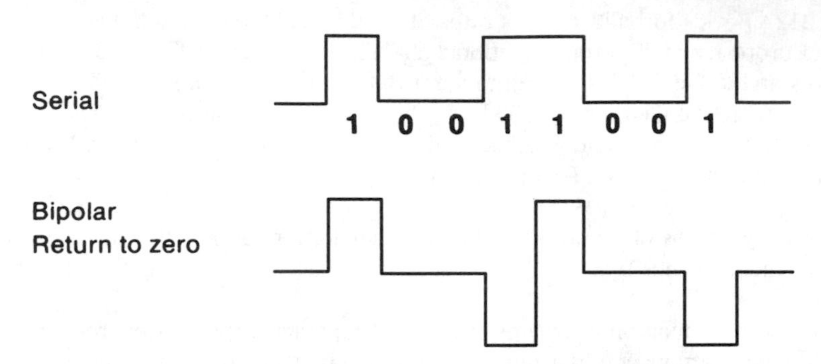

5.24 A channel service unit does not contain circuitry necessary to provide timing recovery and detect or generate DDS network control code, but a digital service unit does.

5.25 For an analog extension to AT&T's DDS, a pair of modems, an analog circuit, and a data auxiliary set are required.

5.26 The analog signal representing a voice conversation is sampled 8000 times per second, with the resulting samples referred to as pulse amplitude modulation (PAM). Each PAM sample is coded as a digital 8-bit PCM byte, which results in a data rate of 64,000 bit/s required to represent one digitized voice conversation. Because the T1 carrier operates at a data rate of 1.544 Mbit/s, a maximum of 24 voice channels can be placed on one T1 carrier circuit.

5.27 As a result of the framing of data on T1 circuits, a frame bit precedes each grouping of 192 bits, which represent digitized 8-bit samples from 24 voice channels. Because the sampling occurs 8000 times per second, a total of 8000 bit/s are required for framing. Adding 8000 bit/s to 1.536 Mbit/s results in the T1 transmission ratio of 1.544 Mbit/s.

Answers to Questions and Problems in Chapter 6

6.1 A tariff is a schedule of charges, practices, services, and other pertinent data concerning communications services offered to the public. Tariffs are important because they regulate the type and scope of service, as well as its cost, permitting the network analyst to plan appropriate action by examining current offerings. Another important tariff feature is that a tariff for interstate service or facility must be approved by the Federal Communications Commission, which normally does not permit subsidized rates on tariffs. This preclusion of subsidized rates prevents large carriers from excluding smaller companies from entering the communications market.

6.2 In the United States, communications common carrier jurisdiction lies with the Federal Communications Commission when communications is interstate, or international with origination in the United States. When communications is intrastate, common carriers might be regulated by state and/or municipal agencies.

6.3 The International Telecommunications Union (ITU) fosters cooperation and compatibility in the areas of radio engineering, data communications, and telephone systems. Several study groups conduct technical and administrative conferences on a periodic basis to develop international operating practices, pricing, policy arrangements, and, most importantly, technical specifications that permit compatibility between telecommunications networks of different nations.

6.4 The Federal Communications Commission's equipment registration program has fostered the proliferation of independent devices that can be connected to the telephone network. By permitting a direct connection via a plug/jack for registered equipment, connection of nontelephone equipment to the telephone network has been simplified.

6.5 Registered equipment can be directly connected to the public switched telephone network via a plug/jack connection, and nonregistered equipment must be connected via a data access arrangement.

Answers to Questions and Problems in Chapter 7

7.1 AT&T divested itself of its 22 Bell operating companies (BOCs). AT&T retained Bell Labs and its parent, Western Electric, which was renamed AT&T Technologies. Long Lines was retained and renamed AT&T Communications. AT&T formed its Information Systems division (ATTIS) to market CPE. The divested BOCs were grouped into seven regional companies (RBOCs): Ameritech, Bell Atlantic, BellSouth, NYNEX, PACTEL, Southwestern Bell, and US West.

7.2 Bell Telephone Laboratories is the research and development arm of AT&T.

7.3 As a result of the 1956 consent decree between the Bell System and the U.S. Department of Justice, the Bell System agreed not to engage in any business activities other than providing communications common carrier offerings. In addition, Western Electric agreed not to manufacture any equipment that was not sold or leased to Bell System operating companies for use in providing common carrier communications services. The establishment of American Bell, and the divestiture by AT&T of operating companies, has largely negated the consent decree.

7.4 Interconnection was the key problem that initially limited the growth of MCI and other specialized carriers. Until a series of legal-ruling-enforced interconnections, telephone companies were reluctant to connect the microwave networks of specialized carriers with their telephone network. This interconnection would cause a loss of revenue for the telephone company, and, in effect, would enhance the competition.

7.5 In the context of a value-added carrier, the term *value-added* means that the carrier uses circuits provided by another carrier and, through special software and hardware devices, improves and adds new capability to the circuit. In most cases, the term *value-added carrier* is synonymous with packet switching.

7.6 In many foreign countries, telecommunications facilities are provided by a department of the government, whereas in the United States such services are provided by independent companies. Another key difference between foreign countries and the United States concerns the area of communications equipment. In the United States, equipment from many commercial vendors can be considered by a communications subscriber, whereas in many foreign countries only one or at best a few devices manufactured by different vendors are approved for consideration. This situation varies from country to country, and is a result of the few suppliers the national transmission company does business with. In 1992, the European community intended to deregulate its telecommunications. Lowered barriers will facilitate acceptance of foreign devices.

Answers to Questions and Problems in Chapter 8

8.1

	Switched	FX	WATS
Traffic volume	Low	Medium-heavy	Medium-heavy
Geographical area of terminals	Distributed	Concentrated	Distributed or concentrated

8.2 The cost per call (applicable factors: distance between stations, time of day, duration of transmission, LEC and IEC charges); number of calls per day; working days per month.

8.3 IEC mileage cost; LEC access charges; conditioning charge.

8.4 The rate step for the applicable band and the cost of that band at the rate step. The monthly cost increments are for the first 25 hours of usage, the next 75 hours, and for each hour over 100. Additional cost items are access line charges and installation.

8.5 When transmission is via Dataphone digital service, the data travels end to end in its original digital format, and the cost and error problems associated with modulation, as required on analog facilities are avoided.

8.6 Mileage between central offices, a fixed mileage-band charge based on the selected service (such as data rate) and the total distance, a central-office connection charge, and channel options and office "functions."

8.7 Number of terminals, calls per day per terminal, average call duration, transmitted characters per call, working days per month, and the average distance between communicating parties.

Answers to Questions and Problems in Chapter 9

9.1 EIA-232-C interface circuits can be arranged into groupings based on ground, data, control, timing, and secondary circuits. Under EIA-232-D, additional interface circuits can be grouped into a testing category, in addition to the previously mentioned EIA-232-C groups.

9.2 Protective ground is used to protect the user from being shocked if he touches a device that has a voltage leak. Signal ground is used to establish a common ground reference potential for all interchange circuits except protective ground.

9.3 The request-to-send signal controls the direction of transmission of the local DCE when half-duplex transmission is used. When the circuit is active, it inhibits receiving of data for half-duplex transmission, whereas it maintains the DCE in the transmit mode for full-duplex transmission.

9.4

Signal	From DCE	To DCE
A. Request to send		X
B. Clear to send	X	
C. Data set ready	X	
D. Received line signal detector	X	
E. Transmit data		X
F. Receive data	X	
G. Data terminal ready		X
H. Ring indicator	X	

9.5 The transmission direction of a reverse channel is opposite that of a primary channel, whereas the transmission direction of a secondary channel is independent of the primary channel.

9.6 Because the bandwidth of a secondary channel is much narrower than that of a primary channel, the transmission speed on the secondary channel is at a much slower rate than that possible on a primary channel.

9.7 The key differences between EIA-232-C and EIA-232-D standards are in the areas of specified connector, test circuits and revised circuit names. EIA-232-D formally specifies the use of 25-pin D connectors which were referenced in an appendix to EIA-232-C and never part of that standard. In addition, EIA-232-D includes three test circuits (test mode, local loopback, and remote loopback) that are not included in EIA-232-**C.** Finally, EIA-232-C renames *data set ready* as *DCE ready* and *data terminal ready* as *DTE ready*.

9.8 The EIA-449 standard provides two connectors, a 37-pin one for primary circuits and a 9-pin one exclusively for secondary channels. In comparison, the EIA-232-C standard requires a 25-pin connector.

9.9 The mode option permits modems to be used as either originate- or answer-type modems. Thus, a modem connected to a terminal would have an originate mode selection, and a modem connected to a dial-in port at a computer site would be placed in the answer mode of operation.

9.10 If automatic answering, NO option is selected for modems at a computer site, and terminal-to-computer transmission must be conducted manually.

9.11 $\frac{200 \times 8}{4800}$ = 333 milliseconds to transmit block

$\left(\frac{80}{80 + 333}\right) \times 100 = 19.37\%$ overhead

9.12 $\left(\frac{10}{10 + 333}\right) \times 100 = 2.9\%$ overhead

9.13 $\frac{300 \times 8}{4800}$ = 500 milliseconds to transmit block

$\left(\frac{80}{80 + 500}\right) \times 100 = 13.8\%$ overhead

9.14 $\frac{200 \times 8}{9600}$ = 167 milliseconds to transmit block

$\left(\frac{80}{80 + 167}\right) \times 100 = 32.4\%$ overhead

Answers to Questions and Problems in Chapter 10

10.1 Three initial LAN design objectives:

1. The LAN's connections should allow data to move directly from one node to another without intervention by immediate host processors.
2. Data should enter the LAN in a standard format.
3. The LAN must be able to allocate transmission capacity on a demand—rather than a partitioned—basis.

10.2 Distinguishing features of the early Cambridge Ring:

1. It consisted of a set of repeaters connected in a loop by a pair of twisted-pair wires.
2. The loop continuously circulated a pair of packets at 10 Mbit/s.
3. Each packet contained two information bytes plus routing and control.

10.3 Ethernet is an example of the bus topology. Some of Ethernet's distinguishing features:

1. All nodes are passive: If one node fails, all other activity remains uninterrupted.
2. Ethernet consists of a single coaxial cable multidropped at its processors.
3. Access contention is by CSMA/CD: carrier sense multiple access with collision detection.

10.4 A third—and possible oldest—LAN topology is the star.

10.5 In Ethernet's CSMA/CD, each station "listens" for the presence or absence of a carrier signal prior to transmitting data. If no carrier signal is sensed, the station is free to transmit.

In token passing access, one or more bits in a field within the "token" frame is used to control a station's decision to transmit data. If the token bit(s) indicate a free (available) token, a station having data to transmit can acquire the token.

10.6 It is difficult to predict a CSMA/CD-based network's performance because of the random time delays that stations incur prior to retransmission attempts. As network utilization increases, a CSMA/CD-based network's performance will worsen more rapidly than that of a token passing network because of the growing collision occurrences, delays introduced by jam signal transmissions, and the aforementioned random time delays that stations incur prior to retransmission attempts.

10.7 The three media alternatives for network wiring of a building are:

1. Twisted pairs (either shielded or unshielded);
2. Coaxial and twinaxial cable;
3. Optical fiber.

The most widely used today is coaxial.

10.8 Optical fiber features immunity to environmental influences and inherent security.

10.9 In a LAN environment, optical fiber is used primarily to extend cabling distances between LAN segments. The electro-optical repeater is the device used to facilitate fiber's use.

10.10 The FDDI network was originally designed to use optical fiber cable as its transmission medium. FDDI's access method is token passing.

10.11 Two wiring-management methods are point-to-point cabling and wiring closets. The wiring closets method lends itself more readily to orderly growth.

10.12 The IBM Cabling System features the star topology.

10.13 RF modems.

10.14 In both Ethernet and IEEE 802.3 networks, the U/L subfield is set to a binary zero to indicate universally administrated addressing; a binary one indicates locally administrated addressing. Under universally administrated addressing, a unique address that is embedded in ROM on each NIC (Network Interface Card) is used for addressing. Under locally administrated addressing, the user is responsible for configuring the source address of each workstation.

10.15 Distinguishing features of the token-ring LAN:

1. The access method is deterministic.
2. There is only one token on the ring at a time.
3. Each ring station can be in either an active or a bypass mode.

10.16 Network control on the token ring is normally distributed among all the nodes. For network malfunctions and recovery, one node is designated the active network monitor.

Answers to Questions and Problems in Chapter 11

11.1 A feasibility study is conducted to determine if there is enough justification to proceed with a project and/or select one alternative from many possible solutions based on performance and economic parameters. A communications plan examines such parameters as the types of transactions to be processed and their urgency, volume, expected growth rates, and other parameters in order to determine a method to implement or revise existing communications. In general, a feasibility study examines data at the macro level in order to obtain a go/no go decision, but a communications plan examines data at a micro level in order to develop a method to implement the feasibility study decision.

11.2 Some of the functions that an intelligent terminal can perform, which normally dumb terminals cannot, include:

- **A.** Formatting input and output—better known as forms mode
- **B.** Data compression/decompression
- **C.** Content-oriented error detection
- **D.** Local editing of data
- **E.** Handling simple computations

11.3 Until the 1980s, a network analyst's efforts were primarily focused on designing an infrastructure for the movement of data from conventional terminal devices to mainframe computers. With the growing acceptance of LANs, the network analyst's job has involved satisfying the requirements of private organizations, government agencies, and academia.

11.4 LANS, WANS, and internetworking.

11.5 If the intelligent terminal preprocesses data input by checking alphabetic fields for numerics and so on, it can reduce the possibility of incorrectly entered data and thereby not only remove this function from the host computer, but also reduce the number and frequency of error messages being transmitted from the host computer to the terminal.

11.6 In general, concentrators combine many low-speed lines into one or more high-speed lines. The functions performed by concentrators range from speed and code conversion to automatic baud recognition, code compression, traffic smoothing, and error control.

11.7 Any of the communications functions listed in the response to 10.4 can be incorporated into front-end processors. In addition, such functions as user password verification can also be incorporated to further reduce the processing burden of the host computer.

11.8 In general, concentrators can physically support more terminal devices than can multiplexers, and are user programmable. Multiplexers are usually programmed by the vendor at the factory.

11.9 Both concentrators and multiplexers combine data from many low- to medium-speed devices into one or more high-speed composite output data channels.

11.10 The restrictions unique to a LAN are caused by its cabling restraints, including the number of allowable stations, a station's connecting-cable length, and the distances between cable connections.

11.11 The repeater functions as a data regenerator. Because the repeater takes a finite time to sample a pulse rise and to regenerate the received pulse, the device introduces a slight pulse delay—known as *jitter*. As jitter accumulates, it adversely affects the ability of network stations to receive data.

11.12 The NIC is responsible for the transmission and reception of data to and from the LAN. The type of NIC used is based on its LAN and computer workstation interfaces and the LAN transport protocol support needed.

11.13 A file server executes the NOS (network operating system) and functions as a repository for application programs and data.

11.14 The primary rationale for using one or more individual print servers, instead of operating print server software on a file server, is based on performance and security. Performance: the buffering and print-control operations can adversely affect the file server's ability to handle network-related requests. Security: most organizations prefer to place their file servers at locations that are not readily accessible because of critical information that can be stored on that device. Because it is desirable to have printing results readily accessible to network users, the operation of separate print servers facilitates accessibility without compromising network file server security.

11.15 A bridge operates at the data link layer. Operating at the network layer permits a router to use network addresses to make routing decisions, which a bridge cannot do at the data link layer. The result is that a router can direct data over different paths (enabling load balancing), make decisions about the use of different paths, and perform other routing decisions for the transport of data from one LAN to another.

11.16 A gateway is a communications device that provides a translation service for the data on one network and the format required by a mainframe or minicomputer application on another network. A typical use of a gateway is to provide a connection between token-ring-network workstations and an IBM host computer. Here, the gateway translates token ring frames into SNA SDLC-transported information for processing by the IBM mainframe.

Answers to Questions and Problems in Chapter 12

12.1 Terminal emulation causes the personal computer to operate as if it were a specific type of terminal device. In comparison, data transfer lets the user exchange information.

12.2 Some of the more common user interfaces by which persons interact with programs are command-line entries, interactive prompts and commands, and menus.

12.3 Two common flow-control methods are the use of X-on/X-off for full-duplex transmission and a special handshake character, such as Control-Q, for half-duplex transmission.

12.4 An external modem can be used with any personal computer that has a serial port. In comparison, an internal modem is designed to be installed in the "system" unit of a particular type of personal computer, and its use might not be compatible with other personal computers. Because an internal modem uses the power of the PC, it does not require a separate power supply. In addition to eliminating a power supply, the internal modem obviously does not require a separate housing, resulting in its cost normally being below that of an external modem.

12.5 One example of the use of a script language to automate a previously performed manual operation is the access of electronic mail. By appropriate programming, your communications program could be instructed to dial your electronic mail network at predefined times during the day, check the contents of your in-box, and, if not empty, retrieve and print your messages on your PC's printer or save the messages onto disk.

12.6 The Kermit protocol permits programs to negotiate compression of repeated bytes, which reduces the actual amount of data transmitted. In comparison, XModem lacks a compression option.

12.7 An increasing number of negative acknowledgements signifies a deteriorating line condition. In this situation, reducing packet length reduces the time required to retransmit a packet containing one or more bit errors. Conversely, as the number of negative acknowledgements decreases, packet length might be increased. This dynamic altering of packet length boosts throughput with improving line conditions, while minimizing packet retransmission delays associated with deteriorating line conditions. ZModem will dynamically alter the length of packets based on the ratio of positive to negative acknowledgements.

12.8 The following questions should be answered prior to selecting communications software to transfer files:

- Can the program transfer binary files?
- Does the program have error-detection-and-correction capability?
- How does the program implement error detection and correction (e.g., what protocol or protocols does it support)?
- Can a group of files be sent in a single operation?
- Can a file transfer be cleanly interrupted and resumed from the point of interruption?

Answers to Questions and Problems in Chapter 13

13.1 Five data sources that could be candidates for data-stream multiplexing are

1. Data streams from modems connected to leased lines
2. Direct-connect terminals
3. Data from digital service units
4. Data from modems connected to the switched network
5. Data streams from other multiplexers

13.2 The two basic techniques used for multiplexing are frequency division and time division. In frequency-division multiplexing (FDM), the bandwidth is the frame of reference, and it is subdivided into segments with each frequency segment assigned to an output channel. In time-division multiplexing (TDM), time is the frame of reference—input data is positioned into an output channel by time. At any point in time, the output channel holds data for only one input line in TDM, whereas the output channels can hold data for numerous lines when the FDM technique is used.

13.3 When multiplexers use a bit interleaving technique, minimal buffer storage is required because each incoming data bit is combined into the high-speed output data stream as the multiplexer receives the bit. In character interleaving, the multiplexer must assemble bits into a character before multiplexing, hence circuitry and buffering is more complex.

13.4 Over head $= \dfrac{SYNC}{SYNC + DATA} = \dfrac{SYNC}{TOTAL}$

A. $\dfrac{2}{2+2} = 50\%$ **C.** $\dfrac{2}{2+6} = \dfrac{2}{8} = 25\%$

B. $\dfrac{2}{2+4} = 33.3\%$ **D.** $\dfrac{2}{2+8} = \dfrac{2}{10} = 20\%$

13.5 When a multiplexer is used as a front-end substitute, special host software must be developed to make the computer compatible with the multiplexer. Normally, a large quantity of such devices must be used so the software development costs can be spread over many such units.

13.6
$$\begin{array}{r} 4800 \\ \times\ {+4} \\ \hline 19{,}200 \text{ bit/s} \end{array} \quad \text{reverse compression}$$

$$\frac{19{,}200}{1200} = 16$$

13.7 The Adaptive Differential PCM module results in a voice digitization rate one-half of the PCM rate. This permits the T1 multiplexer to support twice as many voice channels.

13.8 Two features of optical cable that encourage the multiplexing of data onto that cable are its bandwidth and electrical-hazard immunity. Its high bandwidth permits extremely high data rates to be transmitted in the form of light energy. Because light and not electricity flows on the cable, most building codes permit its installation without the use of a conduit, resulting in a considerable economic savings over the use of conventional copper wire-based cable.

13.9

	Multiport modem	Multiplexers
Data rates supported	Multiple of 2.4 kbit/s	Unlimited
Transmission mode	Synchronous only	Asyn/sync
Number of channels	Usually fixed	Not fixed

13.10 Both concentrators and remote network processors perform data concentration; however, only a remote network processor can concurrently perform local batch and remote batch processing.

13.11 A modem-sharing unit permits many terminals to share the common modem in a poll-and-select environment. A multiplexer permits many types of digital data sources, including directly connected terminals and the output data streams of modems, to be combined and to share a common transmission medium.

13.12 A line-sharing unit is normally used within a computer facility, whereas a modem-sharing unit is normally used at remote locations where terminals are clustered.

Answers to Questions and Problems in Chapter 14

14.1 As explained in Chapter 14, a good starting point in network design is to construct a table similar to Table 14.4.3, using in the example 22 working days/month. You can refer to Chapter 8 for relevant cost figures, specifically Tables 8.2.1 and 8.2.4 for switched- and leased-line costs. Completing the table allows you to determine on a per-terminal basis the most economic method of connecting each terminal to the computer. For this particular problem, at all four locations, the switched network was determined to be the most economical method to use on an individual basis. The completed typical table is shown:

LOCATION	DAILY CONNECT TIME (MINUTES)	MONTHLY CONNECT TIME		MILEAGE TO COMPUTER CENTER	COST/CALL DAY RATE		MONTHLY SWITCHED NETWORK COST/TERMINAL	RATE CENTER CATEGORY	MONTHLY LEASED LINE COST	MONTHLY STATION TERMINAL COST	TOTAL MONTHLY LEASED LINE COST	MOST ECONOMICAL	
		NO. OF FIRST MINUTE	NO. OF ADDITIONAL MINUTES		INITIAL MINUTE	EACH EXTRA MINUTE						TYPE	MONTHLY COST
1	5	22	88	750	.62	.43	51.48	A-A	863.92	72.1	936.02	SWITCHED	51.48
2	20	22	418	400	.59	.42	188.54	B-A	628.83	72.1	700.93	SWITCHED	188.54
3	20	22	418	375	.59	.42	188.54	A-A	511.42	72.1	583.52	SWITCHED	188.54
4	50	22	1,078	1,040	.64	.44	488.40	B-A	1,322.58	72.1	1,394.68	SWITCHED	488.40

Next, you can examine the network layout, denote the number of terminals at each location, and try to ascertain if any type of data concentration can be used to reduce costs. The network layout with the number of terminals indicated at each site and the type and cost of the most economical transmission method to the computer center is shown.

Next, you can begin to examine the utilization of the data concentration equipment listed as available for consideration in solving this problem—4- and 8-channel multiplexers. Because there are only single terminals at locations 1 and 3 and the cost of transmission from those locations to the computer is less than the leased-line cost of linking either of those locations to any other location (see the interconnection distances in the problem), you can exclude those two locations not only from being multiplexer locations, but also from being connected to multiplexers that are located at other points in the network.

Based on the preceding analysis, only locations 2 and 4 can be considered for the installation of multiplexers. Because location 4 is further from the computer site, consider that location first as a candidate for multiplexing.

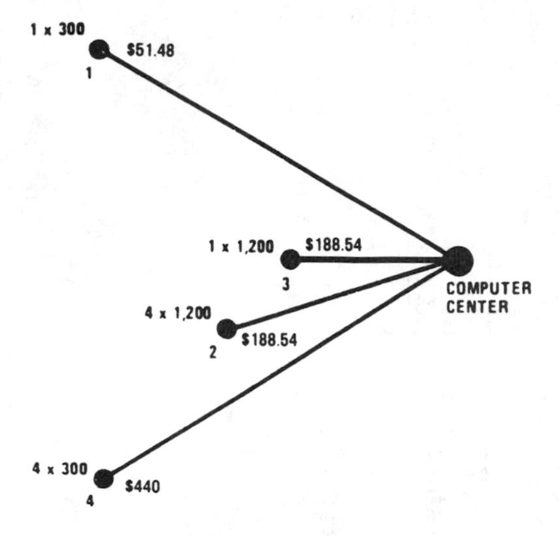

The cost of a 4-channel multiplexer is $100 and a 4800-bit/s modem rents for $135 per month, resulting in a total cost of $235 per month at location 4 (if you assume that the terminals at the location can be directly cabled to the multiplexer). Because similar equipment is required at the computer site, the modem and multiplexer cost is $470 per month. Adding the cost of a leased line, which from the table was computed as $1394.68 per month, the total cost of multiplexing data from the four terminals at location 4 becomes $1864.68, which is less than the switched network cost of 4 × $488.40 = $1953.60 per month. In addition, the multiplexing cost is fixed, whereas the switched cost is variable and will increase if traffic volume and connection time should rise.

Next, similarly examine location 2. Here, the cost of a 4-channel multiplexer and 4800-bit/s modem would be $235 per month. Again, because similar equipment is required at the computer site, the cost for both sites would be $470 per month. Adding the cost of the leased line from location 2 to the computer site (from the table it was $700.93 per month), the total multiplexing cost becomes $1170.93 per month. Because the previously computed most-economical cost pointed to the switched network at $188.54 per month per terminal, for the four terminals such service would cost $754.16—which is less than the cost of multiplexing data.

Notice that an 8-channel statistical multiplexer was not considered since it rents for $75 per month more than a 4-channel conventional multiplexer and would reduce the modem cost by only $70 per month ($135 for a 4800-bit/s modem vs. $65 for a 2400-bit/s modem).

14.2 Sizing is the process of ensuring that the configuration of a selected device will provide the desired level of service.

14.3 One erlang (a dimensionless unit) of traffic intensity on one traffic circuit represents a continuous occupancy of that circuit. One erlang equals 3600 call-seconds.

Answers to Questions and Problems in Chapter 15

15.10 Solution: Using 4-channel TDM

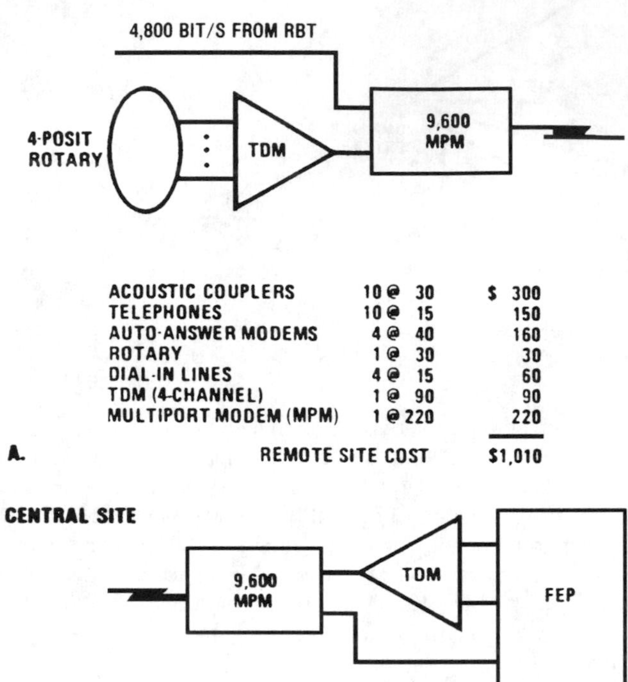

	ACOUSTIC COUPLERS	10 @ 30	$ 300
	TELEPHONES	10 @ 15	150
	AUTO-ANSWER MODEMS	4 @ 40	160
	ROTARY	1 @ 30	30
	DIAL-IN LINES	4 @ 15	60
	TDM (4-CHANNEL)	1 @ 90	90
	MULTIPORT MODEM (MPM)	1 @ 220	220
A.	REMOTE SITE COST		$1,010

	MULTIPORT MODEM	1 @ 220	$ 220
	TDM (4-CHANNEL)	1 @ 90	90
	COMPUTER PORTS	5 @ 35	175
B.	CENTRAL SITE COST		$ 485
C.	LEASED-LINE COST		1,873
D.	PRODUCTIVITY COST		400
	TOTAL COST (A + B + C + D)		$3,768

Solution: Using 8-channel TDM. No additional interactive terminals can be serviced because any increase in terminals would exceed the capacity of the high-speed line to Greensboro.

Solution: Using a statistical multiplexer

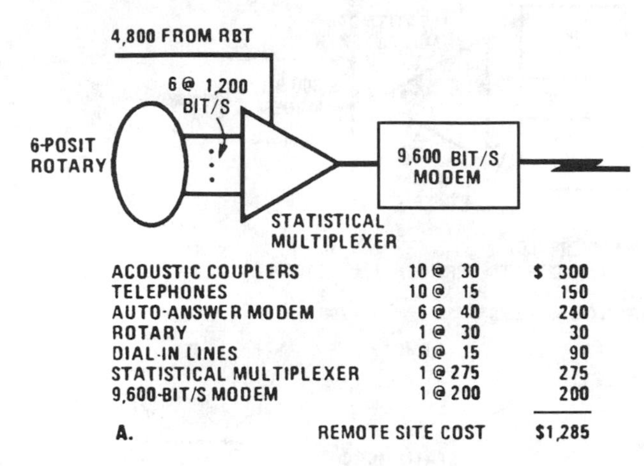

ACOUSTIC COUPLERS	10 @ 30	$ 300
TELEPHONES	10 @ 15	150
AUTO-ANSWER MODEM	6 @ 40	240
ROTARY	1 @ 30	30
DIAL-IN LINES	6 @ 15	90
STATISTICAL MULTIPLEXER	1 @ 275	275
9,600-BIT/S MODEM	1 @ 200	200
A.	REMOTE SITE COST	$1,285

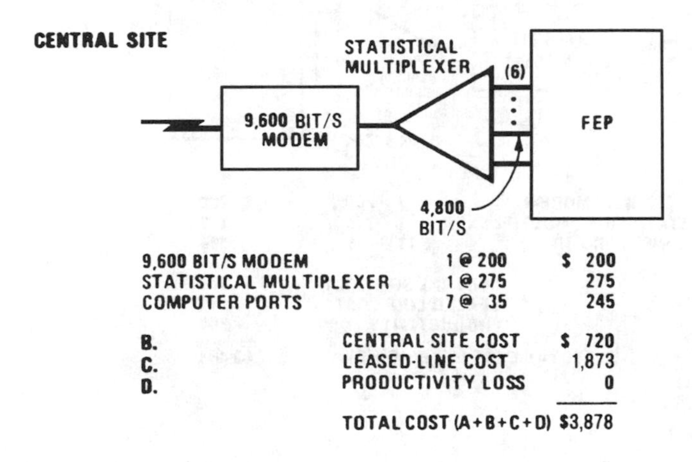

9,600 BIT/S MODEM	1 @ 200	$ 200
STATISTICAL MULTIPLEXER	1 @ 275	275
COMPUTER PORTS	7 @ 35	245
B.	CENTRAL SITE COST	$ 720
C.	LEASED-LINE COST	1,873
D.	PRODUCTIVITY LOSS	0
	TOTAL COST (A+B+C+D)	$3,878

Solution: Using line drivers and a statistical multiplexer

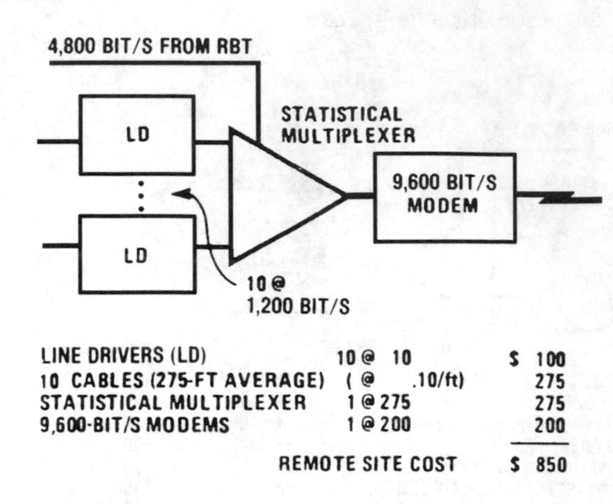

	LINE DRIVERS (LD)	10 @ 10	$ 100
	10 CABLES (275-FT AVERAGE)	(@ .10/ft)	275
	STATISTICAL MULTIPLEXER	1 @ 275	275
	9,600-BIT/S MODEMS	1 @ 200	200
A.		REMOTE SITE COST	$ 850

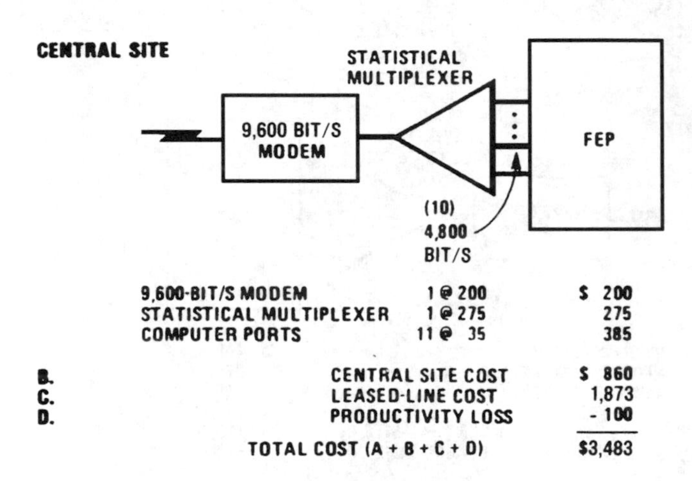

	9,600-BIT/S MODEM	1 @ 200	$ 200
	STATISTICAL MULTIPLEXER	1 @ 275	275
	COMPUTER PORTS	11 @ 35	385
B.		CENTRAL SITE COST	$ 860
C.		LEASED-LINE COST	1,873
D.		PRODUCTIVITY LOSS	- 100
		TOTAL COST (A + B + C + D)	$3,483

Solution: Summary

COST	4-CHNL TDM SERVING 4 TTY	STAT MUX SERVING 6 TTY	LINE DRIVERS W/STAT MUX SERVING 10 TTY
REMOTE SITE	$1,010	$1,285	$ 850
CENTRAL SITE	485	720	860
LEASED LINE	1,873	1,873	1,873
PRODUCTIVITY	400	0	−100
TOTAL	3,768	3,878	3,483

15.11 Solution: Multiplexing new traffic with 2nd shift RBT

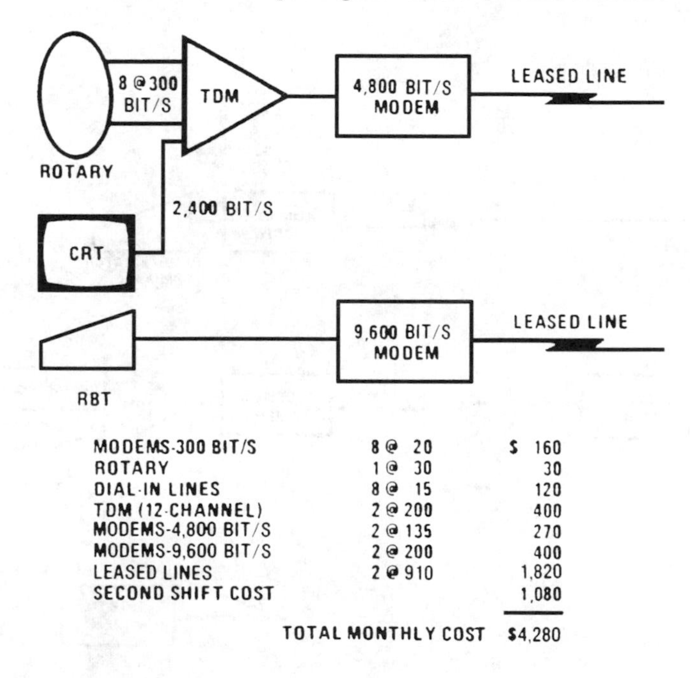

MODEMS-300 BIT/S	8 @ 20	$ 160
ROTARY	1 @ 30	30
DIAL-IN LINES	8 @ 15	120
TDM (12-CHANNEL)	2 @ 200	400
MODEMS-4,800 BIT/S	2 @ 135	270
MODEMS-9,600 BIT/S	2 @ 200	400
LEASED LINES	2 @ 910	1,820
SECOND SHIFT COST		1,080
	TOTAL MONTHLY COST	$4,280

Solution: Pseudo Wideband utilization

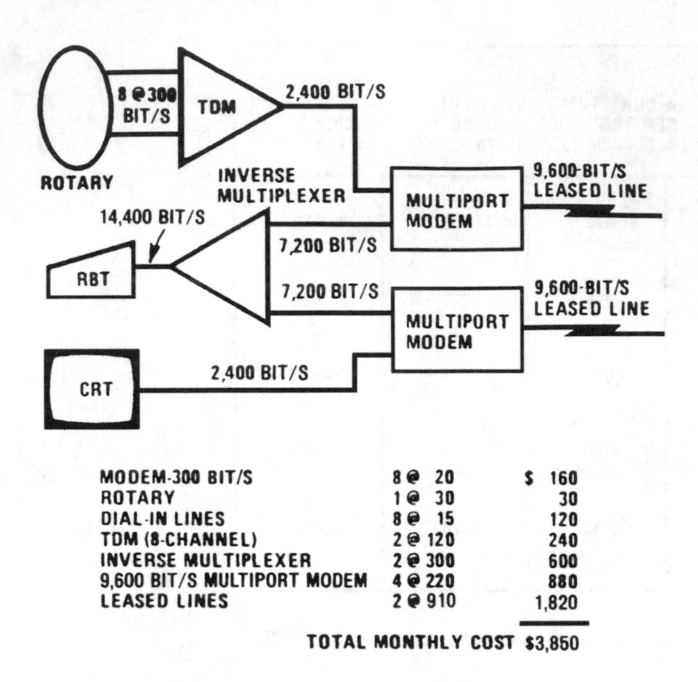

MODEM-300 BIT/S	8 @ 20	$ 160
ROTARY	1 @ 30	30
DIAL-IN LINES	8 @ 15	120
TDM (8-CHANNEL)	2 @ 120	240
INVERSE MULTIPLEXER	2 @ 300	600
9,600 BIT/S MULTIPORT MODEM	4 @ 220	880
LEASED LINES	2 @ 910	1,820
	TOTAL MONTHLY COST	$3,850

15.12 Solution: Using TDMs

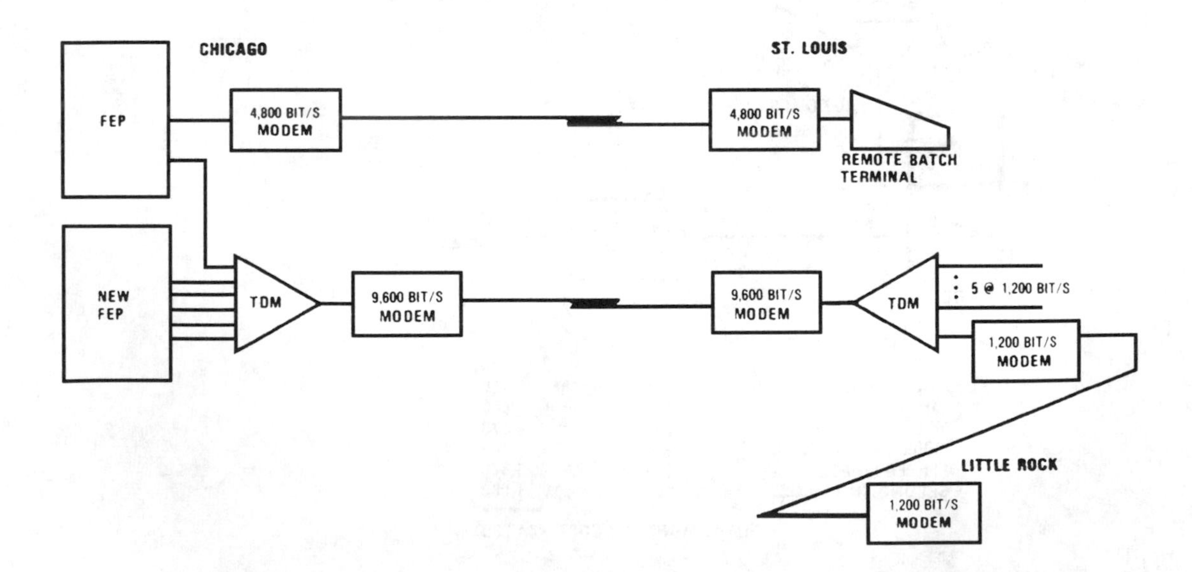

Monthly cost using TDMs

LINE COST		
ST LOUIS – CHICAGO	2 @ 404	$808
LITTLE ROCK – ST. LOUIS	1 @ 432	250
MODEM COST		
1.200 BIT/S MODEMS	2 @ 40	80
4,800 BIT/S MODEMS	2 @ 120	240
9,600 BIT/S MODEMS	2 @ 200	400
FEP COST		
CHANNEL (PORT) COST	7 @ 35	245
ADDITIONAL FEP		1,500
TERMINAL COST		
1,200 BIT/S	6 @ 100	600
REMOTE BATCH TERMINAL	1 @ 500	500
MULTIPLEXERS		
8-CHANNEL TDM	2 @ 160	320
	TOTAL MONTHLY COST	$5,125

Clustering the terminals

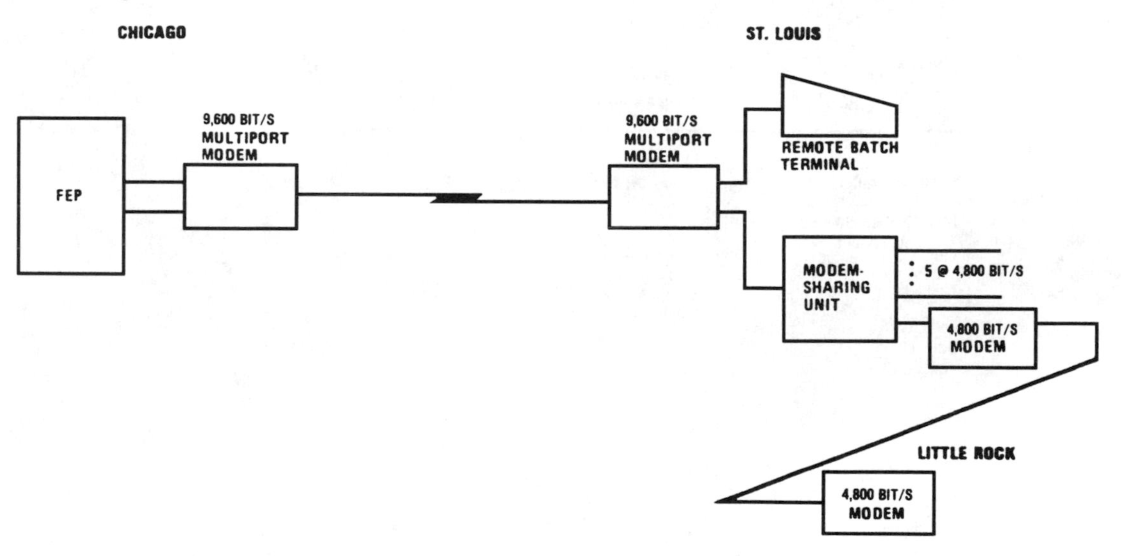

Monthly cost of clustering

LINE COST		
ST. LOUIS – CHICAGO	1 @ 404	$404
ST. LOUIS – LITTLE ROCK	1 @ 432	432
MODEM COST		
4,800 BIT/S MODEMS	2 @ 120	240
9,600 BIT/S MULTIPORT MODEMS	2 @ 220	440
FEP COST		
CHANNEL (PORT) COST	2 @ 35	70
MODEM-SHARING UNIT	1 @ 35	35
TERMINAL COST		
4,800 BIT/S	6 @ 125	750
REMOTE BATCH TERMINAL	1 @ 500	500
	TOTAL MONTHLY COST	$2,871

Cost comparison

	TDM APPROACH	CLUSTER APPROACH
MONTHLY COST	$5,125	$2,871
IF FEP CHANNELS AVAILABLE	–1,500	
	$3,625	$2,871

15.13 Solution: Switching at the remote site

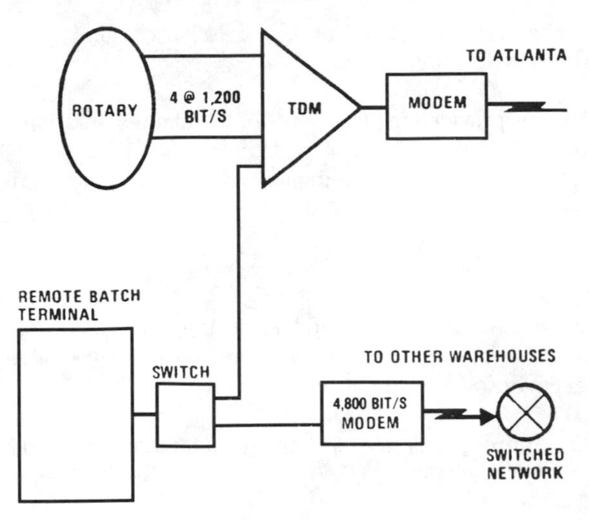

COST

DIAL-UP

5 WAREHOUSES X 12 CALLS X $6.00	$ 360.00
5 MODEMS @ 120	600.00
5 FALLBACK SWITCHES @ $10.00	50.00
MONTHLY COST	$1,010.00

Solution: Switching at the central site

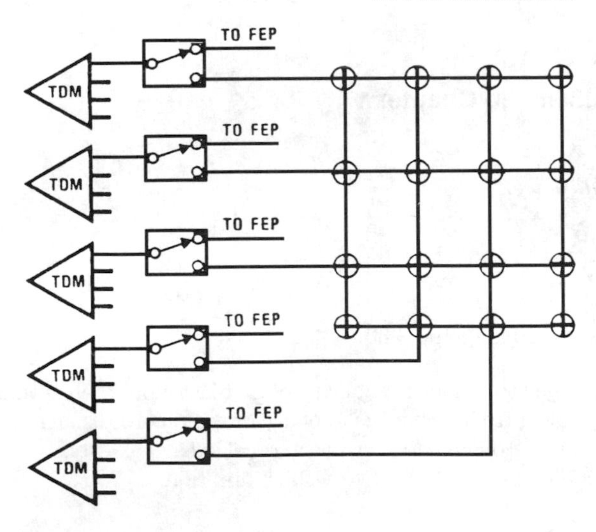

COST

DIAL-UP COORDINATION

5 WAREHOUSES X 12 CALLS X 50 CENTS	$ 30
5 FALLBACK SWITCHES @ $10	50
1 MATRIX SWITCH @ $200	200
MONTHLY COST	$280

Answers to Questions and Problems in Chapter 16

16.1 Videoconferencing is the real-time, two-way transmission of digitized video images, with sound, between two or more locations.

16.2 The reserved offering was provided in increments of 30 minutes duration.

16.3 The costs of Accunet 1.5 Reserved service depend on mileage, time of day, and usage time.

16.4 56 kbps and above.

16.5 H.320 is the dominant videoconferencing standard. It describes videoconferencing terminals. However, H.320 has also come to represent a suite of specifications for enabling compatible videoconferencing sessions.

16.6 An MCU or multipoint control unit is video switching equipment. It is used in multipoint videoconferencing.

16.7 Whiteboarding, also known as *document conferencing*, enables people to work jointly on an electronic document.

16.8 30 frames per second.

16.9 An interim alternative to the digital MCU is the analog bridge. Here, a digital image and its accompanying audio are converted to analog via a codec, then sent over a bridge to another codec, which digitizes the signals for its receiving device.

Answers to Questions and Problems in Chapter 17

17.1 $\text{dB} = 10 \log_{10} \dfrac{\textit{output power}}{\textit{input power}}$

A. $\text{dB} = 10 \log_{10} \dfrac{10}{100} = 10 \log_{10} .1 = -10$

B. $\text{dB} = 10 \log_{10} \dfrac{100}{1} = 10 \log_{10} 100 = 20$

17.2 *Dropouts*. Sudden, large reductions in signal level that last more than several ms.
Phase hits. Sudden, uncontrolled changes in phase of the received signal.
Gain hits. Sudden, uncontrolled increases in the received signal level.
Impulse noise. Sudden noise “spikes” of very short duration.

17.3 Equalization can be used to compensate for changes in amplitude and phase response of modulated signals. This equalization can be done by the telephone company when it conditions a circuit, as well as by circuitry in high-speed modems.

17.4 In digital monitoring, be concerned with the control signals at the RS-232-C interface and the trapping and display of data and control characters between the data circuit-terminating equipment and the data terminal equipment. In analog monitoring, be concerned with the interface between the DCE and the line, examining the voltage, current, and phase levels of the circuit, as well as other unique line parameters.

17.5 Protocol analyzer/digital test set features one should consider include:

- **A.** *Code level*. Does the device support 5, 6, 7, or 8 bits per character?
- **B.** *Stop-length element*. Can the device support 1, 1.42, 1.5, or 2-bit stop intervals?
- **C.** *HDLC/SDLC*. Can the device support new data link control protocols?
- **D.** *Stored-message generator*. Does the device have a stored-message generator to assist in testing?
- **E.** *Character-error counter*. Does the device have the capability to count errors as they occur, and accumulate totals by time intervals?

17.6 Analog monitoring is specifically concerned with measuring dropouts, phase hits, gain hits, impulse noise, envelope delay, and other communications-line parameters. Because the control of a line normally rests with the communications carrier, most organizations rely on the carrier test center to perform line testing.

17.7 A patchfield provides the means for monitoring lines, injecting signals, and reconfiguring a network, as well as providing access to local equipment interfaces. Usually rack mounted, a patchfield consists of a series of plugs, cords, and jacks that permit the interconnection of components and test set interfaces from one device to another.

17.8 Because the cabling associated with a LAN is usually localized to within a building, the effects of inclement weather, sunspot activity, and of similar problem sources, which normally affect a WAN, have minimal effect on a LAN.

17.9 Violations of equipment cabling specifications, electromagnetic interference (EMI), and utilization.

17.10 To monitor network traffic, set network alarms, set workstation options.

Answers to Questions and Problems in Chapter 18

18.1 Three information collection methods include

- **A.** *Personnel interviews*. Most costly and time consuming, but ferret out most items other methods normally miss.
- **B.** *Checklist*. Usually rigid with little or no room for elaboration of answers.
- **C.** *Survey*. Most flexible method.

18.2 Problems that can occur when a checklist (menu) is used include no response to a particular question and no means for users to elaborate on answers.

18.3 Advantages of interviews are the personnel contact and ability to ferret out information that can only be gained by face-to-face exposure. Disadvantages include the cost and time required to use this technique.

18.4 An expansion table can permit you to determine rapidly the capability and cost of upgrading equipment.

18.5 Knowing the protocols supported by a front-end processor will enable the network analyst to consider such line-sharing techniques as multipoint circuits, as well as to plan for the orderly growth of the network to include types of terminals that should be considered based on protocol support of the front end.

18.6 The assignment of additional software modules to a front-end processor can reduce terminal buffer availability size and thus increase time delays in servicing (responding to) all terminals connected to that device.

18.7 Techniques to alleviate front-end-processor port constraints include the use of line- and modem-sharing units, multipoint circuits, and the operation of a port selector.

18.8 The two primary methods of cutover are parallel and serial. In the parallel cutover method, you have to duplicate costs for a period of time. However, you also have a network to fall back to if the new one does not perform as expected and more developmental time becomes necessary.

In the serial cutover technique, duplicate costs are minimized because, as soon as the new network is ready, the old one is eliminated. Unfortunately, if the new one is not error free or does not work as expected, there is no old one to fall back to.

Answers to Questions and Problems in Chapter 19

19.1 Sole-source procurements are less time consuming than competitive procurements; however, competitive procurements normally result in lower prices when large quantities of an item are to be obtained.

19.2 A mandatory feature is one essential for the operation of the device. If the device does not have the feature, it is excluded from consideration. A desirable feature is nice to have, but if the device does not have the feature, it is still considered for selection.

19.3 Mandatory features are normally evaluated on a pass-fail basis. For desirable features, a dollars worth is normally assigned to each item, and the price that the vendor bids is adjusted by the value of all the desirable features to obtain the cost-value bid.

19.4

	Vendor A	Vendor B	Vendor C
Bid	$8750	$8727	$8905
Desirable features	$ 350	$ 500	$ 150
Cost-value	$8400	$8327	$8755

Here, Vendor B is selected.

19.5 Assume no effect as a result of the number of terminals.

$$\text{Dept 1: } \frac{450}{1300} + \frac{750}{1125} = 1.013$$

$$\text{Dept 2: } \frac{850}{1300} + \frac{375}{1125} = \frac{.987}{2.000}$$

$$\text{Dept 1: } \frac{1.013}{2} \times \$4500 = \$2279.25 \text{ pro rata charge}$$

$$\text{Dept 2: } \frac{0.987}{2} \times \$4500 = \$2220.75 \text{ pro rata charge}$$

Information References and Bibliography

Associations

American Federation of Information Processing Societies.
American National Standards Institute.
Electronic Industries Association.
Institute of Electrical and Electronics Engineers.
International Federation of Information Processing Societies.
International Organization for Standardization.
International Telecommunications Union (ITU).
ITU-T (Telecommunication Standardization Sector).

Periodicals

Data Communications
Communications Week

Texts

Abbatiello, Judy and Sarch, Ray, editors, *Telecommunications and Data Communications Factbook*, New York, McGraw-Hill, 1987.

Bracker, William E. Jr. and Sarch, Ray, editors, *Cases in Network Design*, New York, McGraw-Hill, 1985.

Datacomm for the Businessman, Cherry Hill, N.J., Management Information Corp., latest edition.

Held, Gilbert, *Communicating with the IBM PC Series*, New York, John Wiley, 1988.

Held, Gilbert, *Data Communication Networking Devices*, New York, John Wiley, 1984.

Held, Gilbert, *Data Compression*, 3rd Edition, New York, John Wiley, 1991.

Sarch, Ray, ed., *Basic Guide to Data Communications*, New York, McGraw-Hill, 1985.

Sarch, Ray, ed., *Integrating Voice & Data*, New York, McGraw-Hill, 1987.

Stallings, William, *Handbook of Computer-Communications Standards*, 3 Volumes, New York, Macmillan, 1987.

Stallings, William, *Data and Computer Communications*, 2nd Edition, New York, Macmillan, 1988.

Turpin, John and Sarch, Ray, editors, *Beyond the Basics*, New York, McGraw-Hill, 1986.

Index

Illustrations are indicated in **boldface**.

D

U

V

W

X

Y

Z